NONSTANDARD METHODS IN STOCHASTIC ANALYSIS AND MATHEMATICAL PHYSICS

This is a volume in
PURE AND APPLIED MATHEMATICS
A Series of Monographs and Textbooks
Editors: SAMUEL EILENBERG AND HYMAN BASS

The complete listing of books in this series is available from the Publisher upon request.

NONSTANDARD METHODS IN STOCHASTIC ANALYSIS AND MATHEMATICAL PHYSICS

Sergio Albeverio

Institute of Mathematics
Ruhr University Bochum
Bochum, West Germany

Raphael Høegh-Krohn

Institute of Mathematics
University of Oslo
Oslo, Norway

Jens Erik Fenstad

Institute of Mathematics
University of Oslo
Oslo, Norway

Tom Lindstrøm

Institute of Mathematics
University of Trondheim-NTH
Trondheim, Norway

1986

ACADEMIC PRESS, INC.

Harcourt Brace Jovanovich, Publishers
Orlando   San Diego   New York   Austin
Boston   London   Sydney   Tokyo   Toronto

ACADEMIC PRESS, INC.
Orlando, Florida 32887

United Kingdom Edition published by
ACADEMIC PRESS INC. (LONDON) LTD.
24–28 Oval Road, London NW1 7DX

Library of Congress Cataloging in Publication Data
Main entry under title:

Nonstandard methods in stochastic analysis and
    mathematical physics.

    (Pure and applied mathematics ;   )
    Includes bibliographies and index.
    1. Stochastic analysis.   2. Mathematical physics.
I. Albeverio, Sergio.   II. Series: Pure and applied
mathematics (Academic Press) ;
QA3.P8 [QA274.2]   510 s [519.2]   85-21442
ISBN 0–12–048860–4 (alk. paper)
ISBN 0–12–048861–2 (pbk. : alk. paper)

PRINTED IN THE UNITED STATES OF AMERICA

86 87 88 89   9 8 7 6 5 4 3 2 1

CONTENTS

Chapter 3.   Probability

Part II.   SELECTED APPLICATIONS

Chapter 4.   Stochastic Analysis

Chapter 7.   Hyperfinite Lattice Models

PREFACE

This is a book on applied nonstandard analysis. It is divided into two parts. The *basic course* (see accompanying chart) on calculus, topology and linear spaces, and probability gives a complete and self-contained introduction to nonstandard methods. The purpose of this part is expository, but we have tried to enliven the text with some substantial examples in differential equations, linear space theory, and Brownian motion.

The second part presents *selected applications* to stochastic analysis and mathematical physics. Some of the applications of this part are new. Some represent a re-analysis of known results, but from a novel point of view. We hope through this diversity of examples to convince the reader that nonstandard analysis is a viable tool in many parts of the mathematical sciences.

The collaboration that led to this book started in the fall of 1976. At that time P. Loeb and R. Anderson had established the basic facts of nonstandard measure theory and hyperfinite probability theory with applications to Brownian motion and stochastic integration. Our point of entry was somewhat different. In the summer of 1976, E. Nelson had given his AMS lecture on nonstandard theory with application to a singular perturbation problem. This was a field of interest for two of us (S. A. and R. H.-Kr.). They had several "heuristic" calculations and wondered if the "new" numbers of nonstandard theory could make this sound. This led to a joint seminar in 1976–1977 by three of us (S. A., J. E. F., and R. H.-Kr.) and to the subsequent work on singular perturbations (see Chapter 6).

The seminar was very lively and we had the active participation of some of our Oslo colleagues, in particular, Bent Birkeland and Dag Normann. In fact, as the reader will see, they have contributed substantially to the book. At that time the fourth author (T. L.) was a beginning graduate student. We gave him a standard text on stochastic analysis and asked him to do better using the new tools of nonstandard theory. This led to his thesis (see Chapter 4) and to a continued collaboration on the book project. He spent two years in Madison and the contact established with H. J. Keisler has proved important for our project. In fact, Keisler's monograph *An Infinitesimal Approach to Stochastic Analysis,* in various stages of publication, has been a source of much inspiration and insight for us.

Many others have contributed to our project. In Bochum, Albeverio has worked in close contact with C. Kessler and A. Stoll; some of their work is reported in this book. L. Arkeryd in Gothenburg and N. Cutland in Hull have been other friends in this enterprise. In Chapter 6 we report on Arkeryd's new results on the Boltzmann equation and in Chapter 4 we have included some of Cutland's contributions to control theory. The contacts with Arkeryd and Cutland have been very stimulating and we have learned much from both.

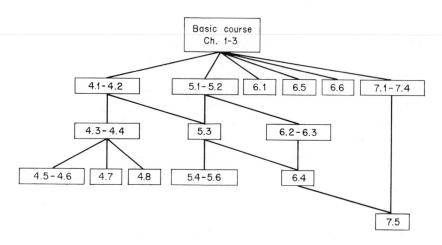

Chart of Main Dependencies

There are certain omissions in this chart, which are all rather innocent. A typical example is that the dependence of Theorems 5.3.9 and 6.4.5 on Proposition 4.8.5 is not indicated since the proof of 4.8.5 can be read independently of the rest of Section 4.8. In the other direction, none of the later chapters really depends on Sections 1.5, 2.4, and 3.5, which have been included to enliven the text by illuminating and worthwhile applications of the basic techniques.

The book has evolved over a number of years and we have benefited through this period from a close interaction with a numbers of friends and colleagues. In addition to those mentioned above we should like to thank Ph. Blanchard, S. Fajardo, W. Henson, H. Holden, P. Loeb, W. Luxemburg, D. Miller, H. Osswald, E. Perkins, D. Ross, K. Rumberger, R. Seising, K. Stroyan, P. Suppes, T. T. Wu, and R. Živaljevič for helpful advice and comments on successive versions of the manuscript.

We also want to express our sincere thanks to Signe Cordtsen for her expert typing and our editors at Academic Press for their extensive help in transforming the manuscript into a printed book.

Part I

BASIC COURSE

CHAPTER 1

CALCULUS

The basic insight or, if you prefer, discovery of nonstandard theory is that the geometric line or continuum can support a point set richer than the standard reals. This, among other things, gives us a framework for a geometric analysis of physical phenomena on many scales, and of physical phenomena that are too singular to fit in a direct way into the standard framework.

Let us elaborate a bit. Take the usual axiomatization of the affine plane. There are two basic categories of objects, *lines* and *points*. Any point lies on a line, but a line is not a set of points. Axioms with a "true" geometric content allow us to introduce coordinates from a field. But we need the less elementary Archimedean axiom to tell us that the field of an ordered geometry is isomorphic to a subfield of the reals.

Is the Archimedean axiom a "true" geometric fact? What is given in our immediate experience is a *limited* part of the geometric line with at most a *finite* number of points marked on it, representing, e.g., the result of some physical measurement. The rest is an extension, ideal or real.

In the orthodox view the real numbers are created from the rationals by a limit construction in which we adjoin points representing certain equivalence classes of convergent sequences. The claim is that the "ideal elements" thus created fill up the line. But in this process we do not pay any attention to the rate of convergence. If we do this and add "witnesses" to distinguish between different convergence behavior, we are led to a richer point set on the line. And if we also care about the asymptotic behavior of sequences we are ineluctably led to the full notion of hyperreals or nonstandard reals. Nonstandard analysis asserts that the geometric line can support this richer set. And we shall see how this gives a frame for rescaling arguments, i.e., for a geometric analysis of phenomena on many scales.

Ours is a book on applied nonstandard analysis and we have not set ourselves the task of writing the history of infinitesimals and nonstandard theory. But two names should be mentioned.

The "founding" memoir of the nonstandard theory is the 1934 paper by T. A. Skolem, "Über die Nichtcharakterisierbarkeit der Zahlenreihe mittels endlich oder abzählbar unendlich vieler Aussagen mit ausschliesslich Zahlenvariablen" (Skolem, 1934).

Skolem's aim was, as the title indicates, in a certain sense destructive. But a deep insight cannot be only negative. In 1960 Abraham Robinson turned the nonstandard method into a new and efficient technique in mathematics. This was a truly remarkable step forward. He extended Skolem's analysis from arithmetic to the reals and saw how the nonstandard version could provide a suitable framework for the development of analysis by means of the infinitely small and infinitely large numbers; see his book (Robinson, 1966) and Volume 2 of his collected works (Robinson, 1979).

Some have seen a vindication of the Leibnizian infinitesimals in nonstandard analysis. There certainly are similarities. But one should be careful in claiming that novel developments prove the correctness of older ideas.

The history of infinitesimals, however, is fascinating; see Chapter X of Robinson (1966) and the many historical remarks in Laugwitz (1978).

We now turn to the text proper. In Sections 1.1 and 1.2 we lay the foundation of the infinitesimal method. In Sections 1.3 and 1.4 we discuss briefly the basic notions of elementary calculus. And in Section 1.5 we present some more substantial examples from differential equation theory.

1.1. INFINITESIMALS

If you are firmly convinced that infinitesimals, and hence their inverses, the infinitely large, exist, then you need not read this section. You may jump directly to the next section, the extended universe.

But if you have some lingering doubts and need to be reassured, we have added some introductory remarks to demonstrate that the real line can be extended to accommodate both the infinitely small and the infinitely large.

The real numbers $\mathbb{R}$ can be constructed from the rationals $\mathbb{Q}$ in various ways. One method is to add to $\mathbb{Q}$ new points to represent limits of convergent sequences of rational numbers. In this process one has to identify sequences converging to the "same point" in $\mathbb{R}$.

We shall follow the same procedure in adding infinitesimals and infinitely large numbers. But we shall be more careful in the process of identifying sequences. If we care about rate of convergence and asymptotic properties, we will identify fewer sequences, hence end up with a richer set of points on the line, the nonstandard extension.

We turn to the construction. Let $\mathbb{N}$ be the set of natural numbers, $\mathbb{N} = \{0, 1, 2, \ldots, n, \ldots\}$. Following usual set-theoretic notation let $\mathbb{R}^{\mathbb{N}}$ be the *direct product* of $\mathbb{N}$ copies of $\mathbb{R}$, i.e., $\mathbb{R}^{\mathbb{N}}$ is the set of all real sequences $\langle a_n \rangle_{n \in \mathbb{N}}$, or, equivalently, the set of all functions $f : \mathbb{N} \to \mathbb{R}$. The direct product has a natural algebraic structure; we add and multiply two sequences by adding and multiplying in each coordinate separately, i.e., let $f, g \in \mathbb{R}^{\mathbb{N}}$, then we define $f + g$ and $f \cdot g$ by the coordinate equations

$$(f + g)(i) = f(i) + g(i), \qquad (f \cdot g)(i) = f(i) \cdot g(i).$$

We are interested in studying the behavior of sequences "in the limit," i.e., the asymptotic properties. As a first requirement we certainly want to identify sequences which agree on a *cofinite* set, where a set $E \subseteq \mathbb{N}$ is cofinite if it is of the form $E = \mathbb{N} - \{m_1, \ldots, m_k\}$. Note that every $m > \max\{m_1, \ldots, m_k\}$ belongs to E.

Two sequences $f, g \in \mathbb{R}^{\mathbb{N}}$ agree on the cofinite set E iff $f(m) = g(m)$, for all $m \in E$. Obviously, if $f, g \in \mathbb{R}^{\mathbb{N}}$ agree on a cofinite set, then they have the same behavior "in the limit." Conversely, if the set $\{i \in \mathbb{N} | f(i) \neq g(i)\}$ is cofinite, then we shall want to distinguish the limiting behavior of f and g, i.e., add different points for f and g in the extension of $\mathbb{R}$.

But cofinite sets are not good enough. It is well known from algebra that the direct product construction may introduce zero divisors, i.e., we may have elements $a \neq 0$ and $b \neq 0$ such that nevertheless $a \cdot b = 0$. The natural candidate for zero in $\mathbb{R}^{\mathbb{N}}$ is the sequence $\mathbf{0} = \langle c_n \rangle_{n \in \mathbb{N}}$, where $c_n = 0$, for all $n \in \mathbb{N}$. Let $f = \langle a_n \rangle_{n \in \mathbb{N}}$ be any sequence such that $a_n = 0$ iff n is even, and $g = \langle b_n \rangle_{n \in \mathbb{N}}$ any sequence such that $b_n = 0$ iff n is odd. Then $f \cdot g = \mathbf{0}$, but $f \neq \mathbf{0}$ and $g \neq \mathbf{0}$. Identifying sequences modulo cofinite sets does not help the situation, neither f nor g is equal to the sequence $\mathbf{0}$ on a cofinite set.

The existence of zero divisors destroys usual algebra, hence we need something more refined than cofinite sets in our construction.

A *filter* $\mathcal{F}$ is a family of subsets of $\mathbb{N}$ satisfying the following properties:

(1) $\qquad\qquad \mathbb{N} \in \mathcal{F}, \qquad \varnothing \notin \mathcal{F}.$

(2) $\qquad\qquad A_1, \ldots, A_n \in \mathcal{F} \Rightarrow A_1 \cap \cdots \cap A_n \in \mathcal{F}.$

(3) $\qquad\qquad A \in \mathcal{F}, \qquad A \subseteq B \Rightarrow B \in \mathcal{F}.$

The first property is a condition of nontriviality. The second is the important *finite intersection property*, which will turn up in a number of connections. The third is a convenient closure property. The class of cofinite sets, Cof, is an example of a filter.

(4) A filter $\mathscr{F}$ on $\mathbb{N}$ is called *free* if it contains no finite set.

We shall be particularly interested in free filters extending the filter of cofinite sets.

(5) A filter $\mathscr{U}$ is called an *ultrafilter* over $\mathbb{N}$ if for all $E \subseteq \mathbb{N}$ either $E \in \mathscr{U}$ or $\mathbb{N} - E \in \mathscr{U}$.

Notice that from the nontriviality condition (1) it follows that if $\mathscr{U}$ is an ultrafilter on $\mathbb{N}$ and $E \subseteq \mathbb{N}$, then exactly one of the sets E and $\mathbb{N} - E$ belongs to $\mathscr{U}$. Also observe that if a finite union $E_1 \cup \cdots \cup E_n$ belongs to $\mathscr{U}$, then at least one E_i belongs to $\mathscr{U}$.

1.1.1. ULTRAFILTER THEOREM. There exist free ultrafilters $\mathscr{U}$ on $\mathbb{N}$ extending the filter of cofinite sets.

The proof is an easy exercise in the use of transfinite methods in set theory, namely Zorn's lemma. Let Δ be the set of all filters $\mathscr{F}$ extending the filter Cof of cofinite sets. Then Δ is nonempty and is closed under union of chains. By Zorn's lemma, any such family has maximal elements. We shall prove that any maximal element of Δ is an ultrafilter extending Cof. So let $\mathscr{U}$ be a maximal element of Δ. $\mathscr{U}$ is clearly a filter; let us show that given any $E \subseteq \mathbb{N}$, either $E \in \mathscr{U}$ or $\mathbb{N} - E \in \mathscr{U}$. There are two cases: (1) assume that $E \cap F$ is infinite for all $F \in \mathscr{U}$; then the set $\mathscr{V} = \{D \subseteq \mathbb{N} \mid D \supseteq E \cap F$, for some $F \in \mathscr{U}\}$, is a filter in Δ which extends $\mathscr{U}$ and $E \in \mathscr{V}$. By maximality $\mathscr{U} = \mathscr{V}$, thus $E \in \mathscr{U}$. (2) Assume that $E \cap F$ is finite for some $F \in \mathscr{U}$. If $(\mathbb{N} - E) \cap D$ is finite for some $D \in \mathscr{U}$, then both $(\mathbb{N} - E) \cap D \cap F$ and $E \cap F \cap D$ will be finite, hence $F \cap D$ will also be finite; but this is a contradiction since both $F, D \in \mathscr{U} \supseteq$ Cof. Thus $(\mathbb{N} - E) \cap D$ is infinite for all $D \in \mathscr{U}$, and we proceed as in case (1).

We can now construct the nonstandard extension. Let $\mathscr{U}$ be a free ultrafilter on $\mathbb{N}$ and introduce an equivalence relation on sequences in $\mathbb{R}^{\mathbb{N}}$ as

$$(6) \qquad f \sim_{\mathscr{U}} g \qquad \text{iff} \quad \{i \in \mathbb{N} \mid f(i) = g(i)\} \in \mathscr{U}.$$

Sets in the ultrafilter are "large" sets. In fact, there is an obvious one-to-one correspondence between ultrafilter on $\mathbb{N}$ and finitely additive 0–1 valued measures on $\mathbb{N}$; with respect to this correspondence the relation in (6) identifies sequences which are equal "almost everywhere," i.e., which are equal on a set of measure 1 in the measure associated to $\mathscr{U}$.

Since $\mathcal{U}$ is free, two sequences that agree on a cofinite set are identified with respect to the relation introduced in (6). But more sequences are identified: since $\mathcal{U}$ is an ultrafilter either the set $\{i \in \mathbb{N} \mid i \text{ even}\}$ or the set $\{i \in \mathbb{N} \mid i \text{ odd}\}$ belongs to $\mathcal{U}$. Thus either the sequence $\langle a_n \rangle_{n \in \mathbb{N}}$, where $a_n = 0$ iff n is even, or the sequence $\langle b_n \rangle_{n \in \mathbb{N}}$, where $b_n = 0$ iff n is odd, will be identified with the zero sequence $\mathbf{0} = \langle c_n \rangle_{n \in \mathbb{N}}$, where $c_n = 0$ for all n. Thus identifying modulo an ultrafilter eliminates the difficulty we noticed above in connection with zero divisors.

$\mathbb{R}^{\mathbb{N}}$ divided out by the equivalence relation $\sim_{\mathcal{U}}$ gives us the nonstandard extension *$\mathbb{R}$, the *hyperreals*; in symbols,

$$(7) \qquad \qquad {}^*\mathbb{R} = \mathbb{R}^{\mathbb{N}} / \mathcal{U}.$$

If $f \in \mathbb{R}^{\mathbb{N}}$, we denote its image in *$\mathbb{R}$ by $f_{\mathcal{U}}$, and, of course, every element in *$\mathbb{R}$ is of the form $f_{\mathcal{U}}$ for some $f \colon \mathbb{N} \to \mathbb{R}$.

For any real number $r \in \mathbb{R}$ let $\mathbf{r}$ denote the constant function with value r in $\mathbb{R}^{\mathbb{N}}$, i.e., $\mathbf{r}(n) = r$, for all $n \in \mathbb{N}$. We then have a natural embedding.

$$(8) \qquad \qquad {}^* \colon \mathbb{R} \to {}^*\mathbb{R}$$

by setting *$r = \mathbf{r}_{\mathcal{U}}$, for all $r \in \mathbb{R}$. We must now lift the structure of $\mathbb{R}$ to the hyperreals *$\mathbb{R}$.

REMARK. What happens if we use in (7) an ultrafilter $\mathcal{U}_0$ that is not free? If $\mathcal{U}_0$ is not free, i.e., does not extend the filter of cofinite sets, then $\mathcal{U}_0$ must contain some finite set; in fact, it is easily seen that in this case there must exist some number $n_0 \in \mathbb{N}$ such that

$$\mathcal{U}_0 = \{E \subseteq \mathbb{N} \mid n_0 \in E\}.$$

In this case the construction of *$\mathbb{R}$ collapses, *$\mathbb{R} = \mathbb{R}^{\mathbb{N}} / \mathcal{U}_0 \cong \mathbb{R}$.

As an algebraic structure, $\mathbb{R}$ is a complete ordered field, i.e., a structure of the form

$$(9) \qquad \qquad \langle \mathbb{R}, +, \cdot, <, 0, 1 \rangle,$$

where $\mathbb{R}$ is the set of elements of the structure, $+$ and $\cdot$ are the binary operations of addition and multiplication, $<$ is the ordering relation, and 0 and 1 are two distinguished elements of the domain. And it is *complete* in the sense that every nonempty set bounded from above has a least upper bound.

The *-embedding of (8) sends 0 to *$0 = \mathbf{0}_{\mathcal{U}}$ and 1 to *$1 = \mathbf{1}_{\mathcal{U}}$. We must lift the operations and relations of $\mathbb{R}$ to *$\mathbb{R}$. We get the clue from (6), which tells us when two elements $f_{\mathcal{U}}$ and $g_{\mathcal{U}}$ of *$\mathbb{R}$ are equal:

$$(10) \qquad \qquad f_{\mathcal{U}} = g_{\mathcal{U}} \qquad \text{iff} \qquad \{i \in \mathbb{N} \mid f(i) = g(i)\} \in \mathcal{U}.$$

In a similar way we extend $<$ to $*\mathbb{R}$ by setting for arbitrary $f_\mathcal{U}$ and $g_\mathcal{U}$ in $*\mathbb{R}$:

(11) $f_\mathcal{U} < g_\mathcal{U}$ iff $\{i \in \mathbb{N} | f(i) < g(i)\} \in \mathcal{U}.$

With this definition of $<$ in $*\mathbb{R}$ we easily show that the extended domain $*\mathbb{R}$ is linearly ordered. As an example we verify transitivity of $<$ in $*\mathbb{R}$: let $f_\mathcal{U} < g_\mathcal{U}$ and $g_\mathcal{U} < h_\mathcal{U}$, i.e.,

$$D_1 = \{i \in \mathbb{N} | f(i) < g(i)\} \in \mathcal{U},$$
$$D_2 = \{i \in \mathbb{N} | g(i) < h(i)\} \in \mathcal{U}.$$

By the finite intersection property (2), $D_1 \cap D_2 \in \mathcal{U}$. If $i \in D_1 \cap D_2$, then $f(i) < g(i)$ and $g(i) < h(i)$; hence by transitivity of $<$ in $\mathbb{R}$, $f(i) < h(i)$. Thus

$$D_1 \cap D_2 \subseteq \{i \in \mathbb{N} | f(i) < h(i)\}.$$

The closure property (3) then tells us that $f_\mathcal{U} < h_\mathcal{U}$. We invite the reader to prove that given any $f_\mathcal{U}, g_\mathcal{U} \in *\mathbb{R}$, then either $f_\mathcal{U} < g_\mathcal{U}$, or $f_\mathcal{U} = g_\mathcal{U}$, or $f_\mathcal{U} > g_\mathcal{U}$.

   REMARK.   The relation $<$ on $*\mathbb{R}$ introduced in (11) extends the relation $<$ on $\mathbb{R}$, i.e., given any $r_1, r_2 \in \mathbb{R}$ we see that $r_1 < r_2$ in $\mathbb{R}$ iff $*r_1 < *r_2$ in $*\mathbb{R}$. This is the reason why we have not decorated the equality introduced in (10) and the ordering in (11) with an asterisk.

   We now have a linear order on $*\mathbb{R}$ and can verify that $*\mathbb{R}$ contains infinitesimals and infinite numbers. A (positive) *infinitesimal* δ in $*\mathbb{R}$ is an element $\delta \in *\mathbb{R}$ such that $*0 < \delta < *r$ for all $r > 0$ in $\mathbb{R}$.

   Infinitesimals exist; let $f(n) = 1/n$ for $n \in \mathbb{N}$. Then $\delta = f_\mathcal{U}$ is a positive infinitesimal. Also notice that $f'(n) = 1/n^2$ introduces another infinitesimal δ' and that $\delta' < \delta$ in $*\mathbb{R}$.

   In the same way $g(n) = n$ and $g'(n) = n^2$ introduce *infinite numbers*, $\omega = g_\mathcal{U}$ and $\omega' = g'_\mathcal{U}$, and we have that $\omega < \omega'$ in $*\mathbb{R}$.

   It remains to extend the operations $+$ and $\cdot$ to $*\mathbb{R}$. Looking back to (10) and (11) we have nothing to do but to set

(12)
$$f_\mathcal{U} + g_\mathcal{U} = h_\mathcal{U}   \text{iff}   \{i \in \mathbb{N} | f(i) + g(i) = h(i)\} \in \mathcal{U},$$
$$f_\mathcal{U} \cdot g_\mathcal{U} = h_\mathcal{U}   \text{iff}   \{i \in \mathbb{N} | f(i) \cdot g(i) = h(i)\} \in \mathcal{U}.$$

With these definitions we can prove easily that $*\mathbb{R}$ is an ordered field extension of $\mathbb{R}$. And these definitions introduce an honest algebra on the infinitesimals and on the infinitely large numbers. As an exercise the reader may wish to verify that if $f_\mathcal{U} < g_\mathcal{U}$ and $*0 < h_\mathcal{U}$, then $f_\mathcal{U} \cdot h_\mathcal{U} < g_\mathcal{U} \cdot h_\mathcal{U}$. One should also notice that for the infinitesimals δ and δ' and the infinite ω and ω' introduced above, we have, e.g., $\omega' = \omega^2$, $\delta \cdot \omega = 1$, $\delta' \cdot \omega$ is infinitesimal, and $\delta \cdot \omega'$ is infinite. (You may directly verify from the definitions that $\delta \cdot \omega'$ is infinite, but you can also calculate $\delta \cdot \omega' = (1/\omega) \cdot \omega^2 = \omega$.) Thus the infinitely small and the infinitely large have a decent arithmetic.

The way we extended the particular operators $+$ and $\cdot$ and the particular relation $=$ from $\mathbb{R}$ to $^*\mathbb{R}$ can be used to extend any function and relation on $\mathbb{R}$ to $^*\mathbb{R}$. Let F be an n-ary function on $\mathbb{R}$, i.e.,

$$F:\underbrace{\mathbb{R} \times \cdots \times \mathbb{R}}_{n \text{ times}} \to \mathbb{R}.$$

We introduce as in (12) the extended function $^*F$ by the equivalence

(13) $\quad ^*F(f^1_{\mathcal{U}}, \ldots, f^n_{\mathcal{U}}) = g_{\mathcal{U}} \qquad$ iff $\quad \{i \in \mathbb{N} \mid F(f^1(i), \ldots, f^n(i)) = g(i)\} \in \mathcal{U}.$

The reader may want to verify that $^*F$ is a function and that $^*F$ really extends F, i.e., $^*F(^*r_1, \ldots, ^*r_n) = ^*r$ iff $F(r_1, \ldots, r_n) = r$. In the same way we extend any n-ary relation S on $\mathbb{R}$ to a relation $^*S$ on $^*\mathbb{R}$. Note that since a subset $E \subseteq \mathbb{R}$ corresponds to an unary relation, we have an extension $^*E$ characterized by the condition

(14) $\qquad\qquad f_{\mathcal{U}} \in ^*E \qquad$ iff $\quad \{i \in \mathbb{N} \mid f(i) \in E\} \in \mathcal{U}.$

Thus if $E = \langle 0, 1]$, then $^*E$ as a subset of $^*\mathbb{R}$ will have every positive infinitesimal as an element, but not ^{*0}, a fact which can be read off immediately from condition (14). We shall later return to some intuitive characterizations of open, closed, and compact sets in $\mathbb{R}$ in terms of their *-extension to $^*\mathbb{R}$.

But first a few elementary observations on the *-extension of subsets of $\mathbb{R}$; $^*\varnothing$ is the empty set in $^*\mathbb{R}$. If $E \subseteq \mathbb{R}$, then $^*r \in ^*E$ for all $r \in E$, but in general (see the example $E = \langle 0, 1]$ above) $^*E$ will contain elements not of the form $^*r$ for any $r \in \mathbb{R}$. Furthermore * is a Boolean homomorphism in the sense that $^*(E_1 \cup E_2) = ^*E_1 \cup ^*E_2$, $^*(E_1 \cap E_2) = ^*E_1 \cap ^*E_2$, for arbitrary sets $E_1, E_2 \subseteq \mathbb{R}$. Finally, we note that $^*E_1 = ^*E_2$ iff $E_1 = E_2$ and $^*r \in ^*E$ iff $r \in E$. Note that * is *not* a σ-homomorphism; see the discussion of Loeb measure in Section 3.1. And, of course, * is *not* onto since there exist sets in $^*\mathbb{R}$ that are not of the form $^*E$ for some $E \subseteq \mathbb{R}$.

Before proceeding we need to discuss the important concept of *standard part*. By virtue of (13) the absolute-value function has an extension to $^*\mathbb{R}$ that we will denote by the usual $|\cdot|$ rather than the "correct" $^*|\cdot|$.

An element $x \in ^*\mathbb{R}$ is called *finite* if $|x| < ^*r$ for some $r > 0$. As we shall see in a moment, every finite $x \in ^*\mathbb{R}$ is infinitely close to some (unique) $r \in \mathbb{R}$ in the sense that $|x - ^*r|$ is either 0 or positively infinitesimal in $^*\mathbb{R}$. This unique r is called the *standard part* of x and is denoted by $\text{st}(x)$ or $^\circ x$.

The proof of existence of the standard part is rather simple. Let $x \in ^*\mathbb{R}$ be finite. Let D_1 be the set of $r \in \mathbb{R}$ such that $^*r < x$ and D_2 the set of $r' \in \mathbb{R}$ such that $x < ^*r'$. The pair (D_1, D_2) forms a Dedekind cut in $\mathbb{R}$, hence determines a unique $r_0 \in \mathbb{R}$. A simple argument shows that $|x - ^*r_0|$ is infinitesimal, i.e., $\text{st}(x) = r_0$.

REMARK. The reader should notice that we did not mess up the above argument with irrelevant details about the construction of *ℝ. It is entirely based upon the properties of the *-embedding that we have already established. This is an important point to which we shall return. The properties of the nonstandard extension *ℝ can be pinned down in a few fundamental principles; see the discussion in Section 1.2. We do the same when we characterize ℝ as a complete ordered field extension of the rationals. And neither for ℝ nor for *ℝ are the basic properties of the extension dependent upon the particular way they are constructed. For ℝ we may use either Dedekind cuts or Cauchy sequences. For *ℝ we can use the ultrafilter construction as above or the Gödel completeness theorem for first-order logic.

The standard part map is well behaved. If $r \in \mathbb{R}$, then $\text{st}(*r) = r$; if x, $y \in *\mathbb{R}$ are both finite, then $\text{st}(x + y) = \text{st}(x) + \text{st}(y)$, $\text{st}(x - y) = \text{st}(x) - \text{st}(y)$; and if $\text{st}(y) \neq 0$, then $\text{st}(x/y) = \text{st}(x)/\text{st}(y)$.

We shall now use the nonstandard extension *ℝ and the standard part map to characterize some topological notions in ℝ. By the *monad* of a real number $r \in \mathbb{R}$, denoted by $\mu(r)$, we understand the set of all $x \in *\mathbb{R}$ such that $\text{st}(x) = r$.

1.1.2. PROPOSITION. Let $E \subseteq \mathbb{R}$; then

  (i) E is open iff $\mu(r) \subseteq *E$ for all $r \in E$;
 (ii) E is closed iff $\text{st}(x) \in E$ for all finite $x \in *E$; and
(iii) E is compact iff for all $x \in *E$, $\text{st}(x)$ exists and is an element of E.

Here (ii) is the dual form of (i) and (iii) follows from (ii), recalling that compact in ℝ means closed and bounded. Since $\mu(x)$ is contained in the *-extension of every standard neighborhood of x, one part of (i) is immediate. Assume now that E is not open; then there exists some $r \in E$ such that every neighborhood of r in ℝ intersects $\mathbb{R} - E$. We shall produce an element $x \in \mu(r)$ such that $x \notin *E$: for each $n \in \mathbb{N}$ pick some $r_n \notin E$ such that $|r - r_n| < 1/n$. Let $f: \mathbb{N} \to \mathbb{R}$ be defined by $f(n) = r_n$. From (14) and the definition of f, we see that $f_\mathcal{U} \notin *E$. But $f_\mathcal{U} \in \mu(r)$ since by construction $|*r - f_\mathcal{U}| < \delta$, where δ is the infinitesimal associated with the sequence $\langle 1/n \rangle_{n \in \mathbb{N}}$.

We hope that the reader by now has been somewhat reassured with respect to his or her doubts about the existence of the infinitely small and the infinitely large. Let us conclude by proving a general result about ultrafilter extensions which draws together in simple terms the many separate insights scattered over the preceding pages.

The result we are hinting at is the *general transfer principle* of nonstandard analysis. We have already noticed that ℝ and *ℝ are similar in many respects;

e.g., both are linearly ordered fields. We shall now make precise in which sense $\mathbb{R}$ and $*\mathbb{R}$ are similar; i.e., we shall specify in more detail which properties of $\mathbb{R}$ transfer to $*\mathbb{R}$.

We consider the reals as a structure

$$(15) \qquad\qquad \langle \mathbb{R}, +, \cdot, <, |\cdot|, 0, 1 \rangle,$$

where, in addition to the information in (9), we have added the absolute value that defines the metric on $\mathbb{R}$. Of course, $|\ |$ is definable in terms of the other entities in (15), but it makes things a bit easier to include it explicitly in the specification.

The structure $\mathbb{R}$ has an associated simple language $L(\mathbb{R})$ that can be used to describe the kind of properties of $\mathbb{R}$ that are preserved under the embedding $*\colon \mathbb{R} \to *\mathbb{R}$.

The *elementary formulas* of $L(\mathbb{R})$ are expressions of the form

(i) $t_1 + t_2 = t_3$,
(ii) $t_1 \cdot t_2 = t_3$,
(iii) $|t_1| = t_2$,
(iv) $t_1 = t_2$,
(v) $t_1 < t_2$,
(vi) $t_1 \in X$,

where t_1, t_2, t_3 are either the constants 0 or 1 or a variable for an arbitrary number $r \in \mathbb{R}$, and X is a variable for a subset $A \subseteq \mathbb{R}$.

From the elementary formulas we generate the class of all *formulas* or *expressions* of $L(\mathbb{R})$ using the *propositional connectives*

$$\wedge \qquad \text{and}$$

$$\vee \qquad \text{or}$$

$$\neg \qquad \text{not}$$

$$\to \qquad \text{if, then}$$

and the *number quantifiers*

$$\forall x \qquad \text{for all} \quad x \quad (\text{in } \mathbb{R})$$

$$\exists x \qquad \text{for some} \quad x \quad (\text{in } \mathbb{R})$$

by the rules:

(vii)   If Φ and Ψ are formulas of $L(\mathbb{R})$, then

$$\Phi \wedge \Psi, \qquad \Phi \vee \Psi, \qquad \neg\Phi, \qquad \Phi \to \Psi$$

are formulas of $L(\mathbb{R})$.

(viii)   If Φ is a formula of $L(\mathbb{R})$ and x is a number variable, then $\forall x \Phi$ and $\exists x \Phi$ are formulas of $L(\mathbb{R})$.

Whenever necessary, we add parentheses to formulas of $L(\mathbb{R})$ to avoid possible ambiguities.

The language $L(\mathbb{R})$ is basically a first-order language; i.e., we allow number quantification but not set quantification. The reader with some knowledge of logic may also be puzzled by our somewhat restricted notion of elementary formula; e.g., $(x + 1) \cdot y = z$ is not elementary in our sense. However, we can easily write a formula in $L(\mathbb{R})$ capturing the meaning of this expression, namely

$$\exists x_1 [x + 1 = x_1 \wedge x_1 \cdot y = z].$$

This trick is universal and shows that $L(\mathbb{R})$ has the intended expressive power.

We give a few examples: in the language $L(\mathbb{R})$ we can write down conditions which express that $<$ is a linear ordering:

transitive	$\forall x \, \forall y \, \forall z [[x < y \wedge y < z] \rightarrow x < z]$
irreflexive	$\forall x \neg [x < x]$
linear	$\forall x \, \forall y [x < y \vee x = y \vee y < x].$

Let us also write down a formula $\Phi(X)$ which expresses that X is an open set

(16) $\Phi(X) =_{\text{def}} \forall y [y \in X \rightarrow \exists z [z > 0 \wedge \forall y_1 [|y - y_1| < z \rightarrow y_1 \in X]]].$

A formula Φ of $L(\mathbb{R})$ is in general of the form

$$\Phi = \Phi(X_1, \ldots, X_m, x_1, \ldots, x_n),$$

where $x_1, \ldots, x_n$ are the free number variables of Φ, i.e., variables not bound by a quantifier $\forall$ or $\exists$, and $X_1, \ldots, X_m$ are the (free) set variables of Φ. Every formula in $L(\mathbb{R})$ has an immediate meaning or interpretation in the structure $\mathbb{R}$; e.g., let $\Phi(X)$ be the formula in (16) and let $A \subseteq \mathbb{R}$, then $\Phi(A)$ expresses the fact that A is open in $\mathbb{R}$.

But formulas of $L(\mathbb{R})$ can also be interpreted in the extended structure $^*\mathbb{R}$. We use (12) to interpret elementary statements of the previous form (i) and (ii). We use the general format of (13) to interpret (iii), and we use (10) and (11) to interpret (iv) and (v), respectively. Finally, we use (14) to interpret (vi). And since the logical symbols have a fixed meaning over any domain, this means that every formula Φ in $L(\mathbb{R})$ has an interpretation in both $\mathbb{R}$ and $^*\mathbb{R}$. Let us once more return to (16); given $A \subseteq \mathbb{R}$ we have already explained what we mean by $\Phi(A)$. Now we can also interpret $\Phi(^*A)$ over $^*\mathbb{R}$ according to the specification given in this paragraph.

REMARKS. (1) As already remarked, not every subset of $^*\mathbb{R}$ is of the form $^*A$ for some subset $A \subseteq \mathbb{R}$. If A_0 is a subset of $^*\mathbb{R}$ that is not of the form $^*A$ for some $A \subseteq \mathbb{R}$, then we have not given any meaning to $\Phi(A_0)$; this follows from the fact that (14) applies only to sets in $^*\mathbb{R}$ of the form $^*A$. (2) Using the general format of (13) we could also have added arbitrary function parameters to our $L(\mathbb{R})$ formulas.

We can now state the main result about ultrafilter extensions, which has the general transfer principle as an immediate corollary.

1.1.3. THEOREM OF ŁOŚ. Let $\Phi(X_1, \ldots, X_m, x_1, \ldots, x_n)$ be a formula of $L(\mathbb{R})$. Then for any $A_1, \ldots, A_m \subseteq \mathbb{R}$ and $f^1_{\mathcal{U}}, \ldots, f^n_{\mathcal{U}} \in {}^*\mathbb{R}$,

$$(17) \qquad \Phi(^*A_1, \ldots, {}^*A_m, f^1_{\mathcal{U}}, \ldots, f^n_{\mathcal{U}}) \qquad \text{iff}$$

$$\{i \in \mathbb{N} \mid \Phi(A_1, \ldots, A_m, f^1(i), \ldots, f^n(i))\} \in \mathcal{U}.$$

The proof is by induction on the number of logical symbols in Φ. If Φ has no logical symbols, it is an elementary formula of the form (i)-(vi), and (17) then reduces to one of (12), (13), (10), (11), or (14). If Φ contains logical symbols, then Φ is of the form $\Phi_1 \wedge \Phi_2$, $\Phi_1 \vee \Phi_2$, $\neg \Phi_1$, $\Phi_1 \to \Phi_2$, $\forall x \Phi_1$, or $\exists x \Phi_1$. The verification of (17) is, by induction, in each case reduced to an elementary property of the ultrafilter $\mathcal{U}$. In fact, we have done bits of the proof in verifying the transitivity of $<$ in $^*\mathbb{R}$; see the paragraph following (11).

Let us, however, comment briefly on a few of the cases. For example, if $\Phi = \Phi_1 \wedge \Phi_2$, (17) follows from the finite intersection property of the ultrafilter. The case $\Phi = \neg \Phi_1$ uses in an essential way that $\mathcal{U}$ is an ultrafilter, namely, that $\mathbb{N} - E \in \mathcal{U}$ iff $E \notin \mathcal{U}$; the reader should recall our discussion of zero divisors above.

Quantifiers offer no special difficulties, but let us spell out the case $\Phi = \exists x \Phi_1$ in detail. For simplicity let Φ have one free variable; we shall prove

$$\Phi(f_{\mathcal{U}}) \qquad \text{iff} \qquad \{i \in \mathbb{N} \mid \Phi(f(i))\} \in \mathcal{U},$$

where $\Phi(f_{\mathcal{U}})$ is of the form $\exists x \Phi_1(x, f_{\mathcal{U}})$. Now $\Phi(f_{\mathcal{U}})$ is true in $^*\mathbb{R}$ iff there is some $g_{\mathcal{U}} \in {}^*\mathbb{R}$ such that $\Phi_1(g_{\mathcal{U}}, f_{\mathcal{U}})$ is true in $^*\mathbb{R}$. By the induction hypothesis this means that $\{i \in \mathbb{N} \mid \Phi_1(g(i), f(i))\} \in \mathcal{U}$. But if $\Phi_1(g(i), f(i))$ is true in $\mathbb{R}$, then $\exists x \Phi_1(x, f(i))$ is also true in $\mathbb{R}$, i.e., $\{i \in \mathbb{N} \mid \Phi_1(g(i), f(i))\} \subseteq \{i \in \mathbb{N} \mid \exists x \Phi_1(x, f(i))\}$. From the property (3) of filters it follows that $\{i \in \mathbb{N} \mid \Phi(f(i))\} \in \mathcal{U}$.

To prove the converse, assume that $\{i \in \mathbb{N} \mid \Phi(f(i))\} \in \mathcal{U}$. For each i in this set choose some $a_i \in \mathbb{R}$ such that $\Phi_1(a_i, f(i))$. Let $g \in \mathbb{R}^{\mathbb{N}}$ be a function g such that $g(i) = a_i$ for all i in the set we started with and $g(i) = a'$

otherwise, where a' is some arbitrary element of $\mathbb{R}$. Then we have $\Phi_1(g(i), f(i))$ for a set of indices in $\mathcal{U}$; hence by the induction hypothesis we have $\Phi_1(g_\mathcal{U}, f_\mathcal{U})$ in *$\mathbb{R}$, i.e., we have $\Phi(f_\mathcal{U}) = \exists x \Phi_1(x, f_\mathcal{U})$ in *$\mathbb{R}$.

The theorem of Łoś has the transfer principle as an immediate corollary.

1.1.4. TRANSFER PRINCIPLE. Let $\Phi(X_1, \ldots, X_m, x_1, \ldots, x_n)$ be a formula of the language $L(\mathbb{R})$. Then for any $A_1, \ldots, A_m \subseteq \mathbb{R}$ and $r_1, \ldots, r_n \in \mathbb{R}$, $\Phi(A_1, \ldots, A_m, r_1, \ldots, r_n)$ is true in $\mathbb{R}$ iff $\Phi(*A_1, \ldots, *A_m, *r_1, \ldots, *r_n)$ is true in *$\mathbb{R}$.

The proof is indeed immediate. From (17) we get at once $\Phi(*A_1, \ldots, *A_m, *r_1, \ldots, *r_n)$ iff $\{i \in \mathbb{N} | \Phi(A_1, \ldots, A_m, r_1, \ldots, r_n)\} \in \mathcal{U}$. But the set $\{i \in \mathbb{N} | \Phi(A_1, \ldots, A_m, r_1, \ldots, r_n)\}$ is equal to $\mathbb{N} \in \mathcal{U}$ if Φ is true of $A_1, \ldots, A_m, r_1, \ldots, r_n$ in $\mathbb{R}$, and is equal to $\varnothing \notin \mathcal{U}$ if Φ is not true of $A_1, \ldots, A_m, r_1, \ldots, r_n$. Thus $\Phi(A_1, \ldots, A_m, r_1, \ldots, r_n)$ is true in $\mathbb{R}$ iff $\Phi(*A_1, \ldots, *A_m, *r_1, \ldots, *r_n)$ is true in *$\mathbb{R}$.

The existence of the embedding $*: \mathbb{R} \to *\mathbb{R}$ satisfying the transfer principle, 1.1.4, is all we need for elementary nonstandard analysis. To make this point let us briefly return to the proof of Proposition 1.1.2. The essential point was to prove that E is open if $\mu(r) \subseteq *E$, for all $r \in E$. In the previous proof we worked inside the model, here we use transfer: let $a \in E$; since $\mu(a) \subseteq *E$ the following statement

$$\Phi(X, x) = \exists z [0 < z \wedge \forall y [|x - y| < z \to y \in X]]$$

is true in *$\mathbb{R}$ when X is interpreted as *E and x as *a. By transfer $\Phi(E, a)$ is true in $\mathbb{R}$, which means that some standard neighborhood of a is contained in E, i.e., E is open.

The reader will find it instructive to compare the two proofs.

Let us round off this introductory discussion by the following more general comments. Starting from the rationals $\mathbb{Q}$ we can form the ultra-product

$$(18) \qquad\qquad *\mathbb{Q} = \mathbb{Q}^\mathbb{N} / \mathcal{U},$$

where $\mathcal{U}$ is some ultrafilter on $\mathbb{N}$ extending the filter of cofinite sets. As in the case of *$\mathbb{R}$ we have the notions of *finite, infinite,* and *infinitesimal* in *$\mathbb{Q}$. But this time we must be a bit careful with the notion of *standard part,* since $\mathbb{Q}$ is not complete in the standard metric uniformity. Let $\mathbb{Q}_f$ denote the finite points of *$\mathbb{Q}$ and $\mathbb{Q}_i$ the set of infinitesimals. It is easy to prove that there is an onto map $\pi: \mathbb{Q}_f \to \mathbb{R}$, obtained by identifying modulo infinitesimals, i.e.,

$$\mathbb{R} \cong \mathbb{Q}_f / \mathbb{Q}_i.$$

This gives perhaps the most direct way of constructing the reals from the rationals; *$\mathbb{Q}$ is an elementary extension of $\mathbb{Q}$, the substructure $\mathbb{Q}_f$ is seen to be an integral domain, and $\mathbb{Q}_i$ is a prime ideal in $\mathbb{Q}_f$. Thus the quotient is a field, namely the reals.

This example can be generalized; $\mathbb{R}$ is usually regarded as the completion of the normed algebraic structure $\mathbb{Q}$. In general, given a normed algebraic structure E, we can construct both the standard completion $\hat{E}$ and the nonstandard extension $*E$. In $*E$ we can distinguish the set of pre-nearstandard (or *bounded*) points E_f. There is a map $\pi : E_f \to \hat{E}$ such that the following diagram is commutative:

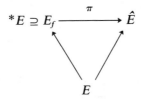

[See Fenstad and Nyberg (1970) for a detailed discussion.] The map π has the property that whenever F is complete and the map $f : E \to F$ has an extension to a map $\hat{f} : \hat{E} \to F$, then

$$\mathrm{st}(*f(x)) = \hat{f}(\pi(x))$$

for all $x \in E_f$.

It is this commutative diagram which "explains" the successful use of nonstandard methods in many situations. It is well known that the algebraic structure of E does not always extend to $\hat{E}$. By transfer it does extend to $*E$. The map $\pi : *E \to \hat{E}$ "confuses" points that from the algebraic point of view should be kept distinct. It is this richness of $*E$ which adds power and intuition to the nonstandard methods.

REMARK. We should admit that, whereas $\hat{E}$ is a "good" extension of E from the topological point of view, $*E$ does not admit a nice intrinsic topological structure; see Zakon (1969). But we are not in general interested in doing topology inside $*E$. The topological aspect will enter the situation in different ways, e.g., via the commutative diagram above; see also Proposition 1.1.2 and the discussion of measure extensions in Sections 3.4 and 3.5.

1.2. THE EXTENDED UNIVERSE

What we have learned so far is that the complete ordered field $\mathbb{R}$ of *real numbers* has a proper (but not unique) ordered field extension $*\mathbb{R}$, the *hyperreals*. The extension is *elementary*, which means that it preserves the true statements about $\mathbb{R}$ which are expressible in the language $L(\mathbb{R})$.

The structure $\mathbb{R}$ is not a large enough domain for the development of classical mathematics. We need an extended universe that, in addition to numbers and functions, also contains sets of functions, sets of spaces of functions, etc.; i.e., we need the finite type structure over $\mathbb{R}$. In more generality, given any set S we introduce the *superstructure* $V(S)$ over S as follows.

1.2.1. DEFINITION. For any set S let

$$V_1(S) = S,$$
$$V_{n+1}(S) = V_n(S) \cup \{X \mid X \subseteq V_n(S)\},$$
$$V(S) = \bigcup_n V_n(S).$$

Thus we see that the superstructure over S is obtained by iterating the power-set operator countably many times. Classical analysis lives inside $V(\mathbb{R})$. The extended universe of nonstandard analysis will be obtained by postulating an extension $*\mathbb{R} \supseteq \mathbb{R}$ and postulating an embedding

$$(1) \qquad\qquad * : V(\mathbb{R}) \to V(*\mathbb{R})$$

that will have properties similar to the embedding $* : \mathbb{R} \to *\mathbb{R}$ constructed in Section 1.1. First of all we assume the following principle.

1.2.2. EXTENSION PRINCIPLE. $*\mathbb{R}$ is a proper extension of $\mathbb{R}$ and $*r = r$ for all $r \in \mathbb{R}$.

In the model of Section 1.1 this means that we identify $\mathbb{R}$ with its $*$-image in $*\mathbb{R}$. We shall now extend the ultrafilter construction to demonstrate that superstructure embeddings of the type (1) satisfying a transfer principle analogous to 1.1.4 exist. This is, of course, not part of our systematic development, but simply a proof that our axioms are nontrivial, i.e., they admit a model.

The construction proceeds in two stages. First we construct a *bounded ultrapower* of $V(\mathbb{R})$ using a free ultrafilter $\mathcal{U}$ on $\mathbb{N}$. This is similar to the construction of $*\mathbb{R}$ in Section 1.1. Then we map the bounded ultrapower into the superstructure $V(*\mathbb{R})$ in such a way that the embedding (1) obeys the transfer principle.

I. *Constructing the Bounded Ultrapower.* A sequence $\langle A_1, A_2, \ldots \rangle$ of elements of $V(\mathbb{R})$ is *bounded* if there is a fixed n such that each $A_i \in V_n(\mathbb{R})$. Two bounded sequences A and B are equivalent with respect to the free ultrafilter $\mathcal{U}$, in symbols $A \sim_{\mathcal{U}} B$, iff $\{i \in \mathbb{N} \mid A_i = B_i\} \in \mathcal{U}$. We let $A_{\mathcal{U}}$ denote the equivalence class of A and define the bounded ultrapower by

$$(2) \qquad V(\mathbb{R})^{\mathbb{N}}/\mathcal{U} = \{A_{\mathcal{U}} \mid A \text{ is a bounded } V(\mathbb{R})\text{-sequence}\}.$$

We define the membership relation $\in_{\mathcal{U}}$ in the ultrapower by

(3) $\qquad\qquad A_{\mathcal{U}} \in_{\mathcal{U}} B_{\mathcal{U}} \qquad$ iff $\quad \{i \in \mathbb{N} \,|\, A_i \in B_i\} \in \mathcal{U}.$

The reader will notice how this extends definition (14) of Section 1.1.
    There is a natural proper embedding

(4) $\qquad\qquad\qquad i: V(\mathbb{R}) \to V(\mathbb{R})^{\mathbb{N}}/\mathcal{U},$

namely let $i(A) = \langle A, A, \ldots \rangle_{\mathcal{U}}$, the equivalence class corresponding to the constant sequence; this is similar to the embedding (8) of Section 1.1.

    II. *Embedding* $V(\mathbb{R})^{\mathbb{N}}/\mathcal{U}$ *into* $V(^*\mathbb{R})$. $^*\mathbb{R}$ is the (bounded) ultrapower $\mathbb{R}^{\mathbb{N}}/\mathcal{U}$. But $V(\mathbb{R})^{\mathbb{N}}/\mathcal{U}$ will not be the same as the full superstructure $V(^*\mathbb{R})$. We shall now construct an embedding

(5) $\qquad\qquad\qquad j: V(\mathbb{R})^{\mathbb{N}}/\mathcal{U} \to V(^*\mathbb{R})$

such that (i) j is the identity on $^*\mathbb{R}$ and (ii) if $A_{\mathcal{U}} \notin {}^*\mathbb{R}$, then $j(A_{\mathcal{U}}) = \{j(B_{\mathcal{U}}) \,|\, B_{\mathcal{U}} \in_{\mathcal{U}} A_{\mathcal{U}}\}$. This means that the relation $\in_{\mathcal{U}}$ in the ultrapower is mapped into the ordinary membership relation in $V(^*\mathbb{R})$.
    The embedding j is constructed in stages. Let

$$V_k(\mathbb{R})^{\mathbb{N}}/\mathcal{U} = \{A_{\mathcal{U}} \,|\, A \text{ is a sequence from } V_k(\mathbb{R})\}.$$

Then the bounded ultrapower is the union of the chain

$$^*\mathbb{R} = V_1(\mathbb{R})^{\mathbb{N}}/\mathcal{U} \subseteq \cdots \subseteq V_k(\mathbb{R})^{\mathbb{N}}/\mathcal{U} \subseteq \cdots,$$

and we can define j by induction. For $k = 1$, j must be the identity. If $A_{\mathcal{U}} \in V_{k+1}(\mathbb{R})^{\mathbb{N}}/\mathcal{U}$ and $A_{\mathcal{U}} \notin {}^*\mathbb{R}$, we simply set $j(A_{\mathcal{U}}) = \{j(B_{\mathcal{U}}) \,|\, B_{\mathcal{U}} \in_{\mathcal{U}} A_{\mathcal{U}}\}$. This makes sense: if $B_{\mathcal{U}} \in_{\mathcal{U}} A_{\mathcal{U}}$, it follows from (3) that $\{i \in \mathbb{N} \,|\, B_i \in V_k(\mathbb{R})\} \in \mathcal{U}$, i.e., $B_{\mathcal{U}} \in V_k(\mathbb{R})^{\mathbb{N}}/\mathcal{U}$, which means that $j(B_{\mathcal{U}})$ is defined at a previous stage of the inductive construction.
    Combining i and j we get *a model of the extended nonstandard universe*

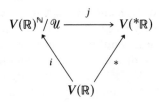

where $^*A = j(i(A))$, for any $A \in V(\mathbb{R})$. The reader may verify that this *-embedding extends the *-embedding constructed in Section 1.1.
    Here $V(\mathbb{R})$ and $V(^*\mathbb{R})$ are connected by a transfer principle generalizing 1.1.4 of Section 1.1. The structure $\mathbb{R}$ has an associated elementary language $L(\mathbb{R})$, which we used to give the necessary precision to the transfer principle. We need a similar formal tool to state the extended transfer principle.

1.2.3. THE RESTRICTED LANGUAGE L(V(ℝ)). The language $L(V(\mathbb{R}))$ will be
an extension of the language $L(\mathbb{R})$. We add to our stock of elementary
formulas [see (i)-(vi) in Section 1.1] expressions of the form

(6) $X \in Y,$

where X and Y are variables for arbitrary sets in $V(\mathbb{R})$. In (vi) of Section
1.1 we restricted ourselves to $t \in Y$, t being a variable for elements of $\mathbb{R}$.

We keep the logical symbols of $L(\mathbb{R})$, but in addition to the number
quantifiers we add *bounded set quantifiers*

(7)
$$\forall X \in Y \qquad \text{for all sets } X \text{ element of } Y,$$
$$\exists X \in Y \qquad \text{for some set } X \text{ element of } Y.$$

Formulas Φ of $L(V(\mathbb{R}))$ are then constructed in exactly the same way as
formulas of $L(\mathbb{R})$.

A formula Φ of $L(V(\mathbb{R}))$ can be interpreted in a natural way in any of
the structures $V(\mathbb{R})$, $V(\mathbb{R})^{\mathbb{N}}/\mathcal{U}$, and $V(^*\mathbb{R})$; note that in $V(\mathbb{R})$ and $V(^*\mathbb{R})$
we have the standard interpretation of the $\in$ symbol, in $V(\mathbb{R})^{\mathbb{N}}/\mathcal{U}$ we use
$\in_{\mathcal{U}}$ as introduced in (3) to interpret membership. Given any formula

$$\Phi = \Phi(X_1, \ldots, X_n)$$

with $X_1, \ldots, X_n$ as the only *free* set parameters, and given sets $A_1, \ldots, A_n \in$
$V(\mathbb{R})$, we mean by $\Phi(A_1, \ldots, A_n)$ the statement about $V(\mathbb{R})$ obtained by
giving the variables $X_1, \ldots, X_n$ the values $A_1, \ldots, A_n$, respectively. In a
similar way we interpret $\Phi(^*A_1, \ldots, ^*A_n)$ as a condition about $V(^*\mathbb{R})$
obtained by giving each X_k the value $^*A_k = j(i(A_k))$, $k = 1, \ldots, n$.

1.2.4. TRANSFER PRINCIPLE. Let $A_1, \ldots, A_n \in V(\mathbb{R})$. Any $L(V(\mathbb{R}))$ statement Φ that is true of $A_1, \ldots, A_n$ in $V(\mathbb{R})$ is true of $^*A_1, \ldots, ^*A_n$ in $V(^*\mathbb{R})$,
and conversely.

Transfer will be used over and over again. We grant that it may take
some time to be completely free and easy in manipulating elementary, i.e.,
$L(V(\mathbb{R}))$, statements. But it is our experience that it is at most a temporary
stumbling block. Here we discuss a simple example.

Let $A \subseteq {}^*\mathbb{R}$, then $A \in V_2(^*\mathbb{R})$. The embedding (1) maps $V_2(\mathbb{R})$ to a set
$^*(V_2(\mathbb{R}))$ in $V(^*\mathbb{R})$. Will A belong to this set? Not necessarily, as we shall
show below (see the remark after 1.2.6). But if $A \in {}^*B$ for some $B \in V(\mathbb{R})$,
then $A \in {}^*(V_2(\mathbb{R}))$.

REMARK. This can be "proved" in two ways. First we can show that it
is true in the model, which means that the result is *consistent* with our
axiomatic principles, extension and transfer. But we can also prove the fact
directly from the two basic principles, which means that it is *true*. We shall

do the latter and leave the former to the reader. Perhaps we should at this point recall the two "proofs" of 1.1.2(i).

We thus want to prove that

(8) $$\forall A[A \in {}^*B \wedge A \subseteq {}^*\mathbb{R} \to A \in {}^*V_2(\mathbb{R})].$$

As it stands, (8) is not an $L(V(\mathbb{R}))$ formula. However, it is equivalent to

(9) $$(\forall A \in {}^*B)[\forall r \in A(r \in {}^*\mathbb{R}) \to A \in {}^*V_2(\mathbb{R})].$$

This is genuine $L(V(\mathbb{R}))$; i.e., we have only bounded set quantifiers. With some experience one always sees how a condition can be rewritten in correct $L(V(\mathbb{R}))$ form. So in the future we will write formulas in the style of (8) rather than always insist on the correct (9).

Now (9) is a condition $\Phi({}^*B, {}^*\mathbb{R}, {}^*V_2(\mathbb{R}))$, which by transfer is true in $V({}^*\mathbb{R})$ iff the corresponding $\Phi(B, \mathbb{R}, V_2(\mathbb{R}))$ is true in $V(\mathbb{R})$. But the latter condition is trivially true. Thus we have shown that if a subset of ${}^*\mathbb{R}$ is an element of some ${}^*B$ in $V({}^*\mathbb{R})$, then it is already an element of the *-image of $V_2(\mathbb{R})$.

REMARK. We shall comment briefly on the proof of 1.2.4. In the ultra-power model there are three structures involved, $V(\mathbb{R})$, $V(\mathbb{R})^\mathbb{N}/\mathcal{U}$, and $V({}^*\mathbb{R})$. Given any $L(V(\mathbb{R}))$ formula $\Phi(X, Y)$, we have explained how to interpret it in the three structures. Now Łoś' theorem, 1.1.3, immediately extends to the *bounded* ultrapower $V(\mathbb{R})^\mathbb{N}/\mathcal{U}$ by exactly the same proof; i.e., for any $A_\mathcal{U}, B_\mathcal{U} \in V(\mathbb{R})^\mathbb{N}/\mathcal{U}$ we have

(10) $$\Phi(A_\mathcal{U}, B_\mathcal{U}) \qquad \text{iff} \qquad \{i \in \mathbb{N} \,|\, \Phi(A_i, B_i)\} \in \mathcal{U},$$

from which transfer follows between $V(\mathbb{R})$ and $V(\mathbb{R})^\mathbb{N}/\mathcal{U}$ exactly as in 1.1.4.

But Principle 1.2.4 asserts transfer between $V(\mathbb{R})$ and $V({}^*\mathbb{R})$. And in order to prove this we need to replace (10) by

(11) $$\Phi(j(A_\mathcal{U}), j(B_\mathcal{U})) \qquad \text{iff} \qquad \{i \in \mathbb{N} \,|\, \Phi(A_i, B_i)\} \in \mathcal{U}.$$

But this is a rather immediate extension which follows from the fact that every element of, say, $j(A_\mathcal{U})$ in $V({}^*\mathbb{R})$ is of the form $j(A'_\mathcal{U})$ for some $A'_\mathcal{U}$ in $V(\mathbb{R})^\mathbb{N}/\mathcal{U}$; see the construction of the j-map.

And once we have (11) the Transfer Principle 1.2.4 follows by the same argument as in 1.1.4.

All properties of ${}^*\mathbb{R}$ discussed in Section 1.1 continue to hold for ${}^*\mathbb{R}$ in $V({}^*\mathbb{R})$. Elements of ${}^*\mathbb{R}$ are either *finite* or *infinite*, and every finite element in ${}^*\mathbb{R}$ has a unique *standard part* in $\mathbb{R}$. Infinitesimals are the inverses of the infinite numbers in ${}^*\mathbb{R}$.

But there are further distinctions to make in the extended universe.

1.2.5. DEFINITION.   Let $A \in V(^*\mathbb{R})$, then

(i) A is called *standard* if $A = {}^*B$ for some $B \in V(\mathbb{R})$,

(ii) A is called *internal* if $A \in {}^*B$ for some $B \in V(\mathbb{R})$, and

(iii) A is called *external* if A is not internal.

It is easy to see that every standard set is internal and that every element of an internal set is internal. The latter fact follows from our previous discussion: let $A \in {}^*B$; then, for some k, $A \subseteq V_k(^*\mathbb{R})$, thus $A \in {}^*(V_{k+1}(\mathbb{R}))$. But it follows from 1.2.1 and transfer that if $A_1 \in A \in {}^*(V_{k+1}(\mathbb{R}))$, then $A_1 \in {}^*(V_k(\mathbb{R}))$; the stages are in set-theoretic terminology *transitive*.

REMARK.   Because of their importance we will describe in detail the internal sets in the model. Let A be internal; thus $A \in {}^*(V_{k+1}(\mathbb{R}))$ for some $k \geq 1$. This means that A will be of the form $A = j(A_\mathcal{U})$, for some $A_\mathcal{U}$. By the construction of j, we then get

$$A \in {}^*(V_{k+1}(\mathbb{R})) \qquad \text{iff} \quad j(A_\mathcal{U}) \in j(i(V_{k+1}(\mathbb{R}))),$$

$$\text{iff} \quad A_\mathcal{U} \in_\mathcal{U} i(V_{k+1}(\mathbb{R})),$$

where i is the embedding of $V(\mathbb{R})$ into the ultrapower. The definition of $\in_\mathcal{U}$ then gives

$$A \in {}^*(V_{k+1}(\mathbb{R})) \qquad \text{iff} \quad \{i \in \mathbb{N} \mid A_i \in V_{k+1}(\mathbb{R})\} \in \mathcal{U},$$

where $\langle A_1, A_2, \ldots \rangle$ is the bounded sequence defining $A_\mathcal{U}$. Thus the *internal* sets are precisely the objects we obtain by starting with an arbitrary bounded sequence $\langle A_1, A_2, \ldots \rangle$; the *standard* objects are obtained by starting from a constant sequence $\langle A, A, \ldots \rangle$.

As an example consider the sequence $\langle A_1, A_2, \ldots, A_n, \ldots \rangle$, where $A_n = [0, n] \subseteq \mathbb{R}$. The internal (but not standard) set defined by this sequence is the interval $[0, \omega] \subseteq {}^*\mathbb{R}$, where ω is the infinite number in $^*\mathbb{R}$ defined by the sequence $\langle 1, 2, \ldots, n, \ldots \rangle$; i.e., a number $x \in {}^*\mathbb{R}$ belongs to $[0, \omega]$ iff $0 \leq x \leq \omega$. We have the following important property of internal sets.

1.2.6. PROPOSITION.   (i)   Every nonempty internal subset of $^*\mathbb{N}$ has a least element.

(ii)   Every nonempty internal subset of $^*\mathbb{R}$ with an upper bound has a least upper bound.

The proof is a simple exercise in the use of the transfer principle. We prove (i), so let $A \subseteq {}^*\mathbb{N}$ be internal. Then $A \in {}^*(V_2(\mathbb{R}))$; see (9). We can express the fact that an internal subset of $^*\mathbb{N}$ has a least element by the condition

$$\Phi =_{\text{def}} \forall X \in {}^* V_2(\mathbb{R})[X \neq \varnothing \wedge X \subseteq {}^*\mathbb{N} \to X \text{ has a } <\text{-least element}].$$

To be exact we must write out in detail the condition that X has a $<$-least element, which is

$$\exists x \in X[\forall y \in X \neg (y < x)],$$

but expanding such clauses into "correct" $L(V(\mathbb{R}))$ statements is a matter of routine and will be largely omitted.

We thus have a condition Φ such that $\Phi(\mathbb{N}, V_2(\mathbb{R}))$ is true in $V(\mathbb{R})$. By transfer $\Phi(*\mathbb{N}, *V_2(\mathbb{R}))$ is true in $V(*\mathbb{R})$, proving (i) of 1.2.6.

REMARK. It follows from this that $*\mathbb{N} - \mathbb{N}$ is external since there is no $<$-least element in $*\mathbb{N} - \mathbb{N}$; if $x \in *\mathbb{N} - \mathbb{N}$, then also $x - 1 \in *\mathbb{N} - \mathbb{N}$. We also see that $\mathbb{N}$ is external; thus $\mathbb{N} \in V_2(*\mathbb{R}) - *V_2(\mathbb{R})$. From (ii) it follows that $\mathbb{R}$ as a subset of $*\mathbb{R}$ is external. Note that Proposition 1.2.6 is valid only for internal sets; the positive infinitesimals in $*\mathbb{R}$ is bounded but has no least upper bound.

We spell out a few important corollaries of 1.2.6.

1.2.7. PROPOSITION. (i)   If A is internal and $\mathbb{N} \subseteq A$, then A contains some infinite natural number, i.e., an element of $*\mathbb{N} - \mathbb{N}$.

(ii)   If A is internal and every infinite $n \in *\mathbb{N}$ belongs to A, then A contains some standard $n \in \mathbb{N}$.

(iii)   If an internal set A contains every positive infinitesimal, then A contains some positive standard real.

(iv)   If an internal set A contains every standard positive real, then A contains some positive infinitesimal.

The first two parts of this proposition are often referred to as *overflow* and *underflow*. The proofs are immediate from 1.2.6; e.g., if an internal set A contained every positive infinitesimal, but did not contain a positive standard real, then A would be an internal subset of $*\mathbb{R}$ with an upper bound, but with *no* least upper bound.

How do you recognize a set as internal? The following simple principle is quite powerful.

1.2.8. INTERNAL DEFINITION PRINCIPLE. Let $A_1, \dots, A_n$ be internal sets in $V(*\mathbb{R})$ and let $\Phi(X_1, \dots, X_n, x)$ be an $L(V(\mathbb{R}))$ statement. Then the set

$$\{x \in A_1 | \Phi(A_1, \dots, A_n, x)\}$$

is internal.

The proof goes as follows. Since $A_1, \dots, A_n$ are internal, there must be some integer m such that $A_1, \dots, A_n \in *(V_m(\mathbb{R}))$. In $V(\mathbb{R})$ we have the truth of the following set existence or comprehension principle:

$$\forall X_1 \cdots \forall X_n \in V_m(\mathbb{R}) \, \exists y \in V_{m+1}(\mathbb{R})[y = \{x \in X_1 | \Phi(X_1, \dots, X_n, x)\}],$$

where $y = \{x \in X_1 | \Phi(X_1, \ldots, X_n, x)\}$ is an abbreviation of the formula $\forall x(x \in y \leftrightarrow x \in X_1 \wedge \Phi(X_1, \ldots, X_n, x))$; thus x is a bound variable of the term $\{x \in X_1 | \Phi(X_1, \ldots, X_n, x)\}$.

Using transfer and the fact that $A_1, \ldots, A_n \in {}^*(V_m(\mathbb{R}))$, we conclude that the set $\{x \in A_1 | \Phi(A_1, \ldots, A_n, x)\}$ is in ${}^*V_{m+1}(\mathbb{R})$, hence internal.

1.2.9. REMARK. Together, 1.2.7 and 1.2.8 carry more punch than you would initially guess. Let the functions f and g be internal, i.e., internal as sets in $V({}^*\mathbb{R})$. Suppose that you have proved that for all $t \in {}^*[0, 1]$ there is some infinitesimal $\delta_t > 0$ such that $|f(t) - g(t)| < \delta_t$. Can we then find a *uniform* estimate δ such that $|f(t) - g(t)| < \delta$ for all $t \in {}^*[0, 1]$?

The answer is yes and follows immediately from 1.2.7(iv) and 1.2.8. The set

$$(12) \qquad A = \{r \in {}^*\mathbb{R} \,|\, r > 0 \wedge \forall t \in {}^*[0, 1](|f(t) - g(t)| < r)\}$$

is internal by 1.2.8. We easily see that A contains every standard positive real, hence by 1.2.7 (iv) it contains some positive infinitesimal δ, which is the uniform estimate we asked for.

This concludes our general description of the extended universe. We have introduced the superstructure $V(\mathbb{R})$ and postulated a *superstructure embedding* (1)

$$* : V(\mathbb{R}) \to V({}^*\mathbb{R})$$

satisfying *extension* (1.2.2) and *transfer* (1.2.4). We have constructed a model of the superstructure embedding via the bounded ultrapower. Thus the axioms are consistent; i.e., there is a coherent conception of the infinitely small and the infinitely large. This conception is ruled by the transfer principle, which will be our main tool in the sections to come.

1.2.10. REMARK. We have restricted ourselves to $V(\mathbb{R})$; in some cases it may be more natural to work inside a different superstructure. For instance, let E be a linear normed space over the complex number $\mathbb{C}$. To apply the present machinery we must assume that E and $\mathbb{C}$ are objects in $V(\mathbb{R})$. However, from a certain point of view it would have been as natural to work in the superstructure $V(E \cup \mathbb{C})$, i.e., to regard E and $\mathbb{C}$ as basic or irreducible objects. The rest would be a set-theoretic construction from the set of "urelements" $E \cup \mathbb{C}$.

The ultrapower construction can easily be adapted to construct an *embedding*

$$(13) \qquad * : V(E \cup \mathbb{C}) \to V({}^*E \cup {}^*\mathbb{C}),$$

where $^*E$ is a normed linear space over $^*\mathbb{C}$, satisfying

(14)

> *Extension*: $^*E$ and $^*\mathbb{C}$ are proper extensions of E and $\mathbb{C}$, respectively, and $^*x = x$ for all $x \in E \cup \mathbb{C}$,

and

(15)

> *Transfer*: Let $A_1, \ldots, A_n \in V(E \cup \mathbb{C})$. Any $L(V(E \cup \mathbb{C}))$ statement Φ which is true of $A_1, \ldots, A_n$ in $V(E \cup \mathbb{C})$ is true of $^*A_1, \ldots, ^*A_n$ in $V(^*E \cup ^*\mathbb{C})$,

where $L(V(E \cup \mathbb{C}))$ is the language of the structure E as a normed linear space over $\mathbb{C}$ augmented by the $\in$-relation and bounded set quantification.

This setting would be the natural framework for our discussion of linear operators in Section 2.2; for other purposes we may choose a different base structure S for $V(S)$.

1.3. LIMITS, CONTINUITY, AND THE DERIVATIVE

We continue our basic course with a brief discussion of limits, continuity, and the derivative.

A sequence $\langle a_n \rangle_{n \in \mathbb{N}}$ is a map $a : \mathbb{N} \to \mathbb{R}$ and, as such, has an extension to a map $^*a : ^*\mathbb{N} \to ^*\mathbb{R}$. For any $n \in {}^*\mathbb{N}$ we write $a_n = {}^*a(n)$. We use $\langle a_n \rangle_{n \in {}^*\mathbb{N}}$ to denote the extended sequence.

For any elements a, $a' \in {}^*\mathbb{R}$ we shall write $a \approx a'$ to mean that the difference $a - a'$ is infinitesimal.

1.3.1. PROPOSITION. $\lim_{n \to \infty} a_n = a$ iff $a_\omega \approx a$ for all $\omega \in {}^*\mathbb{N} - \mathbb{N}$.

Here the left-hand side of the equivalence has its standard meaning inside $V(\mathbb{R})$. The right-hand side is a statement about the extended universe $V(^*\mathbb{R})$. Thus 1.3.1 *characterizes* the limit notion in nonstandard terms. But it could also have been taken as a *definition* of what it means for a standard sequence to converge.

If $\lim a_n = a$, then given any $\varepsilon > 0$ there is some $n \in \mathbb{N}$ such that the following statement is true in $V(\mathbb{R})$:

$$\forall m \in \mathbb{N}(m \geq n \to |a - a_m| < \varepsilon).$$

By transfer (1.2.4) the statement

$$\forall m \in {}^*\mathbb{N}(m \geq n \to |a - a_m| < \varepsilon)$$

is true in $V(^*\mathbb{R})$. If $\omega \in {}^*\mathbb{N} - \mathbb{N}$, then $|a - a_\omega| < \varepsilon$ is true in $V(^*\mathbb{R})$. Since this is true for all standard $\varepsilon > 0$, it means that the difference $a - a_\omega$ is infinitesimal, i.e., $a_\omega \approx a$.

We present two versions of the proof of the converse: (i) If $a \approx a_\omega$ for all $\omega \in {}^*\mathbb{N} - \mathbb{N}$, then, in particular, the following statement is true in $V({}^*\mathbb{R})$, where $\varepsilon > 0$ is some standard real:

$$\Phi(\varepsilon, {}^*\mathbb{N}) \qquad \text{iff} \qquad \exists n \in {}^*\mathbb{N} \, \forall m \in {}^*\mathbb{N}(m \geq n \to |a - a_m| < \varepsilon).$$

By transfer (1.2.4) $\Phi(\varepsilon, \mathbb{N})$ is true in $V(\mathbb{R})$. Since ε is arbitrary, this means that $\lim a_n = a$ in the standard sense.

(ii)   Once more let $a \approx a_\omega$ for all $\omega \in {}^*\mathbb{N} - \mathbb{N}$ and fix an $\varepsilon > 0$ in $\mathbb{R}$. The set

$$A = \{n \in {}^*\mathbb{N} \, \big| \, |a - a_m| < \varepsilon, \text{ all } m \geq n, \, m \in {}^*\mathbb{N}\}$$

is *internal* by 1.2.8. By 1.2.7 (ii) A contains some finite $n_\varepsilon \in \mathbb{N}$. Once more convergence in the standard sense follows.

We could go on from here to develop a nonstandard version of the theory of limits. For instance, Cauchy's criterion tells us that the standard sequence $\langle a_n \rangle_{n \in \mathbb{N}}$ converges iff $a_\omega \approx a_\lambda$ for all $\omega, \lambda \in {}^*\mathbb{N} - \mathbb{N}$, and that the limit is $\text{st}(a_\omega)$ for all $\omega \in {}^*\mathbb{N} - \mathbb{N}$. In general, $\text{st}(a_\omega)$, if it exists, defines a limit point of the sequence, and every limit point is of this form. But we shall leave the general theory to the reader and only concentrate on one particular point which will be of great importance in Chapter 7—double limits.

Given a double sequence $\langle a_{m,n} \rangle_{m,n \in \mathbb{N}}$, we will let $\langle a_{m,n} \rangle_{m,n \in {}^*\mathbb{N}}$ denote the extension ${}^*(\langle a_{m,n} \rangle_{m,n \in \mathbb{N}})$. By analogy with Proposition 1.3.1 it is natural to guess that

$$(1) \qquad\qquad \lim_{n \to \infty} \lim_{m \to \infty} a_{m,n} = \text{st}(a_{\omega, \lambda})$$

for all $\omega, \lambda \in {}^*\mathbb{N} - \mathbb{N}$. But then

$$\lim_{n \to \infty} \lim_{m \to \infty} a_{m,n} = \text{st}(a_{\omega, \lambda}) = \lim_{m \to \infty} \lim_{n \to \infty} a_{m,n},$$

and we know that this is not true in general; the order of the limits matters. We will show that (1) is true if we are more careful in our choice of ω and λ; roughly speaking, we will show that (1) holds if ω is "large" relative to λ, but that if λ is "large" relative to ω, we have instead

$$(2) \qquad\qquad \lim_{m \to \infty} \lim_{n \to \infty} a_{m,n} = \text{st}(a_{\omega, \lambda}).$$

Let us first illustrate this by a simple example. Define $\langle a_{m,n} \rangle_{m,n \in {}^*\mathbb{N}}$ by

$$(3) \qquad\qquad a_{m,n} = \begin{cases} a & \text{if} \quad m > n, \\ b & \text{if} \quad m < n, \\ c & \text{if} \quad m = n, \end{cases}$$

where a, b, and c are real numbers. Thus

$$(4) \qquad \lim_{n \to \infty} \lim_{m \to \infty} a_{m,n} = a,$$

$$(5) \qquad \lim_{m \to \infty} \lim_{n \to \infty} a_{m,n} = b.$$

By transfer, (3) remains true for infinite integers, and we get

$$a = \mathrm{st}(a_{\omega,\lambda})$$

if $\omega > \lambda$, but

$$b = \mathrm{st}(a_{\omega,\lambda})$$

if $\omega < \lambda$. Note that if $\omega = \lambda$, then $c = \mathrm{st}(a_{\omega,\lambda})$, and hence there may be infinite integers ω, λ where the standard part is different from both limits.

1.3.2. PROPOSITION. Let $\langle a_{m,n} \rangle_{m,n \in \mathbb{N}}$ be a standard sequence. (i) Assume that $\lim_{n \to \infty} \lim_{m \to \infty} a_{m,n} = a$ and $\lim_{m \to \infty} \lim_{n \to \infty} a_{m,n} = b$. For each $m \in {}^*\mathbb{N} - \mathbb{N}$ there are numbers n_1, $n_2 \in {}^*\mathbb{N} - \mathbb{N}$ such that $a_{m,n} \approx a$ when $n \leq n_1$ and $a_{m,n} \approx b$ when $n \geq n_2$. Similarly, for each $n \in {}^*\mathbb{N} - \mathbb{N}$ there are m_1, $m_2 \in {}^*\mathbb{N} - \mathbb{N}$ such that $a_{m,n} \approx b$ when $m \leq m_1$ and $a_{m,n} \approx a$ when $m \geq m_2$.

(ii) We have $a_{m,n} \approx a$ for all m, $n \in {}^*\mathbb{N} - \mathbb{N}$ iff for each $\varepsilon \in \mathbb{R}_+$ there is an $N \in \mathbb{N}$ such that $|a_{m,n} - a| < \varepsilon$ whenever m, $n > N$.

Part (i) of the proposition is illustrated in Fig. 1.1. Before we turn to the proof, let us just remark that the nonstandard condition in (ii) does not necessarily hold even when $\lim_{n \to \infty} \lim_{m \to \infty} a_{m,n} = \lim_{m \to \infty} \lim_{n \to \infty} a_{m,n}$; to see this just choose $a = b \neq c$ in the example above.

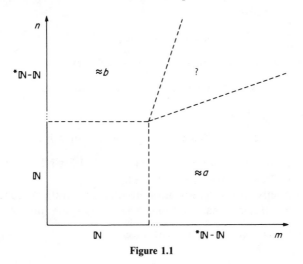

Figure 1.1

To prove (i), we first observe that by applying Proposition 1.3.1 to each of the sequences $\{a_{m,n}\}_{m \in \mathbb{N}}$, $n \in \mathbb{N}$, we get that $a_{m,n} \approx a$ whenever $m \in {}^*\mathbb{N} - \mathbb{N}$, $n \in \mathbb{N}$. Fixing an $m \in {}^*\mathbb{N} - \mathbb{N}$, this implies that the internal set

$$A_m = \{n \in {}^*\mathbb{N} \,|\, \forall k \le n(|a_{m,k} - a| < 1/k)\}$$

contains all of $\mathbb{N}$, and hence has an infinite element n_1. By definition n_1 has the property we want.

To find n_2, we first define

$$a_m = \lim_{n \to \infty} a_{m,n},$$

and note that $a_m \approx b$ when $m \in {}^*\mathbb{N} - \mathbb{N}$. Since

$$\forall m \in \mathbb{N} \; \forall \varepsilon \in \mathbb{R}_+ \; \exists n_2 \in \mathbb{N} \; \forall n \ge n_2(|a_{m,n} - a_m| < \varepsilon)$$

holds in the standard universe, the transferred statement

$$\forall m \in {}^*\mathbb{N} \; \forall \varepsilon \in {}^*\mathbb{R}_+ \; \exists n_2 \in {}^*\mathbb{N} \; \forall n \ge n_2(|a_{m,n} - a_m| < \varepsilon)$$

is true in the nonstandard universe. The n_2 we get by choosing m infinite and ε infinitesimal is the one we want.

The second part of (i) is the same as the first part with the roles of m and n interchanged, and it need not be proved separately.

The proof of (ii) is almost a copy of the proof of Proposition 1.3.1. If $a_{m,n} \approx a$ for all $m, n \in {}^*\mathbb{N} - \mathbb{N}$, then for all $\varepsilon \in \mathbb{R}_+$ the statement

$$\exists N \in {}^*\mathbb{N} \; \forall m, n \ge N(|a_{m,n} - a| < \varepsilon)$$

is true, and by transfer so is

$$\exists N \in \mathbb{N} \; \forall m, n \ge N(|a_{m,n} - a| < \varepsilon).$$

For the converse we note that if $N \in \mathbb{N}$ is such that

$$\forall m, n \in \mathbb{N}(m, n \ge N \to |a_{m,n} - a| < \varepsilon),$$

then by transfer

(6) $$\forall m, n \in {}^*\mathbb{N}(m, n \ge N \to |a_{m,n} - a| < \varepsilon).$$

If $m, n \in {}^*\mathbb{N} - \mathbb{N}$, the condition in (6) is satisfied for all standard ε. Thus $a_{m,n} \approx a$, and Proposition 1.3.2 is proved.

We turn briefly to the topics of continuity and uniform continuity. Let $f : I \to \mathbb{R}$ be a standard, real function, where I is some interval in $\mathbb{R}$.

1.3.3. PROPOSITION. (i) f is continuous at $a \in I$ iff $f(x) \approx f(a)$ for all $x \in {}^*I$ such that $a \approx x$.

(ii) f is uniformly continuous in I iff $f(x) \approx f(y)$ for all $x, y \in {}^*I$ such that $x \approx y$.

The reader should appreciate the difference. $f(x) = 1/x$ is continuous for every $x \in \langle 0, 1 \rangle$, but f is not uniformly continuous. Let ω be an infinite number in ${}^*\mathbb{N}$; then $1/\omega \approx 1/\omega^2 \approx 0$, but $f(1/\omega) = \omega$ and $f(1/\omega^2) = \omega^2$ are far from being close. Here $1/\omega$ and $1/\omega^2$ represents two different ways of converging to 0.

Proofs are again simple; e.g., assume that $f(a) \approx f(x)$ for all $x \in {}^*I$ such that $a \approx x$. Fix an $\varepsilon \in \mathbb{R}_+$, i.e., $\varepsilon > 0$. The set

$$A = \{\delta \in {}^*\mathbb{R} \mid \forall x \in {}^*I (|a - x| < \delta \to |f(a) - f(x)| < \varepsilon)\}$$

is *internal* by 1.2.8 and contains every positive infinitesimal. By 1.2.7(iii) A must then contain some positive $\delta \in \mathbb{R}$; continuity in the standard sense follows.

1.3.4. REMARKS. Again we could develop a nonstandard theory of continuity, uniform continuity, uniform convergence, equicontinuity, and so on. This we shall not do, but we cannot resist inserting the following painless way of proving that if $f : I \to \mathbb{R}$ is continuous and I is compact, then f is uniformly continuous on I. The proof is an immediate combination of Propositions 1.1.2 and 1.3.3. Let $x, y \in {}^*I$ and assume that $x \approx y$. Compactness of I tells us that $\mathrm{st}(x)$, $\mathrm{st}(y)$ exist and are elements of I; obviously $\mathrm{st}(x) = \mathrm{st}(y)$. Since f is continuous in I, $f(x) \approx f(\mathrm{st}(x)) = f(\mathrm{st}(y)) \approx f(y)$.

We round off this section by introducing the nonstandard version of the derivative.

1.3.5. THE DERIVATIVE. Let I be an open interval in $\mathbb{R}$ and let $f : I \to \mathbb{R}$. Let dx be an infinitesimal different from zero. Then we call $dy = f(a + dx) - f(x)$ a *differential* of f at $a \in I$. If the standard part of dy/dx exists and is the same for all nonzero dx, then f has a derivative at a and

$$f'(a) = \mathrm{st}(dy/dx).$$

The reader may amuse himself by proving the standard results. Is, e.g., the chain rule nothing but the calculation

$$\frac{dy}{dx} = \frac{dy}{dx}\frac{du}{du} = \frac{dy}{du}\frac{du}{dx}?$$

1.4. THE INTEGRAL

A general theory of measure and integration will be developed in Chapter 3. Here we shall, in as simple as possible a setting, make sense of the heuristic idea that the integral is an infinite sum of infinitesimal parts.

Thus for this section let $f: I \to \mathbb{R}$ be a positive continuous function, where I is some interval in $\mathbb{R}$. Let $[a, b] \subseteq I$ and let Δx be a positive real. The *Riemann sum* is defined as

(1) $$\sum_a^b f(x)\, \Delta x = f(x_0)\, \Delta x + f(x_1)\, \Delta x + \cdots + f(x_{n-1})\, \Delta x$$
$$+ f(x_n)(b - n\, \Delta x),$$

where n is the largest integer such that $a + n\, \Delta x \le b$ and where $x_0 = a$, $x_1 = a + \Delta x, \ldots, x_n = a + n\, \Delta x$. Note that it may happen that $n\, \Delta x < b < (n + 1)\, \Delta x$. [Since f is positive and continuous we have formed the Riemann sum as the sum of the rectangles over each subinterval with height equal to the value of $f(x)$ at the left end of the base of the rectangle.]

The Riemann sum for fixed a, b is a function of Δx. By extension and transfer this function is also defined for positive infinitesimals dx. We get a corresponding hyperfinite sum

$$\sum_a^b f(x)\, dx,$$

where the number n in (1) is now an infinite number. The reader should notice that the Riemann sum is a finite hyperreal number; thus it has a standard part.

1.4.1. DEFINITION. Let $[a, b] \subseteq I$ and let dx be a positive infinitesimal. The *definite integral* of f from a to b with respect to dx is the standard part of the Riemann sum,

$$\int_a^b f(x)\, dx = \mathrm{st}\left(\sum_a^b f(x)\, dx \right).$$

This definition depends upon the choice of infinitesimal dx. But it can be immediately proved that if dx and du are two positive infinitesimals, then

$$\int_a^b f(x)\, dx = \int_a^b f(u)\, du.$$

Note that the x in $f(x)$ and u in $f(u)$ are dummy variables; the dx and the du are not. The equality above is thus not a matter of typographic convention but an assertion to be verified.

From Definition 1.4.1 we can now develop the elementary theory of integration. We mention one result which "justifies" the intuitive idea of calculating an integral by taking a typical infinitesimal element and adding up. We follow the exposition in Keisler (1976).

First a bit of notation. Given a nonzero infinitesimal Δx, we write

$$u \approx v \qquad \text{(compared to } \Delta x)$$

to mean that $u/\Delta x \approx v/\Delta x$.

1.4.2. INFINITE SUM THEOREM. We assume that (i) h is a real function continuous on $[a, b]$;

(ii) $B(u, w)$ is a real function with the additive property

$$B(u, w) = B(u, v) + B(v, w)$$

for $u < v < w$ in $[a, b]$.

(iii) For any infinitesimal subinterval $[x, x + \Delta x]$ of $*[a, b]$

$$B(x, x + \Delta x) \approx h(x)\, \Delta x \qquad \text{(compared to } \Delta x\text{)}.$$

Then

$$B(a, b) = \int_a^b h(x)\, dx.$$

The proof is rather simple. Choose $\omega \in {}^*\mathbb{N} - \mathbb{N}$ and set $\Delta x = (b - a)/\omega$. We show that for any $r > 0$, $r \in \mathbb{R}$,

$$\sum_a^b (h(x) - r)\, \Delta x < B(a, b) < \sum_a^b (h(x) + r)\, \Delta x,$$

which proves the result. In order to prove the last inequality above, we note that if $B(a, b) \geq \sum_a^b (h(x) + r)\, \Delta x$, then by the transfer principle there is an x such that $a \leq x < x + \Delta x \leq b$ and such that $B(x, x + \Delta x) \geq (h(x) + r)\, \Delta x$. Since $r \in \mathbb{R}_+$, this contradicts the fact that $\Delta B = B(x, x + \Delta x) \approx h(x)\, \Delta x$ (compared to Δx).

A typical application of the infinite sum theorem is the following volume determination (see Fig. 1.2). Let D be the region

$$D = \{(x, y) \mid a \leq x \leq b, 0 \leq y \leq g(x)\}.$$

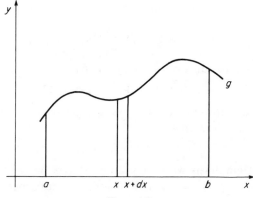

Figure 1.2

The solid generated by revolving D about the x axis has volume

$$V = \int_a^b \pi(g(x))^2 \, dx.$$

This is an immediate consequence of 1.4.2 since $\Delta V \approx \pi(g(x))^2 \Delta x$ (compared to Δx).

1.5. DIFFERENTIAL EQUATIONS

We close our calculus course by giving some examples of the application of nonstandard methods to differential equations. We first present the nonstandard proof of Peano's existence theorem for ordinary differential equations.

1.5.1. THEOREM. Let $f : [0, 1] \times \mathbb{R} \to \mathbb{R}$ be continuous and bounded. Let $u_0 \in \mathbb{R}$ be given. Then there exists a solution $u : [0, 1] \to \mathbb{R}$ such that $u(0) = u_0$ and

$$du(t)/dt = f(t, u(t)).$$

The key to the nonstandard proof consists in noticing that the following procedure is given by an $L(V(\mathbb{R}))$ condition on the parameters involved:

(1)

For all $n \in \mathbb{N}$ divide $[0, 1]$ into equal parts $t_0 = 0, t_1 = 1/n, \ldots, t_n = 1$, and define $v_n(t)$ over $[0, 1]$ inductively by first setting

$$v_n(0) = u_0, \qquad \frac{v_n(t_{i+1}) - v_n(t_i)}{t_{i+1} - t_i} = f(t_i, v_n(t_i))$$

and extend to all of $[0, 1]$ by linear interpolation.

Since the procedure is elementary, let us also carry it through for some $\lambda \in {}^*\mathbb{N} - \mathbb{N}$. A simple calculation shows that for $1 \leq \omega \leq \lambda$

$$(2) \qquad v_\lambda(t_\omega) = u_0 + \sum_{i=1}^{\omega-1} f(t_i, v_\lambda(t_i))(t_{i+1} - t_i).$$

Introduce a standard function $u : [0, 1] \to \mathbb{R}$ by

$$u(t) = \mathrm{st}(v_\lambda({}^*t)), \qquad t \in [0, 1].$$

Elementary considerations (using the compactness of $[0, 1]$ and the boundedness of f) show that

$$(3) \qquad f(t, {}^*u(t)) \approx f(t, v_\lambda(t)),$$

$t \in {}^*[0, 1]$. (For emphasis we have put an asterisk on the extended u, but in conformity with previous conventions we have dropped the asterisk on the extended f.)

Here (3) means that for all $t \in *[0, 1]$ there is some infinitesimal $\delta_t > 0$ such that

$$|f(t, *u(t)) - f(t, v_\lambda(t))| < \delta_t.$$

By Remark 1.2.9 the choice of δ_t can be made uniform (the internal definition principle); i.e., there exists some infinitesimal $\delta > 0$ such that

$$|f(t, *u(t)) - f(t, v_\lambda(t))| < \delta$$

for all $t \in *[0, 1]$. This gives the uniformity needed to pass from (2) to

$$(4) \qquad u(t) \approx u_0 + \sum_{i=1}^{\omega-1} f(t_i, u(t_i))(t_{i+1} - t_i),$$

where ω is such that $t \approx t_\omega$. However (4) is nothing but the standard

$$(5) \qquad u(t) = u_0 + \int_0^t f(s, u(s)) \, ds;$$

see Definition 1.4.1. This completes the proof.

Let us at this point add a mildly polemical remark. It has often been held that nonstandard analysis is highly nonconstructive, thus somewhat suspect, depending as it does upon the ultrapower construction to produce a model; see Section 1.1. On the other hand, nonstandard *praxis* is remarkably constructive; having the extended number set we can proceed with explicit calculations. A case in point is the existence theorem (Theorem 1.5.1). In the standard approach one uses in the final step the Ascoli lemma, which asserts that every bounded equicontinuous sequence of functions on a bounded interval I has a uniformly convergent subsequence. This part of the argument is lacking in the nonstandard proof, which makes it more direct. And indeed it is, in the following precise sense. It is possible to recast the nonstandard proof to give a proof of the Peano existence theorem where the only nonrecursive element is the weak König's lemma asserting that every infinite binary tree, i.e., infinite tree of sequence of 0's and 1's, has an infinite path. And this is a principle which is provably weaker, i.e., more constructive, than the Ascoli lemma (Simpson, 1984). On the other hand, adding saturation (see Chapter 2) adds real power to the nonstandard calculations, which explains the many successes of nonstandard methods in stochastic analysis and related fields (Henson *et al.*, 1984; Henson and Keisler, 1985).

In the generality of 1.5.1 there is no uniqueness of solution. The procedure described in (1) gives one. Is it possible through some "infinitesimal" variation of the data to obtain all solutions via a nonstandard difference equation?

1.5.2. EXAMPLE. The equation

(6) $u' = 3u^{2/3}$

with initial condition $u(0) = 0$ has for each $a \in [0, 1]$ a solution

$$u(t) = 0,   \qquad 0 \le t < a,$$
$$u(t) = (t - a)^3, \qquad a \le t \le 1.$$

The family of solutions can also be parameterized by the values $u(1) = r$, $r \in [0, 1]$; i.e., to each $r \in [0, 1]$ there is a unique solution u of (6) that satisfies $u(0) = 0$ and $u(1) = r$. We will show how to obtain all these solutions from the difference equation of (1),

(7) $v_\lambda(t_{i+1}) = v_\lambda(t_i) + (1/\lambda)3v_\lambda(t_i)^{2/3},$

where $\lambda \in {}^*\mathbb{N} - \mathbb{N}$, by imposing different initial conditions $v_\lambda(0) = \delta$, where δ is a positive infinitesimal.

Let v_δ be the unique solution of (7) with initial condition $v_\lambda(0) = \delta$, and let u_δ be the corresponding solution of (6). Clearly u_0 is the constant zero and is the solution obtained in 1.5.1.

We note that for any standard real $\delta > 0$ equation (6) has a unique solution with initial condition $u(0) = \delta$. This solution is the standard part of the solution v_δ obtained from (7). We observe that $v_\delta(1) > 1$. Thus the *internal* set

$$\{\delta \in {}^*\mathbb{R} \,|\, \delta > 0 \wedge v_\delta(1) > 1\}$$

contains every standard real >0; hence it contains some positive infinitesimal $\delta_0 > 0$.

The function $f(\tau) = v_\tau(1)$ is internal and continuous in the sense of $V({}^*\mathbb{R})$. We know that $f(0) = 0$ and $f(\delta_0) > 1$. Thus for every $r \in {}^*[0, 1]$ there is some $\delta \in [0, \delta_0] \subseteq {}^*\mathbb{R}$ such that $f(\delta) = r$, i.e., such that $v_\delta(1) = r$. Since u_δ is the standard part of v_δ we conclude that *every* solution of (6) with initial value $u(0) = 0$ can be obtained as the standard part of some solution v_δ of (7) by choosing the infinitesimal δ suitably.

This observation also has a numerical content. We can carry out an approximate calculation of the solution corresponding to a given a by choosing an initial value bearing the same relationship to a and the step length as δ has to a and n. It is, of course, a well-known fact in numerical analysis that various approximations, step lengths, often must be chosen to depend on each other in quite specific ways in order to exhibit a particular phenomenon. It could be that the nonstandard theory is the right way to discuss this.

This example is taken from Birkeland and Normann (1980), where the reader can find a more general discussion. We mention some of their other results.

First, if the solutions of the equation $u' = f(t, u)$ are *unique to the left* in the sense that if u_1 and u_2 are solutions and $u_1(t) = u_2(t)$ for some $t \in \langle 0, 1]$, then $u_1(s) = u_2(s)$ for all $s \in [0, t]$, then all solutions u of the equation with initial condition $u(0) = 0$ can be obtained from procedure (1) by choosing a suitable infinitesimal δ and starting from the initial condition $v_\lambda(0) = \delta$.

In the general case it does not suffice to perturb the initial condition only, we must perturb the whole equation. We call g a δ *perturbation* of f, where δ is some infinitesimal, if g is internal, continuous in the sense of $V(*\mathbb{R})$, and $\|f - g\|_\infty < \delta$. We let X_δ denote the set of δ perturbations of f. We further denote by v_g the unique solution of the difference equation in (1) with fixed initial condition $v_\lambda(0) = 0$ and with the δ approximation g replacing f in Eq. (1).

One can now prove that given f there is some infinitesimal $\delta > 0$ such that if u is a solution of $u' = f(t, u)$ with initial condition $u(0) = 0$ there is a δ perturbation g of f such that u is the standard part of v_g.

Our final example comes from the study of vector fields or dynamical systems of the form

$$(8) \qquad \dot{x} = f(x, y), \qquad \varepsilon \dot{y} = g(x, y)$$

where ε is a "small" parameter. Here $x = x(t)$, $y = y(t)$ are functions of "time" t, and $\dot{x}$, $\dot{y}$ denote time derivatives.

We are interested in the behavior for small ε. The standard approach is to use asymptotic expansions in powers of ε. This often leads to complicated computations [see Section 2.6 of Cole (1968)], in particular when f and g also depend on some auxiliary parameter. In the nonstandard setting we may use a fixed, infinitesimal ε instead of a moving, standard one.

1.5.3. REMARK. The work we report on was done by a group of French mathematicians, Callot (1981), F. Diener (1981), and M. Diener (1981). For more complete information the reader may consult the surveys Benoit (1980), Cartier (1982), and Zvonkin and Shubin (1984).

We begin with some general remarks on the system (8). For simplicity we assume that f and g are smooth standard functions, and ε is some fixed infinitesimal constant.

At nearstandard points $(x, y) \in *\mathbb{R}^2$ where $g(x, y)/\varepsilon$ is *infinite*, the trajectories of the system (8) are "quasi parallel" to the y axis; i.e., their standard part is parallel to it. At such points the "speed" $(\dot{x}^2 + \dot{y}^2)^{1/2}$ with which trajectories are traversed is infinite; such points are called *fast points*.

The remaining nearstandard points, the *slow points*, at which the trajectories may have other directions and the speed is finite, all lie close to the set $S: g(x, y) = 0$, which is called the *slow manifold.*

The orientation of the vector field at a fast point is determined by the sign of g, while the direction of the flow at a point on the slow manifold is determined by the sign of f at that point.

In particular, near a segment of the slow manifold S where the partial derivative $g_y(x, y)$ is negative, all fast trajectories are directed toward S, while near segments where g_y is positive, they all go away from it. We speak of *stable* (attracting) and *unstable* (repelling) parts of S.

We shall not develop a general theory here (see the references in Remark 1.5.3), but exhibit the rather complex behavior of trajectories in a simple but fairly representative example.

1.5.4. EXAMPLE. Our system will be a van der Pol equation of the form

$$(9) \qquad\qquad \varepsilon\ddot{x} + (x^2 - 1)\dot{x} + x - a = 0,$$

where a is a parameter and $\varepsilon \approx 0$. For reasons of symmetry we may suppose that $a > 0$ and that ε is a positive infinitesimal.

There are many ways of transforming this equation to a system of the form of (8); we find it convenient to first look at the Lienard's substitution $u = F(x) + \varepsilon\dot{x}$, where $F(x) = x^3/3 - x$. This leads to the vector field

$$(10) \qquad\qquad U_a: \quad \begin{aligned} \varepsilon\dot{x} &= u - F(x), \\ \dot{u} &= a - x. \end{aligned}$$

[Note that the use of variables in (10) is not entirely consistent with (8); this time the fast trajectories will be quasi parallel to the x axis.]

From classical theory, with standard x, u, a, ε, we recall the following facts; see, e.g., LaSalle and Lefschetz (1961). The only stationary or equilibrium point is $(a, F(a))$. If $a < 1$ it is stable, and all trajectories tend toward it as $t \to \infty$. If $a < 1$ it is unstable, and there is a stable limit cycle which goes around it and toward which all trajectories tend as $t \to \infty$. As a increases toward 1, the limit cycle shrinks continuously toward the stationary point.

By transfer this description also applies in a nonstandard setting with ε chosen infinitesimal. But now more can be said. For brevity we restrict the discussion to the case $a < 1$.

The slow manifold S_u for the system U_a of (10) is the cubic $u = F(x)$; it is stable when $|x| > 1$, unstable when $|x| < 1$. Trajectories through nearstandard points (x, u) are quasi parallel to the x axis unless $(u - F(x))/\varepsilon$ is finite, and they are oriented toward increasing or decreasing x depending on the sign of $u - F(x)$. Thus the standard part of any trajectory must be a union of parallels to the x axis and of segments of the slow manifold. In

(a)

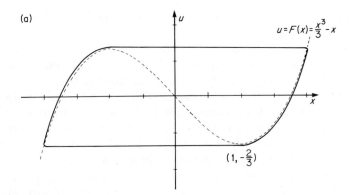

$u = F(x) = \frac{x^3}{3} - x$

$(1, -\frac{2}{3})$

(b)

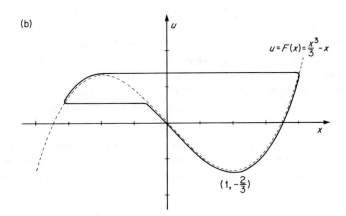

$u = F(x) = \frac{x^3}{3} - x$

$(1, -\frac{2}{3})$

(c)

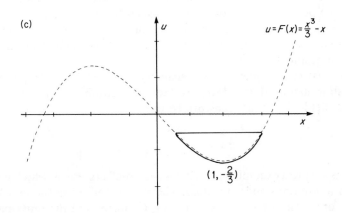

$u = F(x) = \frac{x^3}{3} - x$

$(1, -\frac{2}{3})$

Figure 1.3

particular, the only possible shapes for (the standard part of) a limit cycle are as shown in Fig. 1.3a–c.

Continuity arguments [either by transfer from standard theory or directly for the present nonstandard situation, as in Benoit *et al.* (1980)] show that all three shapes must occur. It is also clear that the cases of Fig. 1.3b,c occur only when $1 - a$ is infinitesimal. The reason for this is simple: if a has a noninfinitesimal distance from 1, a point on a trajectory passing infinitesimally close to the point $(1, -\frac{2}{3})$ on the slow manifold will have a noninfinitesimal distance from the slow manifold at $x = a$, since $\dot{u} < 0$ for $x > a$; this means that we are in the situation of Fig. 1.3a.

We shall be interested in the situations of Fig. 1.3b,c, where a trajectory first follows the stable and then the unstable part of the slow manifold infinitesimally closely, both for some noninfinitesimal distance.

Now numerical experiments (Benoit *et al.*, 1980) have uncovered an unexpected phenomenon: the transition from a cycle of the shape of Fig. 1.3a to an "infinitesimally" small cycle of the shape of Fig. 1.3c is extremely abrupt and takes place well away from $a = 1$; for $\varepsilon = 1/100$ it all takes place within $2 \cdot 10^{-10}$ of the value $a = 0.9987404512$.

Thus we must expect some arithmetic restrictions on the value of a in order to exhibit what we, according to the next definition, will call a "canard."

1.5.5. DEFINITION. A *canard* (duck) for the vector field U_a is a segment of a trajectory which first follows the stable, then the unstable part of the slow manifold S_u infinitesimally closely, both for some noninfinitesimal distance.

The name canard is motivated by the general shape of the cycle in Fig. 1.3b; see Benoit *et al.* (1980). Examples of canards are, of course, segments near the point $(1, -2/3)$ of the limit cycles in Fig. 1.3b,c.

The question now is: For which values of a does the vector field U_a allow canards? Remember that ε is now a positive infinitesimal and a must be infinitesimally close to 1.

To answer this question we "magnify" the immediate neighborhood of the slow manifold $u = F(x)$ by the substitution $y = (u - F(x))/\varepsilon$. The vector field U_a is then transformed to

$$(11) \qquad Y_a: \quad \begin{aligned} \dot{x} &= y, \\ \varepsilon\dot{y} &= a - x - (x^2 - 1)y. \end{aligned}$$

This is a typical example of a "change of scale" argument, which is available to us in the nonstandard setting. The slow manifold in the (x, u) plane is transformed into the x axis in the (x, y) plane, and the finite part of the (x, y) plane represents an infinitesimal strip around the slow manifold in the (x, u) plane. And this rescaling will enable us to answer the question posed above. (See Fig. 1.4.)

Let us note that the field Y_a is of the form (8) with its fast trajectories quasi parallel to the y axis. The slow manifold S_y is defined by $(x^2 - 1)y = a - x$, and it is stable when $|x| > 1$, unstable when $|x| < 1$. Note that when $a \approx 1$ the standard part H of S_y is the union of the straight line $x = 1$ and the hyperbola $y = -1/(x + 1)$.

1.5.6. REMARK. We note that (11) is the usual "phase plane" representation of the van der Pol equation (9). We preferred to start with the Lienard representation (10) because the fast trajectories for (11) for $|x| < 1$ "go off to infinity."

There is one further point we need to discuss before starting to calculate. There is a kind of "transition zone" of points (x, u) in the Lienard plane with $u - F(x)$ infinitesimal but $(u - F(x))/\varepsilon$ infinite. These are fast points in the Lienard plane, but they are also infinitesimally close to the slow

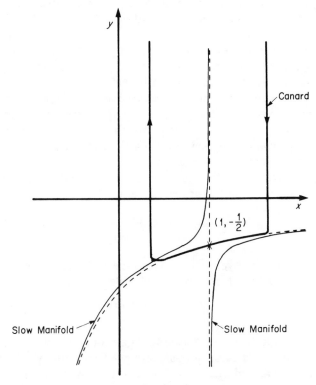

Figure 1.4

manifold; in the phase plane they are infinitely far off. We must show that trajectories cross this zone in a "nice" way.

Consider, for example, the trajectory through a nearstandard point (x_0, y_0) in the phase plane with $x_0 > 1$ and $y_0 > 0$. It is quasi parallel to the y axis, and may therefore be described as the graph of a function $x = \gamma(y)$, defined at least for all finite $y \geq y_0$. But its domain of definition is internal, hence contains some infinite $y_1 > y_0$. Now for all finite y, $\gamma(y) \approx x_0$, hence $\gamma(y) \approx x_0$ on some interval $y_0 \leq y \leq y_2$, with infinite $y_2 \leq y_1$.

1.5.7. REMARK. The existence of y_2 is an exercise in the use of the internal definition principle. Consider the internal set

$$A = \{y \in {}^*\mathbb{R} \mid \forall z \in {}^*\mathbb{R}(y_0 < z < y \rightarrow |\gamma(z) - x_0| < 1/y\}.$$

Since A contains arbitrarily large real numbers, it must have an infinite element y_2.

The point $(\gamma(y_2), y_2)$ corresponds to a point (x_2, u_2) in the Lienard plane with $(u_2 - F(x))/\varepsilon$ infinite, thus to a point on a fast trajectory, quasi parallel to the x axis. In short, the correspondence between trajectories in the two coordinate planes is as nice as one could hope for.

Now let $\gamma: t \rightarrow \gamma(t) = (x(t), u(t))$, $t_1 < t < t_2$, be a canard for the vector field U_a; i.e., γ is a trajectory for U_a, ${}^\circ x(t_1) > 1 > {}^\circ x(t_2)$, and $(u(t) - F(x(t)))/\varepsilon$ is finite for $t_1 \leq t \leq t_2$.

Our change of scale takes γ into a trajectory $\tilde{\gamma}$ for the magnified field Y_a. Since the x coordinate of $\tilde{\gamma}$ is the same as for γ, $\tilde{\gamma}$ also projects onto the same noninfinitesimal interval $(x(t_2), x(t_1))$ of the x axis and it is clear that $\tilde{\gamma}(t)$ is finite for all $t \in [t_1, t_2]$. But this is possible only if $\tilde{\gamma}(t)$ is infinitesimally close to the slow manifold S_y at least for ${}^\circ t_1 < {}^\circ t < {}^\circ t_2$, that is, if $\tilde{\gamma}$ is again a canard (see Fig. 1.4).

For which values of a do canards exist? Heuristically, we may argue as follows: the slope of the vector field Y_a at the point $P = (1, -\frac{1}{2})$ is

$$\frac{dy}{dx} = \frac{1}{\varepsilon}\left(\frac{a-x}{y} - (x^2 - 1)\right)\Bigg|_{\substack{x=1 \\ y=-1/2}} = \frac{2(1-a)}{\varepsilon}.$$

At the same point the hyperbola H has slope $\frac{1}{4}$. If a canard exists, it ought to be quasi parallel to H; i.e., we should expect that $2(1-a)/\varepsilon - \frac{1}{4} \approx 0$, or

$$(12) \qquad\qquad a = 1 - \varepsilon/8 - \varepsilon\eta,$$

for some $\eta \approx 0$.

This result is, in fact, true but it needs a proof. One way is to consider the behavior of Y_a at some suitable straight line l through P with slope between that of Y_a and that of H at P and to show that unless a is of the form (12), all trajectories cross it in the wrong direction.

It is convenient to parameterize l by $x = 1 + \tau$, $y = -\frac{1}{2} + \theta\tau = -\frac{1}{2}(1 - 2\theta\tau)$, and to write $\eta = 2(1 - a)/\varepsilon$. Then on l:

$$\frac{dy}{dx} = \frac{\eta - \tau^2/\varepsilon(1 - 2\theta\tau - 4\theta)}{1 - 2\theta\tau}.$$

Suppose now that $°\eta > \frac{1}{4}$; the case $°\eta < \frac{1}{4}$ is analogous and will be omitted. Choose $\theta \le \frac{1}{2}$ such that $°\eta > °\theta > \frac{1}{4}$ and note that with this choice $(dy/dx) - \theta > 0$ on some standard neighborhood $|\tau| \le \delta$ of P on l (see Fig. 1.5). This implies that all trajectories of Y_a cross this line segment in the downward direction. If a trajectory was a canard for this value of $\eta = 2(1 - a)/\varepsilon$, it would have to "enter" our picture at a point $(1 + \tau, y)$ with $y < -\frac{1}{2} + \theta\tau$. Then in order to be a canard it would have to cross l in an upward direction. But this was impossible on the segment $|\tau| \le \delta$, and since the endpoints of this segment are in the fast part of the plane, we see that no canard is possible. Thus a must necessarily be of the form (12).

We observed above that the canard γ for U_a remained a canard $\tilde{\gamma}$ for the magnified field Y_a. The process can be iterated and eventually leads to

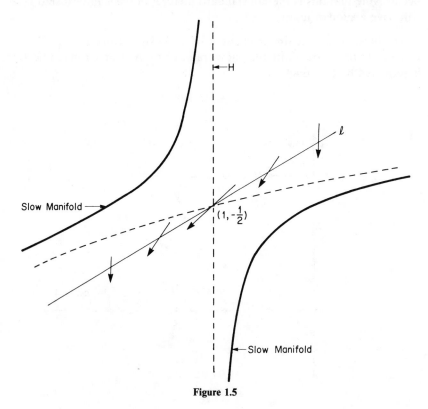

Figure 1.5

the conclusion that the values of a which allow canards are of the form

$$(13) \qquad a = \sum_{j=0}^{N} a_j \varepsilon^j + \varepsilon^N \eta, \qquad \eta \simeq 0,$$

for every standard natural number N. Here the a_j are standard reals depending only on the given vector field U_a; note that $a_0 = \text{st}(a) = 1$, $a_1 = -\frac{1}{8}$. This shows that the values of a for which canards exist all lie in an extremely narrow interval; the reader is asked to recall the numerical experiments reported on above.

1.5.8. REMARK. By the *amplitude* of a limit cycle we understand the length of its projection onto the x axis. In connection with the Lienard plane representation of (9) (see Fig. 1.3), we noted that the transition from a cycle of the shape of Fig. 1.3a to an infinitesimally small cycle of the shape of Fig. 1.3c is rather abrupt. This is illustrated in Fig. 1.6. The amplitude in case of Fig. 1.3a is 4. Near 1 the amplitude is infinitesimal. At $1 - \varepsilon/8$ the curve is extremely steep; this is the region of values of a where canards do occur. Note that this is the nonstandard picture; in the standard setting we will have a sudden jump.

We have seen that linear changes of scale by factors ε^{-n}, $n \in \mathbb{N}$, are insufficient to separate different slow trajectories. A stronger magnification is provided by the function

$$\tau \to \tau^{[\varepsilon]} = |\tau|^\varepsilon \, \text{sgn}(\tau),$$

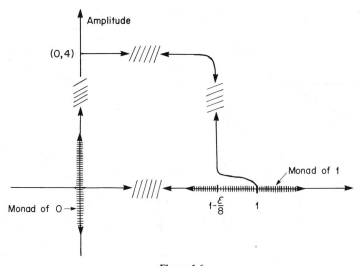

Figure 1.6

i.e., the odd function which is equal to τ^ε for $\tau > 0$. Note that $\tau^{[\varepsilon]} \approx 1$ when ${}^\circ|\tau|$ is finite and noninfinitesimal and even when $\varepsilon^n < |\tau| < \varepsilon^{-n}$ for some $n \in \mathbb{N}$ (since $\varepsilon \ln \varepsilon \approx 0$ when $\varepsilon \approx 0$). And if $0 < {}^\circ|\tau^{[\varepsilon]}| < 1$ then τ is of the form $\tau = \exp(-k/\varepsilon)$ with ${}^\circ k$ finite positive.

Returning to the field U_a, *we consider from now on only that particular value of a for which the limit cycle follows the slow manifold S_u all the way between -1 and 1*, i.e., the border case between parts b and c of Fig. 1.3. This limit cycle is a canard for $-1 \le x \le 2$ and we represent it as a function graph $y = \gamma(x)$, $-1 \le x \le 2$, in the phase plane, Fig. 1.4.

The change of scale

$$w = (y - \gamma(x))^{[\varepsilon]}$$

leads to a new vector field

(14) W_a:
$$\dot{x} = \gamma(x) + w^{[1/\varepsilon]},$$
$$\dot{w} = \frac{(a - x)w}{\gamma(x)}\, w,$$

which in the domain $-1 \le x \le 2$, ${}^\circ|x| < 1$, can be viewed as an infinitesimal perturbation of the elementary integrable vector field

(15) $\tilde{W}$:
$$\dot{x} = -1/(x + 1),$$
$$\dot{w} = (1 - x^2)w.$$

To see this just note that ${}^\circ\gamma(x) = -1/(x + 1)$ and that $w^{[1/\varepsilon]} \approx 0$ when ${}^\circ|w| < 1$. Solution curves for (15) are

(16) $$w(x) = c \exp(\varphi(x)),$$

where c is a constant and $\varphi(x) = x^4/4 + x^3/3 - x^2/2 - x$. Some of the solution curves are sketched in Fig. 1.7.

The important feature of φ is that it has a minimum at $x = 1$ and from there increases monotonically as x increases or decreases. Thus a map ψ is defined by the following prescription: $\psi(1) = 1$, and if $x \ne 1$, then $\psi(x)$ is the only number $\ne x$ such that $\varphi(\psi(x)) = \varphi(x)$. For our purpose both x and $\psi(x)$ must lie in $[-1, 2]$, so we can define ψ only for $-1 \le x \le \psi(-1) = \frac{5}{3}$.

1.5.9. REMARK. It can be proved (see Benoit *et al.*, 1980) that in our domain the trajectories for W_a must lie infinitesimally close to those for $\tilde{W}$.

Further, in analogy to a previous argument, one can show that a trajectory $w(x)$ that reaches 1 (or -1) for some $x = \xi$, with ${}^\circ\xi \in \langle -1, 2\rangle$, reappears in both the phase plane and the Lienard plane as a fast trajectory at abscissas with the same standard part ${}^\circ\xi$.

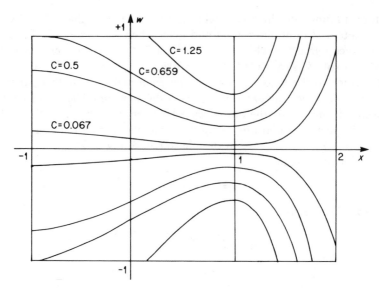

Figure 1.7

We are now in a position to describe the standard parts of the trajectories in the Lienard plane in some detail. We start at a point Q outside the slow manifold S_u. The trajectory through it is quasi horizontal, going to the right if Q is above S_u, to the left if Q is below S_u. It meets S_u in the monad of some standard point $(\xi, F(\xi))$ with $|\xi| > 1$ and from there follows S_u in the direction of decreasing $|x|$. There are now three possibilities:

(i) If $1 < \xi < \frac{5}{3}$, γ reappears in the (x, w) plane as a curve $w(x) = c \exp(\varphi(x))$ with $w(\xi) = 1$ if Q is above S_u, $w(\xi) = -1$ if Q is below S_u. Since $\psi(\xi) > -1$ when $1 < \xi < \frac{5}{3}$, the trajectory seen in the Lienard plane is a canard for values of x such that $\psi(\xi) < x < \xi$; then the trajectory leaves S_u to the same side from which it approached it.

(ii) If $\frac{5}{3} < \xi$, γ reappears in the (x, w) plane as a curve $w(x) = c \exp(\varphi(x))$ with $|w(x)| < 1$ for $-1 < {}^{\circ}x < \min(\xi, 2)$. In the Lienard plane this corresponds to a canard following S_u all the way up to $(-1, \frac{2}{3})$, where it leaves S_u to the upper side, and will then meet S_u at $(2, \frac{2}{3})$.

(iii) If $\xi < -1$, γ follows S_u in the Lienard plane up to $(-1, \frac{2}{3})$, where it leaves S_u and again approaches S_u at $(2, \frac{2}{3})$.

1.5.10. REMARK. It is seen that if the point $Q = (x, u)$ lies in the region $F(x) < u < F(\frac{5}{3}) = -\frac{10}{81}$, $x > 0$, the trajectory through it will exhibit one or more "small" oscillations of the type described under (i) before it enters the limit cycle.

We see that we have a complete description for this particular value of *a*. Similar results hold for other values of *a* that allow canards (see Benoit *et al.*, 1980). It is interesting to note that numerical calculations with finite ε, say $\varepsilon = 0.05$, accord nicely with the descriptions under (i)–(iii).

We have dwelt at length upon this example because it exhibits in a simple situation the richness of *rescaling arguments* that nonstandard analysis makes available to us. This is a theme which will recur throughout this book.

REFERENCES*

S. Albeverio (1984). Nonstandard analysis; polymer models, quantum fields. *Acta Phys. Austriaca, Suppl.* XXVI.

E. Benoit, J.-L. Callot, F. Diener, and M. Diener (1981). Chasse au canard. *Collect. Math.* **32**.

B. Birkeland and D. Normann (1980). A non-standard treatment of the equation $y' = f(y, t)$. *Mat. Sem. Oslo.*

J.-L. Callot (1981). Bifurcations du portrait de phase pour des équations différentielles du second ordre ayant pour type l'equation d'Hermite. Thèse, Strasbourg (1981).

P. Cartier (1982). Perturbations singulières des équations différentielles ordinaires et analyse non-standard. *Sem. Bourbaki*, Astérísque 92–93.

J. D. Cole (1968). *Perturbation Methods in Applied Mathematics*. Ginn (Blaisdell), Boston, Massachusetts.

N. J. Cutland (1983). Nonstandard measure theory and its applications. *Bull. London Math. Soc.* **15**.

M. Davis (1977). *Applied Nonstandard Analysis*. Wiley, New York.

F. Diener (1981). Méthode du plan d'observabilité; développements en ε-ombre. Thèse, Strasbourg.

M. Diener (1981). *Etude générique des canards*. Thèse, Strasbourg.

J. E. Fenstad (1980). Nonstandard methods in stochastic analysis and mathematical physics. *Jber. Deutsch. Math.-Verein.* **82**.

J. E. Fenstad (1985). Is nonstandard analysis relevant for the philosophy of mathematics? *Synthese* **62**.

J. E. Fenstad (1986). Lectures on stochastic analysis with applications to mathematical physics. *Proc. Simposio Chileno Log. Mat.*, Santiago.

J. E. Fenstad and A. Nyberg (1970). Standard and nonstandard methods in uniform topology. *Logic Colloq. 1969*, North-Holland Publ., Amsterdam.

J. M. Henle and E. M. Kleinberg (1979). *Infinitesimal Calculus*. MIT Press, Cambridge, Massachusetts.

C. W. Henson and H. J. Keisler (1985). The strength of nonstandard analysis. *J. Symb. Logic* (to appear).

C. W. Henson, M. Kaufmann, and H. J. Keisler (1984). The strength of nonstandard methods in arithmetic. *J. Symbolic Logic* **49**.

A. Hurd, ed. (1983). *Nonstandard Analysis: Recent Developments, Lecture Notes in Math.* **983**.

A. Hurd and P. A. Loeb (1985). *Introduction to Nonstandard Real Analysis*. Academic Press, New York.

---

* In addition to books and papers explicitly referred to in this chapter we have included a number of references to books and survey papers on nonstandard analysis, which the reader may find useful to consult.

H. J. Keisler (1976). *Foundations of Infinitesimal Calculus.* Prindle, Weber and Schmidt, Boston, Massachusetts.

H. J. Keisler (1984). An infinitesimal approach to stochastic analysis. *Mem. Amer. Math. Soc.* **297**.

D. Laugwitz (1978). *Infinitesimalkalkül.* Bibliographisches Inst., Mannheim.

J. P. LaSalle and S. Lefschetz (1961). *Stability by Liapunov's Direct Method,* Academic Press, New York.

T. Lindstrøm (1986). Nonstandard analysis and perturbations of the Laplacian along Brownian paths. *In Stochastic Processes in Mathematics and Physics.* (S. Albeverio *et al.*, eds.) Proc. of BiBoSI, Lecture Notes in Mathematics, **1158**, Springer-Verlag, Berlin and New York.

P. A. Loeb (1979). An introduction to nonstandard analysis and hyperfinite probability theory. *In Probabilistic Analysis and Related Topics,* ed. A. T. Bharucha-Reid, Vol. 2. Academic Press, New York.

R. Lutz and M. Goze (1981). *Nonstandard Analysis. Lecture Notes in Math.,* **881**, Springer-Verlag, Berlin and New York.

W. A. J. Luxemburg (1973). What is nonstandard analysis. *Amer. Math. Monthly* **80**.

E. Nelson (1977). Internal set theory. *Bull. Amer. Math. Soc.* **83**.

M. M. Richter (1982). *Ideale Punkte, Monaden, und Nichtstandard-Methoden.* Vieweg, Wiesbaden.

A. Robinson (1966). *Non-Standard Analysis.* North-Holland Publ., Amsterdam.

A. Robinson (1979). *Selected Papers,* Vol. 2, North-Holland Publ., Amsterdam.

S. Simpson (1984). Which set existence axioms are needed to prove the Cauchy/Peano theorem for ordinary differential equations? *J. Symbolic Logic* **49**.

T. A. Skolem (1934). Über die Nichtcharakterisierbarkeit der Zahlenreihe mittels endlich oder abzählbar unendlich vieler Aussagen mit ausschliesslich Zahlenvariablen. *Fund. Math.* **33**.

K. D. Stroyan and W. A. J. Luxemburg (1976). *Introduction to the Theory of Infinitesimals.* Academic Press, New York.

K. D. Stroyan and J. M. Bayod (1985). *Foundations of Infinitesimal Stochastic Analysis.* North-Holland Publ., Amsterdam (to appear).

E. Zakon (1969). Remarks on the nonstandard real axis. *In Applications of Model Theory to Algebra, Analysis, and Probability.* Holt, New York.

A. K. Zvonkin and M. A. Shubin (1984). Nonstandard analysis and singular perturbations of ordinary differential equations. *Russian Math. Surveys* **39**.

CHAPTER 2

TOPOLOGY AND LINEAR
SPACES

In this chapter we shall prove the spectral theorem for compact Hermitian operators as an introductory example to a powerful nonstandard technique which we shall apply several times in the second part of this book. But first we need to review a few basic facts about topology and about linear operators in normed spaces. We shall also introduce one new principle of nonstandard theory, *saturation*, in addition to extension and transfer.

2.1. TOPOLOGY AND SATURATION

Our development of the nonstandard theory has so far rested upon two principles, the extension principle, 1.2.2, and the transfer principle, 1.2.4; see also 1.2.10. Now we need to add a new principle.

We recall that a family $\mathcal{F}$ of subsets of some set E has the *finite intersection property* if $X_1 \cap \cdots \cap X_n \neq \varnothing$ for every *finite* set of elements $X_1, \ldots, X_n$ in $\mathcal{F}$. As we noticed in connection with (2) in Section 1.1, a filter as a family of sets has the finite intersection property.

Does the finite intersection property imply that the intersection of all sets in the family is nonempty? Not necessarily, and this deficiency is a basic motivation for the construction of various completions and compactifications. One wishes to adjoin "limit points" to certain families with the finite intersection property, e.g., Cauchy filters.

2.1.1. EXAMPLE. Let $\langle a_n \rangle_{n \in \mathbb{N}}$ be a Cauchy sequence of rationals, i.e., each $a_n \in \mathbb{Q}$. For each $\varepsilon > 0$ there is some $k_\varepsilon \in \mathbb{N}$ such that $|a_{k_\varepsilon} - a_m| < \varepsilon$ for all $m \geq k_\varepsilon$. Thus the set $A_\varepsilon = \{a \in \mathbb{Q} | |a - a_m| < \varepsilon, \text{all } m \geq k_\varepsilon\}$ is a nonempty subset of $\mathbb{Q}$. The family $\mathscr{F} = \{A_\varepsilon\}$ has the finite intersection property, but the intersection $\bigcap A_\varepsilon$ of all sets in the family $\mathscr{F}$ is not necessarily nonempty; observe that $a \in \bigcap A_\varepsilon$ iff $a = \lim_{n \to \infty} a_n$. The completion $\mathbb{R}$ of $\mathbb{Q}$ is introduced exactly to adjoin limit points to families $\mathscr{F}$ of this type.

We shall consider embeddings $*: V(\mathbb{R}) \to V(*\mathbb{R})$ with a certain *saturation property* which in a uniform and general way adjoins limit points to families with the finite intersection property. As an introduction we shall verify the following countable version in the bounded ultrapower model of Section 1.2.

2.1.2. COUNTABLE SATURATION PRINCIPLE. If $A_1 \supseteq A_2 \supset \cdots$ is a countable decreasing chain of nonempty internal sets, then

$$\bigcap_{n \in \mathbb{N}} A_n \neq \varnothing.$$

Each A_n is internal, thus is of the form $A_n = j(A'_n)$ where $A'_n = \langle A^n_1, A^n_2, \ldots \rangle_{\mathscr{U}}$; see the construction of the bounded ultrapower model and the description of internal sets in the model in Section 1.2. We may assume that each $A^n_i \subseteq V_k(\mathbb{R})$ for some fixed $k > 1$. And, for convenience, we adjoin the constant sequence $A'_0 = \langle V_k(\mathbb{R}), \ldots \rangle_{\mathscr{U}}$. Define for $k \geq 0$

$$I_k = \{i \geq k \,|\, A^0_i \supseteq A^1_i \supseteq \cdots \supseteq A^k_i \neq \varnothing\}.$$

We see that: (i) $I_0 = \mathbb{N}$; (ii) $I_k \in \mathscr{U}$, $I_k \supseteq I_{k+1}$, for all $k \geq 0$; and (iii) $\bigcap I_k = \varnothing$.

For each $i \in \mathbb{N}$ let

$$m(i) = \max\{m \,|\, i \in I_m\}.$$

Since $I_0 = \mathbb{N}$ and $\bigcap I_m = \varnothing$, $m(i)$ is well defined. Let B_i be some element in $A^{m(i)}_i$. We shall prove that $B = \langle B_1, B_2, \ldots \rangle_{\mathscr{U}} \in_{\mathscr{U}} A'_k$ for all $k \geq 0$, which, via the j map, verifies 2.1.2 in the model. But $B \in_{\mathscr{U}} A'_k$ will follow if we show that $I_k \subseteq \{i \in \mathbb{N} | B_i \in A^k_i\}$. The latter is, however, immediate since $i \in I_k$ implies that $m(i) \geq k$.

Saturation is a very important uniformity principle which lies behind many mathematical arguments, namely a transition from a quantifier structure $\forall \exists$ (to express the "local" property of finite intersection) to one of the form $\exists \forall$ (to express the "global" or uniform property), which is the heart of many finiteness, compactness, or uniform boundedness arguments.

As a first application of the saturation principle 2.1.2, we prove the following useful extension principle. We know by transfer that any standard sequence $\langle A_n \rangle_{n \in \mathbb{N}}$ has a canonical extension $\langle A_n \rangle_{n \in *\mathbb{N}}$. But what happens if we are given a countable sequence $\langle A_n \rangle_{n \in \mathbb{N}}$ in $V(*\mathbb{R})$?

2.1.3. PROPOSITION. Any bounded countable sequence $\langle A_n \rangle_{n \in \mathbb{N}}$ of internal sets in $V(*\mathbb{R})$ can be extended to an internal sequence $\langle A_n \rangle_{n \in *\mathbb{N}}$ in $V(*\mathbb{R})$.

Notice that the object $\langle A_n \rangle_{n \in \mathbb{N}}$ is *external* in $V(*\mathbb{R})$, even if every element A_n of the sequence is internal. Thus transfer is of no help. But saturation does the trick: define for each $n \in \mathbb{N}$ a set B_n by

$$f \in B_n \qquad \text{iff} \qquad \text{dom} f = *\mathbb{N} \wedge \forall i \le n (f(i) = A_i).$$

Each set B_n is internal. Thus every $f \in B_n$ is internal. We observe by saturation, 2.1.2, that $\bigcap B_n \neq \varnothing$. Any f in this intersection is a suitable extension of the given sequence.

Before turning to topology we shall state the general saturation property. If our base space satisfies some countability condition, e.g., if our base space is a separable metric space, then countable saturation will suffice. For general topological spaces we need something more.

2.1.4. GENERAL SATURATION PRINCIPLE. (1)   Let κ be an infinite cardinal. A nonstandard extension is called κ-*saturated* if for every family $\{X_i\}_{i \in I}$, $\text{card}(I) < \kappa$, with the finite intersection property, the intersection $\bigcap_{i \in I} X_i$ is nonempty, i.e., contains some internal object.

(2)   The embedding $* : V(S) \to V(*S)$ is *saturated* if the extension $V(*S)$ is $\text{card}(*S)$-saturated. The embedding is called *polysaturated* if it is $\text{card}(V(S))$-saturated.

We make a few remarks. If $\kappa = \omega_1$, the first uncountable cardinal, we are back to the countable saturation principle 2.1.2. And we have proved that ω_1-saturated embeddings exist, namely the bounded ultrapower of Section 1.2 is ω_1-saturated. We also notice that by choosing the cardinality of $*S$ large enough saturation implies polysaturation; thus we use "general saturation" to refer to either of the notions in part (2) of 2.1.4. The reader is referred to the literature for a proof that saturated models exist (Chang and Keisler, 1973).

2.1.5. REMARK. For readers who feel some uneasiness about the nonuniqueness of the extension $*\mathbb{R}$, we mention the following *uniqueness theorem for superstructure embeddings*: there is up to isomorphism a unique superstructure embedding $* : V(\mathbb{R}) \to V(*\mathbb{R})$ such that: (i) $*$ satisfies the transfer principle, 1.2.4; (ii) $*$ is saturated in the sense of 2.1.4 (2); (iii) $*\mathbb{R}$ and the set of all internal sets have cardinality equal to the first uncountable inaccessible cardinal [see Keisler (1976) for an excellent discussion].

We now turn to topology. We assume that the topology on the space E is given by the family $\mathcal{O}$ of open sets. For $x \in E$ we let $\mathcal{O}_x$ be the family of open sets containing x. We introduce the *monad* $\mu(x)$ of $x \in E$ by the equation

(1) $$\mu(x) = \bigcap \{*O \mid O \in \mathcal{O}_x\}.$$

The reader will immediately see that this generalizes our previous definition 1.1.2. Furthermore, we have the following extension of Proposition 1.1.2.

2.1.6. PROPOSITION. Let E be a topological space.

(i) E is Hausdorff iff $\mu(x) \cap \mu(y) = \varnothing$ for all $x, y \in E$ such that $x \neq y$.

(ii) A set $A \subseteq E$ is open iff $\mu(x) \subseteq {}^*A$ for all $x \in A$.

(iii) A set $A \subseteq E$ is closed iff for all $x \in E$ and all $y \in {}^*A$, $y \in \mu(x)$ implies that $x \in A$.

(iv) A set $A \subseteq E$ is compact iff for all $x \in {}^*A$ there is a $y \in A$ such that $x \in \mu(y)$.

We call a point $x \in {}^*E$ *nearstandard* iff $x \in \mu(y)$ for some $y \in E$. Thus (iv) of 2.1.6 says that a space E is compact iff every $x \in {}^*E$ is nearstandard. We let $\mathrm{Ns}({}^*E)$ denote the set of nearstandard points in ${}^*E$. Let $x \in \mu(y)$ for some $y \in E$; if E is Hausdorff it follows from 2.1.6(i) that y is unique, and in this case we set $y = \mathrm{st}(x)$. Note that $\mathrm{st} : {}^*E \to E$ is a partial map. For any $A \subseteq {}^*E$ we let

$$\mathrm{st}(A) = \{\mathrm{st}(x) \,|\, x \in A \cap \mathrm{Ns}({}^*E)\}.$$

In the case where E is Hausdorff we have the following useful reformulation of 2.1.6(ii)–(iv):

(ii)′ A set $A \subseteq E$ is open iff $\mathrm{st}^{-1}(A) \subseteq {}^*A$.

(iii)′ A set $A \subseteq E$ is closed iff ${}^*A \cap \mathrm{Ns}({}^*E) \subseteq \mathrm{st}^{-1}(A)$.

(iv)′ A set $A \subseteq E$ is compact iff ${}^*A \subseteq \mathrm{st}^{-1}(A)$.

The map st is called the *standard part map*, and $\mathrm{st}(x)$ will be referred to as the *standard part* of x.

We should comment a bit on the proof of 2.1.6. And for that purpose let us concentrate on the characterization of openness. We recall from the proof of 1.1.2(i) that if $\mu(x) \subseteq {}^*A$, then we could find some standard open neighborhood of x inside A. This was an exercise using transfer. But we had to use a trick. Since $\mu(x)$ is not an internal set, we cannot pull $\mu(x)$ back via the inverse of the *-embedding to a standard neighborhood of x in A. But we can take a positive infinitesimal $\delta > 0$ and consider the internal set $D = \{y \in {}^*\mathbb{R} \,|\, |x - y| < \delta\}$. We now observe that $D \subseteq \mu(x) \subseteq {}^*A$, and D obviously is an element in ${}^*\mathcal{O}_x$. Thus we have the truth in $V({}^*\mathbb{R})$ of the statement

$$\exists D \in {}^*\mathcal{O}_x[D \subseteq {}^*A].$$

By the transfer principle there exists a $D \in \mathcal{O}_x$ such that $D \subseteq A$. This is exactly the proof of 1.1.2(i) (presented immediately after the transfer principle 1.1.4 in Section 1.1), observing that over $\mathbb{R}$ the quantifier $\exists D$ can be reduced to a quantifier $\exists \delta$.

In order to make this proof work in general we need the following fact.

2.1.7. APPROXIMATION LEMMA. For each $x \in E$ there is an internal set $D \in {}^*\mathcal{O}_x$ such that $D \subseteq \mu(x)$.

For each $D \in \mathcal{O}_x$ let $F_D = \{D' \in {}^*\mathcal{O}_x \,|\, D' \subseteq {}^*D\}$. Each set F_D is internal and the family $\{F_D \,|\, D \in \mathcal{O}_x\}$ has the finite intersection property. Hence by general saturation principle 2.1.4, there exists some internal D_0 belonging to F_D for all $D \in \mathcal{O}_x$. Thus $D_0 \subseteq {}^*D$ for all $D \in \mathcal{O}_x$, i.e., $D_0 \subseteq \mu(x)$.

Using 2.1.7, the proof of 2.1.6(ii) is as before. The proof of 2.1.6(i) uses once more Lemma 2.1.7 to approximate $\mu(x)$ from the inside by an internal set. And 2.1.6(iii) is the dual of (ii) and needs no further comment.

We do, however, add a comment on the characterization of compactness. We claim that the space E is compact iff every $x \in {}^*E$ is nearstandard. This is nothing but the well-known ultrafilter characterization of compactness in an easy disguise (see the note Fenstad, 1967). We spell out the details: with every point $x \in {}^*E$ we can associate an ultrafilter $\mathcal{F}_x$ on E in the following way:

$$X \in \mathcal{F}_x \quad \text{iff} \quad x \in {}^*X,$$

i.e., $\mathcal{F}_x$ is the "trace" on E of a principal ultrafilter on ${}^*E$.

Conversely, with every ultrafilter $\mathcal{F}_\lambda$ on E one may associate a point $x_\lambda \in {}^*E$; this is a consequence of general saturation.

It follows that $\mathcal{F}_\lambda = \mathcal{F}_{x_\lambda}$. In fact, if $X \in \mathcal{F}_\lambda$, then by choice $x_\lambda \in {}^*X$, which means that $X \in \mathcal{F}_{x_\lambda}$. As $\mathcal{F}_\lambda$ is maximal, equality follows.

Let $a \in E$ and $b \in {}^*E$. We note that $b \in \mu(a)$ iff $\mathcal{F}_b$ converges as a filter to a, i.e., every open set containing a belongs to $\mathcal{F}_b$.

Combining these observations, the proof of 2.1.6(iv) follows. (i) Let E be compact and $b \in {}^*E$. Compactness means that every ultrafilter on E converges; in particular, there is some $a \in E$ such that $\mathcal{F}_b$ converges to a, i.e., $b \in \mu(a)$. (ii) Conversely, let $\mathcal{F}$ be an ultrafilter on E; then $\mathcal{F}$ is of the form $\mathcal{F} = \mathcal{F}_b$ for some $b \in {}^*E$. Since every point in ${}^*E$ is nearstandard, i.e., $b \in \mu(a)$ for some $a \in E$, we see that $\mathcal{F}_b$ converges.

We shall make one important addition to 2.1.6.

2.1.8. PROPOSITION. Let E be Hausdorff and A an internal set in ${}^*E$. Then $\mathrm{st}(A)$ is closed.

The proof of 2.1.8 is a simple but essential application of the saturation principle. Assume that $O \cap \mathrm{st}(A) \neq \varnothing$ for all $O \in \mathcal{O}_a$, where $a \in E$. If b is any point in $O \cap \mathrm{st}(A)$, then for some $c \in A$, $b = \mathrm{st}(c)$. Since O is open, 2.1.6 tells us that $\mu(b) = \mu(\mathrm{st}(c)) \subseteq {}^*O$, hence $c \in {}^*O \cap A$. Our assumption thus implies that ${}^*O \cap A \neq \varnothing$ for all $O \in \mathcal{O}_a$. Since A is internal saturation implies that there is some point $b_0 \in \mu(a) \cap A$; but this means that $a = \mathrm{st}(b_0) \in \mathrm{st}(A)$. We have thus proved that if a belongs to the closure of $\mathrm{st}(A)$, i.e., $O \cap \mathrm{st}(A) \neq \varnothing$ for all $O \in \mathcal{O}_a$, then $a \in \mathrm{st}(A)$, i.e., $\mathrm{st}(A)$ is closed.

2.1.9. EXAMPLE. The results in 2.1.6–2.1.8 really presuppose that we work in a (poly-)saturated extension. The following is an example of a noncompact space in which every point is nearstandard. Let $\omega_1 = \{0, 1, 2, \ldots, \alpha, \ldots\}$ be the (uncountable) set of all countable ordinals. The structure $\langle \omega_1, < \rangle$, where $<$ is the ordering relation between ordinals, is a complete ordered set. The family of intervals $\langle \alpha, \beta \rangle = \{\gamma \in \omega_1 \mid \alpha < \gamma < \beta\}$, $\alpha, \beta \in \omega_1$, induces a topology on ω_1, and it is easily seen that ω_1 is not compact in this topology.

Let $\mathcal{U}$ be a free ultrafilter on $\mathbb{N}$ and consider

$$^*\omega_1 = \omega_1^{\mathbb{N}}/\mathcal{U}.$$

We claim that every point $\alpha_{\mathcal{U}} \in {}^*\omega_1$ is finite; i.e., there exists some $\alpha \in \omega_1$ such that $\alpha_{\mathcal{U}} < {}^*\alpha$. Now $\alpha_{\mathcal{U}} = \langle \alpha_1, \ldots, \alpha_n, \ldots \rangle_{\mathcal{U}}$. And since a countable set of countable ordinals is bounded in ω_1, there is some $\alpha \in \omega_1$ such that $\alpha_n < \alpha$ for all α_n in the defining sequence for $\alpha_{\mathcal{U}}$. But then $\alpha_{\mathcal{U}} < {}^*\alpha$; i.e., every point in $^*\omega_1$ is finite. But in a complete ordered structure every finite point has a standard part, i.e., $\alpha_{\mathcal{U}}$ is nearstandard for all $\alpha_{\mathcal{U}} \in {}^*\omega_1$.

The extension $^*\omega_1 = \omega_1^{\mathbb{N}}/\mathcal{U}$ satisfies 2.1.2 but not 2.1.4. However, ω_1 does have saturated extensions; see the reference in Remark 2.1.5.

We round off our general topology course by mentioning the characterization of continuity.

2.1.10. PROPOSITION. Let E and F be topological spaces and $f: E \to F$. Then f is continuous at a point $a \in E$ iff $^*f(\mu(a)) \subseteq \mu(f(a))$.

We postpone the few remarks that we shall make on the topic of uniform continuity to our discussion of metric spaces. There is, however, a general discussion of standard versus nonstandard methods in uniform topology in Fenstad and Nyberg (1970), which was behind the commutative diagram in Section 1.1. Suitable references on general topology are Davis (1977) and Stroyan and Luxemburg (1976).

Let us also mention that Richter and Benninghofen have recently introduced an interesting new method to nonstandard topology. Their idea is to extend the transfer principle to a larger class of formulas (allowing certain external quantifiers) in a way which makes it possible to describe monads; see Richter (1982) and Benninghofen and Richter (1983). Benninghofen and Stroyan (1984) have given applications to the bounded weak star topology on linear spaces. Richter's and Benninghofen's ideas were inspired by Nelson's (1977) formulation of nonstandard analysis, but can also be used within the formulation we have described.

2.1.11. REMARK. So far we have been discussing nonstandard characterizations of standard concepts. In 2.1.10 we discussed a map $f: E \to F$ that is a standard object. However, in applications of nonstandard theory we shall be much concerned with internal objects, such as *internal maps* $f: {}^*E \to {}^*F$. For internal maps basic notions such as continuity and differentiability split into two separate concepts. Take as an example continuity; let $f: {}^*\mathbb{R} \to {}^*\mathbb{R}$ and recall that $\mathbb{R}_+$ denotes the set of positive reals:

(i) f is called S-continuous at $a \in {}^*\mathbb{R}$ iff

$$(\forall \varepsilon \in \mathbb{R}_+)(\exists \delta \in \mathbb{R}_+)(\forall x \in \mathbb{R})(|x - a| < \delta \to |f(x) - f(a)| < \varepsilon).$$

(ii) f is called *-continuous at $a \in {}^*\mathbb{R}$ iff

$$(\forall \varepsilon \in {}^*\mathbb{R}_+)(\exists \delta \in {}^*\mathbb{R}_+)(\forall x \in {}^*\mathbb{R})(|x - a| < \delta \to |f(x) - f(a)| < \varepsilon).$$

The reader will appreciate the difference. Let $\omega \in {}^*\mathbb{N} - \mathbb{N}$; then the internal function

$$f(x) = \sin(\omega x)$$

is everywhere *-continuous but nowhere S-continuous. The internal function

$$f(x) = \begin{cases} 1/\omega & \text{if} \quad x \in {}^*\mathbb{Q}, \\ 0 & \text{if} \quad x \notin {}^*\mathbb{Q}, \end{cases}$$

is everywhere S-continuous but nowhere *-continuous.

The difference between S-continuity/differentiability and *-continuity/differentiability will be important in several applications. For instance, the "free energy" is always *-differentiable but not always S-differentiable in an Ising model; see Section 7.2. In the non-S-differentiable case there may be a phase transition.

We now turn for a moment to metric spaces. Let (E, d) be a metric space. The notion of *monad* now makes sense for any $x \in {}^*E$. We set

$$\mu(x) = \{y \in {}^*E \mid d(x, y) \approx 0\}.$$

In analogy with Proposition 2.1.10 we have the following general version of Proposition 1.3.3(ii): a function $f: E \to F$, where E and F are metric spaces, is *uniformly continuous* iff $f(\mu(x)) \subseteq \mu(f(x))$, all $x \in {}^*E$.

In a metric space E we must also carefully distinguish between a notion of *nearstandard* and a notion of *finite* point.

2.1.12. DEFINITION. A point $x \in {}^*E$ is called *nearstandard* if $d(x, y) \approx 0$ for some $y \in E$. The point x is called *finite* if $d(x, y)$ is finite for some $y \in E$.

Trivially, every nearstandard point is finite. The converse is not true. If E is a separable infinite-dimensional Hilbert space there are finite points in $*E$ which are not nearstandard. In order to see this, we let $\langle e_n \rangle_{n \in \mathbb{N}}$ be an orthonormal basis for E. Choose in $*E$ an element e_ω, $\omega \in *\mathbb{N} - \mathbb{N}$. Then e_ω is finite but not nearstandard in the metric derived from the norm of E. We note that for every finite point of $*E$ to be nearstandard it is necessary and sufficient for E to satisfy the condition that every bounded closed set is compact.

Any metric space E is Hausdorff. Thus if $x \in *E$ is nearstandard there is a unique $y \in E$ such that $x \in \mu(y)$; we write $y = {}^\circ x$ or $y = \text{st}(x)$.

We discuss one result that involves the notion of standard part in three spaces: in many applications we have a pair of separable metric spaces E_1, E_2 and the set $C(E_1, E_2)$ of continuous maps $f: E_1 \to E_2$ that in the compact-open topology is a separable metrizable space. Let $F \in *C(E_1, E_2)$; i.e., F is a map from $*E_1$ to $*E_2$. As an element of $*C(E_1, E_2)$, F may have a standard part ${}^\circ F \in C(E_1, E_2)$. How is this notion of standard part related to the notions of standard parts in $*E_1$ and $*E_2$?

2.1.13. PROPOSITION. Let E_1, E_2 be separable metric spaces with E_1 locally compact and let $C(E_1, E_2)$ have the compact-open topology. If $f \in C(E_1, E_2)$ and $F \in *C(E_1, E_2)$, then ${}^\circ F = f$ iff for each nearstandard point $x \in *E_1$, ${}^\circ(F(x)) = f({}^\circ x)$.

We give the proof. Let ${}^\circ F = f$ and let $x \in *E_1$ be nearstandard. E_1 is locally compact, so for every open neighborhood U of $f({}^\circ x)$ we can find a compact neighborhood K of ${}^\circ x$ such that $f(K) \subseteq U$; i.e., $C(K, U)$ is a neighborhood of f. We see that $x \in *K$ and $F \in *C(K, U)$. Thus $F: *K \to *U$ and $F(x) \in *U$. Since this is true for arbitrary open neighborhoods of $f({}^\circ x)$ we conclude that ${}^\circ(F(x)) = f({}^\circ x)$. For the converse, suppose that ${}^\circ(F(x)) = f({}^\circ x)$ for all nearstandard $x \in *E$. Let $C(K, U)$ be any neighborhood of f; we must show that $F \in *C(K, U)$ in order to conclude that ${}^\circ F = f$. Pick any $x \in *K$; K is compact so ${}^\circ x$ exists and is an element of K. Then $f({}^\circ x) \in U$, so by assumption ${}^\circ(F(x)) \in U$, i.e., $F(x) \in *U$. Thus $F: *K \to *U$, i.e., $F \in *C(K, U)$.

We shall prove one more result about metric spaces needed in further applications.

2.1.14. DEFINITION. Let f be a map between metric spaces E, F. f is called *compact* if for every bounded subset A of E, $f(A) \subseteq D$ for some compact subset D of F.

2.1.15. PROPOSITION. $f: E \to F$ is compact iff f maps finite points of $*E$ to nearstandard points of $*F$.

Let f be compact and x finite in $*E$. There is some $r \in \mathbb{R}_+$ such that $d(x, y) \leq r$ for some $y \in E$. The set $A = \{u \in E \mid d(u, y) \leq r\}$ is bounded, hence $f(A) \subseteq D$ for some compact $D \subseteq F$. We observe that $x \in *A$, which implies that $f(x) \in *D$. Since D is compact, $f(x)$ is nearstandard.

For the converse let A be a bounded subset of E. Our candidate for the set D will be the set of all points $y \in F$ such that $y \approx x$ for some $x \in *(f(A))$. We must show that D is compact.

For any $y \in D$ and $\varepsilon \in \mathbb{R}_+$, we have in $V(*\mathbb{R})$ the truth of the statement

$$\exists x \in *f(A)(d(x, y) < \varepsilon).$$

By transfer we see that for all $y \in D$ and all $\varepsilon \in \mathbb{R}_+$ there exists $x \in f(A)$ such that $d(x, y) < \varepsilon$. Thus we have in $V(\mathbb{R})$ the truth of

$$(\forall y \in D)(\forall \varepsilon \in \mathbb{R}_+)(\exists x \in f(A))(d(x, y) < \varepsilon).$$

This means, once more by transfer, that if $y \in *D$, then there is some $x \in *f(A)$ such that $x \approx y$. By assumption A is bounded; thus any point of $*f(A) = f(*A)$ is nearstandard. But if x is nearstandard and $y \approx x$, then y is nearstandard, i.e., D is compact.

2.2. LINEAR SPACES AND OPERATORS

In this section we shall introduce a nonstandard approach to linear spaces and operators. We shall only discuss a few basic facts and refer the reader to the existing literature for further information (Heinrich, 1980; Henson and Moore, 1983). The reader should also preview Chapter 5, where we develop a hyperfinite theory of quadratic forms with applications to stochastic analysis.

Let $(E, \|\cdot\|)$ be a linear normed space. An element $x \in *E$ is called *finite* (*normfinite*) if $\|x\|$ is a finite hyperreal; we let Fin($*E$) denote the finite elements of $*E$. The element $x \in *E$ is *infinitesimal* if $\|x\| \approx 0$; we write $x \approx y$ for $\|x - y\| \approx 0$. Both Fin($*E$) and $\mu(0) = \{x \in *E \mid \|x\| \approx 0\}$ are vector spaces over the same field as E. The quotient space $\hat{E} = \text{Fin}(*E)/\approx$ is also a normed linear space that we call the *nonstandard hull* of E. This notion was introduced by Luxemburg (1969), and there is a large literature on this topic; e.g., see Henson and Moore (1983) for a survey and introduction.

We are working with saturated embeddings, a fact which has important consequences for the study of the nonstandard hull. We shall treat the following somewhat more general situation. Let $(F, \|\cdot\|)$ be an internal normed linear space; in this case we assume that the norm is a map from F into $*\mathbb{R}$. The notions of finite and infinitesimal still make sense; i.e.,

$\text{Fin}(F)/\approx$ is a well-defined normed linear space. We have the following result:

2.2.1. PROPOSITION. The space $\text{Fin}(F)/\approx$ is complete, i.e., is a Banach space.

PROOF. The proof is a variation on Example 2.1.1. Let $\langle \hat{a}_n \rangle_{n \in \mathbb{N}}$ be a Cauchy sequence in $\text{Fin}(F)/\approx$. This sequence comes from a sequence $\langle a_n \rangle_{n \in \mathbb{N}}$ in F; by saturation this sequence can be extended to an internal *Cauchy sequence in F, namely let

$$A_n = \{b \mid b : {}^*\mathbb{N} \to F, b \text{ is }{}^*\text{Cauchy}, \quad \forall i \le n[b_i = a_i]\}.$$

Each A_n is nonempty and internal. By countable saturation 2.1.2, the intersection

$$A = \bigcap_{n \in \mathbb{N}} A_n$$

is nonempty. Let $b \in A$ and let $\eta \in {}^*\mathbb{N} - \mathbb{N}$, then $\hat{b} = b_\eta / \approx$ is an element of $\text{Fin}(F)/\approx$ that is easily seen to be the limit of the given sequence $\langle \hat{a}_n \rangle_{n \in \mathbb{N}}$.

We add the observation that if the norm in F comes from an inner product, then $\text{Fin}(F)/\approx$ will be a Hilbert space.

We now let E be a standard normed linear space. We recall that an element $x \in {}^*E$ is called *nearstandard* if $\|x - y\| \approx 0$ for some $y \in E$. An element $x \in {}^*E$ is called *pre-nearstandard* if for all $\varepsilon \in \mathbb{R}_+$ there is some $y \in E$ such that $\|x - y\| < \varepsilon$; we let $\text{Pns}({}^*E)$ denote the set of pre-nearstandard points of ${}^*E$. By analogy with Proposition 2.2.1 we have:

2.2.2. PROPOSITION. The space $\text{Pns}({}^*E)/\approx$ is the completion of the given space E.

The proof is exactly the same as for 2.2.1. We only add the observation that $\text{Pns}({}^*E)$ is the closure of E in the topology on ${}^*E$ defined by the extended norm; this implies that $\text{Pns}({}^*E)/\approx$ will be the completion of E. We further observe that if E is an inner product space (a pre-Hilbert space) then $\text{Pns}({}^*E)/\approx$ is the Hilbert space completion of E.

Further results can be found in Luxemburg (1969) and in Fenstad and Nyberg (1970) (where "pre-nearstandard" was called "bounded"). Note also that the space E is complete iff every pre-nearstandard point is nearstandard. This generalizes the fact that compactness is equivalent to every point of ${}^*E$ being nearstandard.

Let E be any normed linear space; we denote by $\mathscr{F}_E$ the class of all finite-dimensional subspaces of E. For $F \in \mathscr{F}_E$ let $\dim(F)$ denote the dimension of F. By transfer we obtain an object ${}^*\mathscr{F}_E \in V({}^*\mathbb{R})$ and a map *dim: $*\mathscr{F}_E \to *\mathbb{N}$; in conformity with previous conventions we drop the asterisk and write $\dim(F)$ for all $F \in *\mathscr{F}_E$.

If $F \in *\mathscr{F}_E$ and $\dim(F) = \eta \in *\mathbb{N}$ it follows by transfer that there is an internal sequence $\langle e_i \rangle_{i \leq \eta}$ such that $e_i \in *E$ and

$$F = \left\{ \sum_{i=1}^{\eta} \alpha_i e_i \,\Big|\, \alpha_i \in *K \right\},$$

where $K = \mathbb{R}$ or $\mathbb{C}$ is the field of coefficients of E. The space F is a *hyperfinite-dimensional linear space*. By transfer F has all the elementary properties of finite-dimensional spaces.

2.2.3. PROPOSITION. Let E be a normed linear space. Then there is an $F \in *\mathscr{F}_E$ such that

$$E \subseteq F \subseteq *E.$$

The proof is an exercise in the use of general saturation 2.1.4. For each $x \in E$ let $A_x = \{F \in *\mathscr{F}_E \mid x \in F\}$. The family $\{A_x\}_{x \in E}$ of internal sets has the finite intersection property; hence there is some $F \in *\mathscr{F}_E$ such that $x \in F$ for all $x \in E$.

This is the powerful technique alluded to in the introductory paragraph of this chapter. We imbed the space E into a hyperfinite-dimensional space F. By transfer we prove results about F by proving results about finite-dimensional subspaces of E. Since E sits inside F, we shall use the standard part map to prove results about E. This will be illustrated in the next section. Here we add a few remarks about linear operators.

Let $T: E \to E$ be a *bounded linear operator*. The norm $\|T\| = \sup_{\|x\|=1} \|Tx\|$ is well defined and $\|Tx\| \leq \|T\| \cdot \|x\|$. We have the following nonstandard characterization.

2.2.4. PROPOSITION. Let $T: E \to E$ be a linear operator:
(i) T is bounded iff $*T$ maps finite points to finite points.
(ii) T is compact iff $*T$ maps finite points to nearstandard points.

For the proof note first that (ii) is a corollary of Proposition 2.1.15. For the proof of (i) note that the inequality $\|Tx\| \leq \|T\| \cdot \|x\|$ immediately implies that a bounded T maps finite points to finite points. For the converse we observe that every point in the set $\{Tx \in *E \mid x \in *E, \|x\| = 1\}$ is finite since T maps finite to finite. Since the set is internal and bounded by every hyperfinite integer, there is some finite bound to the set; we conclude that $\|T\| < \infty$.

We now turn to a study of compact symmetric operators in a Hilbert space.

2.3. SPECTRAL DECOMPOSITION OF COMPACT HERMITIAN OPERATORS

In this section we shall illustrate the use of Proposition 2.2.3 by giving a nonstandard proof of the spectral theorem in an arbitrary Hilbert space. We have included this result as a pedagogical example. The result is classical; the nonstandard treatment is due to Robinson (1966), and also consult the exposition in Davis (1977); for a different approach and a more general result see Moore (1976).

2.3.1. THEOREM. If $T \neq 0$ is a compact Hermitian operator in a Hilbert space H, then there exist real numbers $\nu_1, \nu_2, \ldots, \nu_i, \ldots, \nu_i \neq \nu_j, i \neq j$, and finite-dimensional subspaces H_j of H such that for all $x \in H$

$$Tx = \sum_i \nu_i P_{H_i} x,$$

where P_{H_i} is the orthogonal projection onto the subspace H_i. The set of eigenvalues $\{\nu_1, \ldots, \nu_i, \ldots\}$ is either finite or countably infinite. In the latter case $\lim \nu_j = 0$.

We recall that a bounded linear operator T on H is called *Hermitian* or *self-adjoint* if $(Tx, y) = (x, Ty)$ for all $x, y \in H$. A bounded linear operator P on H is called an (orthogonal) *projection* if it is Hermitian and idempotent. There is a one-to-one correspondence between closed linear subspaces E of H and projections P_E on H. We recall that $P_E x$ is the unique point $y \in E$ such that $\|x - y\| \leq \|x - z\|$ for all $z \in E$.

For the proof of Theorem 2.3.1 we use Proposition 2.2.3 to pick a hyperfinite-dimensional space $E \in {}^*\mathscr{F}_H$ such that $H \subseteq E \subseteq {}^*H$. E is a "finite approximation from above" to H. And we shall use "finite-dimensional" linear algebra in E to prove the spectral theorem in H.

Let $P = P_E$ be the projection in ${}^*H$ onto E. We note that if $x \in {}^*H$ is nearstandard, then $Px \approx x$. Let T be a nontrivial compact Hermitian operator on H; we write T' for the operator P^*T restricted to E. T can also be described as P^*TP restricted to E.

We note a few elementary facts about T'. Since T is bounded on H, T' is bounded on E; in fact, $\|T'\| \leq \|T\|$. And if $x \in H$, $T'x = Tx$. We also see that T' is Hermitian on E. The compactness of T immediately implies that if $x \in E$ is finite, then $T'x$ is nearstandard.

Thus we have a compact Hermitian operator T' on the hyperfinite-dimensional space E, and we can use standard linear algebra. By transfer we have a set of eigenvalues $\lambda_1, \ldots, \lambda_\nu, \nu \in {}^*\mathbb{N}$, for T', where $|\lambda_1| \geq |\lambda_2| \geq \cdots \geq |\lambda_\nu|$. We also have a corresponding set of orthonormal eigenvectors

$r_1, r_2, \ldots, r_\nu$ satisfying the eigenvalue equations

$$T'r_i = \lambda_i r_i, \qquad i = 1, 2, \ldots, \nu.$$

By "standard parts" we shall push this back to a result about T on H. First we note that each λ_i is finite; in fact, $|\lambda_i| \leq \|T\|$ for each $i \leq \nu$. This means that r_j is nearstandard for all noninfinitesimal λ_j. The reader should note that this is the point where we use the compactness assumption on T. From the eigenvalue equation $T'r_j = \lambda_j r_j$ and the fact that r_j is normfinite, compactness of T' implies that $\lambda_j r_j$ is nearstandard. So if λ_j is not infinitesimal, then r_j is nearstandard.

We shall need the following lemma, which adds a further ingredient from nonstandard theory.

2.3.2. LEMMA. Let $\varepsilon \in \mathbb{R}_+$ and suppose that $|\lambda_j| \geq \varepsilon$ for all $j \leq k$; then $k \in \mathbb{N}$.

From our remark we see that r_j is nearstandard for all $j \leq k$, and $\|r_i - r_j\| = \sqrt{2}$ for all $i, j \leq k$, $i \neq j$. Suppose to the contrary that $k \in {}^*\mathbb{N} - \mathbb{N}$. Consider the internal sequence

$$s_j = \begin{cases} r_j & \text{if } j \leq k, \\ 0 & \text{if } j > k. \end{cases}$$

This internal sequence has the property that $\|s_i - s_j\| \geq \sqrt{2}$ for $i \neq j$, $i, j \in \mathbb{N}$. Thus the set

$$\{i \in {}^*\mathbb{N} \mid (\forall j < i) \|s_j - s_i\| \geq \sqrt{2}\}$$

contains an $i_1 < k$ in ${}^*\mathbb{N} - \mathbb{N}$.

For $i \in \mathbb{N}$ let $t_i = \operatorname{st}(s_i)$. This is a standard sequence and thus has an extension $\langle t_i \mid i \in {}^*\mathbb{N} \rangle$. Since $t_i \approx s_i$ for all $i \in \mathbb{N}$, there exists an $i_2 \in {}^*\mathbb{N} - \mathbb{N}$ such that $t_i \approx s_i$ for all $i < i_2$.

Choose $i_0 < \min\{i_1, i_2\}$. Then $i_0 < k$ and $t_{i_0} \approx s_{i_0}$ is nearstandard. Therefore $s = \operatorname{st}(t_{i_0}) = \operatorname{st}(s_{i_0}) = \operatorname{st}(r_{i_0})$ is a limit point of the sequence $\langle t_i \mid i \in \mathbb{N} \rangle$. But this is clearly impossible since $\|s - t_i\| \geq \sqrt{2}$ for all $i \in \mathbb{N}$.

From this lemma we see that if we have a block of equal and non-infinitesimal eigenvalues

$$\lambda_{r+1} = \lambda_{r+2} = \cdots = \lambda_{r+k},$$

then the length of the block k is a *finite* integer.

We rewrite the set of eigenvalues as a sequence κ_i, $i \leq \mu \leq \nu$, without repetition. Let the eigenvectors corresponding to κ_i be $r_1^{(i)}, \ldots, r_{m_i}^{(i)}$. Write

$$E_i = \operatorname{span}(r_1^{(i)}, \ldots, r_{m_i}^{(i)}).$$

At this point i and m_i may both be hyperfinite. But transfer of standard matrix theory tells us that we can write

$$E = E_1 \oplus E_2 \oplus \cdots \oplus E_\mu$$

with the associated spectral resolution of T',

$$T' = \kappa_1 P_{E_1} + \kappa_2 P_{E_2} + \cdots + \kappa_\mu P_{E_\mu}.$$

A regularity argument remains; let $W = \{j \in {}^*\mathbb{N} \,|\, \kappa_j \neq 0\}$.

2.3.3. LEMMA. W is a nonempty subset of $\mathbb{N}$.

The main thing to show is that W is nonempty; it then follows from Lemma 2.3.2 that W is already a subset of $\mathbb{N}$. For a contradiction assume that $W = \varnothing$; then $\kappa = |\kappa_1| \approx 0$. Since $T \neq 0$, there is an element $x \in E$ such that $\|x\| < 1$ and $\|T'x\| > 0$. By the above decomposition we can write $x = x_1 + \cdots + x_\mu$ and $T'x = \kappa_1 x_1 + \cdots + \kappa_\mu x_\mu$. Then

$$\|T'x\|^2 = \sum_{i=1}^{\mu} |\kappa_i|^2 \|x_i\|^2 \le \kappa^2 \approx 0.$$

This is the desired contradiction which proves that W is nonempty.

It remains to put the pieces together. If $\lambda_j \neq 0$, then r_j is nearstandard and the pair $\mathrm{st}(r_j)$, $\mathrm{st}(x_j)$ satisfy the eigenvalue equation for T in H. If $j \in W$, then m_j is finite. Let $s_i^{(j)} = \mathrm{st}(r_i^{(j)})$ and $\nu_j = \mathrm{st}(\kappa_j)$ for $j \in W$. Let

$$H_j = \mathrm{span}(s_1^{(j)}, \ldots, s_{m_j}^{(i)}) \subseteq H.$$

We note that if $x \in H$ then $P_{H_j} x \approx P_{E_j} x$. We can now write

$$H = H_0 \oplus H_1 \oplus H_2 \oplus \cdots,$$

where H_0 is the orthogonal complement of $H_1 \oplus H_2 \oplus \cdots$ in H. The sum is finite if W is finite. It is infinite if $W = \mathbb{N}$; both cases may happen. From the spectral decomposition of T' we derive for $x \in H$

$$Tx = \nu_1 P_{H_1} x + \nu_2 P_{H_2} + \cdots.$$

This completes our exposition. We just add that in the case $W = \mathbb{N}$ we immediately see that $\lim_{j \to \infty} \nu_j = 0$. This is so since $|\nu_i| \approx |\kappa_i|$ for all $i \in \mathbb{N}$. Thus for some $\omega \in {}^*\mathbb{N} - \mathbb{N}$, $|\nu_\omega| \approx |\kappa_\omega|$. From the definition of W it follows that $|\kappa_\omega| \approx 0$. Thus 0 is a limit point of the monotone decreasing sequence $\langle \nu_i \,|\, i \in \mathbb{N} \rangle$.

An early success of the technique of approximating a Hilbert space H "from above" by a hyperfinite-dimensional linear space was the solution by Bernstein and Robinson of *the invariant subspace problem*. The result is as follows:

2.3.4. THEOREM. Let T be a bounded linear operator on a separable Hilbert space H. Suppose T is such that for some complex polynomial

$p(\lambda) = c_0 + c_1\lambda + \cdots + c_m\lambda^m$, $c_m \neq 0$, $p(T)$ is compact. Then T leaves invariant at least one closed linear subspace of H other than H or $\{0\}$.

This example is noteworthy; here we have a classical problem whose solution was first given by nonstandard methods. The special case $p(\lambda) = \lambda$ was solved by von Neumann in 1930. Successive extensions are due to Aronszajn, Smith, and Halmos. The problem was finally settled by Bernstein and Robinson in 1966; for an exposition see Robinson (1966) and Bernstein (1973).

The result has subsequently been extended in a standard framework by Lomonozov (1973); for a general exposition see the book by Radjavi and Rosenthal (1973).

However, the nonstandard approach exemplifies a general technique that has proved useful on many other occasions. We start out with a basic finite-dimensional fact, in this case the fact that in a finite-dimensional linear space E of dimension m any linear operator possesses a chain of invariant subspaces

$$E_0 = \{0\} \subseteq E_1 \subseteq \cdots \subseteq E_m = E,$$

where $\dim(E_j) = j$, $0 \leq j \leq m$. The given Hilbert space H will be imbedded into a suitably chosen hyperfinite-dimensional extension H_ν, and one uses the transfer of the above finite-dimensional result in the process of obtaining an invariant subspace for the given operator on H; see Bernstein (1973) and Robinson (1966).

2.4. NONSTANDARD METHODS IN BANACH SPACE THEORY

The *nonstandard hull* construction (see Section 2.2) has had important applications in the study of local properties of Banach spaces. We aim in this book toward other applications, but feel that our introductory account would be incomplete if we did not at least give a glimpse into an area of fruitful applications of nonstandard methods.

Let $(F, \|\cdot\|)$ be an internal normed linear space; in Section 2.2 we introduced the nonstandard hull of F, $\hat{F} = \text{Fin}(F)/\approx$. There is an isometric embedding of F into $\hat{F}$ and we showed in Proposition 2.2.1 that $\hat{F}$ is a Banach space.

2.4.1. EXAMPLE. For each $p \geq 1$ in $^*\mathbb{R}$ and each $n \in {}^*\mathbb{N}$ we can define the internal space $l_p(n)$ in exact analogy with the standard notion. By varying p and n, the nonstandard hulls $\hat{l}_p(n)$ give us many interesting examples of hyperfinite-dimensional spaces. It is easy to see that if p and n are both finite then $\hat{l}_p(n)$ is order isometric to $l_{p'}(n)$, where $p' = \text{st}(p)$. If p is infinite and n finite, then $\hat{l}_p(n)$ is order isometric to $l_\infty(n)$. But new possibilities arise when n is infinite; see Henson and Moore (1983, Section 9) for an introduction.

As remarked in Section 2.2, the notion of nonstandard hull was introduced by Luxemburg (1969). There is a parallel development due to Krivine and continued by Stern using ultraproducts. A comprehensive introduction can be found in the recent survey papers by Heinrich (1980) and by Henson and Moore (1983), who are among the main contributors to this field. As mentioned above, we offer only a brief look and recommend the papers by Heinrich and Henson and Moore for a full account.

We recall that a Banach space E is called *reflexive* if the canonical embedding of E into its second dual E^{**} is surjective. A stronger notion of *superreflexivity* has turned out to be important in many connections. To explain this notion we need the concept of finite representability.

Let E and F be standard Banach spaces; E is said to be *finitely representable* in F if for each finite-dimensional subspace E_0 of E and each positive real number ε there exists a linear transformation T of E_0 into F such that

$$\|x\| \le \|Tx\| \le (1 + \varepsilon)\|x\|$$

for all $x \in E_0$. F is said to be *superreflexive* if E finitely representable in F implies that E is reflexive. Every superreflexive Banach space is reflexive, but the converse is not in general true.

2.4.2. PROPOSITION. A nonstandard hull $\hat{F}$ is reflexive if and only if it is superreflexive.

Let $\hat{F}$ be reflexive; then every separable subspace E of $\hat{F}$ is reflexive. We need to show that every separable space E that is finitely representable in $\hat{F}$ is reflexive. Being finitely representable in a space is not in general the same as being isomorphic to a subspace of it. However, in the case of a nonstandard hull we have the following result, which proves 2.4.2.

2.4.3. PROPOSITION. Let E be a separable Banach space and F an internal Banach space. Then E is finitely representable in $\hat{F}$ if and only if E is isometrically embeddable in $\hat{F}$.

PROOF. Let E be finitely representable in $\hat{F}$. Since E is separable, there is an increasing family $E_1 \subseteq E_2 \subseteq \cdots \subseteq E_n \subseteq \cdots$ of finite-dimensional subspaces of E such that $\dim E_n = n$ and $\bigcup_{n\in\mathbb{N}} E_n$ is dense in E. Let $e_1, \ldots, e_n$ be a basis for E_n; by assumption there exists a linear transformation $T_n : E_n \to \hat{F}$ such that $\|x\| \le \|T_n x\| \le (1 + 1/2n)\|x\|$ for all $x \in E_n$. Pick elements $p_1, p_2, \ldots, p_n \in F$ such that $p_i/\approx = T_n(e_i)$, $i = 1, 2, \ldots, n$. We see that the map T'_n of $*E_n$ onto the linear span [in the sense of $V(*\mathbb{R})$] of $p_1, \ldots, p_n$ in F given by

$$T'_n\left(\sum_{k=1}^n \lambda_k e_k \right) = \sum_{k=1}^n \lambda_k p_k$$

is an *internal*, linear, one-to-one map of $*E_n$ into F satisfying

(*) $(1 - 1/n)\|q\| \le \|T'_n q\| \le (1 + 1/n)\|q\|$

for all $q \in *E_n$ with $\|q\| = 1$.

We are set for an application of the internal definition principle. The set of integers $n \in *\mathbb{N}$ such that there exists an internal, linear, one-to-one mapping T'_n of $*E_n$ into F satisfying (*) is an internal set containing $\mathbb{N}$; hence it contains some $n_0 \in *\mathbb{N} - \mathbb{N}$. The map T defined by

$$T(q/\approx) = T'_{n_0}(q)/\approx$$

defines an embedding of the nonstandard hull $\hat{E}_{n_0}$ of E_{n_0} into $\hat{F}$. We notice that $\bigcup_{n \in \mathbb{N}} E_n$ is in a natural way contained in the hyperfine extension $\hat{E}_{n_0}$; thus the extension to E of the restriction of T to $\bigcup_{n \in \mathbb{N}} E_n$ gives the desired embedding of E into $\hat{F}$.

We can use these ideas to give a simple proof of a result due to Enflo *et al.* (1975). It is well known that reflexivity is a *three-space property*; i.e., if E is a Banach space and F is a closed subspace such that both F and E/F are reflexive, then E is reflexive. Enflo *et al.* showed that superreflexivity is also a three-space property. Rakov observed that using the nonstandard hull construction, one has the following simple proof; see Heinrich (1980) for further references.

2.4.4. PROPOSITION. Let E be a Banach space and F a closed subspace of E. If both F and E/F are superreflexive, then E is superreflexive.

PROOF. A simple argument shows that if a Banach space E is superreflexive, then $\hat{E}$ is reflexive. A further easy calculation shows that the nonstandard hull of E/F is canonically isometric to $\hat{E}/\hat{F}$. It thus follows from the assumptions of 2.4.4 that $\hat{F}$ and $\hat{E}/\hat{F}$ are reflexive. By the classical result $\hat{E}$ is reflexive. By Proposition 2.4.2 it follows that $\hat{E}$ is superreflexive, and so obviously is E since E can be regarded as a subspace of $\hat{E}$.

We stop our account here. The present proof that superreflexivity is a three-space property clearly indicates the strength of the nonstandard hull or ultrapower construction. But we have not in this glimpse touched the really deep and difficult results such as the Kürsten–Stern local duality theorem and Krivine's results on block finite representability of the c_0 and l_p bases. For these and other applications we refer once more to the surveys of Heinrich (1980) and Henson and Moore (1983).

REFERENCES

B. Benninghofen and M. M. Richter (1983). General theory of superinfinitesimals (preprint), RWTH Aachen.

B. Benninghofen and K. D. Stroyan (1984). Bounded-weak-star continuity. Univ. of Iowa.

A. R. Bernstein (1973). Non-standard analysis. *In Studies in Model Theory.* Math. Ass. of Amer.

C. C. Chang and H. J. Keisler (1973). *Model Theory.* North-Holland Publ., Amsterdam.

M. Davis (1977). *Applied Nonstandard Analysis.* Wiley, New York.

P. Enflo, J. Lindenstrauss, and G. Pisier (1975). On the "three space problem". *Math Scand.* **36.**

J. E. Fenstad (1967). A note on "standard" versus "non-standard" topology. *Indag. Math.* **29.**

J. E. Fenstad (1985). Is nonstandard analysis relevant for the philosophy of mathematics? *Synthese* **62.**

J. E. Fenstad and A. Nyberg (1970). Standard versus nonstandard methods in uniform topology. *Logic Colloq. 1969,* North-Holland Publ., Amsterdam.

S. Heinrich (1980). Ultraproducts in Banach space theory. *J. Reine Angew. Math.* **313.**

C. W. Henson and L. C. Moore (1983). Nonstandard analysis and the theory of Banach spaces. *Nonstandard Analysis: Recent developments. Lecture Notes in Math.* **383,** Springer-Verlag, Berlin and New York.

H. J. Keisler (1976). *Foundations of Infinitesimal Calculus.* Prindle, Weber and Schmidt, Boston, Massachusetts.

V. J. Lomonozov (1973). Invariant subspaces for operators commuting with compact operators. *Functional Anal. Appl.* **7.**

W. A. J. Luxemburg (1969). A general theory of monads. *In Applications of Model Theory to Algebra, Analysis and Probability.* Holt, New York.

L. C. Moore (1976). Hyperfinite extensions of bounded operators on a separable Hilbert space. *Trans. Amer. Math. Soc.* **218.**

E. Nelson (1977). Internal set theory. *Bull. Amer. Math. Soc.* **83.**

H. Radjavi and P. Rosenthal (1973). *Invariant subspaces.* Springer-Verlag, Berlin and New York.

M. M. Richter (1982). *Ideale Punkte, Monaden, und Nichtstandard-Methoden.* Vieweg, Wiesbaden.

A. Robinson (1966). *Non-Standard Analysis.* North-Holland Publ., Amsterdam.

K. D. Stroyan and W. A. J. Luxemburg (1976). *Introduction to the Theory of Infinitesimals.* Academic Press, New York.

CHAPTER 3

PROBABILITY

In this chapter we shall give an introduction to *hyperfinite probability theory*, while at the same time presenting the necessary background from the theory of measure and integration. We shall illustrate the general theory by discussing the particular but important case of Brownian motion. In the last two sections we shall present a hyperfinite approach to limit measures and measure extensions.

3.1. THE LOEB MEASURE

Measure theory and probability theory were studied early within the context of nonstandard analysis. However, there were troublesome points, since *-extensions of σ-additive measures are not in general σ-additive in the extended universe $V(*\mathbb{R})$. The breakthrough came with a paper by Loeb (1975). His construction of the *Loeb measure*, which converts an internal measure to a standard σ-additive measure, is the key to the nonstandard approach to stochastic analysis.

Let $(X, \mathcal{A}, \nu)$ be an internal measure space; i.e., X is an internal set in $V(*\mathbb{R})$, $\mathcal{A}$ is an internal algebra of subsets of X, and ν is a finitely additive internal measure on $\mathscr{A}$. It should be noticed that ν takes values in $(*\mathbb{R})_+$. $\mathscr{A}$ being internal means that $\mathscr{A}$ also is closed under *-finite unions; i.e., if $\omega \in *\mathbb{N} - \mathbb{N}$ and $\{A_i\}$ is an internal sequence such that $A_i \in \mathscr{A}$ for each i, then $A_1 \cup A_2 \cup \cdots \cup A_\omega \in \mathscr{A}$.

From the internal algebra $\mathscr{A}$ we can in standard fashion generate the external σ-algebra $\sigma(\mathscr{A})$. Is it possible in some way to obtain from ν on $\mathscr{A}$ a standard σ-additive measure on $\sigma(\mathscr{A})$? Or, more specifically, let $^\circ\nu$, the *standard part* of ν, be the set map

$$^\circ\nu(A) = {}^\circ(\nu(A)), \qquad A \in \mathscr{A}.$$

$^\circ\nu$ is a map from $\mathscr{A}$ to $\mathbb{R}_+ \cup \{\infty\}$. Can $^\circ\nu$ be extended from $\mathscr{A}$ to $\sigma(\mathscr{A})$?

If ν is an internal probability measure, i.e., $\nu(X) = 1$, the Countable Saturation Principle 2.1.2 gives an immediate affirmative answer. In this case $^\circ\nu$ is a finite measure on the algebra $\mathscr{A}$ and it is well known that such a measure can be extended to a σ-additive measure on $\sigma(\mathscr{A})$ if the following continuity property is satisfied: let $A_1, A_2 \ldots \in \mathscr{A}$; if $A_n \downarrow \varnothing$, then $^\circ\nu(A_n) \downarrow 0$. But this is trivially satisfied because of saturation, 2.1.2: if $A_n \downarrow \varnothing$, then $A_n = \varnothing$ for some finite n, and convergence is trivial.

In the general case the Carathéodory extension procedure gives the answer. For the proof we need the following technical lemma, which shows that a countable, infinite union of disjoint nonempty sets in $\mathscr{A}$ is never itself an element of $\mathscr{A}$, which, trivially, implies that $^\circ\nu$ is countably additive on $\mathscr{A}$.

3.1.1. LEMMA. Let $A_n \in \mathscr{A}$ for $n = 0, 1, 2, \ldots$; if $A_0 \subseteq A_1 \cup A_2 \cup \cdots$, then there is an $m > 0$ such that $A_0 \subseteq A_1 \cup A_2 \cup \cdots \cup A_m$.

For the proof, let $\langle A_n \rangle_{n \in *\mathbb{N}}$ be an internal sequence extending $A_0, A_1, A_2, \ldots$, see 2.1.3. The set

$$\left\{ m \in *\mathbb{N} \,\middle|\, A_0 \subseteq \bigcup_{n=1}^{m} A_n \right\}$$

is internal and contains every $m \in *\mathbb{N} - \mathbb{N}$. It follows from 1.2.7 that it must contain some $m \in \mathbb{N}$.

We can now state the main result due to Loeb (1975).

3.1.2. THEOREM. Let $(X, \mathscr{A}, \nu)$ be an internal measure space. The measure $^\circ\nu$ has a unique σ-additive extension, denoted by $L(\nu)$, to the σ-algebra $\sigma(\mathscr{A})$ generated by $\mathscr{A}$. Furthermore, if $^\circ\nu(X) < \infty$, then

   (i)   For each $B \in \sigma(\mathscr{A})$ there exists $A \in \mathscr{A}$ such that $L(\nu)(A \triangle B) = 0$ (where $A \triangle B = (A - B) \cup (B - A)$).

   (ii)   For each $B \in \sigma(\mathscr{A})$ and each $\varepsilon \in \mathbb{R}_+$ there are sets $C, D \in \mathscr{A}$ such that $C \subset B \subset D$ and $L(\nu)(D) - \varepsilon \le L(\nu)(B) \le L(\nu)(C) + \varepsilon$.

The existence is clear from Carathéodory's extension theorem, which also tells us that ν is unique provided $°\nu(X) < \infty$. When ν is infinite, a more refined argument is necessary to establish the uniqueness; see Henson (1979). Since we are mostly interested in probability measure, we omit this part.

We now turn to (ii). By construction of $L(\nu)$ there is a sequence $\langle A_n \rangle_{n \in \mathbb{N}}$ of elements of $\mathscr{A}$ with $A_n \subseteq A_{n+1}$ such that $B \subseteq \tilde{B} = \bigcup_{n \in \mathbb{N}} A_n$ and $L(\nu)$-$(\tilde{B}) < L(\nu)(B) + \varepsilon$. Extend the sequence to an internal sequence $\langle A_n \rangle_{n \in {}^*\mathbb{N}}$. For any $\omega \in {}^*\mathbb{N} - \mathbb{N}$ we clearly have $B \subseteq \tilde{B} \subseteq A_\omega$. We need to choose ω such that $\nu(A_\omega) \leq L(\nu)(B) + \varepsilon$, as we can then put $D = A_\omega$. Let $r = L(\nu)(B)$ and consider the set $\{m \in {}^*\mathbb{N} \mid \nu(A_m) \leq r + \varepsilon\}$. This set is internal and contains $\mathbb{N}$, hence it also contains some $\omega \in {}^*\mathbb{N} - \mathbb{N}$. Note that $L(\nu)$ and B are external objects, but the *number* r is internal. Applying the same argument to $X - B$, we find C.

It is now easy to prove (i). For each n, let C_n and D_n be elements in $\mathscr{A}$ such that $C_n \subset B \subset D_n$ and

$$L(\nu)(D_n) - (1/n) \leq L(\nu)(B) \leq L(\nu)(C_n) + (1/n).$$

We may assume that $\{C_n\}$ is increasing and $\{D_n\}$ decreasing. Extend $\{C_n\}$ and $\{D_n\}$ to an increasing, respectively decreasing, internal sequence such that $C_n \subset D_n$ for all $n \in {}^*\mathbb{N}$, and put $A = C_\omega$ for some $\omega \in {}^*\mathbb{N} - \mathbb{N}$.

3.1.3. EXAMPLE. The space $2^{\mathbb{N}}$ of infinite sequences of 0's and 1's is the standard model of unlimited or infinite coin tossing. More precisely, the model is the measure space $(2^{\mathbb{N}}, \mathscr{B}, \mu)$ defined in the following way: on each factor $2 = \{0, 1\}$ start with the σ-algebra of all subsets and the counting measure giving equal weight to each point in the space. Then $\mathscr{B}$ is the usual product σ-algebra and μ the product measure. In this case existence of $\mathscr{B}$ and μ is trivial; in more general cases there are often problems.

A nonstandard way of modeling the same phenomena would be to choose some $\eta \in {}^*\mathbb{N} - \mathbb{N}$ and consider the space $\Omega = \{0, 1\}^\eta$ of all internal sequences of 0's and 1's of length η. In this case we let $\mathscr{A}$ be the algebra of all internal subsets of Ω and P the associated counting measure. This means that for each $A \in \mathscr{A}$

$$P(A) = |A|/2^\eta,$$

where for any internal set $A \subseteq \Omega$, $|A|$ is the internal cardinality of A, i.e., the number of elements in A, and $2^\eta = |\Omega|$.

Here $(\Omega, \mathscr{A}, P)$ is an internal measure space. Let $L(P)$ be the associated Loeb measure on $\sigma(\mathscr{A})$; $(\Omega, \sigma(A), L(P))$ is a standard measure space; however, on a rather unusual sample space. The possible usefulness of the nonstandard approach lies in the fact that we have two spaces to play with.

The space $(\Omega, \mathscr{A}, P)$ as an internal object is in some sense "finite," which means that we can calculate as if we were in a truly finite situation. On the other hand, the external space $(\Omega, \sigma(\mathscr{A}), L(P))$ has a well-behaved measure and integration theory.

How is $(2^N, \mathscr{B}, \mu)$ related to these structures? Again, in a suitable sense, it should be the "standard part" of the nonstandard constructs. Let $st_\eta : \Omega \to 2^N$ be the restriction map; we would like to assert that for any $A \in \mathscr{B}$

$$\mu(A) = L(P)(st_\eta^{-1}(A)),$$

i.e., that st_η is a measure-preserving map. This is almost true; we need only be a bit careful about the σ-algebras involved.

3.1.4. REMARK. So far we have considered extensions $(\Omega, \sigma(\mathscr{A}), L(P))$ of $(\Omega, \mathscr{A}, P)$. For many purposes it is more convenient to work with the completion of $(\Omega, \sigma(\mathscr{A}), L(P))$. We denote the completion by $(\Omega, L(\mathscr{A}), L(P))$ and call it *the Loeb space* associated with $(\Omega, \mathscr{A}, P)$. Similarly, the completed measure $L(P)$ will be called the *Loeb measure* of P. Referring back to the previous example, we may show that if B is a Borel set in 2^N then $st_\eta^{-1}(B) \in L(A)$ in Ω; see Sections 3.4 and 3.5 for a general discussion.

3.1.5. REMARK. Although we have used Carathéodory's extension theorem to obtain Loeb measures, there is a simple and elegant direct construction which presupposes no measure theory: given an internal measure P on an internal algebra $\mathscr{A}$, define inner and outer measures $\underline{P}$ and $\bar{P}$ by

$$\underline{P}(A) = \sup\{{}^\circ P(B) \,|\, B \in \mathscr{A} \text{ and } B \subset A\}$$

and

$$\bar{P}(A) = \inf\{{}^\circ P(B) \,|\, B \in \mathscr{A} \text{ and } B \supset A\}.$$

Then the Loeb algebra $L(\mathscr{A})$ is just the collection of all sets A such that $\underline{P}(A) = \bar{P}(A)$, and $L(P)(A)$ is just this common value. For the details, see, e.g., Cutland (1983) or Stroyan and Bayod (1985).

A more radically different, Daniell-type approach to nonstandard measure theory has been advocated by Loeb (1983, 1984) [see also Hurd and Loeb (1985)]; it has the advantage of allowing one to develop measure and integration theory simultaneously. There could be interesting connections to vector-valued Loeb measures, a topic which has only recently begun to attract attention (Osswald, 1983, 1985; Zivaljevic, 1985).

3.2. HYPERFINITE PROBABILITY SPACES

We claimed that our internal model $(\Omega, \mathcal{A}, P)$ in a "suitable sense" was finite. The time has come to be more precise.

A subset $A \subseteq \mathbb{N}$ is finite if it is a subset of some proper initial segment of $\mathbb{N}$; i.e., for some $m \in \mathbb{N}$, $A \subseteq \{n \in \mathbb{N} \mid n \le m\}$. An internal subset $A \subseteq {}^*\mathbb{N}$ is called *hyperfinite* (or *-finite) if $A \subseteq \{n \in {}^*\mathbb{N} \mid n \le m\}$, for some $m \in {}^*\mathbb{N}$. In general, an internal set $E \in V({}^*\mathbb{R})$ is called hyperfinite if there is an internal one-to-one map f of a hyperfinite subset of ${}^*\mathbb{N}$ onto E. This is equivalent to the following:

3.2.1. DEFINITION. An internal set $E \in V({}^*\mathbb{R})$ is called *hyperfinite* if there is an internal bijection f of some proper initial segment $\{n \in {}^*\mathbb{N} \mid n \le m\}$ of ${}^*\mathbb{N}$ onto E. The number m is called the *internal cardinality* of E, in symbols $|E| = m$.

REMARK. In the bounded ultrapower model an internal set A was defined from a bounded sequence $\langle A_1, A_2, \ldots, A_n, \ldots \rangle$; see the remark following Definition 1.2.5. If each set A_n is finite, then A will be hyperfinite and the internal cardinality of A will be the hyperfinite integer determined by the sequence $\langle |A_1|, |A_2|, \ldots, |A_n|, \ldots \rangle$ of standard integers, where $|A_n|$ is the number of elements in A_n. Every hyperfinite set in the model can be described in this way.

At this point we should draw the reader's attention to the difference between *internal* and *external* cardinality. The segment $\{n \in {}^*\mathbb{N} \mid n \le n_0\}$, where $n_0 \in {}^*\mathbb{N} - \mathbb{N}$, has the hyperfinite number n_0 as its internal cardinality; i.e., there is no *internal* bijection of this segment onto a smaller one. But seen from the "outside," i.e., within the full structure $V({}^*\mathbb{R})$, the set has uncountably many elements; i.e., there is no (external) map of $\mathbb{N}$ onto $\{n \in {}^*\mathbb{N} \mid n \le n_0\}$. This follows from the general fact that an *internal* set is either finite or uncountable. For let A be an internal set which is countable (but not finite) from the outside; i.e., we have an external enumeration of A, $A = \{a_1, a_2, \ldots, a_n, \ldots\}$, in $V({}^*\mathbb{R})$. We can then define internal sets $A_0 \supset A_1 \supset A_2 \supset \cdots$ by setting $A_0 = A$ and $A_n = A - \{a_1, \ldots, a_n\}$ for $n > 0$. $A_0, A_1, \ldots$ is an external sequence of nonempty internal sets. By the Saturation Principle 2.1.2, $\bigcap A_n \ne \varnothing$; but this is a contradiction! The precise external cardinality of an internal set depends upon the kind of ultrapower construction we use. But it is the internal cardinality which is well behaved and is the important concept in our theory.

Definition 3.2.1 defines the precise sense in which the set $\Omega = \{0, 1\}^\eta$ of Example 3.1.3 is "finite," namely hyperfinite. Thus we see that the definition of $P(A) = |A|/|\Omega|$ is the exact analog of the definition of counting measures in finite probability spaces.

In the following we shall call an internal probability space $(\Omega, \mathscr{A}, P)$, where Ω is hyperfinite, a *hyperfinite probability space*.

3.2.2. EXAMPLE. *The hyperfinite time line* will be another important example of a hyperfinite probability space.

The idea is to replace the time interval $[0, 1]$ by a hyperfinite set T. To this end choose some $\eta \in {}^*\mathbb{N} - \mathbb{N}$, such that $\Delta t = \eta^{-1}$ is a positive infinitesimal. Let

$$T = \{0, \Delta t, 2\,\Delta t, \ldots, \eta\,\Delta t = 1\},$$

where T is hyperfinite. We notice that if $\eta = \omega!$ for some $\omega \in {}^*\mathbb{N} - \mathbb{N}$, then every standard rational m/n in $[0, 1]$ belongs to T; they are all of the form $\lambda\,\Delta t$ for some $\lambda \leq \eta$. The standard part map $°: T \to [0, 1]$ is onto: no irrational number r in $[0, 1]$ is an element of T, but given an irrational r there is a unique $t \in T$ such that $t < r < t + \Delta t$.

We let P be the counting measure on T; i.e., for any internal set $A \subseteq T$ we set

$$P(A) = |A|/|T|.$$

As we shall see below $(T, \mathscr{A}, P)$, where $\mathscr{A}$ is the internal algebra of all internal subsets of T, and the associated Loeb space $(T, L(\mathscr{A}), L(P))$ are our versions of the usual Lebesgue space $([0, 1], \mathscr{B}, \mu)$, where $\mathscr{B}$ denotes the Lebesgue-measurable subsets of $[0, 1]$ and μ is the standard Lebesgue measure.

Spaces and measures give the necessary setting. The mappings and the processes are the objects of central importance. A *standard stochastic process* is a map

$$x : E \times [0, 1] \to \mathbb{R},$$

where $(E, \mathscr{B}, \mu)$ is some probability space. The value space here is $\mathbb{R}$ but could, e.g., be some suitable separable metric space.

A *hyperfinite stochastic process* is an internal map

$$X : \Omega \times T \to {}^*\mathbb{R},$$

where T is a hyperfinite time line and $(\Omega, \mathscr{A}, P)$ is some hyperfinite probability space.

From X we want to derive a standard process x by taking "standard parts." And conversely, if we let our initial data be an external, i.e., standard process $x : \Omega \times [0, 1] \to \mathbb{R}$ with respect to $(\Omega, L(\mathscr{A}), L(P))$, is it possible to approximate it by a hyperfinite process $X : \Omega \times T \to {}^*\mathbb{R}$? Processes are maps of two variables; thus we need to approximate external objects by internal objects in each factor. This leads to the notion of liftings.

3.2.3. DEFINITIONS. (1) Let $f: \Omega \to \mathbb{R}$ and $F: \Omega \to {}^*\mathbb{R}$. F is called a *lifting* of f if F is internal and

$$°F(\omega) = f(\omega)$$

for almost all $\omega \in \Omega$ with respect to the Loeb measure $L(P)$ on Ω.

(2) Let $f: [0, 1] \to \mathbb{R}$ and $F: T \to {}^*\mathbb{R}$. F is called a *lifting* of f if F is internal and

$$°F(t) = f(°t)$$

for almost all $t \in T$ with respect to the Loeb measure $L(P)$ on T.

We have the following important result:

3.2.4. THEOREM. (1) Let $(\Omega, \mathcal{A}, P)$ be a hyperfinite probability space and $(\Omega, L(\mathcal{A}), L(P))$ its associated Loeb space. A function $f: \Omega \to \mathbb{R}$ is Loeb-measurable iff f has a lifting $F: \Omega \to {}^*\mathbb{R}$.

(2) Let $([0, 1], \mathcal{B}, \mu)$ be the standard Lebesgue space and $(T, L(\mathcal{A}), L(P))$ the Loeb space associated to the hyperfinite time line T. A function $f: [0, 1] \to \mathbb{R}$ is Lebesgue-measurable iff f has a lifting $F: T \to {}^*\mathbb{R}$.

We note that $\mathbb{R}$ can be replaced here by any separable metric space, e.g., $C(\mathbb{R}^n, \mathbb{R}^m)$ in the compact-open topology.

Now we turn to the proof of (1). Let F be a lifting of f. Let $N_{1/n}(r) = \{r' \in \mathbb{R} \mid |r - r'| < 1/n\}$; i.e., $N_{1/n}(r)$ is a standard neighborhood of $r \in \mathbb{R}$. We must prove that $f^{-1}(N_{1/n}(r))$ is Loeb-measurable. Since F lifts f the set

$$U = \{\omega \in \Omega \mid °F(\omega) = f(\omega)\}$$

has $L(P)$ measure 1. Let $\omega \in U$; then

$$f(\omega) \in N_{1/n}(r) \qquad \text{iff} \quad |r - °F(\omega)| < 1/n,$$

$$\text{iff} \quad °|r - F(\omega)| < 1/n,$$

$$\text{iff} \quad |r - F(\omega)| \le 1/n - 1/m, \qquad \text{some} \quad m \in \mathbb{N}.$$

Here the first equivalence is by definition and the second comes from the continuity of the absolute-value function. The importance of the third equivalence lies in the fact that an *external* condition $°|r - F(\omega)| < 1/n$ is replaced by an *internal* condition $|r - F(\omega)| \le (1/n) - (1/m)$. Thus the set $\{\omega \in \Omega \mid |r - F(\omega)| \le (1/n) - (1/m)\} \in \mathcal{A}$. Therefore $\bigcup_{m \in \mathbb{N}} \{\omega \in \Omega \mid |r - F(\omega)| \le (1/n) - (1/m)\} \in L(\mathcal{A})$. Hence

$$U \cap f^{-1}(N_{1/n}(r)) \in L(\mathcal{A}),$$

which suffices to show that f is Loeb measurable.

For the converse let $N_1, N_2, \ldots$ be a countable open base for $\mathbb{R}$ and set $U_n = f^{-1}(N_n) \in L(\mathcal{A})$. By Theorem 3.1.2 we can find internal sets $A_{n,m}$ such that $L(P)(U_n) \leq P(A_{n,m}) + 1/m$ and such that $A_{n,m} \subseteq A_{n,m+1} \subseteq U_n$. We calculate that

$$L(P)(U_n - \bigcup_m A_{n,m}) = 0.$$

Thus the set $U = \Omega - \bigcup_m (U_n - \bigcup_m A_{n,m})$ has Loeb measure 1.

For each $n \in \mathbb{N}$ let $\mathcal{G}_{n,m}$ be the internal set of all internal functions $F : \Omega \to {}^*\mathbb{R}$ such that $F(A_{k,l}) \subseteq {}^*N_k$ for all $k \leq n$, $l \leq m$. Each $\mathcal{G}_{n,m}$ is nonempty. Hence by the Saturation Principle 2.1.2 there exists an internal $F \in \bigcap \mathcal{G}_{n,m}$. For $\omega \in U$ we see that

$$F(\omega) \in \bigcap \{ {}^*N_n \mid f(\omega) \in N_n \}.$$

Since $N_1, N_2, \ldots$ is an open base for $\mathbb{R}$, we conclude that $^\circ F(\omega) = f(\omega)$ for all $\omega \in U$. Thus F lifts f.

For the proof of part (2) of 3.2.4 we need the following result.

3.2.5. PROPOSITION. A set $A \subseteq [0, 1]$ is Lebesgue measurable iff the set $\mathrm{st}^{-1}(A) = \{ t \in T \mid {}^\circ t \in A \}$ is Loeb measurable. In this case we have $\mu(A) = L(P)(\mathrm{st}^{-1}(A))$.

The reader should at this point recall our discussion in Example 3.1.3. Granted Proposition 3.2.5, the proof of 3.2.4(2) reduces to the case 3.2.4(1). Given $f : [0, 1] \to \mathbb{R}$, define $f_1 : T \to \mathbb{R}$ by setting $f_1(t) = f(^\circ t)$. For any open $U \subseteq \mathbb{R}$, 3.2.5 tells us that $f^{-1}(U)$ is Lebesgue measurable iff $f_1^{-1}(U)$ is Loeb measurable. Pick by (1) a lifting F of f_1, then F also lifts f.

The intuition behind 3.2.5 should be clear. The Lebesgue measure is the uniform measure on $[0, 1]$, i.e., the continuous version of the counting measure. The corresponding Loeb measure is the counting measure on the hyperfinite approximation T to $[0, 1]$. So everything is all right, except for the fact that there are σ-algebras involved.

Half of the proof is rather immediate. Namely, let A be Lebesgue measurable. It follows from general constructions that we may restrict ourselves to the case $A = [a, b\rangle$, where a and b are rationals. By our choice of Δt, cf. 3.2.2, we have a, $b \in T$ and $\mathrm{st}^{-1}(A) = \bigcup_m \bigcap_n \{ t \in T \mid a - (1/n) \leq t < b - (1/m) \}$. Then $\mathrm{st}^{-1}(A) \in L(\mathcal{A})$, and quite clearly $L(P)(\mathrm{st}^{-1}(A)) = b - a$.

For the converse we use the fact that if B is an internal subset of T in ${}^*\mathbb{R}$, then $\mathrm{st}(B)$ is a closed, hence compact, subset of $[0, 1]$ in $\mathbb{R}$; see Proposition 2.1.8. Now let $A \subseteq [0, 1]$ and assume that $\mathrm{st}^{-1}(A) \in L(\mathcal{A})$. We must show that A is Lebesgue measurable. Given any $\varepsilon \in \mathbb{R}_+$ we may, by Theorem 3.1.2, find an internal $B \subseteq \mathrm{st}^{-1}(A)$ such that $P(B) > L(P)(\mathrm{st}^{-1}(A)) - \varepsilon$. $C = \mathrm{st}(B) \subseteq A$ is compact in $[0, 1]$, hence Lebesgue measurable. By the first part of 3.2.5 we know that $\mathrm{st}^{-1}(C) \in L(A)$ and $\mu(C) = L(P)(\mathrm{st}^{-1}(C)) \geq {}^\circ(P(B)) \geq L(P)(\mathrm{st}^{-1}(A)) - \varepsilon$. By a dual argument we can find for any $\varepsilon \in \mathbb{R}_+$ an open set $D \supseteq A$ such that $\mu(D) \leq L(P)(\mathrm{st}^{-1}(A)) + \varepsilon$. This suffices to show that A is Lebesgue measurable, and $\mu(A) = L(P)(\mathrm{st}^{-1}(A))$.

3.2.6. REMARK. The arguments above generalize in such a way that they can be used to show that any Radon probability space can be obtained from a hyperfinite probability space via a measure isomorphism. In our case the standard part map $\mathrm{st}: T \to [0, 1]$ is a measure isomorphism between $(T, L(\mathscr{A}), L(P))$ and $([0, 1], \mathscr{B}, \mu)$; for a full discussion see Anderson (1977, 1982) and also Sections 3.4 and 3.5.

The theory of integration is particularly simple in a hyperfinite probability space.

3.2.7. DEFINITION. Let $(\Omega, \mathscr{A}, P)$ be a hyperfinite probability space with P the counting measure and $\mathscr{A}$ the algebra of all internal sets. Let $F: \Omega \to {}^*\mathbb{R}$ be an internal function. The *expectation* $E(F)$ of F is defined as

$$E(F) = \int_\Omega F(\omega)\, dP = \sum_{\omega \in \Omega} F(\omega) \frac{1}{|\Omega|}.$$

Some remarks may be in order. First, the hyperfinite sum $\sum_{\omega \in \Omega}$ exists by transfer, thus $E(F)$ is a well-defined hyperreal number.

We have restricted ourselves here to integration with respect to the counting measure P. We could also assign different weights to the points of Ω. Let $\langle a_\omega | \omega \in \Omega \rangle$ be an internal sequence such that $\sum a_\omega = 1$. In a standard way this defines a hyperfinite probability on Ω with associated expectation

$$E(F) = \sum_{\omega \in \Omega} F(\omega) a_\omega.$$

We could also develop the theory with respect to an arbitrary internal algebra $\mathscr{A}$. We would then have to make appropriate mention of $\mathscr{A}$-measurability; e.g., in Theorem 3.2.4(1) we would get that $f: \Omega \to \mathbb{R}$ is Loeb measurable [i.e., $L(\mathscr{A})$-measurable] iff f has an $\mathscr{A}$-measurable lifting $F: \Omega \to \mathbb{R}$.

We shall now relate the hyperfinite expectation with the standard Lebesgue integral on $\mathbb{R}$. First we state a definition:

3.2.8. DEFINITION. Let $(\Omega, \mathscr{A}, P)$ be an internal probability space and $F: \Omega \to {}^*\mathbb{R}$ an $\mathscr{A}$-measurable internal function. F is called *S-integrable* if

(i) $E(|F|)$ is a finite hyperreal;
(ii) $A \in \mathscr{A}$ and $P(A) \approx 0$, then $\int_A |F(\omega)|\, dP \approx 0$.

The space $(\Omega, \mathscr{A}, P)$ has an associated Loeb space $(\Omega, L(\mathscr{A}), L(P))$ with a standard integration theory with respect to the σ-algebra $L(\mathscr{A})$ and the σ-additive measure $L(P)$. We have the following not unexpected result.

3.2.9. THEOREM. (1) Let $(\Omega, \mathcal{A}, P)$ be a hyperfinite probability space and $(\Omega, L(\mathcal{A}), L(P))$ its associated Loeb space. A function $f: \Omega \to \mathbb{R}$ is Loeb integrable iff f has an S-integrable lifting $F: \Omega \to {}^*\mathbb{R}$. In this case

$$E(F) \approx \int_\Omega f(\omega) \, dL(P)(\omega).$$

(2) Let $(\Omega, \mathcal{A}, P)$ be a hyperfinite probability space and $(\Omega, L(\mathcal{A}), L(P))$ its associated Loeb space. Let $F: \Omega \to {}^*\mathbb{R}$ be internal and non-negative. Then F is S-integrable iff
  (i) ${}^\circ F$ is Loeb integrable;
  (ii) ${}^\circ E(f) = \int_\Omega {}^\circ F \, dL(P)$.

(3) Let $([0, 1], \mathcal{B}, \mu)$ be the Lebesgue space and $(T, \mathcal{A}, P)$ a hyperfinite time line. A function $f: [0, 1] \to \mathbb{R}$ is Lebesgue integrable iff f has an S-integrable lifting $F: T \to {}^*\mathbb{R}$. In this case

$$E(F) \approx \int_0^1 f(r) \, d\mu(r).$$

Note that if $f: [0, 1] \to \mathbb{R}$ is continuous, then ${}^*f$ restricted to T is a lifting of f. Thus (3) above shows that the Riemann integral as defined in Section 1.4 is a special case of hyperfinite integration.

For the proof of 3.2.9 note first that part (2) is a variation on (1); it does, however, require a proof! Part (3) follows from the proof of (1) and the fact that st: $T \to [0, 1]$ is a measure isomorphism.

We are thus left with part (1). First note that if F is a *finite function*, i.e., $F: \Omega \to {}^*[-n, n]$ for some $n \in \mathbb{N}$, then a very simple approximation argument shows that

$$E(F) \approx \int_\Omega {}^\circ F(\omega) \, dL(P)(\omega).$$

The general case will follow from the following characterization:

3.2.10. PROPOSITION. A function $F: \Omega \to {}^*\mathbb{R}$ is S-integrable iff there exists a sequence $\langle F_n \mid n \in \mathbb{N} \rangle$ of finite functions such that

$${}^\circ E(|F - F_n|) \to 0 \qquad \text{as} \quad n \to \infty.$$

We give a fairly complete proof of 3.2.10 since it shows us exactly the role of S-integrability. The remaining part of the proof of 3.2.9 is rather standard and is left to the reader. [If needed one may consult Loeb (1975), Anderson (1976), and also Henson (1979) for the case of unbounded Loeb measures; see also the forthcoming book Stroyan and Bayod (1985).]

Let $F : \Omega \to {}^*\mathbb{R}$ be S-integrable. For each $n \in {}^*\mathbb{N}$ define a function $F_n : \Omega \to {}^*\mathbb{R}$ by setting $F_n(\omega) = F(\omega)$ if $|F(\omega)| \le n$, $F_n(\omega) = n$ if $F(\omega) > n$, and $F_n(\omega) = -n$ if $F(\omega) < -n$. Then F_n is a finite function if $n \in \mathbb{N}$. For $m \in {}^*\mathbb{N} - \mathbb{N}$

$$P(\{\omega \in \Omega \,|\, |F(\omega)| > m\}) \le 1/m E[|F|] \approx 0$$

by Chebyshev's inequality. The last part follows since by S-integrability $E(|F|) < \infty$. For $m \in {}^*\mathbb{N} - \mathbb{N}$,

$$E(|F - F_m|) \le \int_{|F(\omega)| > m} |F(\omega)| \, dP.$$

Since $P(\{\omega \in \Omega \,|\, |F(\omega)| > m\}) \approx 0$, S-integrability at once implies that the integral ≈ 0.

For the converse let $\langle F_n \rangle_{n \in \mathbb{N}}$ be given; we assume that $\sup_\omega |F_n| < n$. Obviously $°E(|F|) < \infty$ follows from the approximation $\lim_{n \to \infty} °E(|F - F_n|) = 0$.

In order to verify 3.2.8(ii), let $\varepsilon \in \mathbb{R}_+$ be given. Choose some $n \in \mathbb{N}$ such that $°E(|F - F_n|) < \varepsilon/2$. Let $A \in \mathcal{A}$ be such that $P(A) < \varepsilon/2n$. Then

$$\int_A |F| \, dP \le \int_A |F_n| \, dP + \int_A |F - F_n| \, dP < \varepsilon.$$

Thus if $A \in \mathcal{A}$ and $P(A) \approx 0$, then $\int_A |F| \, dP \approx 0$.

We shall give a brief introduction to *conditional expectations*. Let $(\Omega, \mathcal{A}, P)$ be a hyperfinite probability space where $\mathcal{A}$ is the internal algebra of internal subsets of Ω. Any internal subalgebra $\mathcal{B}$ of $\mathcal{A}$ is generated by a hyperfinite partition $\{\Omega_1, \ldots, \Omega_\eta\}$, $\eta \in {}^*\mathbb{N}$, of the set Ω; this follows by transfer from the finite case.

3.2.11. DEFINITION. Let $F : \Omega \to {}^*\mathbb{R}$ be an internal function and $\mathcal{B}$ a subalgebra of A generated by the internal partition $\{\Omega_1, \Omega_2, \ldots, \Omega_\eta\}$, $\eta \in {}^*\mathbb{N}$, of Ω. The *conditional expectation* $E(F | \mathcal{B})$ is defined as

$$E(F | \mathcal{B})(\omega) = P(\Omega_n)^{-1} \sum_{\omega' \in \Omega_n} F(\omega') P(\omega'),$$

$$\text{for} \quad \omega \in \Omega_n, \qquad n = 1, 2, \ldots, \eta.$$

We note that the function $E(F | \mathcal{B}) : \Omega \to {}^*\mathbb{R}$ is $\mathcal{B}$-measurable and that

$$E(E(F | \mathcal{B})) = E(F).$$

The following proposition relates the hyperfinite concepts to the standard one.

3.2.12. PROPOSITION. Let $(\Omega, \mathcal{A}, P)$ be a hyperfinite probability space, and let $F : \Omega \to {}^*\mathbb{R}$ be S-integrable. For any internal subalgebra $\mathcal{B}$ of $\mathcal{A}$ we let $E(F|\mathscr{B})$ denote the (hyperfinite) conditional expectation of F with respect to $\mathscr{B}$ and $E({}^{\circ}F|L(\mathscr{B}))$ denote the standard conditional expectation of ${}^{\circ}F$ with respect to the sub-σ-algebra $L(\mathscr{B})$ of $L(\mathscr{A})$. Then $E(F|\mathscr{B})$ is S-integrable and

$$^{\circ}E(F|\mathscr{B}) = E({}^{\circ}F|L(\mathscr{B})), \qquad L(P)\text{-a.e.}$$

We begin the proof by observing that for all $A \in \mathscr{B}$,

$$^{\circ}\int_A E(F|\mathscr{B})\,dP = {}^{\circ}\int_A F\,dP = \int_A {}^{\circ}F\,dL(P) = \int_A E({}^{\circ}F|L(\mathscr{B}))\,dL(P),$$

by S-integrability of F and 3.2.9(2). Taking A such that $P(A) \approx 0$, we get S-integrability of $E(F|\mathscr{B})$.

By 3.1.2 we may for any $B \in L(\mathscr{B})$ find an $A \in \mathscr{B}$ such that $L(P)(A \Delta B) = 0$. Hence

$$\int_B {}^{\circ}E(F|\mathscr{B})\,dL(P) = {}^{\circ}\int_A E(F|\mathscr{B})\,dP$$

$$= \int_A E({}^{\circ}F|L(\mathscr{B}))\,dL(P)$$

$$= \int_B E({}^{\circ}F|L(\mathscr{B}))\,dL(P),$$

by the S-integrability of $E(F|\mathscr{B})$, Theorem 3.2.9(2), and the calculation above. Since ${}^{\circ}E(f|\mathscr{B})$ is $L(\mathscr{B})$-integrable, the result follows.

There is more to measure and integration theory than we have touched upon. Let us add a few remarks on product measures.

Anderson (1976) has shown that if Ω_1 and Ω_2 are hyperfinite probability spaces and $U \subseteq \Omega_1 \times \Omega_2$ is measurable with respect to the product of the Loeb measures on Ω_1 and Ω_2, then U is Loeb measurable on $\Omega_1 \times \Omega_2$, and the measure of U is the same with respect to the two measures.

As first pointed out by Hoover (1982), the converse is false in general. Our exposition follows D. Normann (unpublished).

3.2.13. EXAMPLE. Fix some $m \in {}^{*}\mathbb{N} - \mathbb{N}$. Let Ω_1 be the set of all internal subsets of $\{1, 2, \ldots, m\}$ and $\Omega_2 = \{1, 2, \ldots, m\}$. Let P_1 and P_2 be the counting measures on Ω_1 and Ω_2, respectively.

On $\Omega_1 \times \Omega_2$, let $L(P_1) \otimes L(P_2)$ be the product of the two Loeb measures and $L(P_1 \otimes P_2)$ be the Loeb measure of the counting measure $P_1 \otimes P_2$ on $\Omega_1 \times \Omega_2$. Let $F \subseteq \Omega_1 \times \Omega_2$ be defined as

$$F = \{(x, i) \in \Omega_1 \times \Omega_2 \mid i \in x\},$$

where F is internal in $\Omega_1 \times \Omega_2$ and $|F| = \frac{1}{2} \cdot |\Omega_1| \cdot |\Omega_2|$; $|\cdot|$ denotes internal cardinality. Hence F is $L(P_1 \otimes P_2)$-measurable and $L(P_1 \otimes P_2)(F) = \frac{1}{2}$.

We will show that F is *not* $L_1(P_1) \otimes L(P_2)$-measurable. Assume to the contrary that F is $L(P_1) \otimes L(P_2)$-measurable. By the result of Anderson quoted above $L(P_1) \otimes L(P_2)(F) = \frac{1}{2}$.

By definition of the product measure there will be a family of measurable rectangles $\{A_i \times B_i\}_{i \in \mathbb{N}}$ such that

$$F \subseteq \bigcup_{i \in \mathbb{N}} (A_i \times B_i)$$

and such that

$$\sum_{i=1}^{\infty} L(P_1)(A_i)L(P_2)(B_i) < 1.$$

By construction of the Loeb measure (see 3.1.2), we may assume that each A_i, B_i is internal. Then by 3.1.1 there is a number $n \in \mathbb{N}$ such that

$$F \subseteq \bigcup_{i \leq n} (A_i \times B_i).$$

The complement of $\bigcup_{i \leq n} (A_i \times B_i)$ will also be a finite union of internal rectangles. Thus there are internal sets $A \subseteq \Omega_1$ and $B \subseteq \Omega_2$ such that $F \cap (A \times B) = \varnothing$ and $L(P_1)(A) > 0$ and $L(P_2)(B) > 0$. This is impossible as the following argument shows:

$$|\{x \in \Omega_1 | \forall i \in B(i \notin x)\}| = 2^{m-|B|} = |\Omega_1| 2^{-|B|}.$$

Since $F \cap (A \times B) = \varnothing$ we have that $P_1(A) \leq 2^{-|B|}$. But then either $P_1(A)$ is infinitesimal (if B is infinite) or $P_2(B)$ is infinitesimal (if B is finite); thus either $L(P_1)(A) = 0$ or $L(P_2)(B) = 0$.

Even if the product of Loeb measures is not the Loeb measure of the product, Keisler (1977) has provided the following often useful Fubini-type theorem:

3.2.14. THEOREM. Let Ω_1 and Ω_2 be hyperfinite probability spaces and $f : \Omega_1 \times \Omega_2 \to \mathbb{R}$ a Loeb-integrable function. Then

(i) $f(\omega_1, \cdot)$ is Loeb integrable for almost all $\omega_1 \in \Omega_1$;
(ii) the function $g(\omega_1) = \int f(\omega_1, \omega_2) \, dL(P_2)$ is Loeb integrable on Ω_1;
(iii) $\int f(\omega_1, \omega_2) \, dL(P_1 \otimes P_2) = \int (\int f(\omega_1, \omega_2) \, dL(P_2)) \, dL(P_1)$.

The proof we shall give is due to Loeb. It splits naturally into two parts.

(a)   We shall show that if $A \subset \Omega_1 \times \Omega_2$ has $L(P_1 \otimes P_2)$-measure zero, then for $L(P_1)$-a.a. ω_1, the section $A(\omega_1) = \{\omega_2 | (\omega_1, \omega_2) \in A\}$ has $L(P_2)$-measure zero: pick a decreasing sequence $\{B_n\}$ of internal subsets of $\Omega_1 \times \Omega_2$ such that $A \subset B_n$ and $^\circ P_1 \otimes P_2(B_n) \downarrow 0$. Let $B = \bigcap B_n$. Since $P_1 \otimes P_2(B_n) = \int P_2(B_n(\omega_1)) \, dP_1(\omega_1)$, Theorem 3.2.9 tells us that

$$
\begin{aligned}
L(P_1 \otimes P_2)(B_n) = {}^\circ P_1 \otimes P_2(B_n) &= {}^\circ\!\!\int P_2(B_n(\omega_1)) \, dP_1(\omega_1) \\
&= \int {}^\circ P_2(B_n(\omega_1)) \, dL(P_1(\omega_1)) \\
&= \int L(P_2)(B_n(\omega_1)) \, dL(P_1)(\omega_1).
\end{aligned}
$$

By the monotone convergence theorem, this implies that $L(P_2)(B(\omega_1)) = 0$ a.e., and since $A \subset B$, we have proved our claim.

(b)   Pick an S-integrable lifting F of f, and define $G(\omega_1) = \int F(\omega_1, \omega_2) \, dP_2$. It suffices to show that for almost all ω_1, the function $F(\omega_1, \cdot)$ is an S-integrable lifting of $f(\omega_1, \cdot)$ and that G is an S-integrable lifting of g. By Theorem 3.2.9 this immediately implies the first two parts of the theorem, and the last part follows from the calculation

$$
\begin{aligned}
\int f(\omega_1, \omega_2) \, dL(P_1 \otimes P_2) = {}^\circ\!\!\int F(\omega_1, \omega_2) \, d(P_1 \otimes P_2) &= {}^\circ\!\!\int G(\omega_1) \, dP_1 \\
&= \int g(\omega_1) \, dL(P_1) \\
&= \int \left(\int f(\omega_1, \omega_2) \, dL(P_2) \right) dL(P_1).
\end{aligned}
$$

That $F(\omega_1, \cdot)$ is a lifting of $f(\omega_1, \cdot)$ for almost all ω_1 follows immediately from (a). For the S-integrability we use Proposition 3.2.10. Let

$$
F_m(\omega_1, \omega_2) = \begin{cases} F(\omega_1, \omega_2) & \text{if } -m \leq F(\omega_1, \omega_2) \leq m, \\ m & \text{if } F(\omega_1, \omega_2) > m, \\ -m & \text{if } F(\omega_1, \omega_2) < -m \end{cases}
$$

be a truncation of F for each $m \in \mathbb{N}$. Then

$$
0 \leq \int \left({}^\circ\!\!\int |F - F_m| \, dP_2 \right) dL(P_1) \leq {}^\circ\!\!\int |F - F_m| \, d(P_1 \otimes P_2) \to 0
$$

as $m \to \infty$, implying that $^\circ\!\int |F(\omega_1, \omega_2) - F_m(\omega_1, \omega_2)| \, dP_2 \to 0$ a.e. and thus that $F(\omega_1, \cdot)$ is S-integrable for almost all ω_1.

A consequence of what we just proved is that for almost all ω_1

$$^\circ G(\omega_1) = \,^\circ\!\!\int F(\omega_1, \omega_2)\, dP_2 = \int f(\omega_1, \omega_2)\, dL(P_2) = g(\omega_1),$$

and thus G is a lifting of g. If $G_m(\omega_1) = \int F_m(\omega_1, \omega_2)\, dP_2(\omega_2)$ then G_m is a finite function and

$$0 \leq \,^\circ\!\!\int |G - G_m|\, dP_1 \leq \,^\circ\!\!\int\int |F - F_m|\, dP_1\, dP_2 \to 0$$

as $m \to \infty$. By 3.2.10, G is S-integrable, and the theorem is proved.

Let us conclude with the following brief remark. We shall not always start from a hyperfinite probability space. It may be convenient to let the internal space be the $*$-transform of a standard measure space. For instance, if $(\mathbb{R}, \mathcal{B}, \mu)$ is the standard Lebesgue space on $\mathbb{R}$, our internal starting point could be the internal measure space $(*\mathbb{R}, *\mathcal{B}, *\mu)$. Here $*\mu$ is finitely, hence hyperfinitely, additive on the internal algebra $*\mathcal{B}$; σ-additivity is lost in the transition. However, it is restored by passing to the associated Loeb space. By transfer we can write down "integrals" $\int_A f(r)\, d*\mu$, where $A \in *\mathcal{B}$, which however must be handled with some care: no countable manipulations are allowed.

3.2.15. REMARK. The internal measure space $(*\mathbb{R}, *\mathcal{B}, *\mu)$ has an associated Loeb space $(*\mathbb{R}, L(*\mathcal{B}), L(*\mu))$ and results like Theorem 3.2.9 connecting the two spaces are still true. There is, however, one important point to notice. The measure $*\mu$ is not finite, hence we have to add one clause to the definition, 3.2.8, of S-integrability in order for 3.2.9 to extend to the unbounded case. We state the necessary modifications.

Let $(X, \mathcal{A}, \nu)$ be an internal measure space as in Section 3.1 and let $(X, L(\mathcal{A}), L(\nu))$ be its associated Loeb space. A function $f : X \to *\mathbb{R}$ is *S-integrable* if it is $\mathcal{A}$-measurable and

(i) $\int_X |f|\, d\nu$ is finite,
(ii) if $A \in \mathcal{A}$ and $\nu(A) \approx 0$, then $\int_A |f|\, d\nu \approx 0$.
(iii) if $A \in \mathcal{A}$ and $f(A) \subseteq \mu(0)$, then $\int_A |f|\, d\nu \approx 0$.

Condition (iii) is redundant if ν is finite and, hence, was omitted from Definition 3.2.8.

Theorem 3.2.9 extends; for instance, let $f : X \to *\mathbb{R}$ be $\mathcal{A}$-measurable, then f is S-integrable iff $^\circ f$ is $L(\nu)$-integrable and $^\circ\!\int |f|\, d\nu = \int |^\circ f|\, dL(\nu)$. And a function $g : X \to \mathbb{R}$ is $L(\nu)$-integrable iff g has an S-integrable lifting $f : X \to *\mathbb{R}$; in this case $^\circ\!\int f\, d\nu = \int g\, dL(\nu)$.

We round off this section by mentioning the following elegant application of hyperfinite ideas to ergodic theory.

3.2.16. EXAMPLE. Let $K = \{0, 1, \ldots, k - 1\}$, $k \in {}^*\mathbb{N} - \mathbb{N}$, and consider the Loeb space $(K, L(\mathscr{A}), L(P))$ where $\mathscr{A}$ is the algebra of internal subsets of K and P the counting measure. Let φ be the shift operator on K defined by

$$\varphi(x) = \begin{cases} x + 1 & \text{if} \quad x < k - 1, \\ 0 & \text{if} \quad x = k - 1. \end{cases}$$

Kamae (1982) has proved that any dynamical system of the form $(\Omega, \mathscr{B}, \mu, T)$, where $(\Omega, \mathscr{B}, \mu)$ is a (standard) probability space and T a measure-preserving transformation, is a *factor* of the "hypercycle" K in the sense that there exists a measure-preserving transformation $g : K \to \Omega$ such that $g(\varphi(x)) = T(g(x))$ for almost all $x \in K$.

As Kamae (1982) [see also Katznelson and Weiss (1982)] shows this has an immediate application to the individual ergodic theorem, since this theorem is rather simple to prove for hypercycles. It would be interesting to see if this notion of hypercycle has other applications. The reader should also consult the recent work by Ross (1983, 1984a, 1984b) on measurable transformations on Loeb spaces for additional information.

3.3. BROWNIAN MOTION

Now it is time to come down from the abstract theory to a concrete example. The probabilists know how to construct Brownian motion as a limit of random walks. In an important paper Anderson (1976) constructed Brownian motion as a hyperfinite random walk. See also Keisler (1984), whose version of Anderson's process we shall adopt.

For us Brownian motion will be an internal map

$$B : \Omega \times T \to {}^*\mathbb{R},$$

where T is a hyperfinite time line, $T = \{0, \Delta t, 2 \Delta t, \ldots, 1\}$, as in 3.2.2, and Ω is essentially our model for hyperfinite coin tossing with the minor change that the base space $\{0, 1\}$ is replaced by $\{-1, +1\}$ and $\Omega = \{-1, +1\}^T$ (see 3.1.3).

As a *hyperfinite random walk* B has the following explicit definition

$$(1) \qquad\qquad B(\omega, t) = \sum_0^t \omega(s)\sqrt{\Delta t}, \qquad \omega \in \Omega.$$

Thus between times t_0 and $t_0 + \Delta t$ the "particle" moves a distance $\sqrt{\Delta t}$ either to the "left" or to the "right" independently with probability $\frac{1}{2}$.

REMARK. We use the following convention that if $u \le t$, then

$$\sum_u^t X(\omega, s) = X(\omega, u) + \cdots + X(\omega, t - \Delta t),$$

i.e., $X(\omega, t)$ is *not* included in the sum.

The standard Brownian motion is obtained by setting

$$b(\omega, {}^\circ t) = {}^\circ B(\omega, \tilde{t}),$$

where $\tilde{t}$ is the point in T to the immediate right of t; b will be a standard stochastic process from $\Omega \times [0, 1]$ to $\mathbb{R}$ where Ω has the measure structure given by the Loeb construction $(\Omega, L(\mathcal{A}), L(P))$, where $\mathcal{A}$ is the internal algebra of internal subsets of Ω and P is the hyperfinite counting measure on $\mathcal{A}$. We emphasize that b is a standard process although on a somewhat unusual sample space Ω. Perhaps this Ω is closer to "physical intuition"?

In order to prove that b is a Brownian motion, we must verify the following:

(i) $b(\cdot, t)$ is a measurable function of ω for all $t \in [0, 1]$;
(ii) for $s < t$, $b(\omega, t) - b(\omega, s)$ has a normal distribution with mean 0 and variance $t - s$;
(iii) if $s_1 < t_1 \le s_2 < t_2 \le \cdots \le s_n < t_n$ in $[0, 1]$, then $\{b(\omega, t_1) - b(\omega, s_1), \ldots, b(\omega, t_n) - b(\omega, s_n)\}$ is an independent set of random variables.

By construction (i) is immediate. We turn to the verification of (ii) and (iii).

Before giving the proofs we shall digress briefly to make a few remarks on independence of random variables in the extended universe.

3.3.1. DEFINITION. A collection of internal random variables $\langle X_i \rangle_{i \in I}$ on a hyperfinite probability space $(\Omega, \mathcal{A}, P)$ is called *-*independent* if for every hyperfinite subset $\{X_1, \ldots, X_m\}$, $m \in {}^*\mathbb{N}$, and every internal m-tuple $(\alpha_1, \ldots, \alpha_m) \in {}^*\mathbb{R}^m$,

$$P(\{\omega \in \Omega \,|\, X_1(\omega) < \alpha_1, \ldots, X_m(\omega) < \alpha_m\}) = \prod_{k=1}^m P(\{\omega \in \Omega \,|\, X_k(\omega) < \alpha_k\}).$$

The collection is called *S-independent* if for every finite set $\{X_1, \ldots, X_m\}$, $m \in \mathbb{N}$, and every $(\alpha_1, \ldots, \alpha_m) \in \mathbb{R}^m$ the same product formula holds with $=$ replaced by $\approx$.

Once more we have a *-version and an *S*-version in the extended universe. And it is the *S*-version that has a standard significance.

3.3.2. LEMMA. Let $\langle X_i \rangle_{i \in I}$ be *S*-independent on $(\Omega, \mathcal{A}, P)$; then $\langle {}^\circ X_i \rangle_{i \in I}$ is independent on the associated Loeb space $(\Omega, L(\mathcal{A}), L(P))$.

PROOF. (by calculation).   Let $m \in \mathbb{N}$ and $(\alpha_1, \ldots, \alpha_m) \in \mathbb{R}^m$,

$$L(P)(\{\omega \in \Omega \,|\, {}^\circ X_{i_1}(\omega) < \alpha_1, \ldots, {}^\circ X_{i_m}(\omega) < \alpha_m\})$$

$$= \lim_{n \to \infty} {}^\circ P\left(\left\{\omega \in \Omega \,|\, X_{i_1}(\omega) < \alpha_1 - \frac{1}{n}, \ldots, X_{i_m}(\omega) < \alpha_m - \frac{1}{n}\right\}\right)$$

$$= \lim_{n \to \infty} {}^\circ\left(\prod_{k=1}^{m} P\left(\left\{\omega \in \Omega \,|\, X_{i_k}(\omega) < \alpha_k - \frac{1}{n}\right\}\right)\right)$$

$$= \prod_{k=1}^{m} \lim_{n \to \infty} {}^\circ P\left(\left\{\omega \in \Omega \,|\, X_{i_k}(\omega) < \alpha_k - \frac{1}{n}\right\}\right)$$

$$= \prod_{k=1}^{m} L(P)(\{\omega \in \Omega \,|\, {}^\circ X_{i_k}(\omega) < \alpha_k\}).$$

We have included this calculation not because we expect any reader to have any difficulties in providing the proof, but simply once more in a simple context to demonstrate how an external condition involving $L(P)$ and ${}^\circ X_{i_k}$ is converted to an internal condition involving P and X_{i_k}; see also the proof of 3.2.4.

3.3.3. PROPOSITION (Central Limit Theorem).   Let $\langle X_n \rangle_{n \in {}^*\mathbb{N}}$ be an internal sequence of $*$-independent random variables on $(\Omega, \mathscr{A}, P)$ with a common standard distribution function F and with mean 0 and variance 1. Then for any $m \in {}^*\mathbb{N} - \mathbb{N}$ and any $\alpha \in {}^*\mathbb{R}$

$$P\left(\left\{\omega \in \Omega \,\middle|\, \frac{1}{\sqrt{m}} \sum_{k=1}^{m} X_k(\omega) \leq \alpha\right\}\right) \approx {}^*\Psi(\alpha),$$

where

$$\Psi(\alpha) = (2\pi)^{-1/2} \int_{-\infty}^{\alpha} \exp\left(-\frac{x^2}{2}\right) dx$$

is the standard Gaussian distribution.

For the proof let G be the distribution function of ${}^\circ X_n$ on the Loeb space. Since F is standard, it is easy to show that $G = {}^\circ F$ and $F = {}^* G$. It is also straightforward to calculate that $E({}^\circ X_n) = 0$ and $E({}^\circ X_n^2) = 1$. Hence, by the standard Central Limit Theorem, given $\alpha \in \mathbb{R}$ and $\varepsilon \in \mathbb{R}_+$ there exists $n_0 \in \mathbb{N}$ such that if $m > n_0$, then

$$\left| L(P)\left(\left\{\omega \in \Omega \,\middle|\, \frac{1}{\sqrt{m}} \sum_{k=1}^{m} {}^\circ X_k(\omega) \leq \alpha\right\}\right) - \Psi(\alpha) \right| < \varepsilon.$$

The collection $\langle {}^\circ X_k \rangle$ is independent. Thus the sum $\sum_{k=1}^{m} {}^\circ X_k$ has a distribution function G^m which is the m-fold convolution product of G. We get for $m > n_0$ that $|G^m(\sqrt{m}\,\alpha) - \Psi(\alpha)| < \varepsilon$. Now $F = \,^*G$; thus we may apply the transfer principle to conclude that for any $m \in \,^*\mathbb{N} - \mathbb{N}$ and any $\alpha \in \mathbb{R}$, $F^m(\sqrt{m}\,\alpha) \approx \,^*\Psi(\alpha)$. However, F^m is the distribution function of $\sum_{k=0}^{m} X_k$. Thus for any $\alpha \in \mathbb{R}$

$$P\left(\left\{\omega \in \Omega \,\Big|\, \frac{1}{\sqrt{m}} \sum_{k=0}^{m} X_k(\omega) \le \alpha\right\}\right) \approx \,^*\Psi(\alpha).$$

As the reader will appreciate, the distribution function Ψ is so well behaved (also at $\pm\infty$) that we can extend to all $\alpha \in \,^*\mathbb{R}$.

3.3.4. REMARK. In the proof for Proposition 3.3.3 we used the standard central limit theorem. Perhaps the reader would like to give an alternative proof of this result thinking of the Gaussian distribution as a hyperfinite binomial distribution?

3.3.5. THEOREM. Let B be the hyperfinite random walk on $(\Omega, \mathscr{A}, P)$ and b its standard part. Then b is a Brownian motion on $(\Omega, L(\mathscr{A}), L(P))$.

We have to verify conditions (i)-(iii) listed above. As we previously observed, the proof of (i) is immediate by construction. The proof of (iii) follows easily from Lemma 3.3.2. We prove (ii) in two different ways:

FIRST PROOF OF (ii). We use Proposition 3.3.3

$$L(P)(\{\omega \in \Omega \,|\, b(\omega, {}^\circ t) - b(\omega, {}^\circ s) \le \alpha\})$$

$$= L(P)(\{\omega \in \Omega \,|\, {}^\circ B(\omega, t) - {}^\circ B(\omega, s) \le \alpha\})$$

$$= L(P)\left(\left\{\omega \in \Omega \,\Big|\, {}^\circ\left(\sum_{s}^{t} \omega_k \sqrt{\Delta t}\right) \le \alpha\right\}\right)$$

$$= \lim_{n \to \infty} {}^\circ P\left(\left\{\omega \in \Omega \,\Big|\, \sum_{s}^{t} \omega_k \sqrt{\Delta t} \le \alpha + \frac{1}{n}\right\}\right)$$

$$= \lim_{n \to \infty} {}^\circ P\left(\left\{\omega \in \Omega \,\Big|\, \frac{1}{\sqrt{m}} \sum_{k=1}^{m} \omega_{i_k} \le \frac{\alpha + (1/n)}{\sqrt{t-s}}\right\}\right).$$

Here $t - s = m \, \Delta t$ and $i_1, \ldots, i_k$ enumerate the random variables "between" s and t. We now have a form suitable for an application of 3.3.3 and get

$$L(P)(\{\omega \in \Omega \,|\, b(\omega, {}^\circ t) - b(\omega, {}^\circ s) \le \alpha\})$$

$$= \lim_{n \to \infty} {}^{\circ *}\Psi\left(\frac{\alpha + (1/n)}{\sqrt{t-s}}\right) = \Psi\left(\frac{\alpha}{\sqrt{{}^\circ t - {}^\circ s}}\right).$$

Thus $b(\omega, t) - b(\omega, s)$ for $t, s \in [0, 1]$ has a normal distribution with mean 0 and variance $t - s$.

SECOND PROOF OF (ii). We calculate the Fourier transform

$$\int_\Omega \exp[i(b(\omega, {}^\circ t) - b(\omega, {}^\circ s))z] \, dL(P)$$

$$= {}^\circ\!\int_\Omega \exp[i(B(\omega, t) - B(\omega, s))z] \, dP \qquad \text{(this is a use of 3.2.9),}$$

$$= {}^\circ\!\int_\Omega \exp\left[i\left(\sum_s^t \omega_k\sqrt{\Delta t}\right)z\right] dP,$$

$$= {}^\circ\!\int_\Omega \prod_{k=s}^t \exp(i\omega_k\sqrt{\Delta t}\, z) \, dP,$$

$$= {}^\circ\!\left(\prod_{k=s}^t \int_\Omega \exp(i\omega_k\sqrt{\Delta t}\, z) \, dP\right) \qquad \text{(by the independence),}$$

$$= {}^\circ\!\left(\prod_{k=s}^t \left[\frac{\exp(i\sqrt{\Delta t}\, z) + \exp(-i\sqrt{\Delta t}\, z)}{2}\right]\right.$$

(since $\omega_k = \pm 1$, both with probability $\frac{1}{2}$),

$$= {}^\circ\!\left(\left\{\left[1 - \frac{z^2}{2m} + O\!\left(\frac{z^4}{m^2}\right)\right]^m\right\}^{(t-s)}\right) \qquad \text{(where } m = \Delta t^{-1}\text{),}$$

$$= \exp\left(-\frac{{}^\circ t - {}^\circ s}{2} z^2\right).$$

Once more this proves the result. In this case we made a direct calculation which did not appeal to the Central Limit Theorem. Perhaps we should add a comment on the last equality. We know that if $m \in {}^*\mathbb{N} - \mathbb{N}$, then $(1 + x/m)^m \approx e^x$. Since the standard part operation kills the remainder term $O(z^4/m^2)$, the equality follows.

However, there is more to Brownian motion.

3.3.6. THEOREM. $B(\omega, \cdot)$ is S-continuous for almost all $\omega \in \Omega$; i.e., there is a set Ω' of Loeb measure one such that $B(\omega, s) \approx B(\omega, t)$ whenever $\omega \in \Omega'$ and $s \approx t$. Consequently, $b(\omega, \cdot)$ is continuous for all $\omega \in \Omega'$.

Before proving the theorem, we shall establish two simple identities. First observe that if $\Delta B(s) = B(s + \Delta t) - B(s)$, then $\Delta B(s)^2 = \Delta t$ and thus

$$E(B(t)^2) = E \sum_{s=0}^t ((B(s) + \Delta B(s))^2 - B(s)^2)$$

$$= \sum_{s=0}^t E(2B(s)\,\Delta B(s) + \Delta t) = \sum_{s=0}^t \Delta t = t,$$

where we have used that $E(B(s) \, \Delta B(s)) = 0$ since $\Delta B(s)$ is plus or minus $\sqrt{\Delta t}$ with probability $\frac{1}{2}$ no matter what $B(s)$ is. From this we get

$$E(B(t)^4) = E \sum_{s=0}^{t} ((B(s) + \Delta B(s))^4 - B(s)^4)$$

$$= \sum_{s=0}^{t} E(4B(s)^3 \, \Delta B(s) + 6B(s)^2 \, \Delta B(s)^2$$

$$+ 4B(s) \, \Delta B(s)^3 + \Delta B(s)^4)$$

$$= \sum_{s=0}^{t} E(6B(s)^2 \, \Delta t + \Delta t^2) = \sum_{s=0}^{t} (6s \, \Delta t + \Delta t^2)$$

$$= 3t(t - \Delta t) + t \, \Delta t = 3t^2 - 2t \, \Delta t \le 3t^2,$$

by summing the arithmetic series $\sum_{0}^{t} s \, \Delta t = t(t - \Delta t)/2$. The same argument shows that

$$E((B(t) - B(s))^4) \le 3(t - s)^2.$$

Turning to the proof of the theorem, we define a "bad" set $\Omega_{m,n}$ for each pair $(m, n) \in \mathbb{N}^2$:

$$\Omega_{m,n} = \left\{ \omega \in \Omega \,\middle|\, \exists i < n \, \exists s \in T \cap \left(\frac{i}{n}, \frac{i+1}{n} \right] \left(\left| B(\omega, s) - B\left(\omega, \frac{i}{n} \right) \right| \ge \frac{1}{m} \right) \right\}.$$

Note that the path $B(\omega, \cdot)$ is discontinuous iff $\omega \in \bigcup_m \bigcap_n \Omega_{m,n}$ and that it suffices to show that $^{\circ}P(\Omega_{m,n}) \to 0$ as $n \to \infty$ for all $m \in \mathbb{N}$. Observe also that

$$P(\Omega_{m,n}) \le \sum_{i < n} P\left\{ \omega \,\middle|\, \exists s \in T \cap \left(\frac{i}{n}, \frac{i+1}{n} \right] \left(\left| B(\omega, s) - B\left(\omega, \frac{i}{n} \right) \right| \ge \frac{1}{m} \right) \right\}$$

$$\le 2 \sum_{i < n} P\left\{ \omega \,\middle|\, \left| B\left(\omega, \frac{i+1}{n} \right) - B\left(\omega, \frac{i}{n} \right) \right| \ge \frac{1}{m} \right\},$$

where the second inequality follows from a reflection argument: assume that $|B(\omega, (i + 1)/n) - B(\omega, i/n)| < 1/m$, but that there is an $s \in (i/n, (i + 1)/n]$ such that $|B(\omega, s) - B(\omega, i/n)| > 1/m$. Let s_ω be the smallest such s, and consider the "reflected path" ω' defined by $\omega'(t) = \omega(t)$ for $t < s_\omega$, and $\omega'(t) = -\omega(t)$ for $t \ge s_\omega$. Obviously, $|B(\omega', (i + 1)/n) - B(\omega', i/n)| \ge 1/m$, and since each reflected path corresponds to a unique unreflected path, the inequality follows.

Completing our calculations, we now have

$$P(\Omega_{m,n}) \le 2 \sum_{i < n} P\left\{ \omega \,\middle|\, \left| B\left(\omega, \frac{i+1}{n} \right) - B\left(\omega, \frac{i}{n} \right) \right| \ge \frac{1}{m} \right\}$$

$$\le 2m^4 E\left(\left| B\left(\omega, \frac{i+1}{n} \right) - B\left(\omega, \frac{i}{n} \right) \right|^4 \right)$$

$$\leq 6m^4 \sum_{i<n} \frac{1}{n^2} = \frac{6m^4}{n} \to 0$$

as $n \to \infty$. The theorem is proved.

Just as we could get the Lebesgue measure from the counting measure on the hyperfinite time line, we can now get a measure on $C[0, 1]$ by applying the inverse standard part map to the measure induced by B.

3.3.7. REMARK (on Wiener Measure). The Wiener measure is defined as the unique completed Borel measure on $C[0, 1]$ satisfying the following conditions:

(i)   the measure of $\{f \in C[0, 1] \mid f(t) - f(s) < \alpha\}$ is $\Psi(\alpha/\sqrt{t - s})$;

(ii)   if $s_1 < t_1 \leq s_2 \cdots \leq s_n < t_n$ in $[0, 1]$, then $\{f(t_1) - f(s_1), \ldots, f(t_n) - f(s_n)\}$ is an independent set of random variables.

By using 3.3.5 and 3.3.6, we get the following easy construction of the Wiener measure. Let $\mathscr{B}$ be a σ-algebra on $C[0, 1]$ defined by

(2) $F \in \mathscr{B}$ iff $\{\omega \in \Omega \mid b(\omega, \cdot) \in F\} \in L(\mathscr{A})$,

where $L(\mathscr{A})$ is the Loeb algebra on Ω. A measure W on $\mathscr{B}$ is obtained by

(3) $W(F) = L(P)(\{\omega \in \Omega \mid b(\omega, \cdot) \in F\})$.

The space $(C[0, 1], \mathscr{B}, W)$ is the completion of the Wiener space; this can be proved in exactly the same way as 3.2.5 once we realize that (3) can also be expressed as

(4) $W(F) = L(\hat{P})(\mathrm{st}^{-1}(F))$,

where $\hat{P}$ is the measure on $*C[0, 1]$ defined by

$$\hat{P}(A) = P\{\omega \mid B(\omega, \cdot) \in A\}$$

and st is the standard part map on $*C[0, 1]$.

We shall develop a general theory for such "pushed-down" Loeb measures in the next section, and in Section 3.5 we shall return to have a look at the Wiener measure from a somewhat different point of view.

Thus far we have been looking at Brownian motions with a fixed initial position 0; what happens if we also fix the final position a? In the hyperfinite setting everything is straightforward: let $\bar{a}$ be an element in the monad of a which is hit by a Brownian path at time 1; i.e., $\bar{a} = B(\omega, 1)$ for some $\omega \in \Omega$, and put

$$\Omega_{\bar{a}} = \{\omega \in \Omega \mid B(\omega, 1) = \bar{a}\}.$$

Let $P_{\bar{a}}$ be the normalized counting measure on $\Omega_{\bar{a}}$. As before $P_{\bar{a}}$ induces a completed Borel measure on $C[0, 1]$ by

(5) $W_a(F) = L(P_{\bar{a}})\{\omega \in \Omega_{\bar{a}} \mid B(\omega, \cdot) \in \mathrm{st}^{-1}(F)\}$.

The measures W_a are called *conditional Wiener measures*, and a process with such a distribution is often called a *Brownian bridge*. Given s, $t \in [0, 1]$, $s < t$, and a, $b \in \mathbb{R}$, more general conditional Wiener measures $W_{s,t,a,b}$ can be constructed to model particles starting at a at time s and ending at b at time t. We leave these to the reader.

In a purely standard setting a direct construction of conditional Wiener measures is not so easy since we cannot condition on sets like

$$\{\omega \in \Omega \,|\, b(\omega, 1) = a\},$$

which have measure zero. However, we can list the properties we want W_a to have and then prove that such measures exist. Let $P_t(x, y)$ be the Gaussian kernels:

$$P_t(x, y) = (2\pi t)^{-1/2} \exp(-|x - y|^2)/2t).$$

If $0 = t_0 < t_1 < \cdots < t_n < t_{n+1} = 1$ is an increasing sequence of elements from $[0, 1]$, and $A_1, \ldots, A_n$ are Borel sets in $\mathbb{R}$, we want

$$W_a(A) = \int_{A_1} \cdots \int_{A_n} P_1(0, a)^{-1} P_{t_1}(0, x_1)$$
$$\times P_{t_2 - t_1}(x_1, x_2) \cdots P_{t_{n+1} - t_n}(x_n, a) \, dx_1, \ldots, dx_n,$$

where

$$A = \{f \in C[0, 1] \,|\, f(0) = 0, f(1) = a \text{ and } f(t_i) \in A_i \text{ for all } i\}.$$

It can be proved that a unique Borel measure W_a satisfying this condition exists, and this gives a standard construction of conditional Wiener measures. We leave it to the reader to check that the two definitions agree.

3.3.8. BROWNIAN LOCAL TIME. The notion of local time is important in the study of Brownian motion. Formally the local time $l(t, x)$ is given by

$$\text{(i)} \qquad\qquad l(t, x) = \int_0^t \delta(x - b(s)) \, ds,$$

where b is a Brownian motion and δ the delta function. The idea is that $l(t, x)$ measures the number of times the Brownian particle visits the site x before time t.

One way of making this heuristic idea precise is to show that there exists a jointly continuous process $l(t, x)$ such that

$$\text{(ii)} \qquad\qquad l(t, x) = \frac{d}{dx} \int_0^t I_{(-\infty, x]}(b(s)) \, ds,$$

for almost all $(t, x) \in [0, \infty) \times \mathbb{R}$, where I_A is the characteristic function of the set A. This is the standard, but somewhat indirect, approach.

Thinking of b as the standard part of the hyperfinite random walk gives us a different and more direct approach. Starting from the approximation [use either (i) or (ii)]

(iii)
$$(\Delta x)^{-1} \int_0^t I_{[x,x+\Delta x]}(b(s)) \, ds$$

to $l(t, x)$, we replace the time line $[0, \infty)$ by the hyperfinite discretization $T = \{0, \Delta t, \ldots, n \, \Delta t, \ldots\}$, $n \in {}^*\mathbb{N}$, and the space $\mathbb{R}$ by $\Lambda = \{0, \pm \Delta x, \ldots, \pm n \, \Delta x, \ldots\}$, $n \in {}^*\mathbb{N}$, and introduce the internal process $L: T \times \Lambda \to {}^*\mathbb{R}$ by

(iv)
$$L(t, x) = \sum_{s<t} I_{\{x\}}(B(s))(\Delta t)^{1/2}.$$

We note that in the hyperfinite random walk Δx and Δt are chosen such that $\Delta x = (\Delta t)^{1/2}$.

This definition is due to Perkins (1981), who showed that L has a standard part which is, in fact, a Brownian local time, i.e., satisfies (ii).

Perkins used the hyperfinite representations in (ii) to prove the following global intrinsic characterization of local time.

Let $m(t, x, \delta)$ denote the Lebesgue measure of the set of points within $\delta/2$ of $\{s \le t \mid b(s) = x\}$. Then for almost all ω and each $t_0 > 0$,

(v)
$$\lim_{\delta \to 0^+} \sup_{t \le t_0, x \in \mathbb{R}} |m(t, x, \delta)\delta^{1/2} - 2(2/\pi)^{1/2}l(t, x)| = 0.$$

This characterization of $l(t, x)$ is both intrinsic (i.e., depends only on $\{s \mid b(s) = x\}$) and global (i.e., holds for all x simultaneously). Previously it was known to hold for each x separately.

The reader is referred to Perkins (1981, 1983) for further discussion and proofs. We will return to the idea of local time in Chapters 6 and 7 in connection with stochastic potentials given by local time functionals; for the present, we just mention that Perkins' construction has found numerous applications in both standard and nonstandard contexts (Perkins 1981a,b, 1982a–c, 1983; Greenwood, 1982, 1985). There are other hyperfinite constructions of Brownian motion besides Anderson's; see the paper by Oikkonen (1985) for one of them.

3.4. PUSHING DOWN LOEB MEASURES

In the previous sections we have developed the basic theory for Loeb measures and used it to get easy constructions of the Lebesgue and Wiener measures. Our strategy was the same in both cases: we began by constructing a natural internal measure on the *-version $^*X$ of our space X; taking the

Loeb measure, we obtained a countably additive measure on $^*X$; and pulling this back to X with the inverse standard part map, we got the desired measure on X. This method is useful in other situations as well; assume, e.g., that X is a nice enough topological space, and let $\{\mu_\alpha\}_{\alpha \in I}$ be a weakly convergent net of measures on X. How do we construct the limit measure? Consider $^*(\{\mu_\alpha\}_{\alpha \in I})$—which is a nonstandard net $\{\tilde{\mu}_\alpha\}_{\alpha \in {}^*I}$ of internal measures on $^*X$—and pick $\tilde{\mu}_\omega$ for some infinite $\omega \in {}^*I$. Taking the Loeb measure $L(\tilde{\mu}_\omega)$ and pushing it down to X, we get the limit measure [for details, consult Anderson and Rashid (1978) and Loeb (1979a)].

With these examples in mind we shall now study internal probability spaces $(^*X, \mathscr{A}, P)$, where X is a Hausdorff space, and try to determine when $\mu = \mathrm{st}(L(P))$ [i.e., $\mu(A) = L(P)(\mathrm{st}^{-1}(A))$ for all measurable A] is a reasonable probability measure on X. The exposition may seem unduly technical, but we shall try to convince you in the next section that the machinery we develop here is an extremely efficient tool for constructing limit measures and measure extensions. Before we begin, we should mention that the first use of these techniques seems to have been an application to potential theory in Loeb (1976); the first systematic treatments were by Anderson (1977, 1982)—who used them to obtain hyperfinite representations of Radon spaces—and by Loeb (1979a).

Let us first agree on what a "reasonable" probability measure on X is:

3.4.1. DEFINITION. Let X be a Hausdorff space, and let $\mathscr{C}$ be a family of subsets of X. A function $\nu : \mathscr{C} \to \mathbb{R}_+$ is called *regular* if for all $C \in \mathscr{C}$

$$(1) \qquad \nu(C) = \sup\{\nu(F) \,|\, F \subset C, F \in \mathscr{C} \text{ is closed}\}$$

$$= \inf\{\nu(O) \,|\, O \supset C, O \in \mathscr{C} \text{ is open}\}.$$

A measure ν on X is called a *Radon measure* if it is the completion of a Borel measure, and for all ν-measurable C:

$$(2) \qquad \nu(C) = \sup\{\nu(K) \,|\, K \subset C, K \text{ is compact}\}$$

$$= \inf\{\nu(O) \,|\, O \supset C, O \text{ is open}\}.$$

If ν is finite, the condition on approximation by open sets in the definition of Radon measure is clearly redundant.

We shall be mostly interested in Radon measures, but our first result is on regular measures:

3.4.2. PROPOSITION. Let X be a Hausdorff space, and let $(^*X, \mathscr{A}, P)$ be an internal, finitely additive probability space such that $\mathrm{st}^{-1}(F) \in L(\mathscr{A})$ for all closed F. Assume that $L(P)(\mathrm{Ns}(^*X)) = 1$. Then $\mathrm{st}(L(P))$ is a regular, completed Borel probability measure on X.

PROOF.   Since $\mathrm{st}^{-1}(F)$ is measurable for all closed F, the measure $\mathrm{st}(L(P))$ must clearly be defined on a σ-algebra extending the Borel sets, and since $L(P)$ is complete, so is $\mu = \mathrm{st}(L(P))$. The proposition will follow if we can prove the first equality in (1).

Assume that C is μ-measurable, and let $\varepsilon > 0$ be given. Since $\mathrm{st}^{-1}(C) \in L(\mathscr{A})$, there must be an $A \in \mathscr{A}$, $A \subset \mathrm{st}^{-1}C$, with

$$L(P)(A) > L(P)(\mathrm{st}^{-1}(C)) - \varepsilon = \mu(C) - \varepsilon.$$

Since A is internal, $\mathrm{st}(A)$ is a closed subset of C by 2.1.8. But since $\mathrm{st}^{-1}\mathrm{st}(A) \supset A \cap \mathrm{Ns}(^*X)$, we get

$$\mu(\mathrm{st}(A)) = L(P)(\mathrm{st}^{-1}\mathrm{st}(A)) \geq L(P)(A) \geq \mu(C) - \varepsilon,$$

and the proposition is proved.

This proof is exactly the same as the one we gave for Proposition 3.2.5; Anderson (1977, 1982) attributes it to E. Fisher. We shall be more concerned with the following consequence.

3.4.3. COROLLARY.   Let X be a Hausdorff space, and let $(^*X, \mathscr{A}, P)$ be an internal, finitely additive probability space such that $\mathrm{st}^{-1}(K) \in L(\mathscr{A})$ for all compact K. Assume also that for all positive $\varepsilon \in \mathbb{R}$, there is a compact K_ε with $L(P)(\mathrm{st}^{-1}K_\varepsilon) > 1 - \varepsilon$. Then $\mathrm{st}(L(P))$ is a Radon measure on X.

PROOF.   The corollary follows from the proposition if we can prove that $\mathrm{st}^{-1}(F) \in L(\mathscr{A})$ for all closed F. But

$$F = \bigcup_{n \in \mathbb{N}} (F \cap K_{1/n}) \cup (F - \bigcup_{n \in \mathbb{N}} K_{1/n}),$$

and $\mathrm{st}^{-1}(F)$ is hence a countable union of measurable sets.

To use the last result we need to know that $\mathrm{st}^{-1}(K) \in L(\mathscr{A})$ for all compact sets K. In many applications this is far from obvious, and our next task is to reduce this problem to a much easier one. We begin with a topological lemma.

3.4.4. LEMMA.   Let X be a Hausdorff space and τ a basis for the topology closed under finite unions. If $K \subset X$ is compact, then

$$\mathrm{st}^{-1}(K) = \bigcap\{^*O \mid K \subset O, O \in \tau\}.$$

PROOF.   By definition of the standard part map

$$\mathrm{st}^{-1}(K) \subseteq \bigcap\{^*O \mid K \subset O, O \in \tau\}.$$

To prove the opposite inclusion, assume $y \notin \mathrm{st}^{-1}(K)$. For each $x \in K$, we can find $G_x \in \tau$ such that $x \in G_x$ and $y \notin {}^*G_x$. Obviously $K \subseteq \bigcup_{x \in K} G_x$, and since K is compact, we may find a finite subcovering

$$K \subseteq G_{x_1} \cup \cdots \cup G_{x_n},$$

and thus

$$*K \subseteq *(G_{x_1} \cup \cdots \cup G_{x_n}).$$

Since τ is closed under finite unions, we have $G_{x_1} \cup \cdots \cup G_{x_n} \in \tau$, and since $y \notin *(G_{x_1} \cup \cdots \cup G_{x_n})$, the lemma follows.

3.4.5. PROPOSITION. Let X be a Hausdorff space and τ a basis for the topology closed under finite unions. Let $(*X, \mathcal{A}, P)$ be an internal, finitely additive probability space such that $*O \in L(\mathcal{A})$ for all $O \in \tau$. Then $\mathrm{st}^{-1}(K) \in L(\mathcal{A})$ for all compact sets K, and

$$L(P)(\mathrm{st}^{-1}(K)) = \inf\{L(P)(*O) \mid O \in \tau, K \subset O\}.$$

PROOF. Let K be compact, and put

$$\alpha_K = \inf\{L(P)(*O) \mid O \in \tau, K \subset O\}.$$

Given $O_1, \ldots, O_n \in \tau$ with $K \subset O_1 \cap \cdots \cap O_n$, and $m \in \mathbb{N}$, let

$$A_{O_1,\ldots,O_n,m} = \{B \in \mathcal{A} \mid B \subset *O_1 \cap \cdots \cap *O_n, P(B) > \alpha_K - 1/m\}.$$

Each $A_{O_1,\ldots,O_n,m}$ is nonempty since $*O_1 \cap \cdots \cap *O_n$ is Loeb measurable with measure $\geq \alpha_K$. Applying saturation to the family $\{A_{O_1,\ldots,O_n,m}\}$, we find an internal $B \in \mathcal{A}$ such that $°P(B) \geq \alpha_K$ and $B \subset *O$ for all $O \in \tau$, $O \supset K$. By the lemma $B \subset \mathrm{st}^{-1}(K)$, and the proposition follows from the completeness of the Loeb measure.

Combining Corollary 3.4.3 and Proposition 3.4.5, we get the main result of this section.

3.4.6. THEOREM. Let X be a Hausdorff space and τ a basis for the topology closed under finite unions. Let $(*X, \mathcal{A}, P)$ be an internal, finitely additive probability space such that $*O \in L(A)$ for all $O \in \tau$, and assume that for each $\varepsilon \in \mathbb{R}_+$ there is a compact set K_ε with

$$\alpha_{K_\varepsilon} = \inf\{L(P)(*O) \mid K \subset O, O \in \tau\} > 1 - \varepsilon.$$

Then $\mu = \mathrm{st}(L(P))$ is a Radon probability measure on X, and for all compacts K, we have $\mu(K) = \alpha_K$.

Theorem 3.4.6 has three important advantages over Proposition 3.4.3. The first is that we need no longer check if the *external* sets $\mathrm{st}^{-1}(K)$ are in $L(\mathcal{A})$, but only if the *internal* sets $*O$ are; in most applications they will already be in $\mathcal{A}$, and the checking is trivial. The second advantage is that we only have to show the measurability of sets in a *basis* for the topology; this is important in spaces where the open sets are not countably generated from the basis. Finally, we need not show that there exist compacts with arbitrarily large measure; it is enough to come up with compacts that can be approximated from the outside by only basis elements of large measure. When we turn to applications in the next section, we shall exploit these three points systematically.

The hypothesis in Corollary 3.4.3 that $L(P)(K_\varepsilon) > 1 - \varepsilon$ (and the corresponding one in Theorem 3.4.6, that $\alpha_{K_\varepsilon} > 1 - \varepsilon$) serves two purposes. The first is to ensure that $\mathrm{Ns}(*X)$ is Loeb measurable with measure one—which is necessary for $L(P) \circ \mathrm{st}^{-1}$ to be a well-defined probability measure—and the second is to guarantee the existence of arbitrarily large compacts—which is necessary for $L(P) \circ \mathrm{st}^{-1}$ to be a Radon measure, and also for the proof of 3.4.3 to work. We shall now show that if X is a locally compact space or allows a complete metric, then these conditions are satisfied whenever $\mathrm{Ns}(*X)$ has outer measure one.

3.4.7. PROPOSITION. Let X be a Hausdorff space, and assume that $(*X, \mathscr{A}, P)$ is an internal, finitely additive probability space such that $O \in \mathscr{A}$ for all *-open sets O. If either

(a) X is locally compact, or
(b) X is a complete metric space,

then $\mathrm{Ns}(*X)$ is $L(P)$-measurable and

$$L(P)(\mathrm{Ns}(*X)) = \sup\{L(P)(\mathrm{st}^{-1}(K)) \mid K \text{ compact}\}$$

PROOF.   (a)   Define

$$\alpha = \sup\{L(P)(\mathrm{st}^{-1}(K)) \mid K \text{ compact}\}.$$

If $K_1, \ldots, K_n$ are compact and $m \in \mathbb{N}$, let

$$A_{K_1,\ldots,K_n,m} = \{B \in \mathscr{A} \mid B \supset *K_1 \cup \cdots \cup *K_n \text{ and } P(B) < \alpha + 1/m\}.$$

Since by 2.1.6(iv) we have $*K \subset \mathrm{st}^{-1}(K)$ for all compact sets K, the set $A_{K_1,\ldots,K_n,m}$ is nonempty. Using saturation on the family $\{A_{K_1,\ldots,K_n,m}\}$, we find a set $B \in \mathscr{A}$ such that $^\circ P(B) \leq \alpha$ and $*K \subset B$ for all compact K. Since X is locally compact, any element in $\mathrm{Ns}(*X)$ is in the *-version of some compact, and hence $\mathrm{Ns}(*X) \subset B$. It follows that $\mathrm{Ns}(*X)$ is Loeb-measurable with $L(P)(\mathrm{Ns}(*X)) = \alpha$.

(b)   Let $\overline{L(P)}$ be the outer measure generated by $L(P)$, and put

$$\gamma = \overline{L(P)}(\mathrm{Ns}(*X)).$$

If α is as in part (a), all we have to show is that $\gamma \leq \alpha$. The main idea of the argument is indicated by the following claim.

CLAIM. Let $\varepsilon \in \mathbb{R}_+$. For each $m \in \mathbb{N}$, there is a finite sequence $\{C_i^{(m)}\}_{i \leq n(m)}$ of subsets of X such that each $C_i^{(m)}$ is a closed ball of radius

$3/m$ and

$$\overline{L(P)}\left(\text{Ns}(^*X) \cap \bigcap_{m \in \mathbb{N}} \bigcup_{i=1}^{n(m)} {}^*C_i^{(m)}\right) > \gamma - \varepsilon.$$

Let us first use the claim to prove the proposition.

Define

$$K_\varepsilon = \bigcap_{m \in \mathbb{N}} \bigcup_{i=1}^{n(m)} C_i^{(m)};$$

since it is closed and totally bounded, K_ε is compact. We have

$$\text{st}^{-1}(K_\varepsilon) = \bigcap_{m \in \mathbb{N}} \bigcup_{i=1}^{n(m)} \text{st}^{-1}(C_i^{(m)}) \supseteq \bigcap_{m \in \mathbb{N}} \bigcup_{i=1}^{n(m)} {}^*C_i^{(m)} \cap \text{Ns}(^*X)$$

since $C_i^{(m)}$ is closed. By the claim

$$L(P)(\text{st}^{-1}(K_\varepsilon)) > \gamma - \varepsilon,$$

from which the proposition follows.

We now prove the claim. For each $x \in {}^*X$, $r \in {}^*\mathbb{R}$, let

$$B(x, r) = \{y \in {}^*X \mid d(x, y) \leq r\}.$$

Given $\delta \in \mathbb{R}_+$, $n \in {}^*\mathbb{N}$, let

$$\beta_{n,\delta} = \sup\left\{ L(P)\left(\bigcup_{i=1}^{n} B(x_i, \delta) \right) \middle| x_1, \ldots, x_n \in {}^*X \right\},$$

and put $\beta_\delta = \sup\{\beta_{n,\delta} \mid n \in \mathbb{N}\}$. Observe that the set

$$\left\{ n \in {}^*\mathbb{N} \mid \forall x_1, \ldots, x_n \in {}^*X \left(P\left(\bigcup_{i=1}^{n} B(x_i, \delta) \right) < \beta_\delta + \frac{1}{n} \right) \right\}$$

is internal and contains $\mathbb{N}$. Hence we can find $\eta \in {}^*\mathbb{N} - \mathbb{N}$ such that $\beta_{\eta,\delta} = \beta_\delta$ for all $\delta \in \mathbb{R}_+$.

Let ε be the positive real number in the claim. For each $m \in \mathbb{N}$, there is a finite sequence $x_1^{(m)}, \ldots, x_{n(m)}^{(m)}$ such that

$$L(P)\left(\bigcup_{i=1}^{n(m)} B\left(x_i^{(m)}, \frac{1}{m} \right) \right) > \beta_{1/m} - \frac{\varepsilon}{2^m}.$$

We must have

$$\overline{L(P)}\left(\text{Ns}(^*X) - \bigcup_{i=1}^{n(m)} B\left(x_i^{(m)}, \frac{1}{m} \right) \right) < \frac{\varepsilon}{2^m}.$$

The reason is as follows: extend $x_1^{(m)}, \ldots, x_{n(m)}^{(m)}$ to an internal sequence $x_1^{(m)}, \ldots, x_\eta^{(m)}$ containing all standard points; this can be done by saturation. By definition of η, the set

$$B_m = \bigcup_{i=1}^{\eta} B\left(x_i^{(m)}, \frac{1}{m}\right) - \bigcup_{i=1}^{n(m)} B\left(x_i^{(m)}, \frac{1}{m}\right)$$

has measure less than $\varepsilon/2^m$, and since the sequence $x_1^{(m)}, \ldots, x_\eta^{(m)}$ contains all standard points,

$$\text{Ns}(^*X) - \bigcup_{i=1}^{n(m)} B\left(x_i^{(m)}, \frac{1}{m}\right) \subset B_m.$$

We shall now replace the sequence $\{B(x_i^{(m)}, 1/m)\}_{i \leq n(m)}$ by a sequence $\{C_i^{(m)}\}_{i \leq n(m)}$ of standard sets satisfying the claim. If $B(x_i^{(m)}, 1/m) \cap \text{Ns}(^*X) = \varnothing$, let $C_i^{(m)} = \varnothing$. If the intersection is nonempty, we can find a standard element $y_i^{(m)} \in X$ such that

$$B(x_i^{(m)}, 1/m) \subset {}^*C_i^{(m)},$$

where

$$C_i^{(m)} = \{x \in X \mid d(x, y_i^{(m)}) \leq 3/m\}.$$

It follows that

$$\overline{L(P)}\left(\text{Ns}(^*X) - \bigcup_{i=1}^{n(m)} {}^*C_i^{(m)}\right) < \frac{\varepsilon}{2^m},$$

and hence

$$\overline{L(P)}\left(\text{Ns}(^*X) \cap \bigcap_{m \in \mathbb{N}} \bigcup_{i=1}^{n(m)} {}^*C_i^{(m)}\right) > \gamma - \varepsilon.$$

Throwing out the empty $C_i^{(m)}$'s, we prove the claim (and the proposition).

REMARK. The proof of (a) is essentially due to Loeb (1984), while the idea in (b) is based on a well-known argument showing that all probability measures on Polish spaces are tight [see, e.g., Billingsley (1968)]. It should be pointed out that the proof of (b) can be considerably simplified if we assume that X is separable, but since Proposition 3.4.7 will play an important part in Chapter 5, we have decided to present the general case. Note that (a) remains true if we weaken the measurability condition on *-open sets to just demanding that $^*O \in L(\mathscr{A})$ for all open sets O; the same is true of (b) when X is separable.

In many situations it is much easier to show that the nearstandard points have outer measure one than to prove the existence of large compacts, and what Proposition 3.4.7 tells us is that for certain spaces this is sufficient to allow us to apply Theorem 3.4.6. As a matter of fact, in these spaces Theorem 3.4.6 can be used even when $Ns(*X)$ does not have outer measure one, but $L(P) \circ st^{-1}$ will then no longer be a probability measure.

When we have used Theorem 3.4.6 to construct a measure, we often want to compare the result with some given set function; e.g., if we want to extend a measure to a larger algebra, we would like the new measure to agree with the old one where the old one is defined. The next result is an efficient tool for checking this; it goes back to Anderson (1977, 1982).

3.4.8. PROPOSITION. Let X be a Hausdorff space, and $(*X, \mathcal{A}, P)$ an internal, finitely additive probability space with $L(P)(Ns(*X)) = 1$. Let $\mathscr{C}$ be a family of subsets of X, and let $\nu : \mathscr{C} \to \mathbb{R}_+$ be a regular set function such that

$$(3) \qquad \nu(C) = L(P)(*C) \qquad \text{for all} \quad C \in \mathscr{C}.$$

Then $\mu = st(L(P))$ is an extension of ν.

PROOF. Let $C \in \mathscr{C}$ and $\varepsilon \in \mathbb{R}_+$ be given, and choose $F, O \in \mathscr{C}$ closed and open, respectively, such that $F \subset C \subset O$ and

$$(4) \qquad \nu(O) - \nu(F) < \varepsilon,$$

by the regularity of ν. Obviously $*F \subset *C \subset *O$, and by the nonstandard characterization of closed and open sets (see 2.1.6)

$$*F \cap Ns(*X) \subset st^{-1}(F) \subset st^{-1}(C) \subset st^{-1}(O) \subset *O.$$

Combining (4) with (3) applied to O and F, and remembering that $L(P)(Ns(*X)) = 1$, we get

$$L(P)(*O) - L(P)(*F \cap Ns(*X)) < \varepsilon.$$

Since ε is arbitrary, and $*C \cap Ns(*X)$ and $st^{-1}(C)$ are both squeezed between $*F \cap Ns(*X)$ and $*O$, we must have

$$(5) \qquad L(P)(st^{-1}(C) \Delta *C) = 0,$$

and hence

$$\mu(C) = L(P)(st^{-1}(C)) = L(P)(*C) = \nu(C),$$

and the proposition is proved.

As a consequence of the last proposition, we get a nonstandard version of Lusin's theorem due to Anderson (1977, 1982); the result is interesting in its own right and also useful in the construction of liftings.

3.4.9. COROLLARY. Let $(X, \mathscr{C}, \nu)$ be a Radon probability space, and let $f : X \to Y$ be a measurable map into a Hausdorff space with countable basis. Then $°(*f(x)) = f(°x)$ for $L(*\nu)$-a.a. x in $*X$.

PROOF. Let $\{U_n\}_{n \in \mathbb{N}}$ be a countable basis for Y with $U_1 = Y$. If $°(*f(x)) \neq f(°x)$, we must have

$$x \in \bigcup_{n \in \mathbb{N}} \{(f \circ \text{st})^{-1}(U_n) \, \Delta \, *f^{-1}(*U_n)\},$$

and we only have to show that each of the sets in the union has measure zero. But

$$(f \circ \text{st})^{-1}(U_n) \, \Delta \, *f^{-1}(*U_n) = \text{st}^{-1}(f^{-1}(U_n)) \, \Delta \, *(f^{-1}(U_n)),$$

and applying formula (5) with $P = *\nu$ and $C = f^{-1}(U_n)$, we see that the set on the right has measure zero.

Anderson's Lusin theorem tells us that $*f$ is a lifting of f with respect to all *standard* measures $*\nu$. A second and just as useful consequence of 3.4.8 is the following representation theorem, also due to Anderson (1977, 1982):

3.4.10. COROLLARY. Let μ be a Radon probability measure on a Hausdorff space X. Then there exist a hyperfinite subset Y of $*X$ and an internal measure P on Y such that $\mu = \text{st}(L(P))$.

PROOF. Given a finite family $O_1, O_2, \ldots, O_n$ of open sets in X, let $\xi_{O_1, O_2, \ldots, O_n}$ be the set of all hyperfinite partitions of $*X$ into $*\mu$-measurable sets such that each $*O_i$ is a union of partition classes. The family $\{\xi_{O_1, O_2, \ldots, O_n}\}$ has the finite intersection property, and thus there is a hyperfinite partition of $*X$ into $*\mu$-measurable sets such that the *-version of any open set is a union of partition classes. Fix one such partition E; let Y consist of one element from each partition class of E, and if $y \in Y$ and $[y]$ is its partition class, put

$$P\{y\} = *\mu([y]).$$

Since each open set is a union of partition classes, we clearly have

$$\mu(O) = *\mu(*O) = P(*O) = L(P)(*O)$$

for all open sets O. By Proposition 3.4.8, the measures μ and $\text{st}(L(P))$ agree on the open sets, and since they both are Radon measures, they must be equal.

Corollary 3.4.10 states one of the basic facts of nonstandard measure theory, and it will be used repeatedly in the sequel, often without an explicit reference.

3.5. APPLICATIONS TO LIMIT MEASURES
AND MEASURE EXTENSIONS

We shall now apply the techniques developed in the last section to some questions in standard measure and probability theory. Our aim is to show how these techniques form a strong and flexible tool for constructing different kinds of measures. A natural first step is to see what happens when we apply Theorem 3.4.6 to the *-version of a standard space.

3.5.1. THEOREM. Let X be a Hausdorff space, τ a basis for the topology closed under finite unions, and let ν be a regular, finitely additive probability measure defined on an algebra $\mathscr{C}$ extending τ. Assume that for each positive $\varepsilon \in \mathbb{R}_+$, there is a compact K_ε with

(1) $$\beta_{K_\varepsilon} \equiv \inf\{\nu(O)|O \in \tau, O \supset K_\varepsilon\} > 1 - \varepsilon.$$

Then ν has a unique extension to a Radon measure μ on X, and for all compact K, $\mu(K) = \beta_K$.

PROOF. To construct μ, just apply Theorem 3.4.6 to $(*X, *\mathscr{C}, *\nu)$, and put $\mu = \mathrm{st}(L(*\nu))$. It follows immediately from Proposition 3.4.8 that μ is an extension of ν.

For the uniqueness, assume that $\bar{\mu}$ is another Radon extension of ν. Since $\mu(K) = \beta_K$, we must have $\mu(K) \geq \bar{\mu}(K)$ for all compact sets K. Since $\bar{\mu} \neq \mu$, there must be a set B such that $\mu(B) < \bar{\mu}(B)$, and since $\bar{\mu}$ is Radon there is a compact $K \subset B$ with $\bar{\mu}(K) > \mu(B)$. But $\mu(B) \geq \mu(K)$, and hence $\bar{\mu}(K) > \mu(K)$ and we have a contradiction.

To illustrate the strength of this theorem, we shall use it to obtain two famous results of probability theory. The first is concerned with projective limits of measures, and is what Schwartz (1972) calls Prohorov's theorem; the reader is warned that this result is different from the one that is usually known by this name [as a matter of fact, the theorem seems to be due to Kisyński (1969)]. The theorem plays a fundamental role in the theory for cylindrical measures and generalized stochastic processes, but for these applications we can only refer the reader to Schwartz's book. Our second result is another proof of the existence of Brownian motion, from which the Levy modulus of continuity follows immediately.

Before we can state Prohorov's theorem, we need some definitions. A *projective system* of topological spaces is a directed family $\{(X_i, \tau_i)\}_{i \in I}$ of Hausdorff spaces, together with a family $\{\pi_{ij}\}_{i<j}$ of continuous mappings $\pi_{ij}: X_j \to X_i$, satisfying $\pi_{ik} = \pi_{ij} \circ \pi_{jk}$ whenever $i < j < k$, $i, j, k \in I$. The projective limit (X, τ) of $\{(X_i, \tau_i)\}_{i \in I}$ is the space

$$X = \{(x_i)_{i \in I} : \forall i < j (x_i = \pi_{ij}(x_j))\}$$

with the weakest topology making all the maps $\pi_j((x_i)_{i \in I}) = x_j$ continuous; (X, τ) is clearly a Hausdorff space.

A family $\{\mu_i\}_{i \in I}$ of Radon measures on the spaces X_i is a *projective system of measures* if $\mu_i = \pi_{ij}(\mu_j)$ whenever $i < j$. The question is: When does a projective system of measures give rise to a limit Radon measure μ on X such that $\mu_i = \pi_i(\mu)$ for all $i \in I$?

3.5.2. PROHOROV'S THEOREM. Let $(X_i, \tau_i, \mu_i)_{i \in I}$ be a projective system of Hausdorff spaces and Radon probability measures. The following is a necessary and sufficient condition for the existence of a Radon probability measure μ on the projective limit X such that $\mu_i = \pi_i(\mu)$ for all $i \in I$:

(2)

> For all $\varepsilon > 0$, there is a compact $K_\varepsilon \subset X$ such that for all $i \in I$ we have $\mu_i(\pi_i(K_\varepsilon)) > 1 - \varepsilon$.

The limit is unique when it exists.

PROOF. The necessity is almost trivial; since μ is Radon there exists a compact K_ε with $\mu(K_\varepsilon) > 1 - \varepsilon$, and hence

$$\mu_i(\pi_i(K_\varepsilon)) = \mu(\pi_i^{-1}(\pi_i(K_\varepsilon)) \geq \mu(K_\varepsilon) > 1 - \varepsilon.$$

To prove the sufficiency, let τ' be the basis for the topology given by $\pi_i^{-1}(O)$ for all open sets $O \in \tau_i$ and all $i \in I$. Let ν be the finitely additive measure defined on the algebra $\mathscr{C}$ generated by this basis, by $\nu(\pi_i^{-1}(B)) = \mu_i(B)$. Since each μ_i is Radon, and the inverse image of a compact set is closed, ν is regular. Condition (1) of 3.5.1 follows immediately from (2), and hence 3.5.1 gives us the existence of a unique Radon extension of ν. This proves the theorem.

Notice that (2) only allows us to calculate the measure of base elements containing the sets K_ε; thus the extra strength of 3.4.6 was crucial in this application.

We now turn to our next example—another construction of Brownian motion—and in this case we shall see how the equality $\beta_K = \mu(K)$ of 3.5.1 can be used. Let us recall what we are trying to show. Let $C[0, 1]$ be the set of real-valued, continuous functions on $[0, 1]$, and π the uniform topology on $C[0, 1]$. A *cylinder set* C is a subset of $C[0, 1]$ of the form

$$C = \{\omega \in C[0, 1] \mid \omega(t_1) \in A_1, \ldots, \omega(t_n) \in A_n\},$$

where $t_1 < t_2 < \cdots < t_n$ are elements of $[0, 1]$, and $A_1, \ldots, A_n$ are Borel sets in $\mathbb{R}$. Let ν be the finitely additive measure defined on the cylinder sets by

$$\nu(C) = \int_{A_n} \cdots \int_{A_1} \prod_{i=1}^n \frac{\exp[-(y_i - y_{i-1})^2 / 2(t_i - t_{i-1})]}{\sqrt{2\pi(t_i - t_{i-1})}} \, dy_i,$$

where $t_0 = 0$. We shall prove

3.5.3. THEOREM. The finitely additive measure ν has a unique extension to a Borel measure W on $(C[0, 1], \pi)$. Moreover,

$$(3) \qquad W\left\{\omega \in C[0, 1]\Big|\lim_{\substack{0 \le t_1 < t_2 \le 1 \\ t = t_2 - t_1 \downarrow 0}} \frac{|\omega(t_2) - \omega(t_1)|}{(2t \ln(1/t))^{1/2}} = 1\right\} = 1.$$

PROOF. Let σ be the topology of point-wise convergence in $C[0, 1]$. If τ consists of all finite unions of cylinder sets C with open sections $A_1, \ldots, A_n$, then τ is a basis for σ closed under finite unions. We want to apply Theorem 3.5.1 to the space $(C[0, 1], \sigma)$. The regularity of ν is obvious, and we only have to prove (1). Consider the following sets:

$$K_n^C = \Big\{\omega \,\big|\, \forall t_1, t_2 \in [0, 1]$$

$$\left(0 < t_2 - t_1 \le \frac{1}{n} \to \frac{|\omega(t_2) - \omega(t_1)|}{\{2C(t_2 - t_1)\ln[1/(t_2 - t_1)]\}^{1/2}} \le 1\right)\Big\},$$

for $C \in \mathbb{R}_+$, $n \in \mathbb{N}$. The standard part of an element in $^*K_n^C$—in both the σ- and the π-topology—is obviously in K_n^C, and hence K_n^C is compact in both topologies. A little thought will convince the reader that

$$\beta_{K_n^C} = \inf\left\{\prod_{i=0}^{m-1}\left(\int_{-[2C\Delta t_i \ln(1/\Delta t_i)]^{1/2}}^{[2C\Delta t_i \ln(1/\Delta t_i)]^{1/2}} \frac{\exp(-x^2/(2\Delta t)_i)}{\sqrt{2\pi \Delta t_i}}\, dx\right)\right\},$$

where the infimum is over all partitions $0 = t_0 < t_1 < t_2 < \cdots < t_m \le 1$, where $\Delta t_i = t_{i+1} - t_i$ and $\max_{i<m} \Delta t_i \le 1/n$.

Introducing a new variable $y = x/\sqrt{\Delta t_i}$ in the ith integral above, the expression for $\beta_{K_n^C}$ becomes

$$(4) \qquad \beta_{K_n^C} = \inf\left\{\prod_{i=0}^{m-1}\left(\int_{-[2C \ln(1/\Delta t_i)]^{1/2}}^{[2C \ln(1/\Delta t_i)]^{1/2}} \frac{\exp(-y^2/2)}{\sqrt{2\pi}}\, dy\right)\right\}.$$

Let us first consider the case $C > 1$. By using L'Hôpital's rule, we see that

$$\lim_{x \to \infty} \frac{\int_x^\infty (2/\sqrt{2\pi}) \exp(-y^2/2)\, dy}{\exp(-x^2/2)} = 0,$$

and for n large enough, we thus have

$$\beta_{K_n^C} \ge \inf \prod_{i=0}^{m-1}\left\{1 - \exp\left[-\frac{2C \ln(1/\Delta t_i)}{2}\right]\right\} = \inf \prod_{i=0}^{m-1}(1 - \Delta t_i^C)$$

$$= \inf\{\exp[\sum \ln(1 - \Delta t_i^C)]\}$$

$$\ge \inf[\exp(-2\sum \Delta t_i^C)] \ge \exp\left[-2\left(\frac{1}{n}\right)^{C-1}\right],$$

where the inf's are over the same set as previously. This gives us condition
(1) in Theorem 3.5.1, and ν thus has a unique extension to a Radon measure
W on $(C[0, 1], \sigma)$. Since the K_n^C's also are π-compact, W must also be a
Radon measure on $(C[0, 1], \pi)$, and that it is the only π-Radon extension
of ν follows exactly as the uniqueness part of 3.5.1. Since all completed Borel
measures on complete, separable metric spaces are Radon, W is also unique
as a Borel measure on $(C[0, 1], \pi)$.

It remains to prove (3). Let

$$K^C = \left\{ \omega \in C[0, 1] \Big| \varlimsup_{\substack{0 \leq t_1 \leq t_2 \leq 1 \\ t = t_2 - t_1 \downarrow 0}} \frac{|\omega(t_1) - \omega(t_2)|}{[2Ct \ln(1/t)]^{1/2}} \leq 1 \right\};$$

then $K^C \supseteq \bigcup_{n \in \mathbb{N}} K_n^C$ and $K^C \subseteq \bigcup K_n^{C'}$ for all $C' > C$. From what we have
just proved plus the fact that $W(K_n^C) = \beta_{K^C}$, it follows that $W(K^C) = 1$
for $C > 1$. To prove (3) it clearly suffices to show that $W(K^C) = 0$ for
$C < 1$, and this follows if we can show that $\beta_{K_n^C} = 0$ for all $n \in \mathbb{N}$ and $C < 1$.

Assume that $C < 1$, and pick $\alpha > 1$ such that $\alpha C < 1$. We have

$$\lim_{x \to \infty} \frac{\int_x^\infty (2/\sqrt{2\pi}) \exp(-y^2/2) \, dy}{\exp(-\alpha x^2/2)} = \infty,$$

and for all large enough m, we thus have

$$\beta_{K_n^C} \leq \prod_{i=0}^{m-1} \left\{ 1 - \exp\left[\frac{2\alpha C \ln(1/m)}{2} \right] \right\} = \left(1 - \left(\frac{1}{m} \right)^{\alpha C} \right)^m,$$

by applying (4) to the partition $t_i = i/m$. Since $\alpha C < 1$, the limit of $(1 - (1/m)^{\alpha C})^m$ as m goes to infinity is zero, and hence $\beta_{K_n^C} = 0$ for all $C < 1$.
This completes the proof of the theorem.

Equation (3) is the famous Levy modulus for the continuity of the
Brownian sample path; if we only wanted to prove the existence of Wiener
measure, we could get away more easily by choosing simpler compacts. It
is also worthwhile to notice how we used the $W(K) = \beta_K$ part of 3.5.1 to
get $W(K^C) = 1$ for $C > 1$.

For the next application we take our leave of Theorem 3.5.1, and return
to the basic techniques of Section 3.4. Let H be a real separable Hilbert
space; $\mathscr{F}$ the class of finite-dimensional subspaces of H; and $\mathscr{P}$ the class
of finite-dimensional projections in H. If $E, F \in \mathscr{F}, E \subset F$, we let $P_{E,F}$
denote the projection from F to E, and P_E the projection from H to E.

We consider a projective system $(E, \mu_E, P_{E,F})_{E, F \in \mathscr{F}}$ of Radon measures,
and let μ be the finitely additive measure on H defined by $\mu(P_E^{-1}(A)) = \mu_E(A)$. Using Prohorov's theorem, it is not hard to show that there exists

a Borel measure ν extending μ if and only if

$$\sup_{r \in \mathbb{R}_+} \inf_{E \in \mathscr{F}} \mu_E(B_E(r)) = 1,$$

where $B_E(r)$ is the ball in E of radius r centered at the origin [see, e.g., Lindstrøm (1982) for details.]

What happens when this condition is not satisfied? It turns out that although we no longer have a limit measure on H, we may still have a limit measure on some larger space. From the nonstandard point of view we shall see that this means that even though the natural nonstandard limit measure is not supported on the nearstandard points in the Hilbert space topology, it may be supported on the nearstandard points in some weaker topology. The right way of weakening the topology was discovered by Gross (1967):

3.5.4. DEFINITION. Let $\mu = \{\mu_F\}_{F \in \mathscr{F}}$ be a projective system of measures on H. A norm $|\cdot|$ on H is called μ-*measurable* if it is continuous with respect to the Hilbert space norm $\|\cdot\|$ and for all $\varepsilon > 0$ there is a $P_0 \in \mathscr{P}$ such that

(5) $\qquad \mu\{x \in H \,|\, |Px| > \varepsilon\} < \varepsilon \qquad$ for all $\quad P \in \mathscr{P}, P \perp P_0.$

Gross's theorem says that if $|\cdot|$ is μ-measurable, then μ can be extended to a Borel measure on the Banach space B obtained by completing H with respect to $|\cdot|$. Thus for the cases we are interested in, the Hilbert space norm cannot be measurable. Before we formulate Gross's result more precisely, we shall take a look at an example.

3.5.5. EXAMPLE. For each $F \in \mathscr{F}$ let μ_F be Gaussian distributed with mean zero and covariance matrix dim $F \cdot I$, where I is the identity matrix. It is easy to check that $\{\mu_F\}_{F \in \mathscr{F}}$ is a projective system of measures, and if a limit measure on H existed, this would be the natural infinite-dimensional Gaussian measure. But, as is easily seen, no such limit measure exists. However, Gross's theorem tells us that if we can only find a μ-measurable norm, we can produce a limit measure on a larger space B. Let $\{e_n\}_{n \in \mathbb{N}}$ be an orthonormal basis for H, and let $T: H \to H$ be a linear map. Recall that T is called a *Hilbert–Schmidt* operator if $\sum_{n=1}^{\infty} \|Te_n\|^2 < \infty$, and that $\sum_{n=1}^{\infty} \|Te_n\|^2$ is independent of which orthonormal basis we use. If T is a one-to-one Hilbert–Schmidt operator, let $|\cdot|_T$ be the norm defined by

$$|x|_T = \|Tx\|.$$

It is not hard to show that $|\cdot|_T$ is μ-measurable. Thus μ can be extended to a Borel measure on the Hilbert space generated by $|\cdot|_T$.

This example is of great interest in infinite-dimensional probability theory. We shall return to the problem in Section 4.7, when we discuss Brownian motion on Hilbert spaces. Since Gaussian measures do not exist on the "natural" Hilbert space, the Brownian motion must also live on the "wrong" space. For more information on the standard theory the reader should consult Kuo's Lecture Notes (Kuo, 1975).

Before we can prove Gross's theorem, we must agree on what it should mean for a measure on B to extend μ. Let B^* be the dual of B, and embed B^* in H in the natural way. If $y_1, \ldots, y_n \in B^*$ and A is a Borel set in $\mathbb{R}^n$, then

$$\{x \in B \,|\, (y_1(x), \ldots, y_n(x)) \in A\}$$

is called a *cylinder set* in B. We define a finitely additive measure $\hat{\mu}$ on the cylinder sets by

$$\hat{\mu}\{x \in B \,|\, (y_1(x), \ldots, y_n(x)) \in A\} = \mu\{x \in H \,|\, ((y_1, x), \ldots, (y_n, x)) \in A\}.$$

That a Borel measure on B extends μ we now take to mean that it is an extension of $\hat{\mu}$.

3.5.6. GROSS'S THEOREM. Let $\mu = \{\mu_F\}_{F \in \mathcal{F}}$ be a projective system of measures on H, $|\cdot|$ a μ-measurable norm, and B the completion of H with respect to $|\cdot|$. Then μ has an extension to a Borel measure on B.

Let $\mathrm{st}_{|\cdot|}$ be the standard part map in B with respect to the norm $|\cdot|$, and let $\mathrm{Ns}_{|\cdot|}(^*B)$ be the set of nearstandard elements in $^*B$. The family $\{\mu_E\}_{E \in \mathcal{F}}$ extends to an internal family $\{\tilde{\mu}_E\}_{E \in {}^*\mathcal{F}}$ of nonstandard measures. By saturation we find $E \in {}^*\mathcal{F}$ such that $H \subset E$. We define an internal measure $\tilde{\mu}$ on $^*B$ by

$$\tilde{\mu}(A) = \tilde{\mu}_E(A \cap E).$$

We shall show that $\mathrm{st}_{|\cdot|}(L(\tilde{\mu}))$ is the desired measure. First notice that if $O \subset B$ is open, then since $|\cdot|$ is continuous with respect to $\|\cdot\|$, $^*O$ is $\tilde{\mu}$-measurable. The key step is the following lemma, which allows us to apply Proposition 3.4.7.

3.5.7. LEMMA. If $|\cdot|$ is a μ-measurable norm

$$L(\tilde{\mu})(\mathrm{Ns}_{|\cdot|}(^*B)) = 1.$$

PROOF. Define

$$A_m = \{x \in E \,|\, \exists v \in H(|v - x| < 1/m)\};$$

by the completeness of B we see that $\mathrm{Ns}_{|\cdot|}(^*B) \cap E = \bigcap_{m \in \mathbb{N}} A_m$. Hence it suffices to show that $L(\tilde{\mu}_E)(A_m) = 1$ for all $m \in \mathbb{N}$. But for each $P \in \mathcal{F}$ and each $n > m$ we have

$$A_m \supset \{x \in E \,|\, {}^*P(x) \text{ is nearstandard}\} \cap \{x \in E \,|\, |x - {}^*Px| < 1/n\}.$$

The first set on the right has Loeb measure one since $^*(\mu_{P(H)})$ is nearstandardly concentrated (it is the *-version of a Radon measure on a finitely dimensional space), and since $|\cdot|$ is measurable, the second set has measure larger than $1 - 1/n$. Since n is arbitrarily large, $L(\tilde{\mu}_E)(A_m) = 1$.

By Lemma 3.5.7 and Theorem 3.4.6 we now get that $\mathrm{st}_{|\cdot|}(L(\tilde{\mu}))$ is a Radon measure on B, and it only remains to show that it is an extension of $\hat{\mu}$. Let us first calculate the value of $\mathrm{st}_{|\cdot|}(L(\tilde{\mu}))$ on cylindrical sets:

$$\mathrm{st}_{|\cdot|}(L(\tilde{\mu}))\{x \in B \,|\, (y_1(x), \ldots, y_n(x)) \in A\}$$

$$= L(\tilde{\mu}_E)\{x \in E \,|\, (y_1(\mathrm{st}(x)), \ldots, y_n(\mathrm{st}(x)) \in A\}$$

$$= L(\tilde{\mu}_E)\{x \in E \,|\, (y_1(x), \ldots, y_n(x)) \in \mathrm{st}^{-1}(A)\}$$

$$= L(^*\mu_F)\{x \in {}^*F \,|\, (\langle y_1, x \rangle, \ldots, \langle y_n, x \rangle) \in \mathrm{st}^{-1}(A)\}$$

$$= \mathrm{st}(L(^*\mu_F))\{x \in F \,|\, (\langle y_1, x \rangle, \ldots, \langle y_n, x \rangle) \in A\},$$

where F is the finite-dimensional subspace of H, generated by $y_1, \ldots, y_n$. On the other hand, by definition of $\hat{\mu}$

$$\hat{\mu}\{x \in B \,|\, (y_1(x), \ldots, y_n(x)) \in A\} = \mu_F\{x \in F \,|\, (\langle y_1, x \rangle, \ldots, \langle y_n, x \rangle) \in A\}.$$

We must show that these two expressions are equal. Using F as the X of Proposition 3.4.8, μ_F as the ν, and $^*\mu_F$ as P, the equality follows immediately from that result. This completes the proof of Gross's theorem.

As we have already remarked, we shall return to Gross's theorem in Section 4.7; for the present we just remark that in the standard treatments we know (Gross, 1967; Kallianpur, 1971; Kuo, 1975), the theorem is only proved when the cylindrical measure is Gaussian, and that the nonstandard proof of the general case may be the first.

In our last application we return to the problem of extending measures. What happens in Theorem 3.5.1 if we do not even know that ν is defined on a basis for the topology? We shall give a short proof of the following theorem due to Henry (1969).

3.5.8. HENRY'S THEOREM.   Let $\mathscr{C}$ be an algebra of subsets of a Hausdorff space X, and let ν be a finitely additive probability measure on $\mathscr{C}$ such that for all $C \in \mathscr{C}$

$$(6) \qquad \nu(C) = \sup\{\nu(K) \,|\, K \in \mathscr{C},\ K \subset C,\ K \text{ compact}\}.$$

Then ν can be extended to a Radon measure on X.

PROOF. Let

$$\mathcal{B} = \{{}^*C \mid C \in \mathcal{C} \text{ or } C \text{ is a Borel set}\}.$$

Let $\mathcal{A}$ be a hyperfinite algebra containing $\mathcal{B}$, and put $\mathcal{D} = \mathcal{A} \cap {}^*\mathcal{C}$. Obviously, ${}^*\nu$ is an internal measure on $\mathcal{D}$. If we have two finite algebras and a measure defined on the smallest of them, we can in a trivial way extend this measure to the larger algebra. By transfer, we may thus extend ${}^*\nu \upharpoonright \mathcal{D}$ to $\mathcal{A}$ and get an internal measure P. Applying 3.4.6 to $(X, \mathcal{A}, P)$ we get a Radon measure $\mathrm{st}(L(P))$ on X, and by 3.4.8 this is an extension of ν.

For applications of Henry's theorem, we again refer the reader to Schwartz (1972). We have included this proof to show how embeddings in hyperfinite algebras may be used to produce measure extensions; this kind of trick seems to go back to Loeb (1972).

Our hope is that the depth and variety of the results obtained in this section will convince the reader of the strength of the techniques developed in Section 3.4, and we again urge him or her to study the applications to weak convergence given in Anderson and Rashid (1978) and Loeb (1979a). Theorems 3.5.2 and 3.5.6 were first given a nonstandard treatment in Lindstrøm (1982), while (a more general version of) 3.5.7 will appear in Lindstrøm (1986). Although we have not been able to find the exact statement of Theorem 3.5.1 in the literature, we should mention that a standard proof can be obtained by methods introduced by Kisyński (1969) [e.g., combine Theorem 2.1.4 and Lemma 2.1.9 in Berg *et al.* (1984)]. As for Levy's modulus of continuity, the reader should look up Keisler's (1984) proof based on Anderson's random walk. One natural question we have not investigated is how to push down Loeb measures when X does not carry a nice topological structure; see Anderson (1982), Ross (1983), and Lindstrøm (1986) for three different approaches to this problem.

We shall end our introduction to nonstandard measure and probability theory here. The reader who is interested in the history and development of the subject should consult Cutland's excellent survey paper (Cutland, 1983), which contains an almost complete report of what had happened in the field up till the summer of 1983. In the remainder of this book, we shall mainly concentrate on those aspects of the theory which have to do with stochastic processes and stochastic phenomena in physics. The books by Stroyan and Bayod (1985) and Hurd and Loeb (1985) contain wealths of additional information, and the surveys and introductions by Cutland (1982, 1983), Fenstad (1980, 1985), Lindstrøm (1985), Loeb (1979b, 1983), Nelson (1985), Osswald (1985), and Perkins (1983) approach the theory from different perspectives.

REFERENCES

R. M. Anderson (1976). A nonstandard representation for Brownian motion and Itô integration. *Israel J. Math.* **25**.

R. M. Anderson (1977). Star-finite representations of measure spaces. Ph.D. thesis. Yale Univ., New Haven, Connecticut.

R. M. Anderson (1982). Star-finite representations of measure spaces, *Trans. Amer. Math. Soc.* **271**.

R. M. Anderson and S. Rashid (1978). A nonstandard characterization of weak convergence. *Proc. Amer. Math. Soc.* **69**.

C. Berg, J. P. Reus-Christensen, and P. Ressel (1984). *Harmonic Analysis on Semigroups*. Springer-Verlag, Berlin and New York.

P. Billingsley (1968). *Convergence of Probability Measures*. Wiley, New York.

N. J. Cutland (1982). Infinitesimal methods in measure theory, probability theory and stochastic analysis. *Bull. Inst. Math. Appl.* **18**.

N. J. Cutland (1983). Nonstandard measure theory and its applications. *Bull. London Math. Soc.* **15**.

M. Davis (1977). *Applied Nonstandard Analysis*. Wiley, New York.

J. E. Fenstad (1980). Nonstandard methods in stochastic analysis and mathematical physics. *Jber. Deutsch. Math.-Verein.* **82**.

J. E. Fenstad (1985). Is nonstandard analysis relevant for the philosophy of mathematics. *Synthese* **62**.

P. Greenwood and E. Perkins (1983). A conditional limit theorem for random walk and Brownian local time on square root boundaries. *Ann. Probab.* **11**.

P. Greenwood and E. Perkins (1985). Limit theorems for excursions from a moving boundary. *Theory of Prob. and Appl.* **29**.

L. Gross (1967). Abstract Wiener spaces. *Proc. 5th Berkeley Sym. Math. Statist. Probab.* **2** (1965). Univ. of California Press, Berkeley and Los Angeles.

J. P. Henry (1969). Prolongements de Mesure de Radon. *Ann. Inst. Fourier* (Grenoble) **19**.

C. W. Henson (1979). Unbounded Loeb measures. *Proc. Amer. Math. Soc.* **74**.

D. N. Hoover (1982). A normal form theorem for $L_{\omega_1 P}$ with applications. *J. Symbolic. Logic* **47**.

A. E. Hurd and P. A. Loeb (1985). *Nonstandard Real Analysis*. Academic Press, New York.

G. Kallianpur (1971). Abstract Wiener processes and their reproducing kernel Hilbert spaces. *Z. Wahrsch. Verw. Gebiete* **17**.

T. Kamae (1982). A simple proof of the ergodic theorem using non-standard analysis. *Israel J. Math.* **42**.

Y. Katznelson and B. Weiss (1982). A simple proof of some ergodic theorems, *Israel J. Math.* **42**.

H. J. Keisler (1977). Hyperfinite model theory. In (R. O. Gandy and J. M. E. Hyland eds.), *Logic Colloqium 1976*, North-Holland Publ., Amsterdam.

H. J. Keisler (1984). An infinitesimal approach to stochastic analysis. *Mem. Amer. Math. Soc.* **297**.

J. Kisyński (1969). On the generation of tight measures. *Studia Math.* **30**.

H.-H. Kuo (1975). *Gaussian measures in Banach spaces*. Springer-Verlag, Berlin and New York.

T. Lindstrøm (1982). A Loeb-measure approach to theorems by Prohorov, Sazonov and Gross. *Trans. Amer. Math. Soc.* **269**.

T. Lindstrøm (1985). Nonstandard analysis and perturbations of the Laplacian along Brownian paths. *Proceedings of the First BiBos Symposium*. Lecture Notes **1158** Springer-Verlag, Berlin and New York.

T. Lindstrøm (1986). Weak Loeb-space representations (in preparation).

P. A. Loeb (1972). A nonstandard representation of measurable spaces, L_∞ and L_∞^*. In (W. A. J. Luxemburg and A. Robinson, eds.) *Contributions to Nonstandard Analysis.* North-Holland Publ., Amsterdam.

P. A. Loeb (1975). Conversion from nonstandard to standard measure spaces and applications in probability theory. *Trans. Amer. Math. Soc.* **211**.

P. A. Loeb (1976). Applications of nonstandard analysis to ideal boundaries in potential theory. *Israel J. Math.* **25**.

P. A. Loeb (1979a). Weak limits of measures and the standard part map. *Proc. Amer. Math. Soc.* **77**.

P. A. Loeb (1979b). An introduction to nonstandard analysis and hyperfinite probability theory. *In* A. T. Bharucha-Reid (ed.), *Probabilistic Analysis and Related Topics II.* Academic Press, New York.

P. A. Loeb (1984). A functional approach to nonstandard measure theory. *In* (Beals *et al.*, eds.), *Conference on Modern Analysis and Probability,* Amer. Math. Soc., Providence, Rhode Island.

P. A. Loeb (1983). Measure spaces in nonstandard models underlying standard stochastic processes. *Proc. Inter. Congr. Math. Warsaw.*

E. Nelson (1977). Internal set theory: A new approach to nonstandard analysis. *Bull. Amer. Math. Soc.* **83**.

E. Nelson (1985). Radically elementary probability theory (preprint) Math. dept., Princeton Univ.

J. Oikkonen (1985). Harmonic analysis and nonstandard Brownian motion in the plane. *Math. Scand,* (to appear).

H. Osswald (1983). On hyperfinite integration in reflexive Banach spaces (preprint). Univ. of München.

H. Osswald (1985). On Petti's integrability on Loeb spaces (preprint). Univ. of München.

H. Osswald (1985). Introduction to nonstandard measure theory, I-II. Lecture notes, Univ. of München, 1983–85.

E. Perkins (1981a). A global intrinsic characterization of Brownian local time. *Ann. Prob.* **9**.

E. Perkins (1981b). The exact Hausdorff measure of the level sets of Brownian motion. *Z. Wahrsch. Verw. Gebiete* **58**.

E. Perkins (1982a). Weak invariance principles for local time. *Z. Wahrsch. Verw. Gebiete* **60**.

E. Perkins (1982b). Local time is a semimartingale. *Z. Wahrsch. Verw. Gebiete* **60**.

E. Perkins (1982c). Local time and pathwise uniqueness for stochastic differential equations. *Sem. Probab. XVI, Lecture Notes in Math.* **920** Springer-Verlag, Berlin and New York.

E. Perkins (1983a). On the Hausdorff dimension of the Brownian slow points *Z. Wahrsch. Verw. Gebiete* **64**.

E. Perkins (1983b). Stochastic processes and nonstandard analysis. *In* (A. E. Hurd, ed.), *Nonstandard Analysis—Recent Developments.* Springer-Verlag, Berlin and New York.

D. Ross (1983). Measurable transformations in saturated models of analysis. Ph.D. thesis, Univ. of Wisconsin, Madison.

D. Ross (1984a). Automorphisms of the Loeb algebra (preprint). Univ. of Iowa.

D. Ross (1984b). Completeness theorem for probability logic with function symbols (preprint). Univ. of Iowa.

L. Schwartz (1972). *Radon Measures on Arbitrary Topological Spaces and Cylindrical Measures.* Oxford Univ. Press, London and New York.

K. S. Stroyan and J. Bayod (1985). *Foundations of Infinitesimal Stochastic Analysis.* North-Holland Publ., Amsterdam (to appear).

R. Zivaljevic (1985). Loeb-completion of internal vector-valued measures. *Math. Scand.* (to appear).

Part II

SELECTED APPLICATIONS

CHAPTER 4

STOCHASTIC ANALYSIS

Following up the ideas from the last chapter, we shall now study one of the most lively and active areas of nonstandard research—the theory of stochastic processes and their applications. In this field hyperfinite structures play a particularly interesting and important role, combining in the same model the combinatorial aspects of the discrete theory and the analytic character of the continuous one. We have already seen an example of this interplay between the continuous and the discrete in the construction of Brownian motion in Section 3.3, and we shall now consider it in greater breadth and detail. As our central theme we shall take stochastic integration with its applications to diffusions, control theory, and multiparameter processes.

4.1. THE HYPERFINITE ITÔ INTEGRAL

We shall begin by giving an informal introduction to Anderson's (1976) construction of the standard Itô integral as a hyperfinite Stieltjes sum. In this first section our aim is simply to give the reader a feeling for the basic ideas of the subject, and we have postponed the more technical proofs to the systematic treatment starting in the next section. But to illustrate the power of the method, we have included a complete and, we think, illuminating proof of Itô's lemma.

A. Stochastic Integration

The fundamental problem of stochastic integration is to give sense to integrals of the form $\int x\,dy$, where x and y are stochastic processes. Everybody's first idea is to let x_ω and y_ω denote the functions $x(\omega, \cdot)$ and $y(\omega, \cdot)$, respectively, and consider the Stieltjes integrals

$$(1) \qquad\qquad z_\omega(t) = \int_0^t x_\omega(s)\,dy_\omega(s).$$

The natural candidate for the stochastic integral is then the process $(\omega, t) \mapsto z_\omega(t)$. However, for (1) to make sense as a Stieltjes integral, the path $y(\omega, \cdot)$ must be of bounded variation, and for many naturally occurring processes this is not the case; e.g., almost all Brownian paths are of unbounded variation. Hence a pathwise Stieltjes approach breaks down.

It was Itô (1944)—extending work by Wiener—who discovered a way of defining integrals of the form $\int g\,db$, where b is a Brownian motion and g is a suitable process. We shall sketch Itô's idea below, but first we want to explore a nonstandard approach where indeed a pathwise Stieltjes definition works.

Let $T = \{0, \Delta t, 2\,\Delta t, \ldots, 1\}$ be a *hyperfinite time line* with Δt infinitesimal. Our sample space is a hyperfinite probability space $(\Omega, \mathscr{A}, P)$, where for simplicity we assume that $\mathscr{A}$ is the algebra of all internal subsets of Ω. A *hyperfinite stochastic process* on Ω is an internal map

$$X : \Omega \times T \to M$$

into some internal set M; note that since we are using the algebra of all internal subsets of Ω, no measurability conditions are needed in this definition. We shall write $\Delta X(\omega, t)$ for the forward increment of $X(\omega, \cdot)$ at t, i.e.,

$$(2) \qquad\qquad \Delta X(\omega, t) = X(\omega, t + \Delta t) - X(\omega, t),$$

and use the following convention for sums

$$(3) \qquad\qquad \sum_s^t X(\omega, r) = X(\omega, s) + \cdots + X(\omega, t - \Delta t);$$

hence $X(\omega, t)$ is *not* included in the sum $\sum_s^t (X(\omega, r))$.

The nonstandard stochastic integral is just a Stieltjes sum:

4.1.1. DEFINITION. Let $X, Y : \Omega \times T \mapsto {}^*\mathbb{R}$ be two hyperfinite processes. The *stochastic integral* of X with respect to Y is the process $\int X\,dY$ defined by

(4) $$\left(\int X \, dY \right)(\omega, t) = \sum_0^t X(\omega, s) \, \Delta Y(\omega, s).$$

We shall write $\int_0^t X \, dY$ for the random variable $(\int X \, dY)(\cdot, t)$.

The stochastic integral is well defined for all hyperfinite processes X and Y, but, of course, in this generality it may have strange properties and no standard part. One of our first tasks will be to single out classes of processes X and Y for which the integral is well behaved, and we shall devote the next section to this problem. For the present, we shall be satisfied by getting a reasonably good grasp of the basic ideas. To understand what is going on, let us take a look at a simple example.

4.1.2. EXAMPLE. We are back in the setting of Section 3.3; Ω is the set of all internal functions $\omega : T \mapsto \{-1, 1\}$, and $\chi : \Omega \times T \mapsto {}^*\mathbb{R}$ is Anderson's random walk, $\chi(\omega, t) = \sum_0^t \omega(s)\sqrt{\Delta t}$. Let $X(\omega, s) = \omega(s)$ and consider the stochastic integral $\int X \, d\chi$. We have

$$\int_0^t X \, d\chi(\omega) = \sum_0^t \omega(s)\omega(s)\sqrt{\Delta t} = \sum_0^t \sqrt{\Delta t} = \frac{t}{\sqrt{\Delta t}},$$

which is infinite for all noninfinitesimal t. Hence the integral of the finite function X with respect to the "Brownian motion" χ is infinite.

It is easy to see what went wrong in Example 4.1.2—the increments $\pm\sqrt{\Delta t}$ of a Brownian motion are much larger than one would expect of a finite function, and what keeps the Brownian path finite is the delicate balance between the positive and negative contributions. The integrand $\omega(s)$ upsets this balance by making all increments positive. What we learn from this is that we cannot allow integrands which anticipate the behavior of χ if we want to keep the integral finite.

Let us make this notion of nonanticipation precise. If $\omega \in \Omega$ and $t \in T$, let

(5) $$\omega \upharpoonright t = \langle \omega(s) \, | \, s < t \rangle$$

(we are still working with the Ω of Example 4.1.2). A hyperfinite process $X : \Omega \times T \to {}^*\mathbb{R}$ is *nonanticipating* if $X(\omega, t) = X(\omega', t)$ whenever $\omega \upharpoonright t = \omega' \upharpoonright t$. To show that we are headed in the right direction, we shall prove the following result, where λ denotes the uniform probability measure on T:

4.1.3. PROPOSITION. Assume that X is a nonanticipating process which is square S-integrable with respect to $P \times \lambda$. Then for all t, the stochastic integral $\int_0^t X \, d\chi$ is finite a.e.

PROOF. By simple algebra

$$E\left(\left(\int_0^t X \, d\chi\right)^2\right) = E\left(\left(\sum_0^t X \, \Delta\chi\right)^2\right)$$

(6)
$$= E\left(\sum_0^t X(s)^2 \, \Delta\chi(s)^2\right)$$

$$+ 2 \sum_{r<s} E(X(s)X(r) \, \Delta\chi(s) \, \Delta\chi(r)).$$

Here the last term is zero since $\Delta\chi(s)$ is plus or minus $\sqrt{\Delta t}$ [no matter what $X(s)$, $X(r)$, and $\Delta\chi(r)$ are] with probability one-half. Since $\Delta\chi(s)^2 = \Delta t$, we end up with

$$(7) \quad E\left(\left(\int_0^t X \, d\chi\right)^2\right) = E\left(\sum_0^t X(s)^2 \, \Delta t\right) = \int_{\Omega\times[0,t]} X^2 \, d(P \times \lambda) < \infty,$$

and the proposition follows.

This result is the first indication that we get a reasonable theory for stochastic integration if we restrict our integrands to the class of nonanticipating processes. Without delving deeper into this theory here, we mention that in the next section we shall prove that the integral $\int X \, d\chi$ above is S-continuous a.e. and hence induces a nice standard process.

Before turning to the standard theory of stochastic integration, we shall give an alternative description of nonanticipating processes which is less intuitive, but easier to generalize. For each $t \in T$, let $\mathscr{A}_t$ be the internal algebra on Ω generated by the sets

$$[\omega]_t = \{\omega' \in \Omega \,|\, \omega' \restriction t = \omega \restriction t\}.$$

It is easy to see that a process X is nonanticipating if and only if $X(\cdot, t)$ is $\mathscr{A}_t$-measurable for each t. We call the tuple $(\Omega, \{\mathscr{A}_t\}_{t \in T}, P)$ an *internal filtration*.

Let us now turn to the standard theory of stochastic integration with respect to the standard part b of χ. The standard notion of nonanticipation is based on a filtration $(\Omega, \{\mathscr{B}_t\}_{t \in [0,1]}, L(P))$ generated by the internal filtration $(\Omega, \{\mathscr{A}_t\}, P)$ above. If $\mathscr{N}$ is the class of null sets with respect to $L(P)$, we define for each $t \in [0, 1]$,

$$(8) \qquad\qquad \mathscr{B}_t = \sigma\left(\bigcup_{s \approx t} L(\mathscr{A}_s) \cup \mathscr{N}\right),$$

where the σ means that we take the σ-algebra generated by $\bigcup_{s \approx t} L(\mathscr{A}_s) \cup \mathscr{N}$. Basically, $\mathscr{B}_t$ classifies the events that only depend on what happens up to and including the monad of t; the null sets $\mathscr{N}$ are added just for technical convenience.

Recall that a standard process $x : \Omega \times [0, 1] \mapsto \mathbb{R}$ is *measurable* if it is measurable with respect to the completed product of the Loeb measure on Ω and the Lebesgue measure on $[0, 1]$. We say that x is *adapted* to the filtration $(\Omega, \{\mathcal{B}_t\}, L(P))$ if it is measurable and if $x(\cdot, t)$ is $\mathcal{B}_t$-measurable for each $t \in [0, 1]$. Adapted is clearly a standard counterpart of the nonstandard notion of nonanticipating. When it is clear which filtration $(\Omega, \{\mathcal{B}_t\}, L(P))$ we have in mind, we suppress the explicit mention of it, and only refer to x as an adapted process.

In view of Proposition 4.1.3, it is natural to assume that standard stochastic integrals $\int x \, db$ will be defined for all adapted processes x which are square integrable with respect to $L(P) \times m$ (where m is the Lebesgue measure). We first define the integral when x is an adapted step function, i.e., when there is a partition $0 = t_0 < t_1 < \cdots < t_k = 1$ such that $x(\omega, s) = x(\omega, t_i)$ whenever $s \in [t_i, t_{i+1})$, and $x(\cdot, t_i)$ is bounded and $\mathcal{B}_{t_i}$-measurable for each t_i. Let

$$\int_0^1 x(\omega, s) \, db(\omega, s) = \sum_{j=0}^{k-1} x(\omega, t_j)(b(\omega, t_{j+1}) - b(\omega, t_j)).$$

Itô's observation was that since

$$E\left(\left(\int_0^1 x(\omega, s) \, db(\omega, s)\right)^2\right) = E\left(\left(\sum_{j=0}^{k-1} x(t_j) \, \Delta b(t_j)\right)^2\right)$$

$$= E\left(\sum_{j=0}^{k-1} x(t_j)^2 \, \Delta b(t_j)^2\right) + 2 \sum_{j<l} E(x(t_j)x(t_l) \, \Delta b(t_j) \, \Delta b(t_l))$$

$$= E\left(\int_0^1 x(t_j)^2 \, dt\right),$$

where $\Delta b(t_j) = b(t_{j+1}) - b(t_j)$, the map $x \to \int_0^1 x \, db$ is norm-preserving from $L^2(L(P) \times m)$ to $L^2(L(P))$. Since the adapted step functions are dense in the set of adapted processes in $L^2(L(P) \times m)$, we can extend $x \to \int_0^1 x \, db$ to an isometry which we shall also denote by $\int_0^1 x \, db$. If $1_{[0,t]}$ is the indicator function of the interval $[0, t]$, the *stochastic integral* is the process

(9) $$\left(\int x \, db\right)(\omega, t) = \int_0^1 1_{[0,t]} x \, db.$$

Note that since $\int_0^1 g \, db$ is defined as an element in $L^2(L(P))$, the stochastic integral is only determined up to equivalence.

B. Liftings

We have described two stochastic integrals—Anderson's hyperfinite Stieltjes sum (4) and Itô's classical integral (9). As in Section 3.2, the standard and nonstandard theories are connected through the notion of a lifting. Recall that λ is the normalized counting measure on T.

4.1.4. DEFINITION. Let $x : \Omega \times [0, 1] \mapsto \mathbb{R}$; a hyperfinite process $X : \Omega \times T \mapsto {}^*\mathbb{R}$ is a *lifting* of x (with respect to $P \times \lambda$) if

$$(10) \qquad\qquad {}^\circ X(\omega, t) = x(\omega, {}^\circ t)$$

almost surely in $L(P \times \lambda)$.

Notice how this definition combines the two parts of Definition 3.2.3.

Call a process x *almost surely adapted* if there is an adapted process y such that $x(\omega, s) = y(\omega, s)$ for almost all (ω, s). Anderson (1976) proved the "only if" and Keisler (1984) the "if" part of the following theorem.

4.1.5. THEOREM. A stochastic process $x : \Omega \times [0, 1] \mapsto \mathbb{R}$ is almost surely adapted if and only if it has a nonanticipating lifting.

This result is a consequence of the lifting Theorem 3.2.4 for random variables; all that needs to be checked is that the classes of almost surely adapted and nonanticipating sets fit together correctly. This is slightly technical and not very surprising, and we postpone the proof till the systematic treatment of liftings in Section 4.3. It follows easily from 4.1.5 that if x is square-integrable with respect to $L(P) \times m$, then we may choose the lifting to be square S-integrable with respect to $P \times \lambda$.

The next theorem establishes the relationship between Itô integration and internal Stieltjes integration (Anderson, 1976).

4.1.6. THEOREM. Let x be a square-integrable, adapted process, and X a nonanticipating, square S-integrable lifting of x. Then for all $t \in T$

$$(11) \qquad \int_0^{\circ t} x(\omega, s)\, db(\omega, s) = \int_0^{\circ t} X(\omega, s)\, d\chi(\omega, s) \qquad \text{for } L(P)\text{-a.a. } \omega.$$

In this case also the proof is quite simple; we first establish (11) when x is an adapted step function, and then extend to the general case by a limit argument, using the definition of the Itô integral. The full details will be given in Section 4.4.

A consequence of 4.1.6 is that if X_1 and X_2 are two square S-integrable liftings of x, then $\int X_1\, d\chi \approx \int X_2\, d\chi$ almost surely. To give a little more of the flavor of the theory, we shall give a proof of this fact from scratch. Let $X = X_1 - X_2$; then $\int X\, d\chi = \int X_1\, d\chi - \int X_2\, d\chi$ and by (7)

$$E\left(\left(\int_0^t X\, d\chi\right)^2\right) = \int_{\Omega \times [0, t]} X^2\, d(P \times \lambda) \le \int (X_1 - X_2)^2\, d(P \times \lambda) \approx 0.$$

Hence $\int_0^t X_1\, d\chi \approx \int_0^t X_2\, d\chi$ a.e.

C. Itô's Lemma

This is a result of great importance in stochastic analysis. We shall give a nonstandard proof which in a simple way converts the usual heuristics into a precise argument.

Let $\Omega : \mathbb{R}^n \times [0, 1] \mapsto \mathbb{R}$. We shall use ∇ and Δ for space derivatives and ∂_t for time derivative. If b is an n-dimensional Brownian motion [i.e., $b(\omega, t)$ is a vector $(b_1(\omega, t), \ldots, b_n(\omega, t))$ where $b_1, \ldots, b_n$ are independent, one-dimensional Brownian motions], then Itô's lemma in its differential form asserts that

$$(12) \qquad d\varphi(b(t)) = \nabla \varphi(b(t)) \, db + (\tfrac{1}{2}\Delta\varphi + \partial_t\varphi)(b(t)) \, dt,$$

which is just a convenient abbreviation for

$$(13) \quad \varphi(b(t)) - \varphi(b(0)) = \int_0^t \nabla \varphi(b(s)) \, db(s) + \int_0^t (\tfrac{1}{2}\Delta\varphi + \partial_t\varphi)(\underline{b}(s)) \, ds.$$

Itô's lemma is the fundamental theorem of stochastic calculus, and it is particularly useful for calculating with functions φ such that $\partial_t\varphi = -\tfrac{1}{2}\Delta\varphi$; e.g., see Simon (1979) for a proof of the Feynman–Kac formula using this technique.

There are more general versions of the result than (13), and we refer the reader to Anderson (1976), Lindstrøm (1980a), and Section 4.4 for more extensive nonstandard treatments. For simplicity we shall write down the proof only in the one-dimensional case. Assume that $\varphi : \mathbb{R} \times [0, 1] \mapsto \mathbb{R}$ has continuous first and second derivatives; it suffices to show that

$$(14) \qquad \begin{aligned} \varphi(\chi_t, t) - \varphi(\chi_0, 0) &\approx \int_0^t \nabla\varphi(\chi_s, s) \, d\chi + \frac{1}{2}\int_0^t \Delta\varphi(\chi_s, s) \, ds \\ &\quad + \int_0^t \partial_t\varphi(\chi_s, s) \, ds \end{aligned}$$

since $\nabla\varphi(\chi_s, s)$ is a lifting of $\nabla\varphi(b_s, s)$. The left-hand side of (14) can be rewritten as a hyperfinite sum

$$\varphi(\chi_t, t) - \varphi(\chi_0, 0)$$

$$= \sum_{s=0}^t [\varphi(\chi_{s+\Delta t}, s + \Delta t) - \varphi(\chi_s, s)]$$

$$= \sum_{s=0}^t \nabla\varphi(\chi_s, s) \, \Delta\chi(s)$$

$$+ \sum_{s=0}^t [\varphi(\chi_{s+\Delta t}, s) - \varphi(\chi_s, s) - \nabla\varphi(\chi_s, s) \, \Delta\chi(s)]$$

$$+ \sum_{s=0}^{t} [\varphi(\chi_{s+\Delta t}, s + \Delta t) - \varphi(\chi_{s+\Delta t}, s)].$$

This is a simple algebraic reformulation suggested by Taylor's formula. By definition, the first sum is the correct hyperfinite stochastic integral. To treat the middle sum note that

$$|\varphi(x) - \varphi(y) - \nabla\varphi(y)(x - y) - \tfrac{1}{2}\Delta\varphi(y)(x - y)^2| \leq C|x - y|^3$$

for a finite constant C. Thus the second sum can be replaced up to $\approx$ by

$$\tfrac{1}{2} \sum \Delta\varphi(\chi(s), s) \, \Delta\chi(s)^2.$$

But, and here is the point where the usual heuristics becomes exact, $\Delta\chi(s)^2 = \Delta t$. Hence the middle sum can be replaced by

$$\tfrac{1}{2} \int_0^t \Delta\varphi(\chi(s), s) \, ds.$$

To handle the last term we have the inequality

$$|\varphi(x, t) - \varphi(x, s) - \partial_t\varphi(x, s)(t - s)| \leq C(t - s)^2,$$

which tells us that up to $\approx$ the last sum can be replaced by

$$\int_0^t \partial_t\varphi(\chi(s + \Delta t), s) \, ds \approx \int_0^t \partial_t\varphi(\chi(s), s) \, ds,$$

and the proof of (14) is complete; no further limit arguments have to be made.

We have purposely modeled our proof of Itô's lemma on the standard argument in Simon (1979) to make it easier for the reader to compare the two approaches. Two aspects of the nonstandard proof are particularly noteworthy; one is the advantage of the pathwise definition of the stochastic integral, the other is the exact relation $\Delta\chi(t)^2 = \Delta t$, where Δt is infinitesimal.

Our aim so far has been to introduce the most important ideas of hyperfinite stochastic calculus with a minimum of technical details. In the sections to come we will be taking a closer look at the themes that this informal account has suggested, according to the following plan. The next section develops a theory for hyperfinite stochastic integrals $\int X \, dM$, where X is nonanticipating and M is a square-integrable martingale; we will be proving regularity properties for the paths of such integrals and will take a look at an importance criterion for S-continuity. In Section 4.3, we turn to lifting theorems and establish not only a generalized version of Theorem 4.1.5, but also other lifting results which will be important in later applications. We then study the relationship between standard and nonstandard stochastic integration; Section 4.4 contains a more general version of 4.1.6

plus a discussion of other aspects of this question. Applications to stochastic differential equations and stochastic control theory are the themes of Sections 4.5 and 4.6—here the ideas used in the proof of Itô's lemma are more fully exploited. The last two sections of this chapter extend the theory in different directions; in 4.7 we treat stochastic integration in infinite-dimensional spaces with applications to stochastic partial differential equations, and in 4.8 we consider a generalization of Brownian motion to several time parameters.

4.2. GENERAL THEORY OF STOCHASTIC INTEGRATION

It is time to begin the systematic development of the ideas we sketched in the last section. Although the hyperfinite integral $\int X \, dY$ is defined for all $^*\mathbb{R}$-valued, internal processes X and Y, we have seen that certain restrictions on X and Y are necessary to get a decent theory. The appropriate nonanticipation conditions on the integrand X have already been discussed in some detail, but we have so far simply assumed that the integrator process Y is a hyperfinite random walk. In this section we shall allow more general integrators; basically, we shall be working with the class of processes called martingales.

In addition to X being nonanticipating and Y being a martingale, we need to impose certain integrability conditions on the two processes. Here we are given a certain freedom. Originally, standard stochastic integration had a strong L^2 flavor, and both X and Y were assumed to be locally square-integrable (see, e.g., Kunita and Watanabe, 1967). However, by using more sophisticated techniques and stronger inequalities, it has been possible to extend the theory to the L^1 case; thus X and Y need only be assumed locally integrable [see Meyer (1976) and Metivier and Pellaumail (1980) for expositions]. On the nonstandard side, the L^2 and the L^1 theory were developed independently and almost simultaneously by Lindstrøm (1980a–c) and Hoover and Perkins (1983a,b). We have decided to restrict ourselves to the less general L^2 approach in this book as we feel that the extra technicalities needed for the L^1 case tend to blur the very simple ideas underlying the whole theory, and also because the extra generality is not needed for our applications.

A. Internal Martingales

As in the last section we shall be studying hyperfinite stochastic processes $X : \Omega \times T \mapsto {}^*\mathbb{R}$, where $(\Omega, \mathscr{A}, P)$ is a hyperfinite probability space, but we shall allow T to be slightly more general. A *hyperfinite time line* will be a hyperfinite set

$$T = \{t_0, t_1, \ldots, t_\xi\},$$

where $0 = t_0 < t_1 < \cdots < t_\xi = 1$ and $t_{i+1} - t_i \approx 0$ for each i. We write $\Delta X(\omega, t_i)$ for $X(\omega, t_{i+1}) - X(\omega, t_i)$ and use the same convention for sums as in (4.1.3); i.e., if $s = t_i$, $t = t_j$, then

$$(1) \qquad \sum_{r=s}^{t} X(\omega, r) = X(\omega, t_i) + X(\omega, t_{i+1}) + \cdots + X(\omega, t_{j-1});$$

thus the term $X(\omega, t)$ is *not* included in the sum. Given two internal processes $X, Y : \Omega \times T \to {}^*\mathbb{R}$, the stochastic integral is defined as before:

$$(2) \qquad \int_0^t X \, dY = \sum_{s=0}^{t} X(s) \, \Delta Y(s).$$

To define martingales and nonanticipating processes, we must first introduce the notion of a filtration.

4.2.1. DEFINITION. Let T be a hyperfinite time line and $(\Omega, \mathcal{A}, P)$ a hyperfinite probability space. An *internal filtration* on Ω indexed by T is a tuple $\langle \Omega, \{\mathcal{A}_t\}_{t \in T}, P \rangle$, where $\{\mathcal{A}_t\}_{t \in T}$ is an increasing internal sequence of internal algebras on Ω.

Since we are always assuming that $\mathcal{A}$ is the internal power set of Ω, all the $\mathcal{A}_t$'s are automatically subalgebras of $\mathcal{A}$. A natural example of an internal filtration was given in Section 4.1; it gives substance to the assertion that $\mathcal{A}_t$ is supposed to represent the information we have about the stochastic system at time t.

4.2.2. DEFINITION. An internal process $X : \Omega \times T \to {}^*\mathbb{R}$ is *nonanticipating* with respect to the filtration $(\Omega, \{\mathcal{A}_t\}_{t \in T}, P)$ if $\omega \mapsto X(\omega, t)$ is $\mathcal{A}_t$-measurable for all $t \in T$.

There is another way of describing nonanticipating processes that is both more intuitive and closer to the definition we gave for a special case in the previous section. For each $t \in T$, introduce an equivalence relation $\sim_t$ on Ω by

$$(3) \qquad \omega \sim_t \omega' \qquad \text{iff} \qquad \forall A \in \mathcal{A}_t(\omega \in A \Leftrightarrow \omega' \in A).$$

It is easy to see that X is nonanticipating if and only if $X(\omega, t) = X(\omega', t)$ whenever $\omega \sim_t \omega'$.

4.2.3. DEFINITION. An internal process $M : \Omega \times T \to {}^*\mathbb{R}$ is a *martingale* with respect to the filtration $(\Omega, \{\mathcal{A}_t\}_{t \in T}, P)$ if it is nonanticipating and if for all $s, t \in T$, $s < t$, and all $A \in \mathcal{A}_s$,

$$(4) \qquad E(1_A(M_t - M_s)) = 0.$$

If we replace the equality in (4) by the inequality $E(1_A(M_t - M_s)) \geq 0$, then M is called a *submartingale*, and if we replace (4) by the opposite inequality $E(1_A(M_t - M_s)) \leq 0$, then M is called a *supermartingale*. For the final word on this much discussed and somewhat confusing piece of terminology, see Doob 1984 (p. 808).

The easiest way to figure out what Definition 4.2.3 really says is to use the equivalence relation (3). If $[\omega]_t$ is the equivalence class of ω, then a nonanticipating process M is a martingale if and only if

$$(5) \qquad \sum_{\tilde{\omega} \in [\omega]_t} \Delta M(\tilde{\omega}, t) P\{\tilde{\omega}\} = 0$$

for all ω and t. Using this characterization, it is trivial to see that if X is nonanticipating and M is a martingale, then $\int X \, dM$ is also a martingale.

The sample paths of a martingale are usually oscillating wildly—just think of the most studied of all martingales, the Brownian motion, whose paths are not only of unbounded variation, but even nowhere differentiable. To tame such extremely irregular behavior, we introduce an associated increasing process called the quadratic variation.

4.2.4. DEFINITION. By the *quadratic variation* of $X : \Omega \times T \to {}^*\mathbb{R}$, we mean the process $[X] : \Omega \times T \to {}^*\mathbb{R}$ defined by

$$[X](\omega, t) = \sum_{s=0}^{t} \Delta X(\omega, s)^2.$$

A striking example of how much simpler the quadratic variation is than the original process is provided by Anderson's random walk χ:

$$(6) \qquad [\chi](t) = \sum_{s=0}^{t} \Delta \chi(s)^2 = \sum_{s=0}^{t} \Delta t = t.$$

In martingale theory the simplicity of the quadratic variation is exploited systematically by translating problems about martingales into much simpler problems about their quadratic variations. We shall see several examples of this strategy in this and later sections. An important tool in some of these proofs is the following simple identity:

4.2.5. LEMMA. For all hyperfinite processes $X : \Omega \times T \to {}^*\mathbb{R}$,

$$[X](t) = X(t)^2 - X(0)^2 - 2 \int_0^t X \, dX.$$

PROOF. By elementary algebra

$$\Delta[X](t_i) = (X(t_{i+1}) - X(t_i))^2$$
$$= X(t_{i+1})^2 - X(t_i)^2 - 2X(t_i)(X(t_{i+1}) - X(t_i))$$

$$= X(t_{i+1})^2 - X(t_i)^2 - 2 \int_{t_i}^{t_{i+1}} X \, dX.$$

Summing over all $t_i < t$, the lemma follows.

If we apply the formula we just proved to a martingale M, we get

(7) $E(M_t^2) = E(M_0^2 + [M](t))$

since $\int M \, dM$ is a martingale starting at 0 and thus has zero expectation. By connecting the expectations of M^2 and the quadratic variation $[M]$, this formula provides much of the explanation of why the theory of square-integrable martingales is particularly well behaved.

4.2.6. DEFINITION. A hyperfinite martingale M is called a λ^2-*martingale* if $°E(M_t^2) < \infty$ for all $t \in T$.

Note that since $[M](t)$ is an increasing process, (7) tells us that $E(M_t^2)$ is also increasing. To show that M is a λ^2-martingale it thus suffices to check that $E(M_1^2)$ is finite.

Definition 4.2.6 introduces a condition on the size of M. Often we can reduce a martingale to a λ^2-martingale by stopping it before it grows too big. An *internal stopping time* adapted to the filtration $(\Omega, \{\mathscr{A}_t\}, P)$ is an internal mapping $\tau : \Omega \to T$ such that for all $t \in T$, the set $\{\omega \mid \tau(\omega) \leq t\}$ belongs to $\mathscr{A}_t$. In terms of the equivalence relation $\sim_t$ in (3), this just means that if $\tau(\omega) = t$, then $\tau(\omega') = t$ for all $\omega' \sim_t \omega$. The important observation to make is that if τ is a stopping time and M is a martingale, then the *stopped process* M_τ defined by

$$M_\tau(\omega, t) = M(\omega, t \wedge \tau(\omega))$$

is also a martingale.

4.2.7. DEFINITION. An internal martingale M is a *local λ^2-martingale* if there is an increasing sequence $\{\tau_n\}_{n \in \mathbb{N}}$ of internal stopping times such that each M_{τ_n} is a λ^2-martingale, and such that for almost all ω, $\tau_n(\omega) = 1$ for some $n \in \mathbb{N}$. The sequence $\{\tau_n\}_{n \in \mathbb{N}}$ is called a *localizing sequence* for M.

It is the local λ^2-martingales we are going to allow as integrators in our stochastic integrals. The difference between λ^2 and local λ^2 theory is really of little importance as most results in the latter follow immediately from corresponding results in the former, but the extra freedom of localization is handy to have around in applications.

Before we turn to stochastic integration, we would like to show that the local λ^2-martingales are processes with nice standard parts. One thing is clear; if we fix $t \in T$, then M_t is finite almost everywhere. But a problem is that the exceptional set may differ from one t to another, and hence there might conceivably be no ω such that $M(\omega, t)$ is finite for all t. What we would like to know is that $\max_{t \in T} M(\omega, t)$ is finite for almost all ω. To answer "uniform" questions of this type, we have an extremely useful inequality due to Doob (1953).

4.2.8. DOOB'S INEQUALITY. If $X : \Omega \times T \to {}^*\mathbb{R}$ is a positive submartingale, then for all $p > 1$ and all $t \in T$

$$\left\| \sup_{s \le t} X_s \right\|_p \le \frac{p}{p - 1} \|X_t\|_p,$$

where $\| \cdot \|_p$ denotes the L^p norm.

We postpone the proof of this inequality until we have completed our analysis of the paths of M.

To apply 4.2.8 to the problem we are working on, observe that $|M|$ is a positive submartingale. Letting $p = 2$, we get

$$(8) \qquad E(\sup_{s \le t} M_s^2) \le 4E(M_t^2).$$

If M is a λ^2-martingale, the expectation on the right is finite, and hence there is a set of measure one where M_s is finite for all s. Using a localizing sequence, we see that this must also hold for local λ^2-martingales.

But the fact that M is finite a.e. is not sufficient to guarantee that it has a decent standard part; its paths may still be jumping infinitely often between different finite values. We would like M to have one-sided limits in the following sense:

4.2.9. DEFINITION. Let $f : T \to {}^*\mathbb{R}$ be internal. We say that $r \in \mathbb{R}$ is the *S-right limit* of f at $t \in [0, 1]$ if for all standard $\varepsilon > 0$, there is a standard $\delta > 0$ such that if $s \in T$ and $t < {}^\circ s < t + \delta$, then $|f(s) - r| < \varepsilon$. We write $r = S\text{-}\lim_{s \downarrow t} f(s)$. The *S-left limit*, $S\text{-}\lim_{s \uparrow t} f(s)$ is defined analogously.

Note that it is the standard part of s (and not s itself) which should lie between t and $t + \delta$; hence we do not care how f behaves on the upper half-monad of t.

Before we can show that local λ^2-martingales have S-right and S-left limits a.e., we have to know a little more about stopping times. The algebras $\mathscr{A}_t$ in our filtration classify the events which happen before a fixed time t; we shall now generalize this notion and introduce algebras $\mathscr{A}_\tau$ classifying events which happen before a stopping time τ. We first extend the class of equivalence relations $\sim_t$ in (3) by letting

$$(9) \qquad \omega \sim_\tau \omega' \qquad \text{if and only if} \quad \omega \sim_{\tau(\omega)} \omega'$$

when τ is a stopping time. Note that if $\omega \sim_{\tau(\omega)} \omega'$, then $\tau(\omega') = \tau(\omega)$ and hence $\sim_\tau$ is really an equivalence relation. We let $\mathscr{A}_\tau$ be the internal algebra generated by the equivalence classes of $\sim_\tau$.

Assume now that we are given an increasing sequence $\{\tau_n\}_{n \leq \gamma}$ of internal stopping times, and let M_{τ_n} be the random variable $M_{\tau_n}(\omega) = M(\omega, \tau_n(\omega))$. It is trivial to check that $(\omega, n) \mapsto M_{\tau_n}(\omega)$ is a martingale with respect to the filtration $(\Omega, \{\mathscr{A}_{\tau_n}\}, P)$, and hence by (7) we have

$$(10) \qquad E(M_{\tau_\gamma}^2) = E\left(M_0^2 + \sum_{n=0}^{\gamma-1} (M_{\tau_{n+1}} - M_{\tau_n})^2 \right),$$

where $\tau_0 \equiv 0$.

4.2.10. PROPOSITION. If M is a local λ^2-martingale, then almost all M's paths have S-right and S-left limits at each $t \in [0, 1]$.

PROOF. Without loss of generality, we may assume that M is a λ^2-martingale. Since we already know that M is finite almost everywhere, the proposition can only fail if M oscillates too much, i.e., if the set

$$\bigcup_{\substack{a,b \in \mathbb{Q} \\ a < b}} \{\omega \,|\, \text{The path } M(\omega, \cdot) \text{ crosses the interval } [a, b] \text{ infinitely many times}\}$$

has positive probability. Since there are only countably many pairs of rationals, this implies that we can find $a, b \in \mathbb{Q}$, $a < b$, such that M crosses $[a, b]$ infinitely many times with positive probability. Define a sequence $\{\tau_n\}$ of stopping times as follows. Let $\tau_0 = 0$ and for k odd define

$$\tau_k(\omega) = \inf\{t > \tau_{k-1}(\omega) \,|\, X(\omega, t) \leq a\} \wedge 1.$$

Similarly, if k is even, we let

$$\tau_k(\omega) = \inf\{t > \tau_{k-1}(\omega) \,|\, X(\omega, t) \geq b\} \wedge 1.$$

Since the sequence $\{\tau_n\}$ is strictly increasing until it reaches one, we see that if γ is the number of elements in the time line, then τ_γ is identically one. By (10), we thus have

$$E(M(1)^2) = E(M_{\tau_\gamma}^2) = E\left(M_0^2 + \sum_{n=0}^{\gamma-1} (M_{\tau_{n+1}} - M_{\tau_n})^2 \right).$$

The left-hand side of this equation is finite by assumption, while the sum of the right must be infinite on a set of positive measure. Hence we have a contradiction and the theorem is proved.

Hyperfinite processes with one-sided S limits can be turned into standard processes as follows.

4.2.11. DEFINITION. Let $X : \Omega \times T \to {}^*\mathbb{R}$ be a hyperfinite process with S-left and S-right limits a.e. The standard process ${}^\circ X^+ : \Omega \times [0, 1] \to \mathbb{R}$ defined by

$$
{}^\circ X^+(\omega, t) = S\text{-}\lim_{s \downarrow t} X(\omega, s)
$$

is called *the right standard part of X*. The process

$$
{}^\circ X^-(\omega, t) = S\text{-}\lim_{s \uparrow t} X(\omega, s)
$$

is called *the left standard part of X*.

Of these two standard part processes, ${}^\circ X^+$ is by far the most important one and we shall often refer to it simply as the standard part of X. Indeed, Hoover and Perkins (1983a) and Stroyan and Bayod (1985) observed that ${}^\circ X^+$ can be considered as the standard part of X in the Skorohod topology [see, e.g., Billingsley (1968)] on the space D of right continuous functions with left limits. For these reasons most authors denote ${}^\circ X^+$ simply by ${}^\circ X$, but with the notation we have been using, this would make it impossible to distinguish between ${}^\circ X(t)$ (the standard part of the value of X at an internal time t) and ${}^\circ X^+(t)$ (the value of ${}^\circ X^+$ at the standard time t).

Having shown that the local λ^2-martingales have reasonable path properties, we are ready to discuss stochastic integration. However, we still have the postponed proof of Doob's inequality to attend to, and this seems as good a time as any. First a lemma:

4.2.12. LEMMA. Let U and V be two internal, positive maps from Ω to ${}^*\mathbb{R}$. Assume that $p, \alpha \in {}^*\mathbb{R}$ are such that $p > \alpha$, $p > 1$, $\alpha > 0$, and that for all positive $\xi \in {}^*\mathbb{R}$

$$
\tag{11} \xi^\alpha P[U > \xi] \le \int_{\{U > \xi\}} V^\alpha \, dP.
$$

Then

$$
\tag{12} E(U^p) \le \left(\frac{p}{p - \alpha} \right)^{p/\alpha} E(V^p).
$$

PROOF. Let μ be the distribution of U (i.e., $\mu(A) = P[U \in A]$). By straightforward calculations

$$
E(U^p) = \int_0^\infty y^p \, d\mu(y) = \int_0^\infty \left(\int_0^y p\xi^{p-1} \, d\xi \right) d\mu(y)
$$

$$
= \int_0^\infty \left(\int_\xi^\infty p\xi^{p-1} \, d\mu(y) \right) d\xi = \int_0^\infty p\xi^{p-1} \mu[y > \xi] \, d\xi
$$

$$
= \int_0^\infty p\xi^{p-1} P[U > \xi] \, d\xi \le \int_0^\infty p\xi^{p-1-\alpha} \left(\int_{\{U > \xi\}} V^\alpha \, dP \right) d\xi,
$$

where the last step uses (11). Continuing our calculations, we see that

$$E(U^p) \leq \int_0^\infty p\xi^{p-1-\alpha}\left(\int_{\{U>\xi\}} V^\alpha \, dP\right) d\xi = \int \left(\int_0^U p\xi^{p-1-\alpha} \, d\xi\right) V^\alpha \, dP$$

$$= \int \frac{p}{p-\alpha} U^{p-\alpha}V^\alpha \, dP \leq \frac{p}{p-\alpha} E(U^p)^{1-\alpha/p}E(V^p)^{\alpha/p}$$

by Hölder's inequality. Dividing by $E(U^p)^{1-\alpha/p}$ and raising both sides to the p/α-th power, the lemma follows.

To prove Doob's inequality, we apply the lemma with $U = \sup_{s \leq t} X_s$, $V = X_t$ and $\alpha = 1$. All we have to do is to check (11).

Let $\xi > 0$ and define a stopping time τ by

$$\tau(\omega) = \inf\{s \in T \mid X(\omega, s) > \xi\} \wedge 2.$$

Note that since $t < 2$, we have $\{\sup_{s \leq t} X_s > \xi\} = \{\tau \leq t\}$. Thus

$$\xi P[\sup_{s \leq t} X_s > \xi] = \xi P\{\tau \leq t\} \leq \int_{\{\tau \leq t\}} X_\tau \, dP$$

$$= \int (X_{\tau \wedge t} - X_t) \, dP + \int_{\{\tau \leq t\}} X_t \, dP \leq \int_{\{\tau \leq t\}} X_t \, dP,$$

where the last step uses the fact that X is a submartingale. Hence

$$\xi P[\sup_{s \leq t} X_s > \xi] \leq \int_{\{\sup_{s \leq t} X_s > \xi\}} X_t \, dP$$

and Doob's inequality follows.

What we have stated and proved is, of course, a nonstandard version of Doob's result, but we shall not hesitate to use the original, standard version whenever it is convenient.

B. Martingale Integration

In the first section of this chapter we showed that if χ is Anderson's random walk and λ is the normalized counting measure on T, then the integral $\int X \, d\chi$ is well behaved if X is nonanticipating and S-square-integrable with respect to $P \times \lambda$. When we now turn to integration with respect to λ^2-martingales, the integrability condition on the integrand will no longer refer to the measure $P \times \lambda$ but to a measure constructed from the martingale we integrate with respect to. If M is a λ^2-martingale, we let ν_M be the internal measure on $\Omega \times T$ defined by

$$(13) \qquad \nu_M\{(\omega, t)\} = \Delta M(\omega, t)^2 P\{\omega\}.$$

Note that $\nu_M(\Omega \times T) = E([M](1))$ is finite, and that $\nu_\chi = P \times \lambda$ for Anderson's process χ.

Two natural classes of integrands are defined as follows:

4.2.13. DEFINITION. Let M be a λ^2-martingale. A hyperfinite process X belongs to the class $SL^2(M)$ if it is nonanticipating and square S-integrable with respect to ν_M.

If M is a local λ^2-martingale, then X belongs to $SL(M)$ if it is nonanticipating and in $SL^2(M_{\tau_n})$ for all τ_n in a localizing sequence for M.

The reason why ν_M is the right measure to use in this definition should become clear from the proof of the next proposition.

4.2.14. PROPOSITION. If M is a λ^2-martingale and $X \in SL^2(M)$, then $\int X\, dM$ is a λ^2-martingale. If M is a local λ^2-martingale and X is in $SL(M)$, then $\int X\, dM$ is a local λ^2-martingale.

PROOF. The second half of the proposition is an immediate consequence of the first. Thus assume that M is a λ^2-martingale and $X \in SL^2(M)$. Applying (7) to the martingale $\int X\, dM$, we get

$$E\left(\left(\int_0^1 X\, dM\right)^2\right) = E\left(\left[\int_0^1 X\, dM\right]\right) = E\left(\sum_0^1 X^2\, \Delta M^2\right) = \int X^2\, d\nu_M,$$

which is finite by assumption.

Since we already know that local λ^2-martingales are nearstandard, this proposition tells us that we have obtained a reasonable integration theory. Turning to the deeper aspects of this theory, we first recall that an internal function $f: T \mapsto {}^*\mathbb{R}$ is S-continuous if $f(s) \approx f(t)$ whenever $s \approx t$, and each $f(t)$ is nearstandard. A process $X: \Omega \times T \to {}^*\mathbb{R}$ is S-continuous if almost all its paths are. The result we are aiming at is:

4.2.15. THEOREM. If M is an S-continuous local λ^2-martingale and $X \in SL(M)$, then $\int X\, dM$ is also S-continuous.

To prove it, we shall use the following characterization of S-continuous martingales:

4.2.16. THEOREM. A local λ^2-martingale is S-continuous if and only if its quadratic variation is.

This is perhaps the most important example of the interplay between a martingale and its quadratic variation. As a first illustration of its usefulness, we apply it to the hyperfinite random walk χ: since $[\chi](t) = t$ obviously is S-continuous, the theorem tells us that χ itself is S-continuous! A second illustration is provided by the following proof of Theorem 4.2.15:

PROOF OF 4.2.15. It suffices to prove the theorem when M is a λ^2-martingale and $X \in SL^2(M)$. Let us first assume that X is bounded by a real number n. Since M is S-continuous, we know from 4.2.16 that the quadratic variation $[M]$ is S-continuous, and since

$$\left[\int X\,dM\right](t) - \left[\int X\,dM\right](s) = \sum_s^t X^2\,\Delta M^2 \le n^2 \sum_s^t \Delta M^2$$

$$= n^2([M](t) - [M](s)),$$

this implies that $[\int X\,dM]$ is S-continuous. Using 4.2.16 for a second time, we get the S-continuity of $\int X\,dM$.

The idea of the proof is the argument we just gave; to extend the result to general $X \in SL^2(M)$ is just an exercise in measure theory. To carry it out, note that since X is square S-integrable, there is a sequence $\{X_n\}$ of S-bounded functions such that $^\circ\!\int (X - X_n)^2\,d\nu_M \to 0$. By Doob's inequality

$$0 \le E\left(^\circ\!\max_{s\le 1}\left(\int_0^s X\,dM - \int_0^s X_n\,dM\right)^2\right)$$

$$\le 4\,^\circ\!E\left(\left(\int_0^1 (X - X_n)\,dM\right)^2\right) = 4\,^\circ\!\int (X - X_n)^2\,d\nu_M \to 0.$$

Ordinary measure theory tells us that there is a subsequence $\{^\circ\!\max_{s\le 1}(\int_0^s X\,dM - \int_0^s X_{n_k}\,dM)\}_{k\in\mathbb{N}}$ converging to zero almost everywhere. Since each $\int X_{n_k}\,dM$ is S-continuous, the uniform limit $\int X\,dM$ must also be S-continuous, and the proof is complete.

Observe the important part played by the S-integrability of X^2 in this proof; it is not sufficient to assume that $\int X^2\,d\nu_M$ is finite as we need to approximate X by S-bounded functions X_n. Indeed, it is almost trivial to find an example which shows that the theorem is false under the weaker condition. This is rather typical of the theory and is the reason why we require the integrands to be square S-integrable despite the fact that a few simple results (such as 4.2.14) only need the finiteness of $\int X^2\,d\nu_M$.

As almost all our applications will be concerned with continuous processes, Theorems 4.2.15 and 4.2.16 are of the utmost importance to us. Each time we have constructed a process we need to check that it is continuous, and 4.2.15 and 4.2.16 are the perfect tools for this task.

The remainder of this section is devoted to the proof of Theorem 4.2.16. To get the necessary estimates, we shall use the following inequalities.

4.2.17. LEMMA. There exist constants C, $K \in \mathbb{R}_+$ such that for all hyperfinite martingales $M : \Omega \times T \to {}^*\mathbb{R}$ with $M(0) = 0$,

$$(14) \qquad CE(\max_{s\le t} M(s)^4) \le E([M](t)^2) \le KE(\max_{s\le t} M(s)^4).$$

PROOF. By Doob's inequality and simple algebra

$$E(\max_{s \le t} M(s)^4) \le (\tfrac{4}{3})^4 E(M(t)^4)$$

$$= (\tfrac{4}{3})^4 E\left(\sum_{s=0}^{t} \{(M(s) + \Delta M(s))^4 - M(s)^4\}\right)$$

$$= (\tfrac{4}{3})^4 E\left(\sum_{s=0}^{t} \{4M(s)^3 \Delta M(s) + 6M(s)^2 \Delta M(s)^2\right.$$

$$\left. + 4M(s) \Delta M(s)^3 + \Delta M(s)^4\}\right).$$

Using that $E(M(s)^3 \Delta M(s)) = 0$ and $|\Delta M(s)| \le 2 \max_{r \le t} |M(r)|$, we get from this that

$$E(\max_{s \le t} M(s)^4) \le (\tfrac{4}{3})^4 E(6 \max_{s \le t} M(s)^2[M](t) + 8 \max_{s \le t} M(s)^2[M](t)$$

$$+ 4 \max_{s \le t} M(s)^2[M](t))$$

$$= 18(\tfrac{4}{3})^4 E(\max_{s \le t} M(s)^2[M](t))$$

$$\le 18(\tfrac{4}{3})^4 E(\max_{s \le t} M(s)^4)^{1/2} E([M](t)^2)^{1/2},$$

where the last step is Hölder's inequality. Dividing by $E(\max_{s \le t} M(s)^4)^{1/2}$, the first half of (14) follows.

To prove the second half, we first use Lemma 4.2.5 and Hölder's inequality to get

$$E([M](t)^2) = E\left(\left(M_t^2 - 2\int_0^t M \, dM\right)^2\right)$$

$$= E\left(M_t^4 - 4M_t^2 \int_0^t M \, dM + 4\left(\int_0^t M \, dM\right)^2\right)$$

(15)

$$\le E(\max_{s \le t} M_s^4) + 4E(\max_{s \le t} M_s^4)^{1/2} E\left(\left(\int_0^t M \, dM\right)^2\right)^{1/2}$$

$$+ 4E\left(\left(\int_0^t M \, dM\right)^2\right),$$

where we have replaced M_t^4 by the larger quantity $\max_{s \le t} M_s^4$ in the first two terms. Since $\int M \, dM$ is a martingale,

$$E\left(\left(\int_0^t M \, dM\right)^2\right) = E\left(\left[\int M \, dM\right](t)\right) = E\left(\sum_0^t M^2 \Delta M^2\right)$$

$$\le E(\max_{s \le t} M(s)^2[M](t))$$

$$\leq E(\max_{s \leq t} M(s)^4)^{1/2} E([M](t)^2)^{1/2},$$

and putting this into (15) we get

(16)
$$E([M](t)^2) \leq E(\max_{s \leq t} M_s^4) + 4E(\max_{s \leq t} M(s)^4)^{3/4} E([M](t)^2)^{1/4}$$
$$+ 4E(\max_{s \leq t} M(s)^4)^{1/2} E([M](t)^2)^{1/2}.$$

By the part of the lemma we have already proved $E(\max_{s \leq t} M_s^4) \leq (1/C)E([M](t)^2]$, and making use of this, we turn (16) into

(17)
$$E([M](t)^2) \leq \frac{1}{C^{1/2}} E(\max_{s \leq t} M_s^4)^{1/2} E([M](t)^2)^{1/2}$$
$$+ \frac{4}{C^{1/4}} E(\max_{s \leq t} M_s^4)^{1/2} E([M](t)^2)^{1/2}$$
$$+ 4E(\max_{s \leq t} M(s)^4)^{1/2} E([M](t)^2)^{1/2}$$
$$= \left(\frac{1}{C^{1/2}} + \frac{4}{C^{1/4}} + 4 \right) E(\max_{s \leq t} M_s^4)^{1/2} E([M](t)^2)^{1/2}.$$

Dividing by $E([M](t)^2)^{1/2}$, we have proved the lemma.

The inequalities (14) are very simple special cases of the famous Burkholder–Davis–Gundy inequalities, which appear in the literature in various forms. A fairly general version is that

(18)
$$\left\| \sqrt{M_0^2 + [M](t)} \right\|_p \leq 10p \left\| \max_{s \leq t} M_s \right\|_p$$

and

(19)
$$\left\| \max_{s \leq t} M_s \right\|_p \leq p\sqrt{12} \left\| \sqrt{M_0^2 + [M](t)} \right\|_p$$

for all $p \in (1, \infty)$. To prove these results one needs much more sophisticated methods than we used in our proof of 4.2.17; see, e.g., Neveu (1975).

Before we can use our inequalities to prove Theorem 4.2.16, we have one more problem to solve. Assume that we want to stop a martingale before it grows too large. The natural approach is to use a stopping time

$$\tau_K = \min\{t \in T \,||M(t)| \geq K\}$$

for some $K \in \mathbb{R}_+$, but since the last increment before τ_K could be enormous, we might still stop the martingale too late. Let us say that M has *infinitesimal increments* if $\Delta M(\omega, t) \approx 0$ for all ω and t; if this is the case, clearly $^{\circ}|M_{\tau_K}| \leq K$. The next lemma tells us that if one of the processes M, $[M]$ is

S-continuous, then there is a martingale with infinitesimal increments which is almost identical to M.

4.2.18. LEMMA. Let M be a λ^2-martingale such that the set

$$\{\omega \in \Omega \mid \exists t \in T(^\circ \Delta M(\omega, t) \neq 0)\}$$

has Loeb measure zero. Then there is a λ^2-martingale $\tilde{M}$ with infinitesimal increments such that on a set of Loeb measure one

$$\tilde{M}(t) \approx M(t) \quad \text{and} \quad [\tilde{M}](t) \approx [M](t)$$

for all t.

PROOF. For each $n \in {}^*\mathbb{N}$, let $\Omega_n = \{\omega \mid \exists t(|\Delta M(\omega, t)| \geq 1/n)\}$. Since the internal set $A = \{n \in {}^*\mathbb{N} \mid P(\Omega_n) \leq 1/n\}$ contains $\mathbb{N}$, it must have an infinite member η. For $\omega \in \Omega_\eta$, let t_ω be the first t such that $|\Delta M(\omega, t)| \geq 1/\eta$, and put $t_\omega = 1$ for all ω not in Ω_η.

If $\sim_t$ are the equivalence relations in (3), we let $[\omega]_t$ denote the partition class of ω under $\sim_t$. Introduce

$$[\omega]_t^+ = \{\tilde{\omega} \in [\omega]_t \mid t_{\tilde{\omega}} \leq t\},$$

and note that if $t > t_{\tilde{\omega}}$ for some $\tilde{\omega} \in [\omega]_t$, then $[\omega]_t^+ = [\omega]_t$.

We first modify M by cutting away the increments which are larger than $1/\eta$: let K be the internal process defined by $K(0) = M(0)$ and

$$\Delta K(\omega, t) = \begin{cases} \Delta M(\omega, t) & \text{if } t < t_\omega \\ 0 & \text{if } t \geq t_\omega. \end{cases}$$

K is usually not a martingale, but if we add the process N given by $N(0) = 0$,

$$\Delta N(\omega, t) P([\omega]_t) = \int_{[\omega]_t^+} \Delta M(t) \, dP,$$

then $\bar{M} = K + N$ is a martingale.

The crucial observation is that the sum $\sum_{t=0}^1 |\Delta N(\omega, t)|$ is infinitesimal a.e. To see this, introduce the set $\Sigma \subset \Omega \times \Omega \times T$ consisting of those triplets $(\omega, \tilde{\omega}, t)$ such that $t = t_{\tilde{\omega}} < 1$ and $\omega \sim_t \tilde{\omega}$, and make the following computation:

$$E\left(\sum_{t=0}^1 |\Delta N(\omega, t)| \right) = \sum_{\omega \in \Omega} \sum_{t=0}^1 \left| \sum_{\tilde{\omega} \in [\omega]_t^+} \frac{\Delta M(\tilde{\omega}, t) P\{\tilde{\omega}\}}{P([\omega]_t)} \right| P\{\omega\}$$

$$\leq \sum_{(\omega, \tilde{\omega}, t) \in \Sigma} \frac{|\Delta M(\tilde{\omega}, t)| P\{\tilde{\omega}\}}{P([\omega]_t)} P\{\omega\}$$

$$= \sum_{\tilde{\omega} \in \Omega_\eta} \sum_{\omega \in [\tilde{\omega}]_{t_{\tilde{\omega}}}} \frac{|\Delta M(\tilde{\omega}, t_{\tilde{\omega}})| P\{\tilde{\omega}\}}{P([\omega]_{t_{\tilde{\omega}}})} P\{\omega\}$$

$$= \int_{\Omega_\eta} |\Delta M(\tilde{\omega}, t_{\tilde{\omega}})| \, dP \le 2 \int_{\Omega_\eta} \max_{s \le 1} |M(s)| \, dP,$$

which is infinitesimal since $P(\Omega_\eta) = 0$ and $\max_{s \le 1} |M(s)|$ is S-integrable.

On the subset of $\Omega - \Omega_\eta$ where $\sum |\Delta N(s)|$ is infinitesimal, we obviously have $\bar{M}(t) \approx M(t)$ for all t. Moreover, on $\Omega - \Omega_\eta$

$$[\bar{M}](t) - [M](t) = \sum_{s=0}^{t} (2M(s) + \Delta N(s)) \, \Delta N(s),$$

which must be infinitesimal for all t and almost all ω. Since $\bar{M}$ is a λ^2-martingale, it looks like a promising candidate for $\tilde{M}$.

There is one small problem, however; $\bar{M}$ need not have infinitesimal increments since ΔN may be noninfinitesimal. The cure is simple; let γ be an infinite element of the set

$$\left\{ n \in {}^*\mathbb{N} \,\Big|\, P\left\{ \omega \,\Big|\, \sum_{s=0}^{1} |\Delta N(\omega, s)| > \frac{1}{n} \right\} < \frac{1}{n} \right\},$$

and define $\tau : \Omega \to T$ by

$$\tau(\omega) = \min\left\{ t \in T \,\Big|\, \Delta N(\omega, t) > \frac{1}{\gamma} \right\} \wedge 1.$$

Since $\Delta N(\omega, t)$ only depends on the equivalence class $[\omega]_t$, τ is a stopping time. Using that $\tau = 1$ almost everywhere, it is trivial to check that the stopped process $\tilde{M} = \bar{M}_\tau$ satisfies the lemma.

With these preparations it is now quite easy to prove the main theorem.

PROOF OF THEOREM 4.2.16. Without loss of generality, we may restrict ourselves to λ^2-martingales M. By Lemma 4.2.18 it suffices to consider the case where M has infinitesimal increments, and using stopping times

$$\tau_n(\omega) = \min\{t \in T \,|\, |M(\omega, t)| \ge n \text{ or } [M](\omega, t) \ge n\},$$

we may thus assume that M and $[M]$ are S-bounded.

(i) Let us first assume that $[M]$ is S-continuous. For each pair $(m, n) \in {}^*\mathbb{N}$ define a subset $A_{m,n}$ of Ω by

$$A_{m,n} = \left\{ \omega \in \Omega \,\Big|\, \exists i \in {}^*\mathbb{N} \left(\sup_{(\bar{i}/n) \le s \le \overline{(i+1/n)}} \left(M(s) - M\left(\frac{\bar{i}}{n}\right) \right)^4 \ge \frac{1}{m} \right) \right\},$$

where $\bar{r}$ denotes the smallest element in T larger than or equal to r. To prove that M is S-continuous, we must show that $A = \bigcup_{m \in \mathbb{N}} \bigcap_{n \in \mathbb{N}} A_{m,n}$

has measure zero, and for this it is sufficient to show that $P(A_{m,\gamma}) \approx 0$ for all $m \in \mathbb{N}$, $\gamma \in {}^*\mathbb{N} - \mathbb{N}$. But

$$0 \le P(A_{m,\gamma}) \le \sum_{i < \gamma} P\left\{ \omega \,\Big|\, \sup_{(\bar{i}/\gamma) \le s \le \overline{(i+1}/\gamma)} \left(M(s) - M\left(\frac{\bar{i}}{\gamma}\right) \right)^4 \ge \frac{1}{m} \right\}$$

$$\le m \sum_{i < \gamma} E\left(\sup_{(\bar{i}/\gamma) \le s \le \overline{(i+1}/\gamma)} \left(M(s) - M\left(\frac{\bar{i}}{\gamma}\right) \right)^4 \right)$$

$$\le \frac{m}{C} \sum_{i < \gamma} E\left(\left([M]\left(\frac{\overline{i+1}}{\gamma}\right) - [M]\left(\frac{\bar{i}}{\gamma}\right) \right)^2 \right),$$

where the last step uses Lemma 4.2.17 applied to the martingales $M(s) - M(\bar{i}/\gamma)$. Since the quadratic variation is S-continuous and finite, the last expectation is infinitesimal and so is $P(A_{m,\gamma})$.

(ii) We now assume that M is S-continuous and define

$$B_{m,n} = \left\{ \omega \in \Omega \,\Big|\, \exists i \in {}^*\mathbb{N}\left(\left([M]\left(\frac{\overline{i+1}}{n}\right) - [M]\left(\frac{\bar{i}}{n}\right) \right)^2 \ge \frac{1}{m} \right) \right\}.$$

As above it suffices to show that $P(B_{m,\gamma}) \approx 0$ for all $m \in \mathbb{N}$ and $\gamma \in {}^*\mathbb{N} - \mathbb{N}$.

Fix $\gamma \in {}^*\mathbb{N} - \mathbb{N}$, and let N be the restriction of M to the time line $S = \{(\bar{i}/\gamma) \mid i \le \gamma\}$. By stopping M before $[N]$ gets too large if necessary, we may assume that $[N]$ is S-bounded (this uses the fact that M has infinitesimal increments). Using 4.2.17 and Doob's inequality, we have

$$0 \le P(B_{m,\gamma}) \le \sum_{i < \gamma} P\left\{ \omega \,\Big|\, \left([M]\left(\frac{\overline{i+1}}{\gamma}\right) - [M]\left(\frac{\bar{i}}{\gamma}\right) \right)^2 > \frac{1}{m} \right\}$$

$$\le m \sum_{i < \gamma} E\left(\left([M]\left(\frac{\overline{i+1}}{\gamma}\right) - [M]\left(\frac{\bar{i}}{\gamma}\right) \right)^2 \right)$$

$$\le mK \sum_{i < \gamma} E\left(\max_{(\bar{i}/\gamma) \le s \le \overline{(i+1}/\gamma)} \left(M(s) - M\left(\frac{\bar{i}}{\gamma}\right) \right)^4 \right)$$

$$\le mK\left(\tfrac{4}{3}\right)^4 \sum_{i < \gamma} E\left(\left(M\left(\frac{\overline{i+1}}{\gamma}\right) - M\left(\frac{\bar{i}}{\gamma}\right) \right)^4 \right)$$

$$\le mK\left(\tfrac{4}{3}\right)^4 E\left\{ \max_{i < \gamma} \left(\left(M\left(\frac{\overline{i+1}}{\gamma}\right) - M\left(\frac{\bar{i}}{\gamma}\right) \right)^2 \right) [N](1) \right\},$$

which is infinitesimal since $[N]$ is S-bounded and M is S-continuous. This completes the proof.

The theory we have presented above is due independently to Hoover and Perkins (1983a,b) and Lindstrøm (1980a). Our exposition follows Lindstrøm (1980a) quite closely; the Hoover–Perkins approach is based on a systematic use of the Burkholder–Davis–Gundy inequalities and leads to results which are slightly stronger than the ones we have arrived at; basically, they only need local S-integrability where we have required local square S-integrability. Hoover and Perkins also give an alternative formulation of Theorem 4.2.16 which is more convenient to use in certain applications. The continuity theorem 4.2.15 for stochastic integrals has several ancestors and relatives in the literature; Anderson (1976) proved the theorem for M a hyperfinite random walk and X a lifting, and Keisler (1984) removed the lifting hypothesis from X and at the same time estimated the modulus of continuity of the stochastic integral. Indeed, the papers by Anderson and Keisler give a complete theory of stochastic integration with respect to Anderson's random walk, and much of the first part of this section is a straightforward generalization of their work to a martingale setting [see also the thesis of Panetta (1978), which contains the first results on integration with respect to S-continuous martingales]. This generalization would, of course, have been much less "straightforward" if the standard theory of stochastic integration had not told us what to look for and provided us with the necessary inequalities. There are several good introductions to standard stochastic integration available, e.g., Meyer's classic "course on stochastic integrals" (1976) and the books by Metivier and Pellaumail (1980), Chung and Williams (1983), and Ikeda and Watanabe (1981). As for the nonstandard theory, the book by Stroyan and Bayod (1985) gives an account of the Hoover–Perkins approach and also adds many new contributions to the field, while Osswald's (1985) lecture note is a careful and detailed exposition of the approach we have followed.

Martingale theory is much more than stochastic integration; for nonstandard contributions to other aspects of it, see Helms and Loeb (1982), Hoover (1984), Hoover and Keisler (1984), and Perkins (1982) [extending results of Barlow (1981)].

4.3. LIFTING THEOREMS

A lifting theorem gives an internal approximation to an external object and is thus an important technical tool in nonstandard praxis. Lifting theorems also provide simple characterizations of classes of processes and are therefore of equal importance for nonstandard theory. We shall develop both themes.

A. Nonanticipating Liftings

We recall from Theorem 3.2.4 that a function f from a hyperfinite probability space Ω to $\mathbb{R}$ is Loeb measurable if and only if it has a lifting

$F: \Omega \to {}^*\mathbb{R}$, and that a function $f: [0, 1] \to \mathbb{R}$ is Lebesgue measurable if and only if it has a lifting $F: T \to {}^*\mathbb{R}$, where T is a hyperfinite time line.

A stochastic process is a map of two variables

$$x: \Omega \times [0, 1] \to \mathbb{R},$$

and we need to combine the two separate lifting theorems into one result asserting the existence of some suitable hyperfinite process

$$X: \Omega \times T \to {}^*\mathbb{R}.$$

We started this discussion in Section 4.1 and shall continue here. Let us make one preliminary remark; Theorem 4.1.6 would lose some of its force if Theorem 4.1.5 did not assert the existence of "enough" liftings of the appropriate kind.

Before we can begin to prove lifting theorems, we must introduce the basic concepts of stochastic analysis.

A *stochastic filtration* is a tuple $\langle \Omega, \{\mathscr{B}_t\}_{t \in [0,1]}, Q \rangle$, where $\{\mathscr{B}_t\}_{t \in [0,1]}$ is an increasing family of σ-algebras on a set Ω, and Q is a probability measure on $\mathscr{B}_1$. Let $\langle \Omega, \{\mathscr{A}_t\}_{t \in T}, P \rangle$ be an internal filtration as defined in 4.2.3. The *stochastic filtration generated by* $\langle \Omega, \{\mathscr{A}_t\}_{t \in T}, P \rangle$ consists of Ω, the Loeb measure $L(P)$, and the σ-algebras

$$(1) \qquad \mathscr{B}_t = \sigma\left(\bigcup_{s \approx t} L(\mathscr{A}_s) \cup \mathscr{N} \right),$$

where $\mathscr{N}$ consists of the null sets in $L(\mathscr{A})$.

A stochastic filtration $\langle \Omega, \{\mathscr{B}_1\}_{t \in [0,1]}, Q \rangle$ is said to satisfy the *usual conditions* if each $\mathscr{B}_t$ contains all the null sets of $\mathscr{B}_1$, and for all $t \in [0, 1\rangle$

$$(2) \qquad \mathscr{B}_t = \bigcap_{s > t} \mathscr{B}_s.$$

The next lemma implies that all internally generated filtrations satisfy these conditions.

4.3.1. LEMMA. Let $\langle \Omega, \{\mathscr{B}_t\}_{t \in [0,1]}, L(P) \rangle$ be the stochastic filtration generated by $\langle \Omega, \{\mathscr{A}_t\}_{t \in T}, P \rangle$. Then for all $t \in [0, 1]$,

$$(3) \qquad \mathscr{B}_t = \bigcup_{s \approx t} \sigma(L(\mathscr{A}_s) \cup \mathscr{N}).$$

PROOF.   Obviously

$$\bigcup_{s \approx t} \sigma(L(\mathscr{A}_s) \cup \mathscr{N}) \subset \mathscr{B}_t \subset \sigma\left(\bigcup_{s \approx t} \sigma(L(\mathscr{A}_s) \cup \mathscr{N}) \right),$$

and it is thus enough to prove that $\bigcup_{s \approx t} \sigma(L(\mathscr{A}_s) \cup \mathscr{N})$ is a σ-algebra.

Let $\{A_n\}_{n\in\mathbb{N}}$ be a countable family of sets from $\bigcup_{s\approx t}\sigma(L(\mathcal{A}_s)\cup\mathcal{N})$, and assume $A_n\in\sigma(L(\mathcal{A}_{s_n})\cup\mathcal{N})$. The family $S_n=[s_n, s_n+1/n]\cap T$ is countable and has the finite intersection property, hence by saturation $\bigcap_{n\in\mathbb{N}}S_n\neq\varnothing$. If $\tilde{s}\in\bigcap_{n\in\mathbb{N}}S_n$, then $\tilde{s}\approx t$ and $\bigcup_{n\in\mathbb{N}}A_n\in\sigma(L(\mathcal{A}_{\tilde{s}})\cup\mathcal{N})\subset\bigcup_{s\approx t}\sigma(L(\mathcal{A}_s)\cup\mathcal{N})$. This shows that our family is closed under countable unions, and as it obviously has the other properties of a σ-algebra, the lemma is proved.

From (3) and (1), we immediately get

4.3.2. COROLLARY. A stochastic filtration generated by an internal filtration satisfies the usual conditions.

There are several standard attempts to capture the concept of a "nonanticipating" process. Unfortunately, there seems to be no natural candidate; different formulations give rise to different notions and not all are suitable for the same purposes. We shall concentrate on two of these notions—adapted processes and predictable ones.

First recall that a *measurable rectangle* is a subset of $\Omega\times[0,1]$ of the form $B\times[s, t]$, where B is Loeb measurable. A set is *measurable* if it is in the σ-algebra generated by the measurable rectangles. Let $\langle\Omega, \{\mathcal{B}_t\}_{t\in[0,1]}, Q\rangle$ be a stochastic filtration.

4.3.3. DEFINITION. A set $A\subset\Omega\times[0,1]$ is called *adapted* (with respect to $\{\mathcal{B}_t\}$) if A is measurable and each section $A_t=\{\omega\mid(\omega, t)\in A\}$ is $\mathcal{B}_t$-measurable.

4.3.4. DEFINITION. A *predictable rectangle* (with respect to $\{\mathcal{B}_t\}$) is a set of the form $B_s\times(s, t]$, where $B_s\in\mathcal{B}_s$, or $B_0\times[0, t]$, where $B_0\in\mathcal{B}_0$. A set is called *predictable* if it is in the σ-algebra generated by the predictable rectangles.

A process $X:\Omega\times[0,1]\to\mathbb{R}$ is *adapted* if it is measurable with respect to the σ-algebra of adapted sets. Let μ be a measure on $\Omega\times[0,1]$ defined on an extension of the measurable sets. A set A is *almost surely adapted* with respect to μ if there is an adapted set B such that $\mu(A\triangle B)=0$, and a process is *almost surely adapted* if it is equal almost everywhere to an adapted process. We use similar terminology for measurable and predictable sets and processes.

It is clear that all almost surely predictable processes are almost surely adapted. The opposite is not true in general.

We shall be mostly interested in the case where μ is defined from an internal measure ν on $\Omega\times T$ by

$$(4)\qquad\qquad \mu=L(\nu)\circ(\mathrm{id}\times\mathrm{st})^{-1}$$

(here id is the identity map on Ω). The situation we have in mind is $\nu = \nu_M$, where ν_M is the measure derived from a martingale M as in (4.2.13). A special case is $\nu = P \times \lambda$, where λ is the *uniform measure* on T [i.e., $\lambda(\{t_i\}) = t_{i+1} - t_i$], then μ is the completed product of $L(P)$ and the Lebesgue measure. This is the case we discussed in Section 4.1.

For convenience we introduce the notation

(5) $$\text{St} = \text{id} \times \text{st}: \Omega \times T \to \Omega \times [0, 1].$$

From now on $\langle \Omega, \{\mathscr{A}_t\}_{t \in T}, P \rangle$ will be an internal filtration generating $\langle \Omega, \{\mathscr{B}_t\}_{t \in [0,1]}, L(P) \rangle$; ν will be an internal measure defined on all internal subsets of $\Omega \times T$; and $\mu = L(\nu) \circ \text{St}^{-1}$. We shall assume that $\nu(\Omega \times T)$ is finite.

The measure ν is said to be *absolutely continuous* with respect to P if $L(P)(C) = 0$ implies $L(\nu)(C \times T) = 0$, and $\mu(\Omega \times \{0\}) = 0$.

4.3.5. LEMMA. Let ν be absolutely continuous with respect to P. If $B \subset \Omega \times [0, 1]$ is almost surely predictable, then there exists a nonanticipating $A \subset \Omega \times T$ such that $L(\nu)(A \triangle \text{St}^{-1}(B)) = 0$.

PROOF. It is enough to consider the case where B is a predictable rectangle. Assume first that B is of the form $B_s \times (s, t]$. By 4.3.1 we can find an $\tilde{s} \in T$, $\tilde{s} \approx s$, and an $A_{\tilde{s}} \in \mathscr{A}_{\tilde{s}}$ such that $L(P)(B_s \triangle A_{\tilde{s}}) = 0$. By the absolute continuity of ν with respect to P, we get

$$\mu(B_s \times (s, t]) = \lim_{m \to \infty} \lim_{n \to \infty} L(\nu)\left(A_{\tilde{s}} \times \left(s + \frac{1}{n}, t + \frac{1}{m}\right]\right).$$

Thus we can find $n, m \in {}^*\mathbb{N} - \mathbb{N}$ such that $s + 1/n \geq \tilde{s}$ and

$$\mu(B_s \times (s, t]) = {}^\circ \nu(A_{\tilde{s}} \times (s + (1/n), t + (1/m)]).$$

Hence we may take $A = A_{\tilde{s}} \times (s + (1/n), t + (1/m)]$.

To treat the other kind of predictable rectangle $B_0 \times [0, 1]$, we use the second clause in the definition of absolute continuity. The details are left to the reader. $\blacksquare$

In the opposite direction we have

4.3.6. LEMMA. Let $B \subset \Omega \times [0, 1]$ and assume that there is a nonanticipating $A \subset \Omega \times T$ such that $L(\nu)(A \triangle \text{St}^{-1}(B)) = 0$. Then B is almost surely adapted.

PROOF. Let ν_A be the internal measure given by

$$\nu_A(C) = \nu(A \cap C),$$

and put $\mu_A = L(\nu_A) \circ \text{St}^{-1}$. Let g be the Radon–Nikodym derivative

$$g = \partial \mu_A / \partial \mu,$$

and define

$$C = \{(\omega, t) \,|\, g(\omega, t) = 1\}.$$

Since $L(\nu)(A \triangle \mathrm{St}^{-1}(B)) = 0$, we must have $g = 1$ a.e. on B and $g = 0$ a.e. outside B. Hence $\mu(B \triangle C) = 0$, and all we have to do is to find an adapted version of g.

Define internal functions $F, F_A : \Omega \times T \to {}^*\mathbb{R}$ by

$$F(\omega, t) = \sum \{\nu(\omega, s) \,|\, s \le t\}, \qquad F_A(\omega, t) = \sum \{\nu_A(\omega, s) \,|\, s \le t\}.$$

The right standard parts $f = {}^\circ F^+$, $f_A = {}^\circ F_A^+$ are measurable since they are increasing, right continuous processes, and they are obviously $\{\mathscr{B}_t\}$-adapted. But

$$g(\omega, t) = \lim_{h \downarrow 0} \frac{f_A(\omega, t) - f_A(\omega, t - h)}{f(\omega, t) - f(\omega, t - h)}$$

and the lemma is proved.

The proof above has an immediate corollary.

4.3.7. LEMMA. Let ν be absolutely continuous with respect to P. A set $B \subset \Omega \times [0, 1]$ is almost surely measurable if and only if there is an internal subset $A \subset \Omega \times T$ such that $L(\nu)(A \triangle \mathrm{St}^{-1}(B)) = 0$.

PROOF. That an almost surely measurable set can be lifted is proved by a straightforward routine argument, which we leave to the reader.

The proof of the opposite direction is a copy of the proof of 4.3.6; just delete everything which has to do with nonanticipation.

The results in 4.3.5–4.3.7 can be expressed in terms of processes instead of sets. The following important definition generalizes 4.1.1.

4.3.8. DEFINITION. Let $x : \Omega \times [0, 1] \to \mathbb{R}$ be a stochastic process. A *lifting* of x (with respect to ν) is an internal process $X : \Omega \times T \to {}^*\mathbb{R}$ such that

$$^\circ X(\omega, t) = x(\omega, {}^\circ t) \; L(\nu)\text{-a.e.}$$

Of course, the notion of lifting generalizes to processes taking values in arbitrary Hausdorff spaces.

By 3.2.4 (i), the results above can now be reformulated as the following proposition.

4.3.9. PROPOSITION. Assume that ν is absolutely continuous with respect to P, and let $x : \Omega \times [0, 1] \to \mathbb{R}$ be a stochastic process.

  (i)   If x is almost surely predictable, then x has a nonanticipating lifting.

  (ii) If x has a nonanticipating lifting, then x is almost surely adapted.

  (iii) x is almost surely measurable if and only if it has a lifting.

A *lifting theorem* in the sense of Keisler (1984) is a result that characterizes a class of standard processes in terms of what kind of liftings they allow; 4.3.9(iii) is such a characterization of almost surely measurable processes. The two first parts of the proposition do not quite add up to a lifting theorem since we have not determined exactly which standard notion corresponds to the nonstandard one of a nonanticipating process; is it a.s. predictable, a.s. adapted, or something in between? However, in cases where these classes coincide, we do get a lifting theorem from 4.3.9(i) and (ii). We shall take a brief look at one such case.

4.3.10. LEMMA. Let m be the Lebesgue measure on $[0, 1]$, let P be a probability measure on Ω, and put $\mu = P \times m$. A process $x : \Omega \times [0, 1] \to \mathbb{R}$ is almost surely predictable with respect to μ if and only if it is almost surely adapted.

PROOF. It is enough to show that if x is adapted and bounded, then x is almost surely predictable. First observe that if a process is adapted and continuous, then it is almost surely predictable. Hence, taking

$$\delta_\varepsilon(x) = \begin{cases} -(6/\varepsilon^3)x(x - \varepsilon) & \text{for} \quad 0 < x < \varepsilon \\ 0 & \text{elsewhere} \end{cases}$$

as an approximation to the delta function, we see that

$$x_\varepsilon(t) = (x * \delta_\varepsilon)(t) = \int_0^1 x(t - s)\delta_\varepsilon(s) \, ds$$

is a.s. predictable. Since x_ε converges to x in μ-norm as $\varepsilon \to 0$, the original process x must be almost surely predictable.

4.3.11. COROLLARY. If P is an internal probability measure on Ω, λ the uniform measure on T, and $\nu = P \times \lambda$, then a process $x : \Omega \times [0, 1] \to \mathbb{R}$ is almost surely adapted with respect to $\mu = L(\nu) \circ \mathrm{St}^{-1}$ if and only if it has a nonanticipating lifting.

The reader should notice that 4.3.11 is identical to 4.1.5.

The results above are sufficient for our needs in this book; if you are interested in the deeper aspects of the theory, you should consult the work of Stroyan (1985) [see also Stroyan and Bayod (1985)], which contains detailed information on the questions raised by Proposition 4.3.9.

B. Uniform Liftings

We shall now turn our attention to the continuous processes. The appropriate notion of lifting was introduced by Keisler (1984).

4.3.12. DEFINITION. Let E and F be Hausdorff spaces and $x : \Omega \times E \to F$ a stochastic process. An internal process $X : \Omega \times {}^*E \to {}^*F$ is a *uniform lifting* of x if there is a set Ω' of measure one such that

$$ {}^\circ X(\omega, m) = x(\omega, {}^\circ m) $$

for all $\omega \in \Omega'$ and all nearstandard $m \in E$.

Recall that $x : \Omega \times E \to F$ being a stochastic process just means that $x(\cdot, m)$ is measurable for each $m \in E$. In most applications $E = [0, 1]$ and $F = \mathbb{R}^n$, but there are examples in stochastic differential equations and stochastic control theory that make it convenient to consider also the more general situation.

Keisler proved the following result when $E = [0, 1]$ and $F = \mathbb{R}^n$:

4.3.13. PROPOSITION. Assume that E and F are separable metric spaces. A stochastic process $x : \Omega \times E \to F$ is continuous if and only if it has a uniform lifting.

PROOF. Let us do the easy part first. Assume that X is a uniform lifting of x, and fix $m \in E$, $\varepsilon \in \mathbb{R}_+$, and $\omega \in \Omega'$ (where Ω' is the set where x and X agree as in Definition 4.3.12). If $y = x(\omega, m)$ and $\varepsilon \in \mathbb{R}_+$, the internal set

$$ \{ \delta \in {}^*\mathbb{R}_+ \,|\, d(y, X(\omega, m')) < \varepsilon \quad \text{whenever} \quad d(m, m') < \delta \} $$

contains all positive infinitesimals, and hence a noninfinitesimal δ_0. Consequently $d(m, m') \le \delta_0$ implies $'d(x(\omega, m), x(\omega, m')) \le \varepsilon$, and $x(\omega, \cdot)$ is continuous at m.

For the converse, the idea is as follows. Let $C(E, F)$ be the set of continuous functions from E to F, and define $\hat{x} : \Omega \to C(E, F)$ by $\hat{x}(\omega) = x(\omega, \cdot)$. Use Anderson's lifting theorem 3.2.4 for random variables to pick a lifting $\hat{X}$ of $\hat{x}$, and define $X : \Omega \times {}^*E \to {}^*F$ by $X(\omega, m) = \hat{X}(\omega)(m)$. Then X is a uniform lifting of x.

To carry out this plan we need a topology on $C(E, F)$, and in order for Anderson's lifting theorem to work, it has to be second countable. When $E = [0, 1]$, $F = \mathbb{R}^d$, Keisler could use the compact-open topology, but in general this does not work as $C(E, F)$ may fail to be second countable if E is not locally compact. Instead we shall use a topology which can most conveniently be described as follows.

Fix a countable, dense subset E_0 of E, and let $\mathcal{B}(E)$ be the family

$$ \mathcal{B}(E) = \{ B(e, 1/n) \,|\, e \in E_0, n \in \mathbb{N} \} $$

of *closed* balls $B(e, 1/n) = \{ m \in E \,|\, d(e, m) \le 1/n \}$. Let $\mathcal{B}(F)$ be a similarly defined family in F. If $B_1 \in \mathcal{B}(E)$, $B_2 \in \mathcal{B}(F)$, let

$$ O_{B_1, B_2} = \{ f \in C(E, F) \,|\, f[B_1] \subset B_2 \}. $$

Define a topology τ on $C(E, F)$ by letting its open sets be arbitrary unions of finite intersections of the form

(6)
$$O_{B_1^{(1)}, B_2^{(1)}} \cap \cdots \cap O_{B_1^{(m)}, B_2^{(m)}}.$$

Since there are only countably many such sets (6), this topology is second countable, and it is obviously Hausdorff.

As above, define $\hat{x} : \Omega \to C(E, F)$ by $\hat{x}(\omega) = x(\omega, \cdot)$. To lift $\hat{x}$, we must first show that it is measurable with respect to the topology τ. Since there are only countably many basic open sets of the form (6), it suffices to show that each set $\hat{x}^{-1}(O_{B_1, B_2})$ is measurable. Let $\tilde{Q}$ be a countable, dense subset of B_1; then since B_2 is closed

$$\hat{x}^{-1}(O_{B_1, B_2}) = \{\omega \mid \hat{x}(\omega)(q) \in B_2 \quad \text{for all} \quad q \in \tilde{Q}\} = \bigcap_{q \in \tilde{Q}} \{\omega : x(\omega, q) \in B_2\},$$

which is measurable since x is a stochastic process.

We can now pick a lifting $\hat{X}$ of $\hat{x}$, and define X by $X(\omega, m) = \hat{X}(\omega)(m)$. Let Ω' be a set of measure one such that $\hat{x}(\omega) = {}^\circ\hat{X}(\omega)$ for all $\omega \in \Omega'$. It only remains to show that if $\omega \in \Omega'$, then $x(\omega, {}^\circ m) = {}^\circ X(\omega, m)$ for all nearstandard m. Assume not; then $d(x(\omega, {}^\circ m), {}^\circ X(\omega, m))$ is non-infinitesimal, and there is an element $B_2 \in \mathscr{B}(F)$ such that $x(\omega, {}^\circ m)$ belongs to the interior of B_2, but $X(\omega, m) \notin {}^*B_2$. By continuity of $x(\omega, \cdot)$, there is a set $B_1 \in \mathscr{B}(E)$ with m in its interior such that $x(\omega, m') \in B_2$ for all $m' \in B_1$. Hence $\hat{x}(\omega) \in O_{B_1, B_2}$ while clearly $\hat{X}(\omega) \notin {}^*O_{B_1, B_2}$, contradicting the assumption ${}^\circ\hat{X}(\omega) = \hat{x}(\omega)$. The proof is complete.

A special case of the proposition above is that a process $x : \Omega \times [0, 1] \to \mathbb{R}$ is continuous if and only if it has a uniform lifting $X : \Omega \times T \to {}^*\mathbb{R}$. Another situation where the result is useful is when we are considering a function

(7)
$$f : [0, 1] \times C([0, 1], \mathbb{R}^n) \to \mathbb{R}^m,$$

which is measurable in the first coordinate and continuous in the second. If λ is the normalized counting measure on T, Proposition 4.3.13 asserts the existence of a uniform lifting

$$F : T \times {}^*C([0, 1], \mathbb{R}_n) \to {}^*\mathbb{R}^m.$$

Note that in this case

$$C(E, F) = C(C([0, 1], \mathbb{R}^n), \mathbb{R}^m)$$

is not second countable in the compact-open topology, and hence the special choice of the topology in the proof of 4.3.13 was necessary. We shall encounter functions f of the above type in the section on stochastic control theory, where we are interested in expressions of the form $f(t, x(\omega, \cdot))$ for a controlled process x.

A natural question is whether it is possible to combine two already established lifting results to create a new one; e.g., it is tempting to conjecture that a process is continuous and adapted if and only if it has a nonanticipating, uniform lifting. In most cases an extension of this kind is possible, but we may have to formulate it carefully, and the proof is often far from trivial. As an illustration we shall prove a modification of the conjecture we just mentioned.

An internal process $X : \Omega \times T \to {}^*\mathbb{R}$ is an *essentially uniform* lifting of $x : \Omega \times [0, 1] \to \mathbb{R}$ if there is an infinitesimal $\delta \in T$ such that

$$\{\omega \mid \forall t > \delta ({}^\circ X(\omega, t) = x(\omega, {}^\circ t))\}$$

has Loeb measure one; i.e., the uniformity of the lifting may break down on an infinitesimal initial segment. It follows from 4.3.13 that x has an essentially uniform lifting if and only if it is continuous.

4.3.14. THEOREM. A process $x : \Omega \times [0, 1] \to \mathbb{R}$ is continuous and adapted if and only if it has an essentially uniform, nonanticipating lifting.

REMARK. Keisler (1984) originally proved this result in a slightly less general context; the version we are using is taken from Osswald (1985). Another formulation of the theorem, which is more common in the literature, is to claim that the lifting is uniform, but only nonanticipating after an infinitesimal time δ. The statement we get by just removing the word "essentially" is not true as we shall show in Example 4.3.15; hence our conjecture above (i.e., that a process is continuous and adapted iff it has a nonanticipating, uniform lifting) is actually false! False is also the assertion obtained by turning "adapted" into "almost surely adapted"—perhaps somewhat surprisingly at first glance in view of 4.3.9.

4.3.15. EXAMPLE. To construct a continuous, adapted process that does not have a uniform, nonanticipating lifting, we return to the setting of Section 4.1. Hence Ω is the set of all internal maps $\omega : T \to \{-1, 1\}$, and $\mathscr{A}_t$ is the internal algebra generated by the equivalence relation $\omega \sim_t \omega'$ iff $\omega(s) = \omega'(s)$ for all $s < t$. Let x be defined by $x(\omega, t) = \omega(0)$ for all $\omega \in \Omega$, $t \in [0, 1]$. Clearly x is a continuous, $\mathscr{B}_t$-adapted process. Since $\mathscr{A}_0 = \{\varnothing, \Omega\}$, x does not have a uniform, nonanticipating lifting. Note that

$$X(\omega, t) = \begin{cases} 0 & \text{if} \quad t = 0 \\ \omega(0) & \text{for} \quad t > 0 \end{cases}$$

is a nonanticipating, essentially uniform lifting of x.

In the proof of 4.3.14 it will be convenient to work with a standard filtration $\{\mathscr{C}_t\}$ which is slightly coarser than $\{\mathscr{B}_t\}$. To define it, we first introduce an internal equivalence relation $\sim_s$ on Ω for each $s \in T$ by

(8) $\omega \sim_s \omega'$ iff $\forall A \in \mathscr{A}_s (\omega \in A \leftrightarrow \omega' \in A)$.

If $t \in [0, 1]$, let $\equiv_t$ be the (external) equivalence relation given by $\omega \equiv_t \omega'$ if $\omega \sim_s \omega'$ for all s such that $^\circ s \leq t$. The σ-algebra $\mathscr{C}_t$ consists of all Loeb measurable sets C which are closed under $\equiv_t$. Using 4.3.1 and 4.3.2 it is easy to see that $\mathscr{C}_t \subset \mathscr{B}_t$, but that for each $B \in \mathscr{B}_t$ there is a $C \in \mathscr{C}_t$ such that $B \bigtriangleup C$ has Loeb measure zero. It follows from this that for each continuous, $\mathscr{B}_t$-adapted process x, there is a continuous, $\mathscr{C}_t$-adapted process y such that $x(\omega, \cdot) = y(\omega, \cdot)$ for almost all ω. The filtration $\{\mathscr{C}_t\}$ is the one used by Keisler (1984).

PROOF OF THEOREM 4.3.14. Assume that x has an essentially uniform, nonanticipating lifting X. Since it has an essentially uniform lifting, x is continuous. To show that it also is adapted pick $t \in [0, 1]$ and $\tilde{t} \in T$ such that $t = ^\circ\tilde{t}$. Then $^\circ X(\tilde{t}) = x(t)$ almost everywhere, and since $X(\tilde{t})$ is $\mathscr{A}_{\tilde{t}}$-measurable, $x(t)$ must be $\mathscr{B}_t$-measurable.

For the converse, assume that x is continuous and adapted. Let y be a continuous and $\mathscr{C}_t$-adapted process such that $y(\omega, \cdot) = x(\omega, \cdot)$ almost everywhere, and let Y be a uniform lifting of y. We shall modify Y such that it becomes a nonanticipating, essentially uniform lifting of y (and hence of x).

Let Ω' be the set of measure one where Y lifts y, and choose an increasing, internal sequence $\{\Omega_n\}_{n \in {}^*\mathbb{N}}$ such that for each $n \in \mathbb{N}$, $\Omega_n \subset \Omega'$ and $P(\Omega_n) > 1 - 1/n$. Let $\tilde{\Omega} = \bigcup_{n \in \mathbb{N}} \Omega_n$. For each equivalence class $\mathscr{P}$ of the relation $\sim_s$ in (8), choose a representative $\omega_s \in \mathscr{P}$. We pick ω_s such that it belongs to the smallest Ω_n which intersects $\mathscr{P}$, and note that this selection can be carried out in a strictly internal way. To each element $\omega \in \Omega$ and each $s \in T$, we thus associate a representative ω_s (i.e., the representative of the equivalence class of ω under $\sim_s$).

Observe that since we pick ω_s from the smallest possible Ω_n, the representative ω_s is always in $\tilde{\Omega}$ when ω is. This implies that

(9) $\qquad ^\circ Y(\omega, s) = y(\omega, {}^\circ s) \qquad$ and $\qquad ^\circ Y(\omega_s, s) = y(\omega_s, {}^\circ s)$

when $\omega \in \tilde{\Omega}$, since Y lifts y on $\tilde{\Omega}$.

Define $X : \Omega \times T \to {}^*\mathbb{R}$ by

$$X(\omega, s) = Y(\omega_s, s).$$

As X is clearly nonanticipating, it suffices to show that it is an essentially uniform lifting of y. Assume that $\omega \in \tilde{\Omega}$, $s \in T$, and that $^\circ s > 0$. By construction $\omega \sim_s \omega_s$, and hence $\omega \equiv_r \omega_s$ for all $r \in [0, 1]$, $r < {}^\circ s$. Since y is $\mathscr{C}_t$-adapted, this implies that $y(\omega, r) = y(\omega_s, r)$. Taking the limit as r increases to $^\circ s$, we get

(10) $\qquad\qquad\qquad y(\omega, {}^\circ s) = y(\omega_s, {}^\circ s).$

Combining (9) and (10), we see that for all noninfinitesimal $s \in T$ and all $\omega \in \tilde{\Omega}$

(11) $\qquad °X(\omega, s) = °Y(\omega_s, s) = y(\omega_s, °s) = y(\omega, °s),$

which shows that X is a uniform lifting of y on the half-open interval $(0, 1]$. It remains to extend (11) to sufficiently large, infinitesimal s. The trick is as follows: from (11) we get that the internal set

$$\left\{ n \in {}^*\mathbb{N} \,\middle|\, \forall \omega \in \Omega_n \,\forall t \geq \frac{1}{n} \,(|X(\omega, t) - Y(\omega, t)| < \frac{1}{n} \right\}$$

contains $\mathbb{N}$ and thus has an infinite element γ. Since Y is a uniform lifting of y, we see that X must be a uniform lifting of y on $T \cap [1/\gamma, 1]$, and hence an essentially uniform lifting on T.

The basic idea of the proof above is to create a nonanticipating lifting from an ordinary lifting by choosing one element from each equivalence class. In the last result of this section we shall see the same idea at work in a much simpler setting. A function $f : [0, 1] \times C([0, 1], \mathbb{R}^n) \to \mathbb{R}^m$ as in (7) is called *nonanticipating* if $f(t, y) = f(t, y')$ whenever $y(s) = y'(s)$ for all $s \leq t$. We use the same terminology for internal functions $F : T \times {}^*C([0, 1], \mathbb{R}^n) \to {}^*\mathbb{R}^m$. The following result will be useful in Section 4.6.

4.3.16. PROPOSITION. Assume that the nonanticipating function

$$f : [0, 1] \times C([0, 1], \mathbb{R}^n) \to \mathbb{R}^m$$

is measurable in the first coordinate and continuous [w.r.t. the uniform topology on $C([0, 1], \mathbb{R}^n)$] in the second. Then f has a nonanticipating lifting F which is uniform in the second coordinate; i.e., there is a set $T' \subset T$ of Loeb measure one such that $°F(t, y) = f(°t, °y)$ for all nearstandard y and all $t \in T'$.

PROOF. Let $G : T \times {}^*C([0, 1], \mathbb{R}^n) \to {}^*\mathbb{R}^m$ be a lifting of f which is uniform in the second variable. For each $y \in {}^*C([0, 1], \mathbb{R}^n)$ and each $t \in T$, define $y_t \in {}^*C([0, 1], \mathbb{R}^n)$ by

$$y_t(s) = \begin{cases} y(s) & \text{if } s \leq t \\ y(t) & \text{if } s > t, \end{cases}$$

and notice that $(°y)(r) = (°y_t)(r)$ for all $r \in [0, 1]$, $r \leq °t$. Hence

(12) $\qquad f(°t, °y) = f(°t, °y_t).$

Define a nonanticipating $F : T \times {}^*C([0, 1], \mathbb{R}^n) \to {}^*\mathbb{R}^n$ by

(13) $\qquad F(t, y) = G(t, y_t).$

If t is in the set T' where G lifts f, and y is nearstandard, then by (12) and (13)

$$^\circ F(t, y) = G(t, y_t) = f(^\circ t, {}^\circ y_t) = f(^\circ t, {}^\circ y),$$

and thus F is a uniform, nonanticipating lifting of f.

We shall end our survey of lifting theorems here. Although we have at times used a different approach or worked in a more general setting, all the main results in this section are basically due to H. J. Keisler. We urge the reader to consult his monograph Keisler (1984) for further information, e.g., on lifting characterizations of different kinds of Markov processes. Hoover and Perkins (1983a) and Stroyan and Bayod (1985) contain a detailed analysis of liftings of right continuous processes with left limits, and Rodenhausen (1982a,b) has developed an interesting criterion for when a process is a lifting.

4.4. REPRESENTATION THEOREMS

We shall now discuss in more detail the relationship between standard and hyperfinite stochastic integrals. There are two aspects of this relationship. First, is the hyperfinite theory strong enough, i.e., can standard integrals always be reduced in some suitable sense to hyperfinite integrals? And second, does the hyperfinite theory give us something which the standard theory cannot, i.e., are there "meaningful" uses of the hyperfinite integral that cannot be reduced (at least in a simple way) to standard integration theory?

Theorems 4.1.5 and 4.1.6 give a positive answer to the first question in the case of Brownian motion. The nonstandard integral with respect to a hyperfinite random walk gives—via liftings and the standard part map—the classical theory for the Itô integral. In the first part of this section we shall give the proofs we left out in Section 4.1 and at the same time extend the treatment to cover the theory of 4.2; in the second part we shall discuss some results pertaining to the second question.

A. Square S-Integrable Martingales

Let us begin with a brief sketch of the standard theory. An L^2-*martingale* with respect to a stochastic filtration $\langle Z, \{\mathscr{F}_t\}_{t\in[0,1]}, P\rangle$ is just a martingale $N: Z \times [0, 1] \to \mathbb{R}$ such that $E(N_t^2) < \infty$ for all $t \in [0, 1]$. We shall also assume that N is right-continuous with left limits. Most of the results we shall prove about L^2-martingales have trivial extensions to the larger class of local L^2-martingales; we shall just give the definition of this class and leave the extensions to the interested reader. A process $N: Z \times [0, 1] \to \mathbb{R}$

is a *local L^2-martingale* if there exists an increasing sequence $\{\tau_n\}_{n \in \mathbb{N}}$ of stopping times such that each stopped process N_{τ_n} is an L^2-martingale, and for almost all $\omega \in Z$, $\tau_n(\omega) = 1$ for large enough $n \in \mathbb{N}$.

Given an L^2-martingale N, we define a measure ν_N on the σ-algebra of predictable sets by putting

(1) $$\nu_N(A \times (s, t]) = E(1_A(N(t) - N(s))^2)$$

for all predictable rectangles $A \times (s, t]$, and letting

(2) $$\nu_N(A \times \{0\}) = 0.$$

The generated measure ν_N is often called the *Doléans measure* of N (Doléans, 1968; Meyer, 1976).

Stochastic integrals of the form $\int_0^1 X \, dN$ are defined as follows. First, if X is a simple function of the form

(3) $$X = \sum_{i=1}^n a_i 1_{A_i \times (s_i, t_i]}$$

where $A_i \in \mathscr{F}_{s_i}$, the integral is given by

(4) $$\left(\int_0^1 X \, dN \right)(\omega) = \sum_{i=1}^n a_i 1_{A_i}(\omega)(N(\omega, t_i) - N(\omega, s_i)).$$

Observing that $X \mapsto \int_0^1 X \, dN$ is an isometry from $L^2(\nu_N)$ to $L^2(P)$ and that the simple functions of the form (3) are dense in $L^2(\nu_N)$, we extend the mapping $X \mapsto \int_0^1 X \, dN$ to an isometry from all of $L^2(\nu_N)$ into $L^2(P)$. We shall denote this extension also by $\int_0^1 X \, dN$. The stochastic integral as a process is now defined for all $X \in L^2(\nu_N)$ and all $t \in [0, 1]$ by

(5) $$\left(\int X \, dN \right)(\omega, t) = \left(\int_0^1 1_{[0,t]} X \, dN \right)(\omega).$$

Notice that since $X \mapsto \int_0^1 X \, dN$ is given as an L^2 limit, the stochastic integral is only defined up to equivalence. Observe also that the definition above extends the definition of the Itô integral in Section 4.1; we have replaced the measure $L(P) \times m$ by ν_N and restricted the class of integrands from adapted to predictable processes (recall 4.3.10).

We can now formulate the theme of the first half of this section more precisely. Let M be a λ^2-martingale with respect to an internal filtration $\langle \Omega, \{\mathscr{A}_t\}_{t \in T}, P \rangle$. It is easy to check that the standard part $^\circ M^+$ of M is an L^2-martingale with respect to the generated stochastic filtration $\langle \Omega, \{\mathscr{B}_t\}_{t \in [0,1]}, L(P) \rangle$. [A word of caution may be appropriate at this point; it is not true that the standard part of a local λ^2-martingale is always a local L^2-martingale, since we may get problems with the stopping times—see

Lindstrøm (1980b) for an example.] Given an $X \in L^2(\nu_{°M^+})$ we want to show that $\int X \, d°M^+$ can be obtained as a hyperfinite integral of M; i.e., $\int X \, d°M^+ = °(\int Y \, dM)^+$ for suitable $Y \in SL^2(M)$. The idea of the construction is simple; we use Proposition 4.3.9 (i) to pick $Y \in SL^2(M)$ to be a lifting of X with respect to ν_M, and then just check that this Y does the job. However, there is one technical obstacle; Proposition 4.3.9 requires ν to be absolutely continuous with respect to P, and this is not true of ν_M for all λ^2-martingales M. We shall get around this obstacle by restricting our class of martingales slightly.

4.4.1. DEFINITION. An internal martingale $M : \Omega \times T \to {}^*\mathbb{R}$ is called an SL^2-*martingale* if $M_t \in SL^2(\Omega, \mathcal{A}_1, P)$ for all $t \in T$.

We have a corresponding notion of a *local SL^2-martingale*. We also need the following definition.

4.4.2. DEFINITION. An internal process $X : \Omega \times T \to {}^*\mathbb{R}$ is *S-right-continuous* at 0 if $°X(0) = °X^+(0)$ $L(P)$ a.e.

We notice that if M is an SL^2-martingale which is S-right-continuous at 0, then ν_M is absolutely continuous with respect to P.

We shall now interrupt our development of integration theory for a moment while we prove two theorems about SL^2-martingales. These results are included not only to make the exposition self-contained, but also because they provide another illustration of our strategy of reducing questions concerning martingales to questions concerning their quadratic variation. However, a reader primarily interested in a short introduction to stochastic integration may skip the proofs.

4.4.3. PROPOSITION. An internal martingale M is an SL^2-martingale if and only if $M_0^2 + [M](1)$ is S-integrable.

PROOF. Since in either case M is a λ^2-martingale, it is enough to prove the result for such processes.

Recall from Chapter 3 that if $f : \Omega \mapsto {}^*\mathbb{R}$ is internal and non-negative, then $°\int f \, dP \geq \int °f \, dL(P)$ with equality if and only if f is S-integrable. We shall use this several times in the following.

By formula (4.2.7)

(6) $$E(M_0^2 + [M](t)) = E(M_t^2);$$

we shall prove a similar result for the standard part of M. Define a sequence $\{\tau_n\}_{n \in \mathbb{N}}$ of internal stopping times by

$$\tau_n(\omega) = \min\{s \in T \, | \, |M(\omega, s)| \geq n\}.$$

Since

$$E\left(\left(\int_0^t M_{\tau_n}\, dM_{\tau_n}\right)^2\right) \le n^2 E([M](t)) < \infty,$$

$\int M_{\tau_n}\, dM_{\tau_n}$ is a λ^2-martingale, and thus $\int_0^t M_{\tau_n}\, dM_{\tau_n}$ is S-integrable for all t. By the characterization of S-integrability above

$$E\left({}^\circ\!\int_0^t M_{\tau_n}\, dM_{\tau_n}\right) = {}^\circ E\left(\int_0^t M_{\tau_n}\, dM_{\tau_n}\right) = 0.$$

Since

$${}^\circ M_{\tau_n}(0)^2 + {}^\circ[M_{\tau_n}](t) = {}^\circ M_{\tau_n}(t)^2 - 2\int_0^t M_{\tau_n}\, dM_{\tau_n} \qquad \text{a.e.,}$$

we get

(7) $$E({}^\circ M_{\tau_n}(0)^2 + {}^\circ[M_{\tau_n}](t)) = E({}^\circ M_{\tau_n}(t)^2).$$

Obviously, ${}^\circ[M_{\tau_n}](t) \to {}^\circ[M](t)$ and ${}^\circ M_{\tau_n}(t) \to {}^\circ M(t)$ almost everywhere as $n \to \infty$. The sequence $\{{}^\circ[M_{\tau_n}](t)\}$ is bounded by ${}^\circ[M](t)$, which is integrable since

$$E({}^\circ[M](t)) \le {}^\circ E([M](t)) < \infty.$$

Also, ${}^\circ(M_{\tau_n}(t)) \le {}^\circ\max_{s \le t} M_s^2$, and since by Doob's inequality

$$E({}^\circ\max_{s \le t} M_s^2) \le {}^\circ E(\max_{s \le t} M_s^2) \le 4{}^\circ E(M_t^2) < \infty,$$

${}^\circ\max_{s \le t} M_s^2$ is integrable. Applying Lebesgue's convergence theorem to both sides of (7), we get

(8) $$E({}^\circ M(0)^2 + {}^\circ[M](t)) = E({}^\circ M(t)^2).$$

Combining (6) and (8), we see that

$${}^\circ E(M_0^2 + [M](t)) = E({}^\circ M_0^2 + {}^\circ[M](t)) \qquad \text{iff} \qquad {}^\circ E(M_t^2) = E({}^\circ M_t^2).$$

The result now follows from the characterization of S-integrability mentioned at the beginning of the proof.

We mention one more result of the same kind [see Hoover and Perkins (1983a) for a proof]: If M is an SL^2-martingale, then $\max_{s \le 1} M_s^2$ is S-integrable. The next proposition shows that the class of SL^2-martingales is closed under stochastic integration.

4.4.4. PROPOSITION. If M is an SL^2-martingale and $X \in SL^2(M)$, then $\int X\, dM$ is an SL^2-martingale.

PROOF. We first consider the case where X is S-bounded, i.e., $|X| \le n$ for some $n \in \mathbb{N}$. Then

$$0 < \left[\int X \, dM \right](1) = \sum_0^1 X^2 \, \Delta M^2 \le n^2 [M](1),$$

and it follows from 4.4.3 that $\int X \, dM$ is an SL^2-martingale.

Let us consider the general case of $X \in SL^2(M)$. There exists a sequence $\{X_n\}_{n \in \mathbb{N}}$ of S-bounded elements in $SL^2(M)$ such that ${}^\circ\!\int_{\Omega \times T} |X^2 - X_n^2| \, d\nu_M \to 0$. We have

$$0 \le {}^\circ E \left(\left[\int X \, dM \right](1) - \left[\int X_n \, dM \right](1) \right)$$

$$= {}^\circ E \left(\sum_0^1 X^2 \, \Delta M^2 - \sum_0^1 X_n^2 \, \Delta M^2 \right) = {}^\circ\!\int_{\Omega \times T} (X^2 - X_n^2) \, d\nu_M \to 0.$$

Since each $[\int X_n \, dM](1)$ is S-integrable, so is $[\int X \, dM](1)$, and the proposition follows from 4.4.3.

B. Nonstandard Representations of Standard Stochastic Integrals

Having made ourselves a little more familiar with SL^2-martingales, we now return to integration theory. Recall that

$$\text{St} = \text{id} \times \text{st} : \Omega \times T \to \Omega \times [0, 1].$$

4.4.5. LEMMA. Let M be an SL^2-martingale which is S-right-continuous at 0, and let ${}^\circ M^+$ be its standard part. Then $\nu_{{}^\circ M^+}$ is the restriction of $L(\nu_M) \circ \text{St}^{-1}$ to the predictable sets.

PROOF. It suffices to prove that $\nu_{{}^\circ M^+}$ and $L(\nu_M) \circ \text{St}^{-1}$ agree on predictable rectangles. Let $B \in \mathcal{B}_s$, then

$$\nu_{{}^\circ M^+}(B \times (s, t]) = E(1_B({}^\circ M^+(t) - {}^\circ M^+(s))^2).$$

Let A be an internal set such that $L(P)(A \triangle B) = 0$ and $A \in \mathcal{A}_{\tilde{s}}$ for some $\tilde{s} \approx s$; such an A exists by 4.3.1. We get

$$L(\nu_M) \circ \text{St}^{-1}(B \times (s, t]) = L(\nu_M) \circ \text{St}^{-1}(A \times (s, t])$$

$$= \lim_{n \to \infty} \lim_{m \to \infty} {}^\circ\nu_M \left(A \times \left(s + \frac{1}{n}, t + \frac{1}{m} \right] \right)$$

$$= \lim_{n \to \infty} \lim_{m \to \infty} {}^\circ E \left(1_A \left([M] \left(t + \frac{1}{m} \right) - [M] \left(s + \frac{1}{n} \right) \right) \right)$$

$$= \lim_{n \to \infty} \lim_{m \to \infty} {}^\circ E \left(1_A \left(M \left(t + \frac{1}{m} \right) - M \left(s + \frac{1}{n} \right) \right)^2 \right)$$

$$= E(1_B({}^\circ M^+(t) - {}^\circ M^+(s))^2),$$

where the S-integrability of $[M]$ has been used to switch between B and A, and the S-integrability of M^2 to get the standard part inside the expectation. Pulling the limits inside the expectation is justified by a combination of Doob's inequality and Lebesgue's convergence theorem as in the proof of 4.4.3.

It only remains to observe that since M is S-right-continuous at 0,

$$L(\nu_M) \circ \mathrm{St}^{-1}(B \times \{0\}) = 0,$$

and the lemma is proved.

The notion of lifting we shall need is the following.

4.4.6. DEFINITION. Let M be an SL^2-martingale, and let $x : \Omega \times [0, 1] \to \mathbb{R}$ be a predictable process in $L^2(\nu_{\circ M^+})$. A 2-*lifting* of x (with respect to M) is a nonanticipating process $X : \Omega \times T \to {}^*\mathbb{R}$ in $SL^2(M)$ such that ${}^\circ X(\omega, t) = x(\omega, {}^\circ t)$ for $L(\nu_M)$–a.a. (ω, t).

Remembering that we introduced SL^2-martingales which are S-right-continuous at 0 in order to have the conditions of Proposition 4.3.9 satisfied, the following result is not surprising.

4.4.7. LEMMA. Let M be an SL^2-martingale which is S-right-continuous at 0. If $x \in L^2(\nu_{\circ M^+})$, then x has a 2-lifting with respect to M.

PROOF. By 4.3.9 x has a nonanticipating lifting X; we must show that we can choose X in $SL^2(\nu_M)$. For each $n \in \mathbb{N}$, let x_n be the truncation of x given by $x_n = (x \wedge n) \vee (-n)$. If X_n is the corresponding truncation of X, we see that X_n is a nonanticipating lifting of x_n. From 4.4.6 we get

$$\int {}^\circ X_n^2 \, d\nu_M = \int {}^\circ X_n^2 \, dL(\nu_M) = \int x_n^2 \, d\nu_{\circ M^+}.$$

Since $\int x_n^2 \, d\nu_{\circ M^+} \to \int x^2 \, d\nu_{\circ M^+}$, we can find an $\eta \in {}^*\mathbb{N} - \mathbb{N}$ such that

$$\int {}^\circ X_\eta^2 \, d\nu_M = \int x^2 \, d\nu_{\circ M^+}.$$

By Proposition 3.2.10 it follows that $X_\eta \in SL^2(\nu_M)$, and hence that it is a 2-lifting of x.

The next result shows that the particular choice of 2-lifting is irrelevant for stochastic integration.

4.4.8. LEMMA. Let M be an SL^2-martingale and let $x \in L^2(\nu_{\circ M^+})$. If X and Y are 2-liftings of x, then there is a set Ω' of Loeb measure one such that for all $\omega \in \Omega'$ and all $t \in T$

$$^{\circ}\left(\int X \, dM \right)(\omega, t) = {^{\circ}}\left(\int Y \, dM \right)(\omega, t).$$

PROOF. By Doob's inequality

$$E\left(\max_{t \in T} \left(\int_0^t X \, dM - \int_0^t Y \, dM \right)^2 \right) \leq 4E\left(\left(\int_0^1 (X - Y) \, dM \right)^2 \right)$$

$$= 4E\left(\left[\int (X - Y) \, dM \right](1) \right) = 4 \int_{\Omega \times T} (X - Y)^2 \, d\nu_M \approx 0.$$

since $X - Y$ is in $SL^2(\nu_M)$ and is infinitesimal almost everywhere. The lemma follows.

We can now prove the representation theorem.

4.4.9. THEOREM. Let M be an SL^2-martingale that is S-right-continuous at 0, and assume $x \in L^2(\nu_{\circ M^+})$. Then x has a 2-lifting X, and

$$(9) \qquad \int x \, d^{\circ}M^+ = {^{\circ}}\left(\int X \, dM \right)^+.$$

PROOF. All that remains to be done is to prove the equality. We first consider the case where x is a simple process of the form

$$x = \sum_{i=1}^n a_i 1_{B_i \times (s_i, t_i]},$$

where $B_i \in \mathscr{B}_{s_i}$. For each i, choose an $\hat{s}_i \approx s_i$ such that $^{\circ}M(\hat{s}_i) = {^{\circ}}M^+(s_i)$ almost everywhere, and such that there is an $A_i \in \mathscr{A}_{\hat{s}_i}$ with $L(P)(A_i \triangle B_i) = 0$. Pick $\hat{t}_i \approx t_i$ such that $^{\circ}M(\hat{t}_i) = {^{\circ}}M^+(t_i)$ almost everywhere. Let

$$X = \sum_{i=1}^n a_i 1_{A_i \times (\hat{s}_i, \hat{t}_i]};$$

then X is a 2-lifting of x, and (9) obviously holds for this pair.

To prove the general case, we only have to show that the map $x \mapsto {^{\circ}}(\int X \, dM)(1)$ is an isometry from $L^2(\Omega \times [0, 1], \nu_{\circ M^+})$ into $L^2(\Omega, L(P))$. But we have already observed that by 4.4.5

$$\int x^2 \, d\nu_{\circ M^+} = {^{\circ}}\int X^2 \, d\nu_M = {^{\circ}}E\left(\left(\int X \, dM \right)^2 (1) \right) = E\left({^{\circ}}\left(\int X \, dM \right)^2 (1) \right),$$

where we use 4.4.4 in the last step. This proves the theorem.

Recalling Lemma 4.3.10, we now get Theorem 4.1.6 as an immediate corollary of 4.4.9.

In spite of the result above there is a difference between the standard and the nonstandard definition of the stochastic integral that seems worthy of a remark. Since the standard definition only determines the integral process up to a null set for each t, questions about the path properties of this integral does not really make sense; we can not ask whether the integral itself is continuous or has left and right limits, but only whether it has a representative with these properties. By Lemma 4.4.8 the nonstandard approach defines the stochastic integral up to indistinguishability; i.e., if $°(\int X\, dM)^+$ and $°(\int Y\, dM)^+$ are two versions of it, then there exists a null set such that outside this set the two versions agree for all t. Hence in this case it does make sense to ask for path properties, and it is reassuring to notice that the representatives picked by the nonstandard definition always seem to have the nicest properties possible (see, e.g., Proposition 4.2.15).

C. Quadratic Variation and Itô's Lemma

As the quadratic variation of an internal martingale has proved to be such a useful tool, it may be interesting to take a look at the corresponding standard notion. Let $N: \Omega \times [0, 1] \to \mathbb{R}$ be an L^2-martingale. If $\pi = \{0 = t_0 < t_1 < t_2 < \cdots < t_n = 1\}$ is a partition of $[0, 1]$, let $\delta(\pi) = \max_{0 \le i < n}(t_{i+1} - t_i)$ be its mesh. Given a sequence $\{\pi_m\} = \{0 = t_0^m < t_1^m < \cdots < t_{n_m}^m = 1\}$ of such partitions with $\delta(\pi_m) \to 0$, it is natural to define the *quadratic variation* $[N]$ of N by

$$(10) \qquad [N](t) = \lim_{m \to \infty} \sum_{i=0}^{n_m - 1} (N(t_{i+1}^m \wedge t) - N(t_i^m \wedge t))^2,$$

where the limit is taken in the L^1-sense. There are two obvious questions; first whether the limit exists and is independent of the sequence $\{\pi_m\}$, and then whether

$$(11) \qquad [°M^+] = °[M]^+$$

for all internal SL^2-martingales M. Standard probabilists have answered the first question affirmatively; the quadratic variation exists and can be defined through (10) [see, e.g., Metivier and Pellaumail (1980) and Meyer (1976)]. As we have stated it, the answer to the second question is "no"; there are SL^2-martingales M for which (11) does not hold; an example is given in Hoover and Perkins (1983a). However, we shall now introduce a class of SL^2-martingales for which (11) is true.

4.4.10. DEFINITION. An internal function $f: T \to {}^*\mathbb{R}$ is called *well-behaved* if for each $t \in [0, 1]$ there is a $\tilde{t} \in T$, $\tilde{t} \approx t$, such that for all $s \approx \tilde{t}$, $s \le \tilde{t}$,

$$f(s) \approx f(\tilde{t}),$$

and for all $s \approx \tilde{t}$, $s > \tilde{t}$,

$$f(s) \approx f(\tilde{t}^+),$$

where $\tilde{t}^+$ denotes the successor of $\tilde{t}$ in T. A process is *well-behaved* if almost all its paths are.

REMARK. A well-behaved process is simply a process which "jumps" at most once per monad. The nondescript term "well-behaved" was introduced by Lindstrøm (1980b), who could not come up with anything better. Unfortunately, Hoover and Perkins' (1983a) terminology is hardly an improvement—a "well-behaved process" is to them an "SDJ-process"—although it makes a certain sense in their setting. We shall continue to use the worst alternative "well-behaved" in the hope that its sheer awfulness will force somebody to invent a better name for it.

To prove that $[°M^+] = °[M]^+$ for all well-behaved SL^2-martingales M, we need the following lemma.

4.4.11. LEMMA. Let $M : \Omega \times T \to {}^*\mathbb{R}$ be a well-behaved SL^2-martingale. If S is a hyperfinite subline of T and M^S is the restriction of M to S, then

$$(12) \qquad [M](s) \approx [M^S](s) \qquad \text{a.e.}$$

for all $s \in S$.

PROOF. Let $\tau_n(\omega) = \inf\{t \in T \mid M(\omega, t) \geq n\}$. Since $M_{\tau_n} \in SL^2(M_{\tau_n})$, and for a.a. ω we have $\tau_n(\omega) = 1$ for sufficiently large $n \in \mathbb{N}$, it suffices to prove the lemma when $M \in SL^2(M)$.

By Lemma 4.2.5

$$(13) \qquad [M](t) = M(t)^2 - M(0)^2 - 2 \int_0^t M \, dM$$

and

$$(14) \qquad [M^S](t) = M^S(t)^2 - M(0)^2 - 2 \int_0^t M^S \, dM^S.$$

Note that $\int_0^t M^S \, dM^S = \int_0^t \tilde{M}^S \, dM$ where $\tilde{M}^S : \Omega \times T \to {}^*\mathbb{R}$ is defined by letting $\tilde{M}^S(\omega, t)$ equal $M^S(\omega, s)$ for the largest s in S smaller than t. Since M is well behaved, M^S and $\tilde{M}^S$ are both 2-liftings of the left standard part $°M^-$ of M, and hence

$$(15) \qquad \int_0^t M \, dM \approx \int_0^t \tilde{M}^S \, dM = \int_0^t M^S \, dM^S \qquad \text{a.e.}$$

by Lemma 4.4.8. Putting this into (13) and (14), the lemma follows.

The reader may feel that this proof is unnecessarily roundabout. Why are we using Lemma 4.2.5 to reduce the problem to a question about stochastic integrals, and not attacking it directly through the definition of the quadratic variation? The answer is simply that we do not know of a more direct proof; the trick above was used by both Hoover and Perkins (1983a) and Lindstrøm (1980b).

4.4.12. PROPOSITION. If M is a well-behaved SL^2-martingale that is S-continuous at 0, then $[{}^{\circ}M^{+}]$ exists and equals ${}^{\circ}[M]^{+}$. Moreover,

$$(16) \qquad [{}^{\circ}M^{+}](t) = {}^{\circ}M^{+}(t)^2 - {}^{\circ}M^{+}(0)^2 - 2 \int_0^t {}^{\circ}M^{-} \, d{}^{\circ}M^{+}.$$

Before we give the proof, let us say how to interpret the stochastic integral in (16). The point is that since ${}^{\circ}M^{-}$ need not belong to the class $L^2(\nu_{{}^{\circ}M^{+}})$, this integral is not covered by the definition given at the beginning of this section. But the solution is simple: put $\int {}^{\circ}M^{-} \, d{}^{\circ}M^{+} = \lim_{n \to \infty} \int {}^{\circ}M^{-}_{\sigma_n} \, d{}^{\circ}M^{+}$, where $\{\sigma_n\}$ is a sequence of stopping times increasing to 1 such that ${}^{\circ}M^{-}_{\sigma_n} \in L^2(\nu_{{}^{\circ}M^{+}})$ for each n.

To prove the proposition, fix $t \in [0, 1]$ and choose $\tilde{t} \in T$, $\tilde{t} \approx t$, so large that ${}^{\circ}M(\tilde{t}) = {}^{\circ}M^{+}(t)$ almost everywhere. Given a sequence $\{\pi_m\}_{m \in \mathbb{N}}$ of partitions of $[0, 1]$ with mesh going to zero, construct a sequence $\{\tilde{\pi}_m\}_{m \in \mathbb{N}}$ of internal partitions of T by choosing for each t_i^m a $\tilde{t}_i^m \in T$, $\tilde{t}_i^m \approx t_i^m$ such that ${}^{\circ}M(\tilde{t}_i^m) = {}^{\circ}M^{+}(t_i^m)$. Extend $\{\tilde{\pi}_m\}_{m \in \mathbb{N}}$ to an internal sequence $\{\tilde{\pi}_m\}_{m \in {}^{*}\mathbb{N}}$. For each $m \in {}^{*}\mathbb{N} - \mathbb{N}$, we have by the lemma that

$$E\left(\left|[M](\tilde{t}) - \sum_{i=0}^{n_m - 1} (M(\tilde{t}_{i+1}^m \wedge \tilde{t}) - M(\tilde{t}_i^m \wedge \tilde{t}))^2\right|\right) \approx 0,$$

and the first part of the proposition follows.

To prove (16), let $\tau_n(\omega) = \min\{t \in T \mid M(\omega, t) \geq n\}$ and put $\sigma_n = {}^{\circ}\tau_n$. Then $M_{\tau_n} \in SL^2(M_{\tau_n})$ is a 2-lifting of ${}^{\circ}M^{-}_{\sigma_n}$, and thus

$$\int {}^{\circ}M^{-}_{\sigma_n} \, d{}^{\circ}M^{+} = {}^{\circ}\left(\int M_{\tau_n} \, dM_{\tau_n}\right)^{+}.$$

If we combine this with the identity

$$[M_{\tau_n}](\tilde{t}) = M_{\tau_n}^2(\tilde{t}) - M_{\tau_n}^2(0) - 2 \int_0^{\tilde{t}} M_{\tau_n} \, dM_{\tau_n}$$

and take the limit as n goes to infinity, we get (16).

Not much reflection should be need to realize why 4.4.12 fails for SL^2-martingales in general; if M is not well behaved it will have jumps inside monads which will be counted by $[M]$ but missed by $[{}^{\circ}M^{+}]$. That the proposition is useful despite the restriction to well-behaved processes is guaranteed by the following fact: if X is an internal process with S-left and S-right limits, there is a subline S such that the restriction X^S of X to S is well behaved. Thus we can always assume that our processes are well behaved by restricting them to a coarser time line if necessary. We shall not prove this fact here as proofs are easily found in Hoover and Perkins (1983a), Stroyan and Bayod (1985), and Lindstrøm (1980b), but would like to make the following observation first pointed out by Hoover and Perkins: if N is the right standard part of an SL^2-martingale M, we can always choose M to be well behaved. Thus Proposition 4.4.12 establishes the existence of the quadratic variation $[N]$ and the formula (16) for all such processes. As an application of Proposition 4.4.12, we shall prove the following generalization of Itô's lemma.

4.4.13. PROPOSITION. Let $N : \Omega \times [0, 1] \to \mathbb{R}$ be the right standard part of an S-continuous, SL^2-martingale M. If $\varphi : \mathbb{R} \times [0, 1] \to \mathbb{R}$ is twice continuously differentiable in the first variable and once in the second variable, then

(17)
$$
\varphi(N_t, t) - \varphi(N_0, 0) = \int_0^t \frac{\partial \varphi}{\partial t}(N_s, s) \, ds + \int_0^t \frac{\partial \varphi}{\partial x}(N_s, s) \, dN_s
$$
$$
+ \frac{1}{2} \int_0^t \frac{\partial^2 \varphi}{\partial x^2}(N_s, s) \, d[N](s).
$$

PROOF. We have

$$
\varphi(M_t, t) - \varphi(M_0, 0)
$$

$$
= \sum_{s=0}^{t} \{ \varphi(M_{s+\Delta t}, s + \Delta t) - \varphi(M_{s+\Delta t}, s) + \varphi(M_{s+\Delta t}, s) - \varphi(M_s, s) \}
$$

$$
\approx \sum_{s=0}^{t} \left\{ \frac{\partial \varphi}{\partial t}(M_{s+\Delta t}, s) \, \Delta t + \frac{\partial \varphi}{\partial x}(M_s, s) \, \Delta M_s + \frac{1}{2} \frac{\partial^2 \varphi}{\partial x^2}(M_s, s) \, \Delta M_s^2 \right\}
$$

$$
\approx \int_0^t \frac{\partial \varphi}{\partial t}(M_s, s) \, ds + \int_0^t \frac{\partial \varphi}{\partial x}(M_s, s) \, dM + \frac{1}{2} \int_0^t \frac{\partial^2 \varphi}{\partial x^2}(M_s, s) \, d[M].
$$

Each of the three last terms equals the corresponding term on the right-hand side of (17). The last of these equalities uses the fact that $[N] = {}^\circ[M]^+$.

There is a version of Itô's lemma for discontinuous martingales also, but it is more complicated as we must add each jump term separately [see Lindstrøm (1980a) for a nonstandard treatment]. The continuous version above suffices for our purposes in this book.

Let us end our discussion of well-behaved martingales on a cautious note; although "well-behaved" with respect to the quadratic variation, the class is not closed under stochastic integration: if M is a well-behaved martingale and X is a lifting, then $\int X \, dM$ is well behaved, but this is not always true for general $X \in SL^2(M)$ [see Lindstrøm (1980b)]. However, Hoover and Perkins (1983b) have shown that it is possible to find a subline such that the restriction of the martingale to this subline and all its integrals are well behaved.

D. Further Remarks on Nonstandard Representations

We have shown that the standard theory of stochastic integration with respect to the right standard part $°M^+$ of an internal martingale M can be reduced to the nonstandard theory of stochastic integration with respect to M itself. To claim on this basis that standard stochastic integration in general can be reduced to nonstandard stochastic integration is still somewhat premature; what about the martingales that are not the standard parts of internal martingales? A point in case is Proposition 4.4.13; do we really need the awkward formulation "N is the right standard part of an S-continuous martingale," or does it suffice to assume that N itself is a continuous martingale? In this subsection we shall take a brief look at this and a few related questions. The reader who does not care for such niceties, and who believes that the taste is a better proof of the pudding than a close scrutiny of the recipe, may skip this discussion and let the applications in the next sections convince him or her that the nonstandard theory is rich enough for all practical purposes.

By a saturation argument it is quite easy to show that if $(\Omega, \mathscr{A}, P)$ is an internal probability space and $N : \Omega \times [0, 1] \to \mathbb{R}$ is a martingale with respect to a filtration $(\Omega, \{\mathscr{C}_t\}, L(P))$, then N is the standard part of an internal martingale in the following sense. There is an internal filtration $(\Omega, \{\mathscr{A}_t\}_{t \in T}, P)$ and an internal martingale $M : \Omega \times T \to {}^*\mathbb{R}$ adapted to it such that $N = °M^+$. Moreover, $\mathscr{C}_t \subset \mathscr{B}_t$, where $\mathscr{B}_t$ is the standard filtration generated by $\{\mathscr{A}_t\}$. This answers our question about Proposition 4.4.13; it suffices to assume that N is a continuous martingale.

But not all processes live on Loeb spaces; if $N : \Sigma \times [0, 1] \to \mathbb{R}$ is a martingale with respect to some general filtration $(\Sigma, \{\mathscr{F}_t\}, Q)$, what do we then do? It turns out that N can be represented by a martingale N^Θ adapted to a Loeb-space filtration $(\Omega, \{\mathscr{C}_t\}, L(P))$ through a measure-preserving, Boolean σ-homomorphism Θ, mapping Loeb measurable sets into Q-measurable sets. As we have already remarked, N^Θ is the right standard part of an internal martingale M, and by using the map Θ all stochastic integrals with respect to N can be interpreted as nonstandard integrals with respect to M [see Lindstrøm (1980c) for the details]. Although this procedure makes complete the reduction of the standard theory to the nonstandard, it does not seem to be of any great significance. The reason is that probabilists usually do not care which probability space they are working with as long as it is "rich" enough to support the phenomenon they are studying. Keisler (1984, 1985) and Hoover and Keisler (1984) have shown that Loeb spaces are extremely rich in this respect; whatever happens on some probability space also happens on a Loeb space. This universality property implies that we need not care about other spaces; we can always assume that we are working in a Loeb setting. To obtain their results, Hoover and Keisler use a notion of elementary equivalence taken from probability logic, which gives a classification of processes much finer than those habitually used by probabilists. For the reader with some background in logic, we mention that Loeb spaces play much the same role in probability theory as saturated models do in first-order model theory. Additional information about probability logic can be found in Keisler (1977, 1985), Hoover (1978, 1982, 1985), Rodenhausen (1982a), Fajardo (1984, 1985), and Ross (1984).

E. Standard Representations of Nonstandard Stochastic Integrals

If standard stochastic integrals can be obtained from hyperfinite ones, a natural question is whether the converse also holds; if $X \in SL^2(M)$, can we find a standard process N somehow related to M, and an $x \in L^2(\nu_N)$ such that $\int x \, dN = {}^\circ(\int X \, dM)^+$? In the remainder of this section we shall discuss various formulations of this problem. The following example shows that the answer is no if we insist on having $N = {}^\circ M^+$.

4.4.14. EXAMPLE. We are in the setting of Section 4.1; our time line is $\{0, \Delta t, 2\,\Delta t, \ldots, 1\}$, and $\chi : \Omega \times T \to {}^*\mathbb{R}$ is Anderson's random walk. Let $X : \Omega \times T \to {}^*\mathbb{R}$ be given by

$$X(\omega, k\,\Delta t) = (-1)^k;$$

obviously $X \in SL^2(\chi)$. If $\beta = {}^\circ\chi^+$ and $x \in L^2(\nu_\beta)$, we shall show that $\int x \, d\beta \neq {}^\circ(\int X \, d\chi)^+$.

Let Y be a 2-lifting of x; we may assume that $Y(\omega, 2k\,\Delta t) = Y(\omega, (2k+1)\,\Delta t)$ for all k. We have

$$E\left(\left({}^\circ\!\int X \, d\chi^+ - \int x \, d\beta\right)^2\right) = {}^\circ E\left(\left(\int_0^1 X \, d\chi - \int Y \, d\chi\right)^2\right)$$

$$= E\left(\sum_0^1 (X - Y)^2 \, \Delta t\right).$$

The last expression is smallest when Y is identically zero, and hence

$$E\left(\left({}^{\circ}\!\int X \, d\chi^+ - \int x \, d\beta\right)^2\right) \geq 1$$

for all β-integrable x.

As observed by Rodenhausen, this calculation can be used to prove

4.4.15. PROPOSITION. Let M be an SL^2-martingale that is S-right continuous at the origin, and let $X \in SL^2(M)$. If $x \in L^2(\nu_{\circ M^+})$ is such that $\int x \, d^\circ M^+ = {}^\circ(\int X \, dM)^+$, then X is a 2-lifting of x.

PROOF. Let Y be a 2-lifting of x; then $\int_0^1 X \, dM = \int_0^1 Y \, dM$ almost everywhere. Hence

$$\int_{\Omega \times T} (X - Y)^2 \, d\nu_M = E\left(\sum_0^1 (X - Y)^2 \, \Delta M^2\right) = E\left(\left[\int (X - Y) \, dM\right](1)\right)$$

$$= E\left(\left(\left(\int X \, dM\right)(1) - \left(\int Y \, dM\right)(1)\right)^2\right) \approx 0.$$

Since Y is a 2-lifting of x, so is X, and the proposition is proved.

The example and the proposition above show that the class of hyperfinite stochastic integrals with respect to M is much larger than the class of standard stochastic integrals with respect to ${}^\circ M^+$. Before we go on to explore this extra richness, let us introduce a note of caution to the discussion by stating

4.4.16. PROPOSITION. Let $M : \Omega \times T \to {}^*\mathbb{R}$ be an SL^2-martingale that is S-continuous at the origin. Let $N : \Omega \times [0, 1] \to \mathbb{R}$ be an L^2-martingale such that for each hyperfinite subline S of T, there is a process $X^S \in SL^2(M^S)$ such that $N = {}^\circ(\int X^S \, dM^S)^+$. Then there is an $x \in L^2(\nu_{\circ M^+})$ such that $N = \int x \, d^\circ M^+$.

The proof is not hard, and we leave it to the reader. Proposition 4.4.16 may be interpreted as saying that as long as we only use M as an arbitrary representation of its standard part, the extra power of hyperfinite stochastic integration may not be very useful since it depends solely on the representation and not on the original process. However, it is not quite clear that this interpretation is correct since for any hyperfinite representation we get a much richer class of stochastic integrals by allowing integrands that are not liftings, and it may be that it is this extra richness that is important and not the representation of some particular process as a stochastic integral.

Returning to our original question we see that we have no chance of finding an x such that $\int x \, dN = {}^\circ(\int X \, dM)^+$ if $N = {}^\circ M^+$. The next question is what happens if we allow a looser relationship between N and M. We shall answer this question in the case of Brownian motion. From now on we work in the setting of Section 4.1, and let Ω, P, $\{\mathcal{A}_t\}$, $\{\mathcal{B}_t\}$ and χ be as in that section. Our aim is to prove

4.4.17. THEOREM. Let χ be Anderson's random walk and $X \in SL^2(\chi)$. Then there exist a Brownian motion β and an adapted process $x \in L^2(\Omega \times [0, 1])$ such that

$$\int x \, d\beta = {}^{\circ}\left(\int X \, d\chi \right)^{+}.$$

Before we turn to the proof, let us make a few comments. If X is not a lifting, Proposition 4.4.15 tells us that $\beta \neq {}^{\circ}\chi^{+}$. A natural extension of 4.4.17 would be that given an SL^2-martingale M and an $X \in SL^2(M)$, there always exist an L^2-martingale N having the same finite-dimensional distributions as ${}^{\circ}M^{+}$ and an $x \in L^2(\nu_N)$ such that

$$\int x \, dN = {}^{\circ}\left(\int X \, dM \right)^{+}.$$

This turns out to be false even if M is assumed to be continuous [see Lindstrøm (1985) for an example]. On the other hand, the generalization of 4.4.17 to n-dimensional Brownian motions does hold, and with a proof that is basically the same as the one we shall give for the one-dimensional case [see Lindstrøm (1985)].

Let us agree that by a *Brownian motion β adapted to a filtration* $(\Omega, \{\mathcal{F}_t\}, P)$, we mean a continuous martingale with respect to this filtration such that for all $t > s$, $\beta(t) - \beta(s)$ is independent of $\mathcal{F}_s$ and Gaussian distributed with mean zero and variance $t - s$.

Before we can prove 4.4.17 we need a result which can help us in recognizing Brownian motions. The following is a standard result (Lévy, 1948) in hyperfinite disguise:

4.4.18. PROPOSITION. Let $M: \Omega \times T \to {}^{*}\mathbb{R}$ be an S-continuous SL^2-martingale such that $E({}^{\circ}[M]^{+}(t) - {}^{\circ}[M]^{+}(s)|\mathcal{B}_s) = t - s$ for all $t > s$. Then ${}^{\circ}M^{+}$ is a Brownian motion with respect to $(\Omega, \{\mathcal{B}_s\}, L(P))$.

PROOF. Applying 4.4.13, Itô's lemma, to the martingale M and the function $\varphi(x) = e^{iyx}$, we get

$$\exp(iy(M_t - M_s)) \approx 1 + iy \int_s^t \exp(iy(M_r - M_s)) \, dM(r)$$

$$-\frac{y^2}{2} \int_s^t \exp(iy(M_r - M_s)) \, d[M](r).$$

Let $B \in \mathcal{B}_s$. If we take the standard part of the equation above, then multiply by 1_B and take expectations, we get

$$\int_B \exp(iy(^\circ M_t^+ - {}^\circ M_s^+)) \, dL(P)$$

$$= L(P)(B) - \frac{y^2}{2} \int_s^t \int_B \exp(iy(^\circ M_r^+ - {}^\circ M_s^+)) \, dL(P) \, dr.$$

Since the unique solution to the integral equation

$$\varphi(t) = L(P)(B) - \frac{y^2}{2} \int_s^t \varphi(r) \, dr$$

is $\varphi(t) = L(P)(B) \exp[-y^2/2L(P)(B)(t - s)]$, we see that $(^\circ M^+(t) - {}^\circ M^+(s)) \upharpoonright B$ is Gaussian distributed with mean zero and variance $L(P)(B)(t - s)$. This proves the proposition.

We now turn to the proof of 4.4.17. Let

$$M = \int X \, d\chi.$$

First notice that since $[M] = \int X^2 \, dt$, the standard part $^\circ[M]^+$ is increasing and absolutely continuous. Hence

$$f(\omega, t) = \frac{d}{dt} {}^\circ[M]^+(t) = \lim_{h \downarrow 0} \frac{{}^\circ[M]^+(t) - {}^\circ[M]^+(t - h)}{h}$$

exists, and is a non-negative, adapted process with

$$(18) \qquad {}^\circ[M]^+(\omega, t) = \int_0^t f(\omega, s) \, ds \qquad \text{a.e.}$$

Define a new adapted process g by

$$g(\omega, t) = \begin{cases} f^{-1/2}(\omega, t) & \text{if } f(\omega, t) \neq 0, \\ 0 & \text{elsewhere} \end{cases}$$

and let 1_g be the characteristic function of the set $\{(\omega, t) \mid g(\omega, t) = 0\}$. Since by (18)

$$E\left(\int_0^1 g(\omega, s)^2 \, d^\circ[M]^+\right) = E\left(\int_0^1 g(\omega, s)^2 f(\omega, s) \, ds\right) \leq 1,$$

we have $g \in L^2(\nu_{{}^\circ[M]^+})$. Let $G \in SL^2(M)$ be a 2-lifting of g, and 1_G a lifting of 1_g. We may assume that $G \cdot 1_G = 0$. Define

$$(19) \qquad \beta(\omega, t) = {}^\circ\left(\int_0^t G(\omega, s) \, dM(\omega, s) + \int_0^t 1_G(\omega, s) \, d\chi(\omega, s)\right)^+.$$

Since G and 1_G have disjoint supports

$$[\beta](\omega, t) = \,^\circ\left[\int G\,dM\right]^+(\omega, t) + \,^\circ\left[\int 1_G\,d\chi\right]^+(\omega, t)$$

$$= \,^\circ\left(\int G^2\,d[M]\right)^+(\omega, t) + \,^\circ\left[\int 1_G\,dt\right]^+(\omega, t)$$

$$= \int_0^t g^2 f\,ds + \int_0^t 1_g^2\,ds = \int_0^t 1\,ds = t,$$

and combining this with 4.4.18 we see that β is a Brownian motion adapted to $(\Omega, \{\mathscr{B}_s\}, L(P))$.

By (18) we have $f^{1/2} \in L^2(\nu_\beta)$, and

$$\int f^{1/2}\,d\beta = \int f^{1/2} g\,d\,^\circ M^+ + \int f^{1/2} \cdot 1_g\,d\,^\circ\chi^+ = \int f^{1/2} g\,d\,^\circ M^+$$

since $f^{1/2} \cdot 1_g = 0$. It only remains to prove $^\circ M^+ = \int f^{1/2} g\,d\,^\circ M^+$, since we then get 4.4.17 by putting $x = f^{1/2}$. By Doob's inequality

$$E\left(\sup_{\substack{q \leq 1 \\ q \in Q}}\left(^\circ M^+(q) - \int_0^q f^{1/2} g\,d\,^\circ M^+\right)^2\right) \leq 4E\left(\left(^\circ M^+(1) - \int_0^1 f^{1/2} g\,d\,^\circ M^+\right)^2\right)$$

$$= 4E\left(\int_0^1 (1 - f^{1/2} g)^2\,d\,^\circ[M]^+\right)$$

$$= 4E\left(\int_0^1 (1 - f^{1/2} g)^2 f\,dt\right) = 0$$

since $f^{1/2} g = 1$ whenever $f \neq 0$. Theorem 4.4.17 is proved.

This proof is not new; this technique for proving that processes are integrals of Brownian motions goes back to Doob. One way of using 4.4.17 is the following. Assume that we know a certain property holds for all standard integrals of Brownian motions. *A priori* there is no reason to believe that this property should also hold for all hyperfinite integrals, but by 4.4.17 all such integrals are standard integrals, and thus the property extends.

However, Theorem 4.4.17 is not really a standard characterization of nonstandard integrals since it does not tell us which Brownian motions β should be allowed. This question is answered in Lindstrøm (1986) and we shall give a brief account of the main results.

The *absolute joint variation* of two internal processes M_1 and M_2 is defined as

$$(20) \qquad [\![M_1, M_2]\!](\omega, t) = \sum_{s=0}^{t} \varepsilon(\omega, s) \, \Delta M_1(\omega, s) \, \Delta M_2(\omega, s),$$

where $\varepsilon(\omega, s) = \operatorname{sgn} E(\Delta M_1(s) \, \Delta M_2(s) | \mathscr{A}_s)(\omega)$. We say that M_1 and M_2 are *absolutely orthogonal* if $[\![M_1, M_2]\!](1) \approx 0$ a.e. If M_2 is absolutely orthogonal to all S-continuous SL^2-martingales that are absolutely orthogonal to M_1, we say that M_2 is in the *absolute span* of M_1.

4.4.19. PROPOSITION. Let N be an S-continuous SL^2-martingale with $N(0) = 0$. Then $N = \int X \, d\chi$ for some $X \in SL^2(\chi)$ if and only if N is in the absolute span of χ.

The nonstandard version has a standard counterpart which generalizes (10). If $M, N : \Omega \times [0, 1] \to \mathbb{R}$ are continuous L^2-martingales and $\pi = \{0 = t_0 < t_1 < \cdots < t_n = 1\}$ is a partition of $[0, 1]$ we let

$$(21) \qquad [\![M, N]\!]_{\pi} = \sum_{i=0}^{n-1} \varepsilon_{\pi}(\omega, t_i)(M(t_{i+1}) - M(t_i))(N(t_{i+1}) - N(t_i)),$$

where

$$\varepsilon_{\pi}(\omega, t_i) = \operatorname{sgn} E((M(t_{i+1}) - M(t_i))(N(t_{i+1}) - N(t_i)) | \mathscr{B}_{t_i})(\omega).$$

We define the absolute joint variation of M and N to be

$$(22) \qquad [\![M, N]\!] = \lim_{n \to \infty} [\![M, N]\!]_{\pi_n},$$

where $\{\pi_n\}$ is a sequence of partitions with $\delta(\pi_n) \to 0$. A lemma is needed to show that this is a well-defined notion at least for all continuous L^2-martingales with respect to $(\Omega, \{\mathscr{B}_t\}, L(P))$. We have the associated notion of *absolute span*.

4.4.20. THEOREM. Let $\chi : \Omega \times T \to {}^*\mathbb{R}$ be Anderson's random walk. A process $N : \Omega \times [0, 1] \to \mathbb{R}$ is of the form $N = {}^{\circ}(\int X \, d\chi)^{+}$ with $X \in SL^2(\chi)$ iff there is a Brownian motion β in the absolute span of ${}^{\circ}\chi^{+}$ and an adapted $x \in L^2(\Omega \times [0, 1])$ such that $N = \int x \, d\beta$.

The results in the first half of this section are adapted from Lindstrøm (1980b) and Hoover and Perkins (1983a,b), but the treatment is closely modeled on the theory of Anderson (1976) and Keisler (1984) for Brownian motions. Theorem 4.4.17 is from Lindstrøm (1985) and the last two results from Lindstrøm (1986).

4.5. STOCHASTIC DIFFERENTIAL EQUATIONS

We have completed our account of the fundamentals of the hyperfinite theory for stochastic integration and its relationship to the standard theory. In the sections which remain of this chapter, we shall take a look at various applications and extensions of the theory, trying to show by example what we have so far only postulated—namely, the strength and the flexibility of the nonstandard approach.

A. Itô Equations with Continuous Coefficients

In this section we shall discuss the existence of solutions of stochastic differential equations of the form

$$(1) \quad x(\omega, t) = x_0(\omega) + \int_0^t f(s, x(\omega, s)) \, ds + \int_0^t g(s, x(\omega, s)) \, db(\omega, s),$$

where b is an n-dimensional Brownian motion on a suitable Loeb space Ω, and f and g are functions

$$(2) \qquad f : [0, 1] \times \mathbb{R}^n \to \mathbb{R}^n, \qquad g : [0, 1] \times \mathbb{R}^n \to \mathbb{R}^n \otimes \mathbb{R}^n,$$

where $\mathbb{R}^n \otimes \mathbb{R}^n$ is the space of real $n \times n$-matrices.

First of all we must explain what this means. We have thus far restricted our discussion to real-valued processes, since the multidimensional theory for stochastic integration is a trivial extension of the one-dimensional. However, for stochastic differential equations the situation is quite different; here the multidimensional theory is far more complex, and one-dimensional techniques often have no extensions to the general case. To show the strength and applicability of our approach we shall have to work in a multidimensional setting.

Let $T = \{0, 1/\eta, 2/\eta, \ldots, 1\}$ be a hyperfinite time line. As our sample space we shall use $\Omega = \{-1, 1\}^{T \times \{1, 2, \ldots, H\}}$—the set of all internal functions from $T \times \{1, 2, \ldots, H\}$ to $\{-1, 1\}$—for some $H \in {}^*\mathbb{N}$, $H \geq n$. If $\omega \in \Omega$, we often write $\omega_i(t)$ for $\omega(t, i)$.

Let $\{e_1, e_2, \ldots, e_n\}$ be the natural basis for $\mathbb{R}^n$, and define an n-dimensional Anderson process by

$$(3) \qquad \chi(\omega, t) = \sum_{i=1}^n \sum_{s=0}^t \frac{\omega_i(s)}{\sqrt{\eta}} e_i.$$

Notice that χ consists of n independent random walks $\chi_1, \chi_2, \ldots, \chi_n$ running in orthogonal directions. The standard part of χ is an n-dimensional Brownian motion with respect to the Loeb measure $L(P)$ of the normalized counting measure P.

As in Section 4.1 an internal filtration $\{\mathcal{A}_t\}_{t \in T}$ is defined in terms of the equivalence classes

$$[\omega]_t = \{\omega' \in \Omega \,|\, \forall s < t \,\forall i \le H(\omega_i(s) = \omega_i'(s))\},$$

and the standard filtration $\{\mathcal{B}_t\}_{t \in [0,1]}$ is generated from $\{\mathcal{A}_t\}$ in the usual way.

For all $i, j \le n$ let X_{ij} be a nonanticipating process in $SL^2(P \times \lambda)$ (where λ is the normalized counting measure on T); the $*\mathbb{R}^n \otimes *\mathbb{R}^n$-valued process $X(\omega, t) = (X_{ij}(\omega, t))_{i,j \le n}$ is then said to be in $SL^2(\chi)$, and the *stochastic integral* is defined by

$$(4) \qquad \left(\int X \, d\chi \right)(\omega, t) = \sum_{s=0}^{t} X(\omega, s) \cdot \Delta\chi(\omega, s),$$

where $\cdot$ means matrix multiplication.

Notice that $\int X \, d\chi$ is an $*\mathbb{R}^n$-valued process; the ith component is given by

$$(5) \qquad \left(\int X \, d\chi \right)_i (\omega, t) = \sum_{j=1}^{n} \left(\int X_{ij} \, d\chi_j \right)(\omega, t).$$

This gives a definition of $\int X \, d\chi$ in terms of one-dimensional integrals, and the corresponding formula is used to define standard stochastic integrals of multidimensional processes. Because of this reduction all lifting and representation theorems we shall need will carry over from the one-dimensional case.

Having explained what each part of (1) means, let us now define what a solution is:

4.5.1. DEFINITION. Let $(\Omega, \{\mathcal{F}_t\}_{t \in [0,1]}, P)$ be a filtration and b a Brownian motion adapted to it. A *solution x of* (1) *with respect to* $(\Omega, \{\mathcal{F}_t\}, P, b)$ is an adapted process $x : \Omega \times [0, 1] \to \mathbb{R}$ such that $f(s, x(\omega, s))$ is in $L^1(\Omega \times [0, 1])$ and $g(s, x(\omega, s))$ is in $L^2(\Omega \times [0, 1])$, and such that for all $t \in [0, 1]$ equation (1) holds almost surely.

Existence theorems for solutions usually state that for all f in a certain class F and all g in a class G, equation (1) has a solution. Probabilistic jargon classifies such theorems as "weak," "strong," and "strict" according to the dependence of $(\Omega, \{\mathcal{F}_t\}, P, b)$ on f and g. We shall use the following version of this terminology (Barlow, 1982; Cutland, 1982; Keisler, 1984).

Weak solutions. For all $f \in F$, $g \in G$, there exist $(\Omega, \{\mathcal{F}_t\}, P, b)$ (depending on f and g) such that (1) has a solution x with respect to $(\Omega, \{\mathcal{F}_t\}, P, b)$.

Strong solutions. There exist $(\Omega, \{\mathcal{F}_t\}, P, b)$ such that for all $f \in F$, $g \in G$, Eq. (1) has a solution x with respect to $(\Omega, \{\mathcal{F}_t\}, P, b)$.

Strict solutions. Let $\Omega = C([0, 1])$, P Wiener measure, and $\mathcal{F}_t$ the σ-algebra generated by $C([0, t])$. Let b be the coordinate function; $b(\omega, t) = \omega(t)$. For all $f \in F$, $g \in G$, equation (1) has a solution with respect to $(\Omega, \{\mathcal{F}_t\}, P, b)$.

Notice that if an equation has a strict solution, then it also has a solution with respect to any other Brownian motion.

What we get by nonstandard methods are strong solutions living on hyperfinite Loeb spaces. Since most methods only yield weak solutions when the conditions on f and g become fairly general, this is in itself noteworthy. Also, hyperfinite Loeb spaces have the following homogeneity property [in the remainder of this section, unless otherwise specified, $(\Omega, \{\mathcal{A}_s\}, P)$ is a hyperfinite probability space of the form previously described]:

4.5.2. PROPOSITION. Let

$$f : [0, 1] \times \mathbb{R}^n \to \mathbb{R}^n, \qquad g : [0, 1] \times \mathbb{R}^n \to \mathbb{R}^n \otimes \mathbb{R}^n$$

be measurable functions. Suppose that there is a Brownian motion b adapted to $(\Omega, \{\mathcal{B}_t\}, L(P))$ such that for all $\mathcal{B}_0$-measurable initial conditions $x_0 : \Omega \to \mathbb{R}^n$, the equation

$$x(\omega, t) = x_0(\omega) + \int_0^t f(s, x(\omega, s)) \, ds + \int_0^t g(s, x(\omega, s)) \, db(\omega, s)$$

has a solution. Then the equation has a solution for every Brownian motion adapted to $\{\mathcal{B}_s\}$ and every $\mathcal{B}_0$-measurable initial condition.

This result is due to Keisler (1984), and although we shall not prove it here, it will be used below to extend results from one Brownian motion to all.

Proposition 4.5.2 shows that Loeb spaces of the form $\Omega = \{-1, 1\}^{T \times \{1, \dots, H\}}$ are extremely regular; e.g., if we can produce weak solutions on such spaces, we automatically get strong solutions.

To illustrate the hyperfinite approach to stochastic differential equations, we first take a look at a fairly simple, but typical proof from Keisler (1984).

4.5.3. PROPOSITION. Let b be an n-dimensional Brownian motion adapted to $(\Omega, \{\mathcal{B}_s\}, L(P))$, and assume that

$$f : [0, 1] \times \mathbb{R}^n \to \mathbb{R}^n, \qquad g : [0, 1] \times \mathbb{R}^n \to \mathbb{R}^n \otimes \mathbb{R}^n$$

are bounded measurable functions which are continuous in the second variable. Let x_0 be a $\mathcal{B}_0$-measurable initial condition. Then the equation

$$(6) \quad x(\omega, t) = x_0(\omega) + \int_0^t g(s, x(\omega, s)) \, ds + \int_0^t f(s, x(\omega, s)) \, db(\omega, s)$$

has a solution.

PROOF. Choose $\tilde{0} \approx 0$ such that x_0 has an $\mathscr{A}_{\tilde{0}}$-measurable lifting X_0. By Proposition 4.3.13 we can find liftings $F : T \times {}^*\mathbb{R}^n \to {}^*\mathbb{R}^n$ and $G : T \times {}^*\mathbb{R}^n \to {}^*\mathbb{R}^n \otimes {}^*\mathbb{R}^n$ of f and g which are S-bounded and uniform in the second variable. Hence there exists a set $T' \subset T$ of Loeb measure one such that ${}^\circ F(t, y) = f({}^\circ t, {}^\circ y)$ and ${}^\circ G(t, y) = g({}^\circ t, {}^\circ y)$ whenever $t \in T'$ and y is near-standard.

Let us consider the hyperfinite difference equation

$$(7) \quad X(\omega, t) = X_0(\omega) + \sum_{s=0}^{t} F(s, X(\omega, s)) \, \Delta t + \sum_{s=0}^{t} G(s, X(\omega, s)) \, \Delta\chi(\omega, s),$$

where χ is Anderson's random walk. This equation obviously has a unique, nonanticipating solution X defined inductively for all $t \geq \tilde{0}$ [recall that by our convention (4.1.3), the sum $\sum_{s=\tilde{0}}^{t}$ is really a sum $\sum_{s=\tilde{0}}^{t-\Delta t}$]. We extend X to all of T by letting $X(\omega, t) = 0$ for $t < \tilde{0}$.

Let x be the standard part ${}^\circ X^+$ of X; we shall show that x is a solution of the original equation (6) in the case where $b = {}^\circ \chi^+$. It suffices to show that $F(s, X(\omega, s))$ is a lifting of $f(s, x(\omega, s))$ and that $G(s, X(\omega, s))$ is a lifting of $g(s, x(\omega, s))$, as we will then have for all $t \geq \tilde{0}$

$$x({}^\circ t) \approx X(t) = X_0 + \sum_{s=0}^{t} F(s, X(s)) \, \Delta t + \sum_{s=0}^{t} G(s, X(s)) \, \Delta\chi(s)$$

$$\approx x_0 + \int_0^{{}^\circ t} f(s, x(s)) \, ds + \int_0^{{}^\circ t} g(s, x(s)) \, db,$$

which proves the theorem for $b = {}^\circ \chi^+$. The general case then follows from 4.5.2.

To prove that $F(s, X(\omega, s))$ and $G(s, X(\omega, s))$ are liftings of $f(s, x(\omega, s))$ and $g(s, x(\omega, s))$, respectively, we note that since F and G are S-bounded, there is a set $\Omega' \subset \Omega$ of Loeb measure one such that $X(\omega, s)$ is nearstandard for all s when $\omega \in \Omega'$. Thus if $(\omega, s) \in \Omega' \times T'$,

$$(8) \quad {}^\circ F(s, X(\omega, s)) = f({}^\circ s, x(\omega, {}^\circ s)), \qquad {}^\circ G(s, X(\omega, s)) = g({}^\circ s, x(\omega, {}^\circ s)),$$

which completes the proof, since $\Omega' \times T'$ obviously has measure one.

For an application with more sting, we refer the reader to Theorem 5.14 of Keisler (1984). The proof is just a slightly more technical variant of the one above, and the result is a new, strong existence theorem. Even Proposition 4.5.3 is less innocent than the proof might lead you to believe; Barlow (1982) has shown that (6) may have no *strict* solution even in the one-dimensional case! (Warning: Barlow's terminology is slightly different from ours.) This is an indication that the Loeb space is a better setting for stochastic differential equations than standard path space $C[0, 1]$.

B. Itô Equations with Measurable Coefficients

The argument we have just given is typical; by choosing appropriate liftings F and G of f and g, we turn Eq. (1) into a hyperfinite difference equation (7), and the solution of Eq. (1) is obtained as the standard part of the solution of Eq. (7). There are two technical problems that occur; first we have to find the proper liftings F and G, and then we have to prove that $F(s, X(\omega, s))$ and $G(s, X(\omega, s))$ are liftings of $f(s, x(\omega, s))$ and $g(s, x(\omega, s))$, respectively. In the theorem above these problems were relatively easy to solve since f and g were continuous in the space variable. If this is not the case the problems become much harder; the reason is that although we may choose F and G such that they differ from f and g only on a null set in $T \times \Omega$, the process may nevertheless stay in this null set with positive probability, and this will destroy the lifting properties. Following Keisler (1984), we shall see how this problem can be solved under the extra hypothesis that g does not degenerate.

We shall need the following inequality due to Krylov (1974, 1980):

4.5.4. THEOREM. Let $x : \Omega \times [0, 1] \to \mathbb{R}^n$ be a process of the form

$$x(\omega, t) = x_0 + \int_0^t f(\omega, s) \, ds + \int_0^t g(\omega, s) \, db(\omega, s),$$

where b is a Brownian motion adapted to a filtration $(\Omega, \{\mathscr{F}_t\}, P)$;

$$f : \Omega \times [0, 1] \to \mathbb{R}^n \qquad \text{and} \qquad g : \Omega \times [0, 1] \to \mathbb{R}^n \otimes \mathbb{R}^n$$

are bounded, adapted processes; and $x_0 \in \mathbb{R}^n$. For $d \in \mathbb{R}_+$, let

$$\tau_d(\omega) = \inf\{t \in [0, 1] \mid |x(\omega, t)| \geq d\} \wedge 1.$$

Given $d, K \in \mathbb{R}_+$, there is a constant $N = N(n, d, K)$—depending only on d, K, and the dimension n—such that for all adapted f and g bounded by K, and all functions $h \in L^{n+1}([0, 1] \times \mathbb{R}^n)$

$$E\left(\int_0^{\tau_d} (\det g(\omega, t))^{2/(n+1)} |h(t, x(\omega, t))| \, dt \right) \leq N \|h\|_{n+1}.$$

Krylov's inequality is a deep result which we do not prove here, especially as we have no new, nonstandard insight to offer. Instead we refer the reader to Krylov's (1980) book; what we have stated is a special case of his Theorem 2.2.2.

We shall need a hyperfinite version of 4.5.4. If $J \in {}^*\mathbb{R}_+$, a function $h : {}^*([0, 1] \times \mathbb{R}^n) \to {}^*\mathbb{R}^m$ is called J-*Lipschitz* if it is internal, bounded by J, and satisfies

$$\|h(t, x) - h(s, y)\| \leq J \|(t, x) - (s, y)\|$$

for all $(t, x), (s, y) \in {}^*([0, 1] \times \mathbb{R}^n)$.

If χ is Anderson's random walk in $^*\mathbb{R}^n$, and $U : \Omega \times T \to {}^*(\mathbb{R}^n \otimes \mathbb{R}^n)$ is a nonanticipating process such that $U(\omega, t)$ is a unitary matrix for all (ω, t), then the standard part of

$$\chi' = \int U \, d\chi$$

is a Brownian motion. We denote the class of all such χ' by $\mathcal{U}(\chi)$.

4.5.5. PROPOSITION. Let $\chi : \Omega \times T \to {}^*\mathbb{R}^n$ be Anderson's random walk. For all $n \in \mathbb{N}$, D, $K \in \mathbb{R}_+$, there exist constants $N = N(n, D, K) \in \mathbb{R}_+$ and $J \in {}^*\mathbb{N} - \mathbb{N}$ such that if $X : \Omega \times T \to {}^*\mathbb{R}^n$ is of the form

$$(9) \quad X(\omega, t) = X_0 + \int_0^t F(s, X(\omega, s)) \, ds + \int_0^t G(s, X(\omega, s)) \, d\chi'(\omega, s),$$

where

$$F : {}^*([0, 1] \times \mathbb{R}^n) \to {}^*\mathbb{R}^n \qquad G : {}^*([0, 1] \times \mathbb{R}^n) \to {}^*(\mathbb{R}^n \otimes \mathbb{R}^n)$$

are J-Lipschitz and bounded by K, $X_0 \in {}^*\mathbb{R}^n$, and $\chi' \in \mathcal{U}(\chi)$, then

$$(10) \quad {}^\circ E \left(\int_0^{\sigma_D} (\det G(s, X(\omega, s)))^{(2/(n+1))} |H(s, X(\omega, s))| \, dt \right) \leq N^\circ \|H\|_{n+1}$$

for all J-Lipschitz $H : {}^*([0, 1] \times \mathbb{R}^n) \to {}^*\mathbb{R}$, where

$$\sigma_D = \inf\{t \in T \,|\, |X(\omega, t)| \geq D\} \wedge 1.$$

Before we prove the proposition, we observe that we can take J to be independent of $(n, D, K) \in \mathbb{N} \times \mathbb{R}_+ \times \mathbb{R}_+$. If we can find a suitable $J(n, D, K)$ for each such triple, then there is by saturation a $J \in {}^*\mathbb{N} - \mathbb{N}$ smaller than all of them, and this J will work for all choices of n, D, and K.

To prove the theorem, we first consider the case where $J \in \mathbb{N}$. Then F and G are S-continuous and we can define functions f and g by

$$f({}^\circ t, {}^\circ x) = {}^\circ F(t, x); \qquad g({}^\circ t, {}^\circ x) = {}^\circ G(t, x).$$

If x is the standard part of X,

$$(11) \quad x(\omega, t) = x_0 + \int_0^t g(s, x(\omega, s)) \, ds + \int_0^t f(s, x(\omega, s)) \, db'(\omega, s)$$

where $x_0 = {}^\circ X_0$ and $b' = {}^\circ \chi'^+$ (if X_0 is not nearstandard, $\sigma_D = 0$ and there is nothing to prove).

Applying Theorem 4.5.4 to x with $d = D + 1$, we get

$$(12) \quad E \left(\int_0^{\tau_d} (\det g(s, x(\omega, s)))^{2/(n+1)} |h(s, x(\omega, s))| \, ds \right)$$

$$\leq N(n, d, K) \|h\|_{n+1}.$$

Since all J-Lipschitz functions are S-continuous and S-bounded, it follows from (12) that for all J-Lipschitz H:

$$(13) \qquad E\left(\int_0^{\sigma_D} (\det G(s, X(s, \omega)))^{2/(n+1)} |H(s, X(\omega, s))| \, ds \right)$$

$$\leq N(n, D + 1, K) \|H\|_{n+1} + 1/J,$$

where $1/J$ is added to take care of the case where we have equality in (12).

Consider the internal set A defined as $\{J \in {}^*\mathbb{N} \,|\, (13) \text{ holds for all } J\text{-Lipschitz } F, G, H, \text{ all } X_0 \in {}^*\mathbb{R}^d, \text{ and all } \chi' \in \mathcal{U}(\chi)\}$. Since we have just shown that $A \supset \mathbb{N}$, it also contains an infinite J, and the proposition follows.

In order to use Proposition 4.5.5, we must know how to produce functions that are J-Lipschitz. Let $h : {}^*\mathbb{R}^m \to {}^*\mathbb{R}^k$ be any internal, S-bounded function. If $x \in {}^*\mathbb{R}^m$, let $[x]^J$ be the box centered at x with sides of length $1/\sqrt{J}$. If $J \in {}^*\mathbb{N} - \mathbb{N}$, an easy calculation shows that the function h^J defined by

$$h^J(x) = m([x]^J)^{-1} \int_{[x]^J} h(x) \, dm(x)$$

is J-Lipschitz (here m is Lebesgue measure).

Another important aspect of 4.5.5 is the need to keep track of points where ${}^\circ\det(G(s, X(\omega, s))) = 0$ since the estimate does not tell us anything about them. The key observation is the following lemma.

4.5.6. LEMMA. Let $a : [0, 1] \times \mathbb{R}^n \to \mathbb{R}^n \otimes \mathbb{R}^n$ be a bounded and measurable function taking non-negative, symmetric values. Then a^J is a J-Lipschitz lifting of a also taking symmetric and non-negative values. Moreover, for each $M \in \mathbb{R}_+$, there is an $m \in \mathbb{R}_+$ such that if $\det {}^*a(s, y) \geq M$ for all $(s, y) \in [(t, x)]^J$, then $\det a^J(t, x) \geq m$.

PROOF. By Anderson's Lusin theorem 3.4.9, we know that ${}^*a$ is a lifting of a, and hence $a^J = {}^*a^J$ is a J-Lipschitz lifting of a.

To prove the last part of the lemma, we define

$$l(B) = \inf\{\langle Bx, x \rangle \,|\, \|x\| = 1\}$$

for all non-negative, symmetric $n \times n$ matrices B, and note that

$$(14) \qquad\qquad l(B + C) \geq l(B) + l(C).$$

Since $\det B$ is the product of the eigenvalues of B and $l(B)$ is the smallest eigenvalue,

$$(15) \qquad\qquad l(B)^n \leq \det B \leq \|B\|^{n-1} l(B).$$

Returning to our function a, we get from (15) that if $\det {}^*a(s, y) \geq M$ for all $(s, y) \in [(t, x)]^J$, then $l({}^*a(s, y)) \geq M/\|a\|^{n-1}$ for the same (s, y)'s.

From the superadditivity property (14), we get $l(a^J(t, x)) \geq M / \|a\|^{n-1}$, and using (15) again we obtain

$$\det(a^J(t, x)) \geq (M / \|a\|^{n-1})^n,$$

which shows that we can choose $m = (M / \|a\|^{n-1})^n$.

We have now reached the stage where we can prove that solutions to (1) exist even in the case where f and g are not continuous in the space variable, provided that the determinant of g is bounded away from zero. A "weak" solution of this kind was first obtained by Krylov (1974); the "strong" result we shall give is due to Keisler (1984).

4.5.7. THEOREM. Let

$$f : [0, 1] \times \mathbb{R}^n \to \mathbb{R}^n, \qquad g : [0, 1] \times \mathbb{R}^n \to \mathbb{R}^n \otimes \mathbb{R}^n$$

be bounded, measurable functions, and assume that there exists an $\varepsilon \in \mathbb{R}_+$ such that $|\det g(t, y)| > \varepsilon$ for all $(t, y) \in [0, 1] \times \mathbb{R}^n$. For all $\mathcal{B}_0$-measurable random variables x_0, and all Brownian motions b adapted to $(\Omega, \{\mathcal{B}_t\}, L(P))$ the equation

$$(16) \quad x(\omega, t) = x_0(\omega) + \int_0^t f(s, x(\omega, s)) \, ds + \int_0^t g(s, x(\omega, s)) \, db(\omega, s)$$

has a solution.

PROOF. We shall only consider the case where x_0 is constant and b is a particular Brownian motion. The easy task of gluing these processes together to obtain solutions for general x_0's is left to the reader, as is the final appeal to Proposition 4.5.2.

We begin by simplifying the problem further. For each (t, x) let $|g|(t, x)$ be the absolute value of $g(t, x)$, i.e., the unique symmetric, non-negative matrix such that $|g|^2 = g\,'g$, where $'g$ is the transpose of g. Let $u(t, x)$ be the unitary matrix such that $|g| = g \cdot u$. We claim that it suffices to find a solution to the equation

$$(17) \qquad x(\omega, t) = x_0 + \int_0^t f(s, x(\omega, s)) \, ds + \int_0^t |g|(s, x(\omega, s)) \, db,$$

where $|g|$ has replaced g. The reason is simply that since $u(t, x)$ is a unitary matrix, the process $b' = \int u \, db$ is a Brownian motion. Thus $\int |g| \, db = \int g \, db'$, and a solution of (17) is also a solution of (16), although with respect to a different Brownian motion.

To solve (17), we let J be an element of $*\mathbb{N} - \mathbb{N}$ satisfying Proposition 4.5.5, and choose $J_0 \in *\mathbb{N} - \mathbb{N}$ infinitesimal compared to J. The reason why we pass to the second constant will become clear later. Let $F = f^{J_0}$ and

$|G| = |g|^{J_0}$, and observe that by 4.5.6, the standard part of $|G|$ is bounded away from zero.

Let X be the solution of the hyperfinite difference equation

$$(18) \qquad X(\omega, t) = x_0 + \sum_{s=0}^{t} F(s, X(\omega, s)) \, \Delta t + \sum_{s=0}^{t} |G|(s, X(\omega, s)) \, \Delta \chi(\omega, s);$$

we shall show that $x = {}^\circ X^+$ is a solution of (17) when $b = {}^\circ \chi^+$. All we need to prove is that ${}^\circ F(s, X(\omega, s)) = f({}^\circ s, x(\omega, {}^\circ s))$ and ${}^\circ |G|(s, X(\omega, s)) = |g|({}^\circ s, x(\omega, {}^\circ s))$ almost everywhere in $\Omega \times T$. To this end let $B_K = \{(t, x) \in T \times {}^*\mathbb{R}^n \mid \|x\| \le K\}$, and define

$$A_K = \{(t, x) \in B_K \mid {}^\circ F(t, x) \neq f({}^\circ t, {}^\circ x) \text{ or } {}^\circ |G|(t, x) \neq |g|({}^\circ t, {}^\circ x)\}.$$

We shall show that there is J-Lipschitz function H_ε which is 1 on A_K and has ${}^\circ \|H_\varepsilon\|_{n+1} \le \varepsilon$. Applying 4.5.5 to X and H_ε, we get

$$E\left(\int^{\sigma_0} \det|G|(s, X(\omega, s))^{2/(n+1)} |H_\varepsilon(s, X(\omega, s))| \right) ds \le N\varepsilon.$$

Since $\det|G|$ is bounded away from zero, it follows that

$$L(P \times \lambda)\{(\omega, t) \mid (t, X(\omega, t)) \in A_K\} = 0$$

for all $K \in \mathbb{R}_+$, and hence $x = {}^\circ X^+$ is a solution of (17).

To construct H_ε, observe that since F and G are J_0-Lipschitz for a J_0 which is infinitesimal compared to J, the set A_K has the following "openness" property: if $x \in A_K$ and $|x - y| \le 1/J$, then $y \in A_K$. Thus if B is an internal set in the complement of A_K, the function $H_\varepsilon(y) = J \cdot d(y, B) \wedge 1$ is one on A_K. If we also choose B such that its complement has measure less than ε, then $\|H_\varepsilon\|_{n+1} < \varepsilon$ as H_ε is zero on B. But H_ε is obviously J-Lipschitz, and hence the theorem is proved.

Observe that in the proof above we only proved that (16) has a weak solution; Proposition 4.5.2 took care of the rest. Since we have claimed that one of the nicest features of the nonstandard theory is its ability to produce strong solutions, but at the same time have refused to prove 4.5.2, the reader could accuse us of not playing fair. However, by changing the proof above slightly, we can get a strong solution directly. The idea is to replace (17) by

$$(19) \qquad X(\omega, t) = x_0 + \sum_0^t F(s, X(\omega, s)) \, \Delta t$$

$$+ \sum_0^t |G|(s, X(\omega, s)) \, \Delta\left(\int V(s, X(\omega, s)) \, d\chi\right),$$

where V is a suitable lifting of u^{-1}. The proof proceeds along the same lines as above, but becomes a little more technical since we also have to treat the relationship between u and V.

In a certain sense Keisler's result is the best we can expect if we only assume f and g to be measurable. If we allow g to degenerate, all potential solutions of (16) may take values in a null set in $\mathbb{R}^n$, and if f and g are only given as measurable functions, their values on null sets are arbitrary— the equation no longer makes sense. However, by assuming that f and g are continuous on the set where g degenerates, Kosciuk (1982, 1983) obtained the following generalization:

4.5.8. THEOREM. Let $f : [0, 1] \times \mathbb{R}^n \to \mathbb{R}^n$ and $g : [0, 1] \times \mathbb{R}^n \to \mathbb{R}^n \otimes \mathbb{R}^n$ be bounded, measurable functions, and define $N \subset [0, 1] \times \mathbb{R}^n$ by $N = \{(t, x) |$ there is a sequence $(t_n, x_n) \to (t, x)$ such that $\det g(t_n, x_n) \to 0\}$. Assume that the restrictions of f and g to N are continuous, and that if either f or g is discontinuous at a point $(t, x) \in N$, then $\det g$ is bounded away from zero on $V - N$ for some neighborhood V of (t, x). Under these conditions the equation

$$(20) \qquad x(\omega, t) = x_0 + \int_0^t f(s, x(\omega, s)) \, ds + \int_0^t g(s, x(\omega, s)) \, db(s)$$

has a solution for all Brownian motions b adapted to $(\Omega, \{\mathscr{B}_t\}, L(P))$ and all $\mathscr{B}_0$-measurable initial conditions x_0.

PROOF. As in the proof of 4.5.7 we may assume that g is symmetric and non-negative by passing to the absolute value $|g|$ if necessary. We shall also assume that x_0 is constant and that b is the right standard part of Anderson's random walk χ.

Choose a $J \in {}^*\mathbb{N} - \mathbb{N}$ satisfying 4.5.5, and let $F = f^{J_0}$, $G = g^{J_0}$ for a $J_0 \in {}^*\mathbb{N} - \mathbb{N}$ that is infinitesimal compared to J. We shall prove that the standard part x of the process

$$(21) \quad X(\omega, t) = x_0 + \sum_{s=0}^{t} F(s, X(\omega, s)) \, \Delta t + \sum_{s=0}^{t} G(s, X(\omega, s)) \, \Delta\chi(\omega, s)$$

is a solution of (20). As usual, it suffices to show that

$$(22) \qquad\qquad L(P \times \lambda)\{(\omega, t) | (t, X(\omega, t)) \in A_K\} = 0$$

for all $K \in \mathbb{N}$, where

$$(23) \qquad \begin{aligned} A_K = \{(t, x) \mid \|x\| &\le K \text{ and either } {}^\circ F(t, x) \ne f({}^\circ t, {}^\circ x) \\ &\text{or } {}^\circ G(t, x) \ne g({}^\circ t, {}^\circ x)\}. \end{aligned}$$

We shall first show that if $\det G(t, x) \approx 0$, then $(t, x) \notin A_K$. By the conditions on g and Lemma 4.5.6, it is clear that $\det G(t, x)$ can only be infinitesimal when $(t, x) \in \mathrm{st}^{-1}(N)$. Let N_C be the points in N where both

f and g are continuous, and let $N_P = N - N_C$. Assume first that $(t, x) \in$ $\mathrm{st}^{-1}(N_C)$; then f and g are continuous at $(^\circ t, {}^\circ x)$, and (t, x) cannot be in A_K. If, on the other hand, $(t, x) \in \mathrm{st}^{-1}(N_P)$, then $\det G(t, x)$ can only be infinitesimal if

$$(24) \qquad \frac{m([(t, x)]^{J_0} \cap {}^*N)}{m([(t, x)]^{J_0})} \approx 1,$$

since $\det g$ is bounded away from zero on $V - N$. However, since f and g are continuous in N, formula (24) implies that $(t, x) \notin A_K$. Thus we have shown that if $\det G(t, x) \approx 0$, then $(t, x) \notin A_K$.

For each $\varepsilon \in \mathbb{R}_+$, let H_ε be a J-Lipschitz function that is one on A_K and has L^{n+1} norm less than ε, as in the proof of 4.5.7. By 4.5.5

$$E \left(\int_0^{\sigma_D} \det G(s, X(\omega, s))^{2/(n+1)} |H_\varepsilon(s, X(\omega, s))| \, ds \right) \leq N\varepsilon.$$

Since $\det G \not\approx 0$ on A_K, and ε, K, and D are arbitrary elements of $\mathbb{R}_+$, $\mathbb{N}$, and $\mathbb{N}$, respectively, (22) follows. The proof is complete.

We should mention that we have reformulated Kosciuk's result somewhat; in his papers Kosciuk (1982, 1983) does not consider the stochastic differential equation (20), but constructs a solution to the associated martingale problem without using stochastic integration [for information on martingale problems, see Stroock and Varadhan (1979)]. The proof we have given is shorter and fits better into our presentation.

As Kosciuk (1982) pointed out, Theorem 4.5.8 may have interesting applications in biophysics and biochemistry, e.g., to the study of reactions between regulatory proteins and DNA molecules. Here is what is believed to happen (Richter and Eigen, 1974): the protein is diffusing in a solution until it hits a DNA molecule (nonspecific binding); the process then abruptly changes to a diffusion along the surface of the DNA until the protein either dissociates from it or reaches a particular site where it gets trapped (specific binding). The lower-dimensional diffusion along the DNA chain is necessary to explain the high association rate found in experiments; if we use a model with only a three-dimensional diffusion, the value will be much too low.

A mathematical model for the phenomenon was suggested by Berg and Blomberg (1976, 1977), using a coupling of three-dimensional and one-dimensional diffusion processes, but their theory has some unsatisfactory features; e.g., the dissociation of the protein from the DNA can only be obtained by allowing it to jump a finite distance back into the solution. In Theorem 4.5.8 we have a flexible tool for engineering models of diffusions with abrupt changes and degeneracies in the coefficients, and it seems likely that by applying it to proteins and DNA molecules, we would arrive at a better mathematical model. However, it should be pointed out that using their model, Berg and Blomberg were able to obtain connections between the diffusion coefficients and physical properties of the DNA molecule, and that no analysis on this level has yet been attempted using 4.5.8—it remains an interesting challenge for future research.

C. Equations with Coefficients Depending on the Past

As a final example of the hyperfinite approach to stochastic differential equations, we shall take a brief look at equations where the coefficients depend on the past of the process and not only on its present state. Although there are several papers dealing with equations of this sort, we shall concentrate on Hoover and Perkins's (1983b) work on equations where the driving term is a semimartingale, and even in this case restrict ourselves to the main ideas—for a description of the technical machinery, you will have to consult the original paper or Perkins' (1983) survey article.

A semimartingale is a process $z = a + m$ where a has paths of bounded variation and m is a local martingale. Hoover and Perkins studied equations of the form

$$(25) \qquad y(\omega, t) = h(\omega, t) + \int_0^t f(\omega, s, y(\omega, \cdot)) \, dz(\omega, s),$$

where z is a d-dimensional semimartingale, $h : \Omega \times [0, 1] \to \mathbb{R}^n$ is right-continuous with left limits, and $f(\omega, s, y(\omega, \cdot)) \in \mathbb{R}^n \otimes \mathbb{R}^d$ only depends on the values of $y(\omega, \cdot)$ up to time s. The strategy is the same as before; we find a well-behaved (in the technical sense of Section 4.4) lifting H of h, and a hyperfinite process Z such that $^\circ Z^+ = z$ and Z and all its stochastic integrals are well behaved. If we can construct a suitable lifting F of f, we can solve the hyperfinite difference equation

$$(26) \qquad Y(\omega, t) = H(\omega, t) + \sum_{s=0}^{t} F(\omega, s, Y(\omega, \cdot)) \, \Delta Z(\omega, s),$$

and since all processes in (26) are well behaved, it is not hard to see that under reasonable conditions the right standard part y of Y is a solution of (25).

But how do we construct F? If J_1 is the Skorohod topology on the space $\mathcal{D}$ of right-continuous processes with left limits, we can use Proposition 4.3.13 to pick a lifting F of f which is uniformly continuous in the third variable, provided that $f(\omega, s, \cdot)$ is J_1-continuous for all ω and all s. However, Hoover and Perkins only assumed that $f(\omega, s, \cdot)$ was continuous in the uniform topology, and this makes the problem much harder as $\mathcal{D}$ is then no longer separable. Since we only need to lift f on functions y which are possible solutions to (25), Hoover and Perkins could show that we are only interested in a subspace of $\mathscr{D}$ where the two topologies coincide [this is basically because a solution to (25) can only jump when h or z jumps]. Hence we get our lifting F, and the standard part of Y is a solution of (25). For the precise statement of the result, we again refer the reader to Hoover and Perkins (1983b) and Perkins (1983).

Prior to Hoover and Perkins's work, solutions to (25) were only known to exist under Lipschitz conditions on the space variable in f [Doléans–Dade (1976); Protter (1977a,b)], but a variant of the general result was found independently and at about the same time by Jacod and Memin (1981). Other nonstandard studies of stochastic differential equations with dependence on the past include Cutland (1982) and a paper by Osswald (1984). We shall take a closer look at Cutland's ideas when we turn to stochastic control theory in the next section; at this point we only mention an interesting feature of his latest papers, where he works with internal Brownian motions instead of hyperfinite random walks. This *-continuous approach seems to simplify a number of technical problems concerning liftings and transfer of standard results; see his proof (Cutland, 1985e) of Keisler's existence theorem 4.5.7 as an illustrative example.

In this section we have concentrated on existence theorems, although nonstandard methods have also been used successfully to obtain other kinds of results about stochastic differential equations, notably invariance principles (Keisler, 1984; Kosciuk, 1982) and Markov properties (Keisler, 1984). With uniqueness questions there has yet been no progress. For an interesting application to economics, see Keisler (1983).

4.6. OPTIMAL STOCHASTIC CONTROLS

The stochastic differential equations

$$(1) \qquad x(t) = x_0 + \int_0^t f(s, x)\, ds + \int_0^t g(s, x)\, db(s)$$

of the last section have been used to model a wide variety of phenomena in science, engineering, and economics. What many of these phenomena have in common is a feedback mechanism which modifies the process's future behavior on the basis of an observation of its past. The purpose of this modification is usually to optimize a certain outcome of the process. Mathematically the situation can be described as follows: we consider processes

$$(2) \qquad x^u(t) = x_0 + \int_0^t f(s, x^u, u(s, x^u))\, ds + \int_0^t g(s, x^u, u(s, x^u))\, db$$

where f and g are given functions, but where the *control u* may vary. If the outcome of the process is determined by a function h, we define the *cost* of x^u to be

$$(3) \qquad j(u) = E\left(\int_0^1 h(x^u(t))\, dt \right).$$

The idea is to find the control u which minimizes the cost $j(u)$.

It is obvious that the solution to this problem depends on which functions u we allow. Certain restrictions are necessary; e.g., (2) does not make sense unless u is nonanticipating in the sense that $u(s, x)$ only depends on the values of x up to time s. On the other hand, it is not always the case that u should be allowed to depend on the entire past of the process; the feedback mechanism may be based on incomplete information about the system and this will result in a smaller choice of u's. Hence we are led to the subject matter of this section, optimal controls for partially observed stochastic systems.

The nonstandard approach we shall present here has been developed by Cutland (1983a–c, 1985a,b) in a series of papers. Although we have changed the details slightly to fit into our framework, the main ideas are all due to him. Our aim is to prove the existence of optimal controls under quite general conditions, and we shall first take a look at the case where $f(s, x, a)$, $g(s, x, a)$, and $u(s, x)$ only depend on the value of x at time s. In this situation Krylov's inequality and the techniques of the last section can be used. When f, g, and u are allowed to depend on the past of x, the problem becomes more difficult, and we can only solve it under extra hypotheses. The main tool in this case is Girsanov's formula, which we shall develop from scratch.

A. Optimal Controls: The Markov Case

The first type of controlled systems we shall study is

$$(4) \qquad x(t) = \int_0^t f(x(s), s, u(x, s))\, ds + \int_0^t g(x(s), s, u(x, s))\, db(s),$$

where $f : \mathbb{R}^n \times [0, 1] \times K \to \mathbb{R}^n$, $g : \mathbb{R}^n \times [0, 1] \times K \to \mathbb{R}^n \otimes \mathbb{R}^n$ are continuous in the third variable and K is a compact Polish space (i.e., a topological space which allows a complete, separable metric). Since f and g do not depend on the past of x, we shall refer to this rather sloppily as "the Markov case"; strictly speaking this terminology is incorrect, as a certain dependence on the past is introduced through the control u.

The feedback mechanism works as follows. *Observations* are made at fixed times $0 \le t_1 < t_2 < \cdots < t_p < 1$, and the result of the observation at time t_i is recorded as a value $y_i(x(t_i))$, where $y_i : \mathbb{R}^n \to \mathcal{Y}$ is a measurable function into a countable *observation space* $\mathcal{Y}$. Given a path $x : [0, 1] \to \mathbb{R}^n$, we let

$$y(x) = (y_1(x(t_1)), \ldots, y_p(x(t_p)))$$

be the sequence of observations made on x.

4.6.1. DEFINITION. An *admissible control* is a measurable function $u : \mathcal{Y}^p \times [0, 1] \to K$ that is nonanticipating in the following sense: if $t_i \leq t < t_{i+1}$, then $u(y, t)$ depends only on the first i components of $y = (y_1, y_2, \ldots, y_p)$.

Given a bounded, measurable *cost function* $h : \mathbb{R}^n \times [0, 1] \to \mathbb{R}$, the problem is to find a control u which minimizes the *cost*

$$(5) \qquad j(u) = E\left(\int_0^1 h(x^u(t)) \, dt \right),$$

where

$$(6) \qquad x^u(t) = \int_0^t f(x^u(s), s, u(y(x^u), s)) \, ds$$

$$+ \int_0^t g(x^u(s), s, u(y(x^u)s)) \, db(s).$$

For this to make sense, the solutions to (6) must be unique in distribution, and this is guaranteed by assuming that g is Lipschitz continuous in the first variable and that g^{-1} is bounded on compacts. But even when (6) has a unique solution, there may not exist an admissible optimal control. The problem is that since the set of controls has rather weak closure properties, the infimum

$$(7) \qquad \inf\{j(u) \mid u \text{ an admissible control}\}$$

need not be attained. One way of solving—or evading—this difficulty is to allow measure-valued controls. Let $\mathcal{M}(K)$ be the set of Borel probability measures on K with the weak topology.

4.6.2. DEFINITION. An *admissible relaxed control* is a measurable function $u : \mathcal{Y}^p \times [0, 1] \to \mathcal{M}(K)$ that is nonanticipating in the sense of Definition 4.6.1.

Relaxed controls were first introduced in a standard context (Filippov, 1962; McShane, 1967; Warga, 1967, 1972), but we shall see below that they arise naturally in nonstandard theory as the standard parts of internal controls.

We must explain the proper way of interpreting (6) when u is measure valued. The drift term causes no difficulties; just let

(8)
$$\int_r^t f(x^u(s), s, u(y(x^u), s)) \, ds$$

$$= \int_r^t \int_K f(x^u(s), s, a) \, du(y(x^u), s)(a) \, ds.$$

The interpretation of the martingale term is perhaps more surprising; we let

(9)
$$\int_r^t g(x^u(s), s, u(y(x^u), s)) \, db$$

$$= \int_r^t \left(\int_K g(x^u(s), s, a)^2 \, du(y(x^u), s)(a) \right)^{1/2} db.$$

Why this is the appropriate definition will become clear later; it suffices to say here that it is connected to the fact that it is the covariance of the diffusion that determines the dynamics of the system.

Before stating the result we are aiming at, we introduce conditions on f, g, and h. The conditions are stronger than what is strictly necessary, but they make the theory run smoothly without too many technicalities.

4.6.3. CONDITION. Assume that $f : \mathbb{R}^n \times [0, 1] \times K \to \mathbb{R}^n$, $g : \mathbb{R}^n \times [0, 1] \times K \to \mathbb{R}^n \otimes \mathbb{R}^n$, and $h : \mathbb{R}^n \times [0, 1] \to \mathbb{R}$ are bounded, measurable functions with the following properties:

(i) f and g are continuous in the third variable;

(ii) g is uniformly Lipschitz continuous in the first variable; i.e., there is a constant K such that $\|g(x, t, a) - g(y, t, a)\| \leq K \|x - y\|$ for all x, y, t, a;

(iii) the values of g are positive definite, symmetric matrices whose determinants are bounded away from zero, i.e., there is a constant $\varepsilon > 0$ such that $|\det g(x, t, a)| > \varepsilon$ for all x, t, a.

Note that (i) and (ii) imply that g is jointly continuous in the first and the third variable.

4.6.4. THEOREM. Assume that f, g, and h satisfy Condition 4.6.3. Then there exists an admissible relaxed control u that minimizes the cost

(10)
$$j(u) = E \left(\int_0^1 h(x^u(t), t) \, dt \right);$$

i.e., $j(u) \leq j(\tilde{u})$ for all other admissible relaxed controls $\tilde{u}$.

The control u is called an *optimal relaxed control* for the system defined by f, g, and h.

To prove Theorem 4.6.4 we shall apply the following strategy. We first translate the problem into a hyperfinite setting by using Anderson's random walk and suitable liftings of f, g, and h. In the nonstandard problem it will be immediately clear that an optimal control exists. Taking the standard part of this internal optimal control, we obtain the optimal relaxed control for the original system.

In order to carry out this program, we must first study internal controls and their relationship to relaxed admissible controls. Let

(11) $$T = \{0, \Delta t, 2 \Delta t, \ldots, 1\}$$

be an internal time line and

(12) $$0 \le \tilde{t}_1 < \tilde{t}_2 < \cdots < \tilde{t}_p < 1$$

a finite sequence of elements of T with $°\tilde{t}_i = t_i$ for all $i \le p$.

4.6.5. DEFINITION. An *internal control* is an internal function $U : {}^*\mathcal{Y}^p \times T \to {}^*K$ that is nonanticipating in the following sense: if $\tilde{t}_i \le t < \tilde{t}_{i+1}$, then $U(Y, t)$ depends only on the first i components of $Y = (Y_1, \ldots, Y_p)$.

We shall now define the standard part of an internal control to be an admissible relaxed control. The intuitive idea behind the construction is as follows: fix a time $t \in [0, 1]$ and a control parameter $y \in \mathcal{Y}^p$. For each $n \in \mathbb{N}$, let $\nu_n(y, t)$ be the internal measure on ${}^*K$ given by

(13) $$\nu_n(y, t)(A) = 2n \cdot \# \left\{ s \mid t - \frac{1}{n} \le s \le t + \frac{1}{n} \wedge U(y, s) \in A \right\}.$$

If $u_n(y, t) = L(\nu_n(y, t)) \circ \mathrm{st}^{-1}$, the standard part of U is the limit of u_n as n goes to infinity. Thus $u(y, t)$ is the distribution of $U(y, s)$ over the monad of t. The proof of the next lemma gives an alternative construction of u which is perhaps less intuitive, but technically more convenient.

4.6.6. LEMMA. For each internal control U there is an admissible relaxed control u with the following property: assume that $z : [0, 1] \times K \to \mathbb{R}$ is a bounded, measurable function continuous in the second variable, and let $Z : T \times {}^*K \to {}^*\mathbb{R}$ be an S-bounded lifting of z such that for all t outside a set of Loeb measure zero, $°Z(t, a) = z(°t, °a)$ for all $a \in {}^*K$. Then for all a, b, y

(14) $$\int_{°a}^{°b} z(t, u(y, t)) \, dt = °\sum_{a}^{b} Z(t, U(y, t)) \, \Delta t.$$

PROOF. Fix a control parameter y, and let ν_y be the internal measure on $T \times {}^*K$ defined by

$$\nu_y(A) = \#\{t \in T \mid (t, U(y, t)) \in A\} \cdot \Delta t.$$

Put $\mu_y = L(\nu_y) \circ \mathrm{st}^{-1}$.

Note that for each Borel set $B \subset K$, the measure $\mu_{y,B}$ defined on $[0, 1]$ by

$$\mu_{y,B}(C) = \mu_y(C \times B)$$

is absolutely continuous with respect to the Lebesgue measure. We define $u(y, t)(B)$ to be the Radon–Nikodym derivative, i.e.,

$$(15) \qquad \mu_y(C \times B) = \mu_{y,B}(C) = \int_C u(y, t)(B)\, dt.$$

By working on each of the intervals $[t_i, t_{i+1}]$ at a time, it is easy to check that we can choose u nonanticipating. What is intuitively just as obvious is that we may assume $u(y, t)(\cdot)$ to be a measure for each pair (y, t), but this is actually a nontrivial observation [see, e.g., Stroock and Varadhan (1979), Theorem 1.1.6].

It remains to prove that (14) is satisfied. Since Z is a lifting of z with respect to ν_y, we have

$$\int_{^\circ a}^{^\circ b} z(t, u(y, t))\, dt = \int_{^\circ a}^{^\circ b} \left(\int z(t, a)\, du(y, t)(a) \right) dt$$

$$= \int_{[^\circ a, ^\circ b] \times K} z(t, a)\, d\mu_y(t, a)$$

$$= {}^\circ\!\!\int_{[a, b] \times {}^*K} Z(t, a)\, d\nu_y(t, a),$$

and the lemma is proved.

The relaxed control u in the lemma above is called the *standard part* of U.

4.6.7. LEMMA. All admissible relaxed controls are standard parts of internal controls.

PROOF. We shall only sketch this simple construction. Fix an admissible relaxed control u and a control parameter $y \in \mathcal{Y}^p$. Choose $N \in {}^*\mathbb{N} - \mathbb{N}$ such that $1/N$ is infinitely large compared to Δt, and consider the internal measures u_k on ${}^*K$ defined by

$$(16) \qquad u_k(A) = \int_{k/N}^{(k+1)/N} {}^*u(t, y)(A)\, dt.$$

By strictly internal methods we can find an internal function $U_y : T \to {}^*K$ such that

$$u_k(A) \approx N \cdot \#\{t \mid (k/N) \le t < ((k+1)/N) \wedge U_y(t) \in A\}$$

for all internal A, and $U(y, t) = U_y(t)$ is nonanticipating. It is easy to check that u is the standard part of U.

4.6.8. REMARK. The last two results are fundamental to the theory in this section. They were discovered by Cutland, who applied them first in deterministic (Cutland, 1983a) and then later in stochastic control theory (Cutland, 1983b,c, 1985a,b). Recently, a similar connection between internal solutions and "weak" standard solutions has turned up in Arkeryd's work on partial differential equations, but here the relationship seems to be more complicated, especially in the nonlinear case. Arkeryd's interest in these problems grew out of his work on the Boltzmann equation (which we report on in Section 6.5), but in two recent papers (Arkeryd, 1984; Arkeryd and Bergh, 1985), he has given a general treatment in a Sobolev space setting.

We now pick internal functions

$$F : {}^*\mathbb{R}^n \times T \times {}^*K \to {}^*\mathbb{R}^n,$$

$$G : {}^*\mathbb{R}^n \times T \times {}^*K \to {}^*\mathbb{R}^n \otimes {}^*\mathbb{R}^n,$$

$$H : {}^*\mathbb{R}^n \times T \to {}^*\mathbb{R},$$

$$Y_i : {}^*\mathbb{R}^n \to {}^*\mathcal{Y}, \qquad i \le p,$$

and consider the controlled internal process

$$X^U(t) = \int_0^t F(X^U(s), s, U(Y(X^U), s)) \, ds$$

$$+ \int_0^t G(X^U(s), s, U(Y(X^U), s)) \, d\chi,$$

where $\chi : \Omega \times T \to \mathbb{R}^n$ is Anderson's random walk, U is an internal control, and

$$Y(X^U) = (Y_1(X^U(\tilde{t}_1)), \ldots, Y_p(X^U(\tilde{t}_p))).$$

The *cost* of X^U is defined to be

(17) $$J(U) = E\left(\int_0^1 H(X^U(t), t) \, dt \right).$$

We shall assume that F, G, and H are S-bounded.

The next lemma reduces the proof of Theorem 4.6.4 to a question of constructing the right kinds of liftings of f, g, and h. It also explains why we have chosen (9) as the interpretation of $\int g(x^u, s, u(y(x^u), s)) \, db$ when u is measure valued.

4.6.9. LEMMA. Assume that there is a set $N \subset \Omega \times T$ of Loeb measure zero such that if $(\omega, s) \in N$, then

$$°F(X^U(\omega, s), s, a) = f(°X^U(\omega, s), °s, °a) \qquad \text{for all} \quad a \in {}^*K$$

$$°G(X^U(\omega, s), s, a) = g(°X^U(\omega, s), °s, °a) \qquad \text{for all} \quad a \in {}^*K$$

$$°H(X^U(\omega, s), s) = h(°X^U(\omega, s), °s)$$

Assume, moreover, that $Y(X^U(\omega)) = y(°X^U(\omega))$ for $L(P)$ almost all ω, and that f, g, and h satisfy Condition 4.6.3. If u is the standard part of U, then

$$j(u) = °J(U).$$

PROOF. Let x be the right standard part of X^U. We shall show that x is a solution of

$$x(t) = \int_0^t f(x(s), s, u(y(x), s)) \, ds + \int_0^t g(x(s), s, u(y(x), s)) \, db(s)$$

for a suitably chosen Brownian motion b. Since Condition 4.6.3 guarantees that the solution to this problem is unique in distribution, this is sufficient to prove the lemma.

First, observe that by Lemma 4.6.6 and the conditions on F and Y,

$$\int_{°a}^{°b} f(x(s), s, u(y(x), s)) \, ds = °\sum_a^b F(X^U(s), s, U(Y(X^U), s)) \, \Delta t$$

for almost all ω.

In treating the martingale part of our process, we shall first assume for simplicity that the dimension n is equal to one. The quadratic variation is given by

$$\left[\int G(X^U(s), s, U(Y(X^U), s)) \, d\chi \right](t)$$

$$= \int_0^t G^2(X^U(s), s, U(Y(X^U), s)) \, ds.$$

By Lemma 4.6.6 and the conditions on G and Y, the standard part of the integral on the right is equal to

$$\int_0^{°t} g^2(x(s), s, u(y(x), s)) \, ds.$$

Thus by (9)

$$\frac{d}{dt} {}^{\circ}\!\left[\int G(X^U(s), s, U(Y(X^U), s))\, dx\right](t)$$

$$= \int_K g^2(x(t), t, a)\, du(y(x), t)(a).$$

By (the proof of) Theorem 4.4.17, there is a Brownian motion b such that

$${}^{\circ}\!\int G(X^U(s), s, U(Y(X^U), s))\, d\chi$$

$$= \int \left(\int g^2(x(t), t, a)\, du(y(x), t)(a)\right)^{1/2} db.$$

Recalling (9) again, we get

$$x(t) = \int_0^t f(x(s), s, u(y(x), s))\, ds + \int_0^t g(x(s), s, u(y(x), s))\, db$$

which completes the proof of the one-dimensional case.

The proof in higher dimensions is similar and we shall leave it to the reader, with the following comments. If M is an $*\mathbb{R}^n$-valued martingale, its quadratic variation is the $*\mathbb{R}^n \otimes *\mathbb{R}^n$-valued process whose (i, j) component is given by

$$[M]_{i,j}(t) = \sum_0^t \Delta M_i(s)\, \Delta M_j(s),$$

where M_i, M_j are the ith and jth components of M. The proof of 4.4.17 can be extended to the n-dimensional case rather easily [see Lindstrøm (1985), but observe that we are only interested in the nondegenerate case], and the lemma then follows as in dimension one.

We did not use the full force of Condition 4.6.3 in the proof above; essentially all that was needed was the continuity of $f(x(s), s, \cdot)$ and $g(x(s), s, \cdot)$, and the uniqueness in distribution of the solutions to the stochastic differential equations

$$x(t) = \int_0^t f(x(s), s, u(y(x), s))\, ds + \int_0^t g(x(s), s, u(y(x), s))\, db(s).$$

However, it is not easy to see how one should construct the liftings F, G, H, and Y and guarantee the uniqueness, unless something close to 4.6.3 is satisfied. Observe how that condition is used in the proof of the main theorem:

PROOF OF THEOREM 4.6.4. Assume that we can find liftings F, G, H, and Y of f, g, h, and y such that the conditions of Lemma 4.6.9 are satisfied for all internal controls U. By 4.6.6, 4.6.7, and 4.6.9, we then have

$$\inf\{j(u) \mid u \text{ a relaxed, admissible control}\}$$

$$= \inf\{{}^\circ J(U) \mid U \text{ an internal control}\}.$$

If α is this common infimum, the set

$$\{n \in {}^*\mathbb{N} \mid \text{there is an internal control with } J(U) \le \alpha + 1/n\}$$

is internal and contains $\mathbb{N}$, and hence it has an infinite element γ. If U is an internal control such that $J(U) \le \alpha + 1/\gamma$, the standard part u of U is an optimal relaxed control since

$$j(u) = {}^\circ J(U) = \alpha.$$

To complete the proof, it only remains to construct the liftings F, G, H, and Y. Let $J \in {}^*\mathbb{N} - \mathbb{N}$ be the constant occurring in the nonstandard version of Krylov's inequality, Proposition 4.5.5. We choose F, G, and H to be J-Lipschitz liftings of f, g, and h, and we make sure that for almost all nearstandard (x, t), the functions $F(x, t, \cdot)$ and $G(x, t, \cdot)$ are S-continuous. We also require that ${}^\circ\det G$ is bounded away from zero; since we have assumed that g is positive definite and symmetric, this does not interfere with the requirement that G is J-Lipschitz (recall 4.5.6).

Note that if $y \in \mathcal{Y}^p$ is an observation parameter and

$$X(t) = \int_0^t F(X(s), s, U(y, s))\, ds + \int_0^t G(X(s), s, U(y, s))\, d\chi$$

for some internal control U, then by Krylov's inequality

$${}^\circ F(X(\omega, s), s, a) = f({}^\circ X(\omega, s), {}^\circ s, {}^\circ a) \qquad \text{for all} \quad a \in {}^*K,$$

$${}^\circ G(X(\omega, s), \dot{s}, a) = g({}^\circ X(\omega, s), {}^\circ s, {}^\circ a) \qquad \text{for all} \quad a \in {}^*K,$$

$${}^\circ H(X(\omega, s), s) = h({}^\circ X(\omega, s), {}^\circ s),$$

for almost all (ω, s) as needed for Lemma 4.6.9. If we can only show that there is a Y such that

$$Y(X(\omega)) = y({}^\circ X(\omega)) \qquad \text{a.e.}$$

for all controls U, the proof will be finished.

Since $y = (y_1, \ldots, y_p)$, we need only explain how to lift each component $y_i : \mathbb{R}^n \to \mathcal{Y}$. The trouble with y_i is that it depends only on the value $x(t_i)$ of the process at time t_i, and not on the joint distribution $(x(t), t)$ as t varies. This makes it impossible to apply Krylov's inequality to y_i the way we did for f, g, and h. However, there is a similar inequality which holds for fixed t's, namely that for all $t > 0$

(18) $$L(P)\{\omega : x(\omega, t) \in A\} \le Cm(A)^{1/2},$$

where C is a constant depending only on t, the bounds on f, g, and det g, and the Lipschitz coefficient K of g. We omit the proof of this estimate, which is long, dull, and not very informative.

To turn (18) into an internal inequality we argue as follows: if $H \in {}^*\mathbb{N}$, let an *H-element* be a cube in ${}^*\mathbb{R}^n$ of the form

(19) $$\left\{ x \in {}^*\mathbb{R}^n \,|\, \forall i \le n \left(\frac{k_i}{H} \le x_i < \frac{k_i + 1}{H} \right) \right\},$$

where $(k_1, \ldots, k_n) \in {}^*\mathbb{Z}^n$. An *H-set* is an internal union of *H*-elements. For finite H it follows from (19) that

(20) $$P\{\omega \,|\, X(\omega, t) \in A\} \le 3^{d/2} C^* m(A)^{1/2} + 1/H$$

for all *H*-sets A. The factor $3^{d/2}$ is due to the fact that when $X(\omega, t)$ is in one *H*-element, x may be in the same element or any one of its neighbors, and the extra term $1/H$ is included to compensate for the fact that $P\{X(\omega, t) \in A\}$ may be larger than $L(P)\{X(\omega, t) \in A\}$ by an infinitesimal amount.

Since (20) is valid for all finite H, there must be an infinite H_0 for which it also holds. We let Y_i be a lifting of y_i (with respect to the *version of the Lebesgue measure) which is constant on H_0-elements. It follows from (20) that

$$Y_i(X(\omega)) = y_i(^\circ X(\omega)) \qquad \text{a.e.},$$

and this completes the proof of Theorem 4.6.4.

Although much more deserves to be said about the stochastic system we have been discussing, we shall restrict ourselves to a few short remarks. First we would like to mention that Theorem 4.6.4 is a simplified version of the results in Cutland (1985a,b); the original treatment covers a more complicated and general situation which is really an amalgamation of the one we have used above and the one we shall discuss at the end of this section. Next we should admit that the feedback mechanism we have been using is partly dictated by technical considerations. As an example of the problems we run into a more general setting, note that if the observation space $\mathcal{Y}$ is not discrete, then it is not clear that an internal control U has a standard part satisfying anything like Lemma 4.6.6, the problem being that $U(y, t)$ and $U(y', t)$ may be completely different for infinitely close y and y'. Indeed, the existence of optimal relaxed controls for partially observed stochastic systems still seems to be a wide open question in the general case.

But there is more to stochastic control theory than the existence of optimal controls. We shall take a very brief look at one of the other aspects as a further indication of the strength of the nonstandard approach.

If $H \in {}^*\mathbb{N}$, call an internal control U an H-control if $U(y, \cdot)$ is constant on the intervals $[k/H, (k + 1)/H) \cap T$, $k \in \mathbb{Z}$, for all $y \in {}^*\mathcal{Y}$. Recalling the proof of Lemma 4.6.7, we see that if u is an admissible relaxed control and H is infinite, then u is the standard part of an internal H control. If $\varepsilon \in \mathbb{R}_+$ and α is the lowest possible cost, the set

$$\{H \in {}^*\mathbb{N} \,|\, \text{there is an } H \text{ control } U \text{ with } J(U) < \alpha + \varepsilon\}$$

is internal and contains all infinite H, hence it contains a finite element H_ε. The standard part of an H_ε control is an ordinary (not measure-valued) admissible control which is constant on intervals of length $1/H_\varepsilon$. Thus we have shown that the optimal relaxed control can be approximated arbitrarily well by very simple ordinary controls. This argument can be refined by also allowing Δt to become finite, and we can then describe the system as a limit of discrete systems where the Brownian motion is replaced by random walks. Results of this kind are of interest in applications; see Christopheit (1983) for a discussion.

There is one natural question we have not touched on yet: When is there an optimal ordinary control? We shall postpone this until the end of the section.

B. Girsanov's Formula

In the stochastic systems we studied above, the drift coefficient f, the diffusion coefficient g, and the observations y_i all depended only on the present value of the process. For many applications this is too restrictive; f, g, and y_i will also depend on the past; they will be nonanticipating functions

$$f : C[0, 1] \times [0, 1] \times K \to \mathbb{R}^n,$$

$$g : C[0, 1] \times [0, 1] \times K \to \mathbb{R}^n \otimes \mathbb{R}^n,$$

$$y_i : C[0, 1] \to K,$$

where $C[0, 1]$ is short for $C([0, 1], \mathbb{R}^n)$. Looking for liftings of f, g, and y_i, it is clear that they will be functions

$$F : {}^*C[0, 1] \times T \times {}^*K \to {}^*\mathbb{R}^n,$$

$$G : {}^*C[0, 1] \times T \times {}^*K \to {}^*\mathbb{R}^n \otimes {}^*\mathbb{R}^n,$$

$$Y_i : {}^*C[0, 1] \to {}^*K,$$

but the big question is which measure on $*C[0, 1]$ they should be liftings with respect to. What we need, of course, is that F, G, and the Y_i's are liftings with respect to all the measures induced on $*C[0, 1]$ by the controlled processes

$$X^U = \int F(X^U, s, U(Y(X^U), s)) \, ds + \int G(X^U, s, U(Y(X^U), s)) \, d\chi,$$

but it is impossible to define F, G, and Y_i in this way since there are uncountably many of these measures and most of them are mutually singular. In the simpler case we studied above where $\mathbb{R}^n$ replaced $C[0, 1]$, the liftings were picked with respect to the Lebesgue measure, and Krylov's inequality was used to prove that they were also liftings with respect to the measures induced by the X^U's. It is possible to use a similar idea in the present setting, but the price we shall have to pay is that the diffusion coefficient g is no longer allowed to depend on the control u; hence g is a function

$$g: C[0, 1] \times [0, 1] \to \mathbb{R}^n \otimes \mathbb{R}^n.$$

The point is that if G is a lifting of g and

(21) $$Z = \int G(Z, s) \, d\chi,$$

then (under reasonable conditions) all the measures induced by the X^U's are absolutely continuous with respect to the one induced by Z. Hence we can define our liftings with respect to the latter, and the absolute continuity will take care of the rest.

The result we are referring to is a nonstandard version of Girsanov's formula. To prove it, we shall take a look at stochastic differential equations from a slightly different angle than we did in the last section. This change of viewpoint is not, strictly speaking, necessary (see Cutland, 1982), but we find that the alternative approach is not only a convenient way of proving Girsanov's theorem, but also interesting in its own right. The basic idea is that we can deal with the drift term by changing the underlying probability measure rather than changing the paths of the process.

As usual we work with the hyperfinite time line $T = \{0, \Delta t, 2 \, \Delta t, \ldots, 1\}$ and the probability space $\Omega = \{-1, 1\}^{T \times \{1, \ldots, n\}}$ with normalized counting measure P. We let $\{\mathscr{A}_t\}_{t \in T}$ be the natural filtration on Ω and

$$\chi(\omega, t) = \sum_{i=1}^{n} \sum_{s=0}^{t} \omega_i(s) \sqrt{\Delta t} \, \mathbf{e}_i$$

an n-dimensional version of Anderson's random walk. By Σ we shall denote the set of all internal maps $\sigma : T \to {}^*\mathbb{R}^n$ which are zero at the origin. Given a nonanticipating function $H : \Sigma \times T \to {}^*\mathbb{R}^n$, we let $P_{H,I}$ be the probability measure on Ω given by

$$(22) \qquad P_{H,I}(\omega) = \prod_{s=0}^{1} \prod_{i=1}^{n} (\tfrac{1}{2} + \tfrac{1}{2}H_i(\chi(\omega), s) \, \Delta\chi_i(\omega, s)),$$

where H_i and χ_i are the ith components of H and χ, respectively. This is the measure governing a random walk obtained by flipping unfair coins, H being a measure of the unfairness. Note that the density of $P_{H,I}$ with respect to P is

$$(23) \qquad \frac{P_{H,I}}{P}(\omega) = \prod_{s=0}^{1} \prod_{i=1}^{n} (1 + H_i(\chi(\omega), s) \, \Delta\chi_i(\omega, s)).$$

If $E_{H,I}$ denotes expectation with respect to $P_{H,I}$, an easy calculation shows that

$$E_{H,I}(\Delta\chi(t) \,|\, \mathscr{A}_t)(\omega) = H(\chi(\omega), t) \, \Delta t.$$

Hence

$$(24) \qquad W_{H,I}(\omega, t) = \chi(\omega, t) - \sum_0^t H(\chi(\omega), t) \, \Delta t$$

is a martingale with respect to $(\Omega, \{\mathscr{A}_t\}, P_{H,I})$. In fact, it is more than just a martingale:

4.6.10. LEMMA. Assume that $\|H(\sigma, t)\|^2 \, \Delta t \approx 0$ for all σ and t. Then the standard part of $W_{H,I}$ is a Brownian motion with respect to the Loeb measure $L(P_{H,I})$.

PROOF. Let $\Delta W_{H,I}(t)_i$ be the ith component of the increment $\Delta W_{H,I}(t)$. Then

$$
\begin{aligned}
E_{H,I}(\Delta W_{H,I}(t)_i^2 \,|\, \mathscr{A}_t)(\omega) &= (\sqrt{\Delta t} - H_i(\chi(\omega), t) \, \Delta t)^2 (\tfrac{1}{2} + \tfrac{1}{2}H_i(\chi(\omega), t)\sqrt{\Delta t}) \\
&\quad + (-\sqrt{\Delta t} - H_i(\chi(\omega), t) \, \Delta t)^2 \\
&\quad \times (\tfrac{1}{2} - \tfrac{1}{2}H_i(\chi(\omega), t)\sqrt{\Delta t}) \\
&= \Delta t - H_i(\chi(\omega), t)^2 \, \Delta t^2.
\end{aligned}
$$

Since $\|H(\chi(\omega), t)\|^2 \, \Delta t \approx 0$, Proposition 4.4.18 tell us that $({}^\circ W_{H,I}^+)_i$ is a Brownian motion with respect to $L(P_{H,I})$, and since the components of ${}^\circ W_{H,I}^+$ are independent, the lemma follows.

Rewriting (24) as

$$(25) \qquad \chi(\omega, t) = \sum_0^t H(\chi(\omega), s) \, \Delta t + W_{H,I}(\omega, t),$$

we see that if H is a suitable lifting of a standard function h, then $°\chi^+$ is a solution of the stochastic differential equation

$$(26) \qquad x(t) = \int_0^t h(x, t) \, ds + b(t)$$

on $(\Omega, L(P_{H,I}))$. In the terminology of the last section, this is a weak solution as the Brownian motion $b = °W_{H,I}^+$ depends on the coefficient h.

Let us take a brief look at what we have been doing. Starting with a Brownian motion $°\chi^+$ on $(\Omega, L(P))$, we have constructed a new measure $L(P_{H,I})$ such that on $(\Omega, L(P_{H,I}))$ the "same" process is a solution to (26). The connection between $L(P)$ and $L(P_{H,I})$ is given by (23). We shall now explain how the method can be extended to deal with stochastic differential equations of the form

$$(27) \qquad x(t) = \int_0^t f(x, s) \, ds + \int_0^t g(x, s) \, db(s).$$

The idea is the same as before; given a solution of the simpler equation

$$(28) \qquad x(t) = \int_0^t g(x, s) \, db(s),$$

we turn it into a solution of (27) by changing the underlying measure.

Assume that

$$F : \Sigma \times T \to {}^*\mathbb{R}^n, \qquad G : \Sigma \times T \to {}^*\mathbb{R}^n \otimes {}^*\mathbb{R}^n$$

are nonanticipating, and that $G(\sigma, t)$ is invertible for all σ and t. For each $\sigma \in \Sigma$, let σ_G be given by

$$\sigma_G(t) = \sum_0^t G(\sigma_G, s) \, \Delta\sigma(s),$$

and similarly let $\chi_G : \Omega \times T \to {}^*\mathbb{R}^n$ be the process

$$(29) \qquad \chi_G(t) = \sum_0^t G(\chi_G, s) \, \Delta\chi(s).$$

Note that (29) is the nonstandard version of (28).

We define a nonanticipating function $H : \Sigma \times T \to {}^*\mathbb{R}^n$ by

$$(30) \qquad H(\sigma, t) = G^{-1} F(\sigma_G, t).$$

Combining (29), (30), and (25), we get

(31)
$$\chi_G(t) = \sum_0^t G(\chi_G, s) \, \Delta\chi(s) = \sum_0^t G(\chi_G, s)(H(\chi, s) \, \Delta t + \Delta W_{H,I}(s))$$
$$= \sum_0^t F(\chi_G, s) \, \Delta t + \sum_0^t G(\chi_G, s) \, \Delta W_{H,I}(s),$$

which is a nonstandard version of (27).

Let us introduce the following notation. We shall write $P_{F,G}$ and $W_{F,G}$ for $P_{H,I}$ and $W_{H,I}$, respectively, where H is defined as in (30). When we consider χ_G as a stochastic process on $(\Omega, P_{F,G})$, we shall denote it by $\chi_{F,G}$; when we consider it as a process on (Ω, P), we shall keep the old notation χ_G. In this way we avoid having to refer explicitly to the underlying measure every time we mention a process.

By Lemma 4.6.10 and formula (31) we have

4.6.11. LEMMA. Assume that $\|G^{-1}F(\sigma, t)\|^2 \, \Delta t \approx 0$ for all σ and t, and that the sets

(32) $\{(\omega, t) \,|\, {}^\circ F(\chi_{F,G}, t) \neq f({}^\circ \chi_{F,G}^+, {}^\circ t)\}$

(33) $\{(\omega, t) \,|\, {}^\circ G(\chi_{F,G}, t) \neq g({}^\circ \chi_{F,G}^+, {}^\circ t)\}$

have $L(P_{F,G} \times \lambda)$ measure zero, where λ is the normalized counting measure on T. Then ${}^\circ \chi_{F,G}^+$ is a solution to

(34) $$x(t) = \int_0^t f(x, s) \, ds + \int_0^t g(x, s) \, db(s)$$

on $(\Omega, L(P_{F,G}))$.

Recall that we are not interested in the solution of (34) for its own sake, but rather in its density with respect to the solution of

(35) $$x(t) = \int_0^t g(x, s) \, db(s).$$

Since χ_G and $\chi_{F,G}$ have the same paths, this density is easy to calculate; in fact, formula (23) tells us that

(36) $$\frac{P_{F,G}}{P}\{\omega\} = \prod_{s=0}^1 \prod_{i=1}^n (1 + H_i(\chi(\omega), s) \, \Delta\chi_i(\omega, s)).$$

To study the expression on the right-hand side, we introduce a process

$$M : \Omega \times S \to {}^*\mathbb{R}$$

as follows: S is the time line obtained by inserting $n - 1$ new points between t and $t + \Delta t$ for each $t \in T$, i.e.,

$$S = \{0, \Delta s, 2 \Delta s, \ldots, 1\},$$

where $\Delta s = \Delta t/n$. A point in S is thus of the form $s = t + i \Delta s$ for a $t \in T$ and an i between 0 and n. We define M by letting

(37) $M(0) = 1,$

(38) $M(t + (i + 1) \Delta s) = M(t + i \Delta s)(1 + H_i(\chi, t) \Delta \chi_i(t)).$

It is clear that M is a martingale with respect to P and that

(39) $(P_{F,G}/P)(\omega) = M(\omega, 1).$

4.6.12. LEMMA.   If $\| H(\sigma, t) \|^2 \leq K$ for all σ and t, then

(40) $E(M(s)^2) \leq \exp(nK^2 s)$

for all $s \in S$.

PROOF.   First note that if $s = t + i \Delta s$, then

$$E(M(s + \Delta s)^2) = E(M(s)^2(1 + H_i(\chi, t) \Delta \chi_i(t))^2)$$
$$= E(M(s)^2(1 + H_i(\chi, t)^2 \Delta t)) \leq E(M(s)^2)(1 + nK^2 \Delta s).$$

By induction we see that if $f: S \to {}^*\mathbb{R}$ is an internal function such that $f(0) = 1$ and $f(s + \Delta s) \leq (1 + nK^2 \Delta s)f(s)$, then

$$f(s) \leq (1 + nK^2 \Delta s)^{s/\Delta s} \leq \exp(nK^2 s).$$

Putting $f(s) = E(M(s)^2)$, the lemma follows.

If $A \in \Omega$ has infinitesimal P measure, then by (39), Hölder's inequality and the lemma,

$$P_{F,G}(A) = \int_A M(\omega, 1) \, dP \leq P(A)^{1/2} \left(\int_A M(1)^2 \, dP \right)^{1/2} \approx 0.$$

Hence $L(P_{F,G})$ is absolutely continuous with respect to $L(P)$. The next lemma weakens the boundedness assumption on H.

4.6.13. LEMMA.   Let F and G be S-bounded, nonanticipating functions. Assume that for all S-bounded $\sigma \in \Sigma$, we have ${}^\circ|G^{-1}F(\sigma, t)| < \infty$ for all $t \in T$. Then $L(P_{F,G})$ is absolutely continuous with respect to $L(P)$, and

(41) $\dfrac{\partial L(P_{F,G})}{\partial L(P)}(\omega) = {}^\circ\prod_{s=0}^{1} \prod_{i=1}^{n} (1 + (G^{-1}F)_i(\chi_G, s) \Delta \chi_i(s)).$

PROOF.   We already know that the conclusion is true if $|G^{-1}F|$ is uniformly S-bounded. Let

$$B_N = \{\sigma \in \Sigma \mid \sup_{t \in T} |\sigma(t)| \le N\}$$

for each $N \in \mathbb{N}$, and observe that

$$\lim_{N \to \infty} {}^\circ P_{F,G}\{\omega \mid \chi_{F,G}(\omega, \cdot) \in B_N\} = 1.$$

If $L(P_{F,G})$ is not absolutely continuous with respect to $L(P)$, there must be an internal set A contained in some B_N, $N \in \mathbb{N}$, such that $L(P)(A) = 0$ and $L(P_{F,G})(A) > 0$. Define $\tilde{F}: \Sigma \times T \to {}^*\mathbb{R}^n$ by

$$\tilde{F}(\sigma, t) = \begin{cases} F(\sigma, t) & \text{if } |\sigma(s)| \le N \text{ for all } s \le t, \\ 0 & \text{otherwise.} \end{cases}$$

Then ${}^\circ|G^{-1}\tilde{F}|$ is uniformly S-bounded, and $P_{\tilde{F},G}$ and $P_{F,G}$ agree on B_N. Thus $L(P_{\tilde{F},G})(A) = L(P_{F,G})(A) > 0$, while $L(P)(A) = 0$, contradicting the preceding lemma.

The conditions in Lemma 4.6.13 are still far from optimal; e.g., the S-boundedness of F and G can be replaced by bounds of the form

$$|F(\sigma, t)|, |G(\sigma, t)| \le K(1 + \|\sigma\|),$$

where $\|\cdot\|$ is the supremum norm, but we shall leave all further refinements to the reader. However, although we shall not need it, we would like to bring our expression for the density $\partial L(P_{F,G})/\partial L(P)$ to a more familiar form.

4.6.14. NONSTANDARD GIRSANOV FORMULA.   Let F and G be S-bounded, nonanticipating functions, and assume that for all S-bounded $\sigma \in \Sigma$, we have ${}^\circ|G^{-1}F(\sigma, t)| < \infty$ for all $t \in T$. Then $L(P_{F,G})$ is absolutely continuous with respect to $L(P)$ and

$$\frac{\partial L(P_{F,G})}{\partial L(P)}(\omega) = {}^\circ\prod_{s=0}^{1} \prod_{i=1}^{n} (1 + G^{-1}F)_i(\chi_G, s) \, \Delta\chi_i(s))$$

(42)
$$= {}^\circ\exp\left(\int_0^1 {}^tFA^{-1}(\chi_G, s) \, d\chi_G(s) \right.$$

$$\left. -\frac{1}{2}\int_0^1 (G^{-1}F)^2(\chi_G, s) \, ds\right),$$

where $A = G \cdot {}^tG$.

PROOF. It only remains to prove the second equality in (42). We have

$$\prod_{s=0}^{1} \prod_{i=1}^{n} (1 + (G^{-1}F)_i(\chi_G, s) \, \Delta\chi_i(s))$$

$$= \exp \sum_{s=0}^{1} \sum_{i=1}^{n} \ln(1 + (G^{-1}F)_i(\chi_G, s) \, \Delta\chi_i(s))$$

$$\approx \exp \sum_{s=0}^{1} \sum_{i=1}^{n} \left\{ (G^{-1}F)_i(\chi_G, s) \, \Delta\chi_i(s) - \frac{1}{2}(G^{-1}F)_i^2(\chi_G, s) \, \Delta t \right\}$$

$$= \exp\left\{ \int_0^1 {}^t(G^{-1}F)(\chi_G, s) \, d\chi - \frac{1}{2} \int_0^1 (G^{-1}F)^2(\chi_G, s) \, ds \right\},$$

where we have used the Taylor expansion $\ln(1 + x) \approx x - (x^2/2)$. Recalling that by (31) $d\chi = G^{-1}(\chi_G) \, d\chi_G$, we get

$$\exp\left\{ \int_0^1 {}^t(G^{-1}F)(\chi_G, s) \, d\chi - \frac{1}{2} \int_0^1 (G^{-1}F)^2(\chi_G, s) \, ds \right\}$$

$$= \exp\left\{ \int_0^1 {}^tFA^{-1}(\chi_G, s) \, d\chi_G(s) - \frac{1}{2} \int_0^1 (G^{-1}F)^2(\chi_G, s) \, ds \right\},$$

and the theorem is proved.

In our work on stochastic control theory we shall only use the fact that $L(P_{F,G})$ is absolutely continuous with respect to $L(P)$, and not the explicit formula for the Radon–Nikodym derivative given in (42). All we shall need to know is that a lifting with respect to P is also a lifting with respect to $P_{F,G}$. For other applications, however, formula (42) is of the greatest importance; see Stroock and Varadhan (1979) for numerous examples.

C. Optimal Controls: Dependence on the Past

We shall now take a look at a stochastic control problem where the coefficients f, g, and the observations y_i are allowed to depend on the past of the process. This means that f and g will be functions

$$f: C[0, 1] \times [0, 1] \times K \to \mathbb{R}^n, \qquad g: C[0, 1] \times [0, 1] \to \mathbb{R}^n \otimes \mathbb{R}^n,$$

and that for each i, y_i is a function

$$y_i: C[0, 1] \to \mathcal{Y},$$

where $\mathcal{Y}$ is a countable observation space as before. We shall assume that f and g are *nonanticipating* in the sense that for all x, t, a, the values $f(x, t, a)$ and $g(x, t)$ depend only on the values $\{x_s\}_{s \le t}$ of x up to time t. Similarly, if

$$0 \le t_1 < t_2 < \cdots < t_p < 1$$

are the times observations are made, then the ith observation $y_i(x)$ depends only on the values of x up to time t_i. The sequence $(y_1(x), y_2(x), \ldots, y_p(x)) \in \mathcal{Y}^p$ of observations is denoted by $y(x)$.

If u is a relaxed admissible control

$$u : \mathcal{Y}^p \times [0, 1] \to \mathcal{M}(K)$$

as in Definition 4.6.2, we are interested in the process

$$(43) \qquad x^u = \int f(x^u, s, u(y(x^u), s)) \, ds + \int g(x^u, s) \, db(s).$$

Given a *cost function* $h : C[0, 1] \to \mathbb{R}$, we want to minimize the *cost*

$$j(u) = E(h(x^u)).$$

We shall use the following conditions on f, g, and h.

4.6.15. CONDITION. Assume that:

(i) h is a measurable, bounded function;
(ii) f and g are bounded, measurable, and nonanticipating, and f is continuous in the third variable;
(iii) g is uniformly Lipschitz in the first variable, i.e., there is a number $L \in \mathbb{R}$ such that $|g(x, t) - g(y, t)| \le L \|x - y\|$ for all x, y, t, where $\| \cdot \|$ denotes the supremum norm;
(iv) for each n, the function $g^{-1}f(x, t, a)$ is bounded on the set

$$A_n = \{(x, t, a) \in C[0, 1] \times [0, 1] \times K \,|\, \|x\| \le n\}.$$

The main result is due to Cutland (1985a):

4.6.16. THEOREM. If Condition 4.6.15 is satisfied there exists an optimal relaxed control.

PROOF. We shall first construct the nonstandard counterparts of f, g, h, and y. Let λ be the normalized counting measure on T. Since g is continuous in the first variable, we can find a nonanticipating lifting $G : \Sigma \times T \to {}^*\mathbb{R}^n \otimes {}^*\mathbb{R}^n$ such that for all t outside a set of measure zero with respect to $L(\lambda)$,

$$(44) \qquad {}^\circ G(\sigma, t) = g({}^\circ \sigma^+, {}^\circ t)$$

for all nearstandard σ (recall Proposition 4.3.16).

Let

$$(45) \qquad \chi_G(t) = \sum_{s=0}^{t} G(\chi_G, s) \, \Delta\chi(s)$$

be the process defined in (29). Since f is continuous in the third variable, it has a nonanticipating lifting $F: \Sigma \times T \times {}^*K \to {}^*\mathbb{R}^n$ such that for $L(P \times \lambda)$ almost all (ω, t), we have

(46) $\qquad {}^\circ F(\chi_G(\omega), t, a) = f({}^\circ\chi_G^+(\omega), {}^\circ t, {}^\circ a)$

for all $a \in {}^*K$. We let H be a lifting of h such that

(47) $\qquad {}^\circ H(\chi_G(\omega)) = h({}^\circ\chi_G^+(\omega))$

for $L(P)$ almost all ω.

We can obviously choose F, G, and H S-bounded. Note that we can also choose F and G such that $|G^{-1}F(\sigma, t, a)|$ is S-bounded when we restrict σ to an S-bounded set.

For each i, choose $\tilde{t}_i \in T$ infinitely close to the observation time t_i, and let

(48) $\qquad Y_i: \Sigma \to {}^*\mathcal{Y}$

be a lifting of y_i such that

(49) $\qquad {}^\circ Y_i(\chi_G(\omega)) = y_i({}^\circ\chi_G^+(\omega))$ $\qquad$ a.e.,

and Y_i depends only on $\chi_G(\omega)$ up to time $\tilde{t}_i$.

We are now ready to define the internal counterparts of the x^u processes. Given an internal control $U: {}^*\mathcal{Y}^p \times T \to {}^*K$, we let $F^U: \Sigma \times T \to {}^*\mathbb{R}^n$ be defined by

(50) $\qquad F^U(\sigma, t) = F(\sigma, t, U(Y(\sigma), t)).$

The process $\chi_{F^U, G}$ [i.e., χ_G considered as defined on $(\Omega, P_{F^U, G})$] is a hyperfinite version of x^u since by (31)

$$\chi_{F^U, G}(t) = \sum_0^t F(\chi_{F^U, G}, s, U(Y(\chi_{F^U, G}), s)) \, \Delta t + \sum_0^t G(\chi_{F^U, G}, s) \, \Delta W_{F^U, G},$$

where ${}^\circ W_{F^U, G}^+$ is a Brownian motion on $(\Omega, L(P_{F^U, G}))$.

Let u be the standard part of U. We shall show that ${}^\circ\chi_{F^U, G}^+$ is a solution of (43). To this end let Ω_0 be the set of all $\omega \in \Omega$ such that (49) holds for all i and (46) holds for almost all t and all a. By Lemma 4.6.6 we see that if $\omega \in \Omega_0$, then

$$\, {}^\circ\sum_0^t F(\chi_G(\omega), s, U(Y(\chi_G(\omega)), s)) \, \Delta t = \int_0^{{}^\circ t} f({}^\circ\chi_G^+(\omega), s, u(y({}^\circ\chi_G(\omega), s)) \, ds.$$

Since Ω_0 has $L(P)$-measure one and $L(P_{F^U, G})$ is absolutely continuous with respect to $L(P)$, the last equation holds $L(P_{F^U, G})$ almost everywhere. By a similar argument

$$\, {}^\circ\sum_0^t G(\chi_G(\omega), s) \, \Delta W_{F^U, G} = \int_0^{{}^\circ t} g({}^\circ\chi_G^+(\omega), s) \, d{}^\circ W_{F^U, G}^+$$

$L(P_{F^U,G})$ almost everywhere. Since χ_G and $\chi_{F^U,G}$ have the same sample paths, it follows that $^\circ\chi^+_{F^U,G}$ is a solution of (43). Moreover, since H is a lifting of h also with respect to $P_{F^U,G}$, we get

$$j(u) = {}^\circ J(U),$$

where

$$J(U) = E_{F^U,G}(\chi_{F^U,G}(\omega))$$

(here we have tacitly assumed that the solutions of (43) are unique in distribution; this follows from the Lipschitz condition on g).

Thus

$$\inf\{j(u)\,|\,u \text{ an admissible relaxed control}\}$$

$$= \inf\{{}^\circ J(U)\,|\,U \text{ an internal control}\}.$$

Since the last infimum is achieved, so is the first, and the theorem is proved.

Cutland (1985a) has extended the result above to a situation which incorporates both the present and the Markov case we looked at before. This is achieved by considering a system consisting of two interacting parts, an "observation" part and a "performance" part, both modeled by stochastic differential equations. Both parts are allowed to depend on the past, but only the noise in the observation part can be controlled. The observation scheme is also slightly different. Without going deeper into this problem, we only mention that Cutland's methods are rather similar to the ones we have used, the main difference being that he works in a *-continuous setting with *-Brownian motions instead of hyperfinite random walks (see the comment at the end of Section 4.5).

We would like to point out that Girsanov's theorem was used in the proof of Theorem 4.6.16 only to show that F, G, H, and Y were liftings of f, g, h, and y with respect to all the measures $P_{F^U,G}$, and not only with respect to P. A natural question is whether absolute continuity is the "best" condition for the existence of simultaneous liftings of this sort, or whether it is possible to find a weaker one. Observe that by Anderson's Lusin theorem, 3.4.9, $^*f$ is a lifting of f with respect to all *standard* measures $^*\mu$, and hence absolute continuity is not always required. However, since the internal measures we construct in nonstandard probability theory are rarely standard, the Lusin theorem is often of little use; what we need is an extension which applies to a larger class of internal measures. No such result seems to be known yet, and we shall make no attempt to prove one here, restricting ourselves to the rather cryptic remark that Anderson's work on standardly distributed measures (Anderson, 1982) may turn out to contain valuable ideas for such a project.

We have promised to say a few words about the existence of optimal ordinary controls. In the setting of Theorem 4.6.16 there is a general strategy for turning an optimal relaxed control into an optimal ordinary control which works under suitable convexity assumptions on the range of f. The idea is that if f has a convex and closed range, then for all x, t and each measure u in $\mathcal{M}(K)$ there is an element $a(x, t, u) \in K$ such that

$$f(x, t, a(x, t, u)) = \int f(x, t, c) \, du(c).$$

Given a relaxed control u, the idea is to choose an ordinary control $\tilde{u}$ by

$$\tilde{u}(y(x), t) = a(x, t, u(y(x), t)).$$

Since y contains only partial information about x, this definition need not make sense as we may have $y(x) = y(x')$ and $a(x, t, u(y(x), t)) \neq a(x', t, u(y(x'), t))$. To circumvent this difficulty, one has to assume that f is of the form $f(x, t, a) = f_1(x, t) \cdot f_2(t, a)$. Once this assumption is made, the difficulties disappear and the argument can be carried through [see Elliott and Kohlmann (1982) for the details].

When the control is also allowed to enter into the martingale term as in Theorem 4.6.4, the situation becomes more complicated. Not only do we need a convexity assumption on g^2 guaranteeing the existence of an element $b(x, t, u) \in K$ satisfying

$$g^2(x, t, b(x, t, u)) = \int g^2(x, t, c) \, du(c),$$

but we must also be able to choose $a(x, t, u) = b(x, t, u)$. The simplest way of ensuring this is perhaps to let f and g be of the form

$$f(x, t, a) = f_1(x, t)h(t, a), \qquad g(x, t, a) = g_1(x, t)h^{1/2}(t, a)$$

for the same function h. We shall leave it to the reader to obtain an optimal ordinary control in this case.

The theory of optimal stochastic controls is a fascinating area both in its own right and as a laboratory for the techniques of nonstandard probability theory. There are books by Krylov (1980), Elliott (1982), and Davis (1984) for those who want to know more about the subject.

4.7. STOCHASTIC INTEGRATION IN INFINITE-DIMENSIONAL SPACES

In the last sections of this chapter we shall extend the theory of Brownian motion and stochastic integration in two directions; first we shall consider processes taking values in infinite-dimensional spaces, and then—in the next section—we shall turn to the case of multidimensional "time" parameters.

Infinite-dimensional Brownian motions are important for the study of stochastic partial differential equations. As an example we shall take a look at a problem from hydrodynamics. Recall that the velocity of a fluid at position x at time t is given by a function $u(x, t)$ satisfying the Navier–Stokes equation

(1) $$\partial_t u + (u \cdot \nabla)u = -\nabla p + \nu \, \Delta u + f(x, t)$$

(where p is the pressure, ν is the viscosity, and f is the external force) and the incompressibility assumption

(2) $$\nabla u = 0.$$

To (1) and (2) one must in each particular case add the appropriate boundary conditions.

The mathematical formulation we have just given is the one most often found in the literature, but from the point of view of integral equations, there is another that is more convenient. Using the incompressibility assumption $\nabla u = 0$, the pressure p can be eliminated from (1), and the system (1), (2) is reduced to the equation

(3) $$\partial_t u + (u \cdot \nabla)u = \nu \, \Delta u + Kf,$$

where K is a certain projection operator. The actual form of K is of no importance to us here, and we only refer the interested reader to Chow (1978) and Vishik *et al.* (1979). We can consider (3) as an infinite-dimensional evolution equation; for each t the solution is a function $u_t(\cdot)$ in a suitably chosen function space.

It is not very realistic to assume full knowledge of the external force f; in many situations it may be reasonable to introduce it as a stochastic term. The natural candidate is white noise, and we are led to the stochastic equation

(4) $$du = \nu \, \Delta u \, dt - (u \cdot \nabla)u \, dt + h \, dt + g \, dw,$$

where h is the deterministic contribution of the external force, and w is a "Brownian motion."

But what kind of a Brownian motion can w be? Since u takes values in an infinite-dimensional space, so does w. In order to understand (4), we must first study infinite-dimensional Brownian motions and their stochastic integrals. This, in fact, is almost all we shall do in this section; only briefly shall we return to (4).

A. Brownian Motions on Hilbert Spaces

Brownian motions are defined in terms of rotationally invariant Gaussian measures. In Section 3.5 (recall Gross's theorem), we noticed that in the infinite-dimensional case such measures exist only in a rather curious sense, not as measures on the original Hilbert space, but rather as measures on a larger Banach space. We also discovered an easy way of constructing these measures by pulling down a hyperfinite-dimensional Gaussian measure using the Banach space topology. The same idea will be used to define Brownian motions on a Hilbert space, starting with a hyperfinite-dimensional random walk and taking the standard part with respect to a measurable norm.

Let us begin by recalling and extending some of the results from Section 2.2. If $(E, |\cdot|)$ is an internal normed linear space, a *standard generating set* for E is a set $\{v_n\}_{n \in \mathbb{N}}$ of elements from E with $0 < {}^\circ|v_n| < \infty$ for all $n \in \mathbb{N}$. An element $u \in E$ is called $|\cdot|$-*pre-nearstandard* with respect to $\{v_n\}$ if for all $\varepsilon \in \mathbb{R}_+$, there is a $v = \sum_{n=1}^{k} a_n v_n$, with $k \in \mathbb{N}$, $a_1, \ldots, a_k \in \mathbb{R}$, such that $|u - v| < \varepsilon$. We denote the set of all pre-nearstandard points by $\mathrm{Pns}(E, |\cdot|)$. Let $\sim_{|\cdot|}$ be the equivalence relation on $\mathrm{Pns}(E, |\cdot|)$ defined by $u \sim_{|\cdot|} v$ if and only if $|u - v| \approx 0$. We let ${}^\circ u$ be the equivalence class of u, and define a norm ${}^\circ|\cdot|$ on ${}^\circ E_{|\cdot|} = \mathrm{Pns}(E, |\cdot|)/\sim_{|\cdot|}$ by ${}^\circ|{}^\circ u| = \mathrm{st}|u|$. The following lemma is just a slight reformulation of Proposition 2.2.2.

4.7.1. LEMMA. $({}^\circ E_{|\cdot|}, {}^\circ|\cdot|)$ is a Banach space, and the set of finite linear combinations $\sum_{k=1}^{n} a_k {}^\circ v_k$ is dense in ${}^\circ E_{|\cdot|}$.

The mapping $\mathrm{st}_{|\cdot|} : \mathrm{Pns}(E, |\cdot|) \to {}^\circ E_{|\cdot|}$ defined by $\mathrm{st}_{|\cdot|}(x) = {}^\circ x$ is called the *standard part map*. This slight abuse of terminology is justified by $\mathrm{st}_{|\cdot|}$ having exactly the same properties as a real standard part function. For example, mimicking the proof of Theorem 3.2.4 (1), we get.

4.7.2. PROPOSITION. Let $(\Omega, \mathscr{A}, P)$ be a hyperfinite probability space, and let $Y : \Omega \to {}^\circ E_{|\cdot|}$ be an $L(\mathscr{A})$-measurable random variable. Then there exists an internal, $\mathscr{A}$-measurable random variable $X : \Omega \to E$ such that $\mathrm{st}_{|\cdot|}(X(\omega)) = {}^\circ X(\omega) = Y(\omega)$ for $L(P)$ almost all ω.

We call X a *lifting* of Y. Using 4.7.2 we can derive the natural counterparts of the lifting results in Section 4.3.

Let $(E, \langle \cdot, \cdot \rangle)$ be a hyperfinite inner product space where the norm is denoted by $\|\cdot\|$. Choose an orthornormal generating set $\{e_n\}_{n \in \mathbb{N}}$ in E, and let $\{e_n\}_{n \leq \gamma}$, $\eta \in {}^*\mathbb{N} - \mathbb{N}$, be an extension of $\{e_n\}_{n \in \mathbb{N}}$ to an orthonormal basis for E. We shall construct a hyperfinite-dimensional random walk on E.

Let $T = \{0, \Delta t, 2\Delta t, \ldots, 1\}$ be a hyperfinite time line, and let $H \in {}^*\mathbb{N} - \mathbb{N}$ be at least as large as the dimension γ of E. The sample space Ω consists of all internal mappings $\omega : \{1, 2, \ldots, H\} \times T \to \{-1, 1\}$, and P is the normalized counting measure on Ω. We shall write $\omega_i(s)$ for $\omega(i, s)$. The process $\chi : \Omega \times T \to E$ is defined analogously to the n-dimensional random walk in Section 4.5;

$$(5) \qquad\qquad \chi(\omega, t) = \sum_{j=1}^{\gamma} \sum_{s=0}^{t} \sqrt{\Delta t}\, \omega_j(s) e_j.$$

We let $(\Omega, \{\mathscr{A}_t\}, P)$ be the natural internal filtration on Ω; i.e., $\mathscr{A}_t$ is generated by the equivalence relation $\omega \sim_t \omega'$ iff $\omega_i(s) = \omega_i'(s)$ for all $s < t$ and all i.

It is easy to see that $E(\|\chi_t\|^2) = \gamma \cdot t$ and that χ_t is not nearstandard in $(E, \|\cdot\|)$. It does, however, have the right finite-dimensional properties:

4.7.3. PROPOSITION. Let $P : E \to E_0$ be the projection onto a finite-dimensional subspace E_0. Then $°(P\chi_t)$ is a Brownian motion on $(°E_{0,\|\cdot\|}, °\|\cdot\|)$.

PROOF. It is not hard to see that $P\chi$ is S-continuous; in fact, this is a special case of Theorem 4.7.6 below, and we postpone the proof until then. To show that $P\chi$ is a Brownian motion, we first check the distributions of one-dimensional projections. Let Q be the projection on a unit vector $e \in E_0$, and let $q_k(e)$ be the Fourier coefficient $\langle e, e_k \rangle$. The quadratic variation of $Q\chi$ is given by

$$E([Q\chi](t) - [Q\chi](s) \mid \mathscr{A}_s) = E\left(\sum_{r=s}^{t} \left(\sum_{i=1}^{\gamma} \sqrt{\Delta t}\, \omega_i(r) q_i(e) \right)^2 \Bigg| \mathscr{A}_s \right)$$

$$= \sum_{r=s}^{t} \Delta t \sum_{i=1}^{\gamma} q_i(e)^2 = t - s,$$

where we have used the independence of $\omega_i(r)$ and $\omega_j(r)$ for $i \neq j$. By Proposition 4.4.18, $°Q\chi$ is a Brownian motion.

To show that $°P\chi$ itself is a Brownian motion, it suffices to show that two one-dimensional projections $°Q\chi$ and $°\tilde{Q}\chi$ along orthonormal axes e and $\tilde{e}$ are independent. Using that $\omega_i(s)$ and $\omega_j(r)$ are independent when $(i, s) \neq (j, r)$, we get

$$E\left(\left(\sum_{s=0}^{t} \sum_{i=1}^{\gamma} \omega_i(s)\sqrt{\Delta t}\, q_i(e) \right) \left(\sum_{r=0}^{t} \sum_{j=1}^{\gamma} \omega_j(r)\sqrt{\Delta t}\, q_j(\tilde{e}) \right) \right)$$

$$= \sum_{s=0}^{t} \Delta t \sum_{i=0}^{\gamma} q_i(e)\, q_i(\tilde{e}) = t \langle e, \tilde{e} \rangle = 0,$$

which shows that $°Q\chi$ and $°\tilde{Q}\chi$ are uncorrelated and hence independent.

Let $(H, \|\cdot\|)$ be the Hilbert space $(°E_{\|\cdot\|}, °\|\cdot\|)$. We have just seen that the hyperfinite random walk χ is in some sense a Brownian motion on H.

But how do we turn χ into a standard process? Since we have already observed that χ does not have a standard part in H, the only hope is to take the standard part with respect to a weaker topology. This is exactly the situation we encountered in connection with Gross's theorem in Section 3.5. In fact, if for each finite-dimensional subspace F of H and each $t \in [0, 1]$ we let μ_F^t be the measure on F induced by the projection ${}^{\circ}P_F\chi$ of χ to F, then $\{\mu_F^t\}_{F \in \mathscr{F}}$ is a cylindrical measure on H for each t. Gross's theorem tells us that if $|\cdot|$ is a norm on H which is measurable with respect to Gaussian measures and B is the completion of H with respect to $|\cdot|$, then $\{\mu_F^t\}$ induces a measure μ^t on B. The proof of Gross's theorem and in particular Lemma 3.5.7 tell us more; they say that χ_t is nearstandard almost everywhere with respect to the norm $|\cdot|$ (note that since we can consider E as subspace of $^*H$, there is a natural extension of $|\cdot|$ to E). Since $B = {}^{\circ}E_{|\cdot|}$, the standard part b of χ is thus a process living in B, and μ^t is nothing but the measure b induces on B.

If $i: H \to B$ is the inclusion map, the triple (i, H, B) is often referred to as an *abstract Wiener space*, and $b: \Omega \times [0, 1] \to B$ is a *Brownian motion* on (i, H, B). From the hyperfinite-dimensional random walk χ, we have thus obtained a standard Brownian motion b on H simply by taking the standard part with respect to a measurable norm. More information about Brownian motions in abstract Wiener spaces can be found, e.g., in Kuo (1975).

From the nonstandard point of view, the last two paragraphs may be considered an unnecessary detour; the random walk χ is a perfectly legitimate mathematical object, and the existence of the standard part b will play no part in our theory for stochastic integration.

B. Infinite Dimensional Stochastic Integrals

Our first task is to define a suitable class of integrands for stochastic integrals with respect to hyperfinite-dimensional random walks. Recall from Section 4.5 that if χ is an n-dimensional random walk, then the integrands are matrix-valued processes. In infinite dimensions one would expect the integrands to be operator valued, and it might be useful to take a brief look at operators between hyperfinite-dimensional spaces before turning to stochastic integration.

Let E and F be hyperfinite-dimensional inner product spaces, and let $\langle \cdot, \cdot \rangle$ and $\|\cdot\|$ denote the inner products and norms in both E and F. Fix two orthonormal generating sets $\{e_n\}_{n \in \mathbb{N}}$ and $\{f_n\}_{n \in \mathbb{N}}$ for E and F, respectively. A linear map $A: E \to F$ is called *S-bounded* if the operator norm $\|A\|$ is finite; it is called *nearstandard* if $A(e_n) \in \text{Pns}(F, \|\cdot\|)$ for all $n \in \mathbb{N}$. Thus, if A is S-bounded and nearstandard, it maps pre-nearstandard points to pre-nearstandard points.

Let $\{e_n\}_{n \leq \gamma}$ and $\{f_n\}_{n \leq \eta}$ be extensions of $\{e_n\}_{n \in \mathbb{N}}$, $\{f_n\}_{n \in \mathbb{N}}$ to internal orthonormal bases for E and F. A linear map $A : E \to F$ is called a *Hilbert–Schmidt operator* if

$$\sum_{n=1}^{\infty} {}^{\circ}\|A^*f_n\|^2 = {}^{\circ}\sum_{n=1}^{\eta} \|A^*f_n\|^2 < \infty,$$

where A^* is the adjoint of A. The sum $\sum_{n=1}^{\eta} \|A^*f_n\|^2$ is independent of the choice of basis $\{f_n\}_{n \leq \eta}$. Moreover,

$$\|A^*f_n\|^2 = \sum_{m=1}^{\gamma} \langle e_m, A^*f_n \rangle^2 = \sum_{m=1}^{\gamma} \langle Ae_m, f_n \rangle^2,$$

and hence

$$\sum_{n=1}^{\eta} \|A^*f_n\|^2 = \sum_{m=1}^{\gamma} \sum_{n=1}^{\eta} \langle Ae_m, f_n \rangle^2 = \sum_{m=1}^{\gamma} \|Ae_m\|^2,$$

which shows that $\sum_{n=1}^{\eta} \|A^*f_n\|^2$ is the square of the usual Hilbert–Schmidt norm for A. We let

$$(6) \qquad\qquad \|A\|_{(2)} = \left(\sum_{n=1}^{\eta} \|A^*f_n\|^2 \right)^{1/2}.$$

All Hilbert–Schmidt maps are obviously S-bounded, and they are also nearstandard:

$${}^{\circ}\sum_{m=k}^{\eta} \langle Ae_n, f_m \rangle^2 = {}^{\circ}\sum_{m=k}^{\eta} \langle e_n, A^*f_m \rangle^2 \leq {}^{\circ}\sum_{m=k}^{\eta} \|A^*f_m\|^2 \to 0$$

as $k \to \infty$ in $\mathbb{N}$.

Letting $\mathscr{L}(E, F)$ denote the space of internal linear maps from E to F, we can now define the set of stochastic integrands we shall be working with as follows.

4.7.4. DEFINITION. A process $X : \Omega \times T \to \mathscr{L}(E, F)$ is said to be in $SL^2(E, F)$ if it is nonanticipating [with respect to $(\Omega, \{\mathscr{A}_t\}P)$] and

$$\lim_{k \to \infty} E\left\{ \int_0^1 {}^{\circ}\sum_{m=1}^{k} \|X(\omega, s)^*(f_m)\|^2 \, ds \right\} = {}^{\circ}E\left\{ \int_0^1 \|X(\omega, s)\|_{(2)}^2 \, ds \right\} < \infty,$$

where an asterisk denotes the adjoint operator, and $\|\cdot\|_2$ is the Hilbert–Schmidt norm.

Notice the position of the standard parts in (4.7.4), inside both integrals on the left and outside both on the right. This guarantees that $X(\omega, s)$ is Hilbert–Schmidt for almost all (ω, s), and that $\|X(\omega, s)\|_{(2)}^2$ is S-integrable in the product measure on $\Omega \times T$.

4.7.5. DEFINITION. Let $\chi: \Omega \times T \to E$ be the random walk defined in (5), and let $X \in SL^2(E, F)$. The *stochastic integral* $\int X \, d\chi$ is the F-valued process given by

$$(7) \qquad \left(\int X \, d\chi \right)(\omega, t) = \sum_{s=0}^{t} X(\omega, s)(\Delta\chi(\omega, s)).$$

This definition is the natural generalization of the ones given in 4.1.1 and equation (4.5.5). We shall show that by restricting the integrands to the class $SL^2(E, F)$, we have obtained an integral with decent properties:

4.7.6. THEOREM. Let $X \in SL^2(E, F)$. Then $\int X \, d\chi$ is pre-nearstandard and S-continuous almost everywhere.

PROOF. Let Y_k be the kth component of $Y = \int X \, d\chi$. We make a preliminary calculation:

$$E(Y_k(t)^2) = E\left(\left(\sum_{s=0}^{t} \langle X_s(\Delta\chi(s)), f_k \rangle \right)^2 \right)$$

$$= E\left(\left(\sum_{s=0}^{t} \sum_{j=1}^{\gamma} \omega_j(s)\sqrt{\Delta t} \langle X_s(e_j), f_k \rangle \right)^2 \right)$$

$$= E\left(\sum_{s=0}^{t} \sum_{j=1}^{\gamma} \Delta t \langle X_s(e_j), f_k \rangle^2 \right)$$

$$= E\left(\int_0^t \sum_{j=1}^{\gamma} \langle e_j, X_s^*(f_k) \rangle \, ds \right)$$

$$= E\left(\int_0^t \| X_s^*(f_k) \|^2 \, ds \right).$$

By Doob's inequality, 4.2.8, we obtain

$$0 \le E\left(\sup_{t \le 1} \sum_{k=m}^{\eta} Y_k(t)^2 \right) \le 4E\left(\int_0^1 \sum_{k=m}^{\eta} \| X(\omega, s)^*(f_k) \|^2 \, ds \right).$$

Since $X \in SL^2(E, F)$, the right-hand side approaches zero as m goes to infinity in $\mathbb{N}$, and hence $\int X \, d\chi$ is nearstandard almost everywhere.

To prove that Y is S-continuous, we shall first prove that each component Y_k is. We remind the reader that according to Theorem 4.2.16, a square-integrable martingale is continuous if and only if its quadratic variation is. To apply this result in the present setting, we use the following trick: construct a new time line T' by dividing each interval $[k \, \Delta t, (k + 1) \, \Delta t)$ of T into γ points (recall that γ is the dimension of E). Introduce a new martingale $\tilde{Y}_k: \Omega \times T' \to {}^*\mathbb{R}$ which makes the jump $\omega_j(t)\sqrt{\Delta t} \langle X_t(e_j), f_k \rangle$ at time $t + ((j - 1)/\gamma)\,\Delta t$. Obviously, T is a subline of T', and Y_k is the restriction of $\tilde{Y}_k$ to T.

The point of this trick is that $\tilde{Y}_k$ has quadratic variation

$$[\tilde{Y}_k](t) = \sum_{s=0}^{t} \sum_{j=1}^{\gamma} \langle X_s(e_j), f_k \rangle^2\, \Delta t = \int_0^t \|X_s^*(f_k)\|^2\, ds.$$

Since $\|X(\omega, s)^*(f_k)\|^2 \le \|X(\omega, s)\|_{(2)}^2$, and $(\omega, s) \to \|X(\omega, s)\|_{(2)}^2$ is S-integrable, it follows that $s \to \|X_s^*(f_k)\|^2$ is S-integrable for almost all ω. Hence $[\tilde{Y}_k]$ is S-continuous almost everywhere, and by 4.2.16 so are $\tilde{Y}_k$ and Y_k.

It still remains to prove the S-continuity of Y. Let Ω' be a set of measure one such that for all $\omega \in \Omega'$, $Y(\omega, t)$ is pre-nearstandard for all t, and $Y_k(\omega, \cdot)$ is S-continuous for all $k \in \mathbb{N}$. Pick $\omega \in \Omega'$, and let s, $t \in T$ be infinitely close. By the continuity of the Y_k's, we have

$$(8) \qquad \sum_{k=1}^{n} (Y_k(\omega, t) - Y_k(\omega, s))^2 < 1/n$$

for all $n \in \mathbb{N}$. But then (8) holds for some $n = n_0 \in {}^*\mathbb{N} - \mathbb{N}$. Since $Y(\omega, t)$ and $Y(\omega, s)$ are pre-nearstandard

$$(9) \qquad \sum_{k=n_0+1}^{\eta} (Y_k(\omega, t) - Y_k(\omega, s))^2 \approx 0.$$

Combining (8) and (9), we get $\|Y(t) - Y(s)\|^2 \approx 0$, and hence $Y(\omega, \cdot)$ is S-continuous when $\omega \in \Omega'$.

We shall now show how the hyperfinite integrals $\int X\, d\chi$ can be used to define standard integrals $\int X\, db$, where $b = \mathrm{st}_{|\cdot|}\chi$. Let $\mathscr{B}_{(2)}(H, K)$ be the set of all Hilbert–Schmidt operators from H to K; with the $\|\cdot\|_{(2)}$-norm this is a separable Hilbert space. As usual $(\Omega, \{\mathscr{B}_s\}, L(P))$ is the external filtration generated by $(\Omega, \{\mathscr{A}_s\}, P)$. The class of b-integrable processes is defined as follows:

4.7.7. DEFINITION. A process $x : \Omega \times [0, 1] \to \mathscr{B}_{(2)}(H, K)$ is in $L^2(H, K)$ if it is adapted to $(\Omega, \{\mathscr{B}_t\}, L(P))$ and

$$E\left(\int_0^1 \|x(t)\|_{(2)}^2\, dt\right) < \infty.$$

As usual $\int x\, db$ will be defined as the standard part of $\int X\, d\chi$ for a suitable lifting X of x, but this time the notion of a lifting is a little more complicated. Recall that $\mathscr{L}(E, F)$ is the set of all internal linear maps from E to F. If $T \in \mathscr{B}_{(2)}(H, K)$, we define $\tilde{T} : E \to F$ by

$$\tilde{T} = P_F {}^* T \restriction E,$$

where $P_F : {}^*K \to F$ is the orthogonal projection. Observe that $\tilde{T}$ is an internal Hilbert–Schmidt operator. Pick an orthonormal basis $\{T_n\}_{n \in \mathbb{N}}$ for $\mathscr{B}_{(2)}(H, K)$, and let $\{\tilde{T}_n\}_{n \in \mathbb{N}}$ be a standard generating set in $(\mathscr{L}(E, F), \|\cdot\|_{(2)})$. The map $h : \mathscr{B}_{(2)}(H, K) \to ({}^\circ\mathscr{L}(E, F)_{\|\cdot\|_2}, {}^\circ\|\cdot\|_{(2)})$ defined by

$$h(T) = {}^\circ(\tilde{T})$$

is an isomorphism, and we shall identify $\mathscr{B}_{(2)}(H, K)$ and ${}^\circ\mathscr{L}(E, F)_{\|\cdot\|_2}$.

4.7.8. DEFINITION. An internal process $X : \Omega \times T \to \mathscr{L}(E, F)$ is called a 2-*lifting* of $x : \Omega \times [0, 1] \to B_{(2)}(H, K)$ if $X \in SL^2(E, F)$ and $\mathrm{st}_{\|\cdot\|_2}(X(\omega, t)) = x(\omega, {}^\circ t)$ almost everywhere in $\Omega \times T$.

4.7.9. LEMMA. Any $x \in L^2(H, K)$ has a 2-lifting.

PROOF. By 4.7.2 and 4.3.11 there is a nonanticipating process $Y : \Omega \times T \to \mathscr{L}(E, F)$ such that $\mathrm{st}_{\|\cdot\|_2}(Y(\omega, t)) = x(\omega, {}^\circ t)$ almost everywhere. We must modify Y to make it an element of $SL^2(E, F)$. For each $m \in {}^*\mathbb{N}$, let Y_m be defined by

$$Y_m(\omega, t) = \begin{cases} Y(\omega, t) & \text{if } \|Y(\omega, t)\|_{(2)} \le m, \\ 0 & \text{otherwise.} \end{cases}$$

Since Y lifts x, we have for $m \in \mathbb{N}$

$$ {}^\circ E\left(\int_0^1 \|Y_m(\omega, t)\|_{(2)}^2 \, dt \right) = E\left(\int_0^1 {}^\circ \|Y_m(\omega, t)\|_{(2)}^2 \, dt \right) $$

$$ \le E\left(\int_0^1 \|x(\omega, t)\|_{(2)}^2 \, dt \right). $$

Thus there exists an $m_0 \in {}^*\mathbb{N} - \mathbb{N}$ such that

$$ {}^\circ E\left(\int_0^1 \|Y_{m_0}(\omega, t)\|_{(2)}^2 \, dt \right) = \lim_{\substack{m \to \infty \\ m \in \mathbb{N}}} {}^\circ E\left(\int_0^1 \|Y_m(\omega, t)\|_{(2)}^2 \, dt \right), $$

which implies that $\|Y_{m_0}(\omega, t)\|_{(2)}^2$ is S-integrable. Since Y_{m_0} is a lifting of x, we get

$$ \lim_{k \to \infty} E\left(\int_0^1 {}^\circ \sum_{n=1}^k \|Y_{m_0}(\omega, s)^*(f_n)\|^2 \, ds \right) $$

$$ = \lim_{k \to \infty} E\left(\int_0^1 \sum_{n=1}^k \|x(\omega, s)^*(f_n)\|^2 \, ds \right) = E\left(\int_0^1 \|x(\omega, s)\|_{(2)}^2 \, ds \right) $$

$$= E\left(\int_0^1 {}^\circ \| Y_{m_0}(\omega, s)\|_{(2)}^2 \, ds\right) = {}^\circ E\left(\int_0^1 \| Y_{m_0}(\omega, s)\|_{(2)}^2 \, ds\right),$$

which proves that Y_{m_0} is a 2-lifting of x.

We shall leave it to the reader to check that if X and Y are two 2-liftings of x, then $\int X \, d\chi \approx \int Y \, d\chi$ almost everywhere. With these two facts in mind, we can make the following definition.

4.7.10. DEFINITION. Let $x \in L^2(H, K)$. Then the *stochastic integral* $\int x \, db$ is defined as the standard part of $\int X \, d\chi$ where X is a 2-lifting of x.

Notice that the stochastic integral is independent of which measurable norm was used in the construction of b; in fact, b itself only enters the theory as a notational device in the expression $\int x \, db$.

We could now go on to prove that the stochastic integral defined in 4.7.10 agrees with the one obtained by standard methods in, e.g., Kuo (1975). This can be done very much along the same lines as the one-dimensional theory in Section 4.4, but we shall not try to carry it out here. The interested reader is referred to Lindstrøm (1983).

The nonstandard approach to the infinite-dimensional stochastic integral has two advantages. First, the integral has a simple, intuitive definition as a pathwise Stieltjes integral; and second, the Brownian motion has a natural construction as a random walk on a hyperfinite-dimensional linear space; no measurable norms or appeals to Gross's theorem are necessary. Also notice that although the Brownian motion b is B-valued, the integrand x takes values in $\mathscr{B}_{(2)}(H, K)$. In the standard approach [see, e.g., Kuo (1975)] this necessitates a rather messy study of the relationship between H and B and the linear maps on the two spaces. The nonstandard theory does not need B, and thus no such study is necessary.

C. A Remark on Stochastic Partial Differential Equations

Let us finally return to partial stochastic differential equations. We shall first take a look at a simplified, linearized version of Eq. (4):

$$(10) \qquad du = (\nu \, \Delta u - (v\Delta) \cdot u + h) \, dt + g \, dw.$$

Here v is a constant equilibrium solution of the deterministic Navier–Stokes equation, and $u(x, t)$ is the fluctuating velocity about this equilibrium due to the random perturbations $g \, dw$. As a model of physical reality, (10) is of limited interest, but it has the mathematical virtue of being exactly solvable.

Observe that (10) is of the form

$$(11) \qquad dx(t) = Ax(t) \, dt + \xi(x(t)) \, db(t),$$

where A is an unbounded, linear operator on a Hilbert space K [in (10), $K = L^2(D)$, where D is the domain of the problem], and b is an infinite-dimensional Brownian motion. We have the following general result:

4.7.11. THEOREM. Let $A: K \to K$ be the densely defined infinitesimal generator of a strongly continuous semigroup $\{T_t\}_{t \in \mathbb{R}_+}$. Let $\xi: [0, 1] \times K \to \mathcal{B}_{(2)}(H, K)$ be measurable, and assume that there is an $L \in \mathbb{N}$ such that $\|\xi(t, u) - \xi(t, v)\|_{(2)} \le L \cdot \|u - v\|$ for all $t,$ $u,$ $v.$ Assume also that $\int_0^1 \|\xi(t, 0)\|_{(2)}^2 \, dt < \infty$. Finally, let x_0 be a $\mathcal{B}_0$-measurable, square-integrable initial condition. Then the equation

$$(12) \qquad dx(t) = Ax(t) \, dt + \xi(t, x(t)) \, db(t), \qquad x(0) = x_0,$$

has a continuous weakened and mild integral solution.

There are several ways of defining a "solution" to (12); we have focused on the following two:

The process x is called a *weakened solution* if

$$(13) \qquad x(t) = x_0 + A \int_0^t x(s) \, ds + \int_0^t \xi(s, x(s)) \, db(s),$$

and it is called a *mild integral solution* if

$$(14) \qquad x(t) = T_t x_0 + \int_0^t T_{t-s} \xi(s, x(s)) \, db(s),$$

where $\{T_t\}$ is the semigroup generated by A.

Results like 4.7.11 are well known from the standard literature, and a general result [see Chojnowskaja–Michalik (1979)] states that a process is a weakened solution if and only if it is a mild integral solution. We shall restrict ourselves to an outline of the strategy for producing solutions of the latter kind.

Embed H and K in hyperfinite-dimensional spaces E and F in the usual way, and let χ be a random walk on E. Write $\{*T_t\}_{t \in *\mathbb{R}_+}$ for $*(\{T_t\}_{t \in \mathbb{R}_+})$, and let $\tilde{T}_{\Delta t}: F \to F$ be defined by

$$\tilde{T}_{\Delta t} = P_F \, {}^*T_{\Delta t} \restriction F,$$

where $P_F: {}^*K \to F$ is the projection. Define an internal semigroup $\{\tilde{T}_t\}_{t \in T}$ by $\tilde{T}_{k \Delta t} = (\tilde{T}_{\Delta t})^k$, and let

$$\tilde{A} = (I - \tilde{T}_{\Delta t})/\Delta t.$$

Let $\Xi: T \times F \to \mathscr{L}(E, F)$ be a uniform lifting of ξ with respect to the normalized counting measure λ on T [i.e., there is a $T' \subset T$ of Loeb measure one such that $\mathrm{st}_{\|\cdot\|_{(2)}} (\Xi(t, v)) = \xi(^\circ t, {}^\circ v)$ whenever $t \in T'$ and v is pre-nearstandard]. We choose Ξ S-Lipschitz-continuous in the second variable, and such that $\|\Xi(t, 0)\|_{(2)}^2$ is S-integrable over T. Finally, we let X_0 be an $\mathscr{A}_\delta$-measurable 2-lifting of x_0 for some $\delta \approx 0$.

Corresponding to (14), we now have the hyperfinite difference equation

$$(15) \qquad X_t = \tilde{T}_t X_0 + \sum_{s=\delta}^{t} \tilde{T}_{t-\Delta t - s} \Xi(s, X_s) \, \Delta \chi(s).$$

This equation obviously has a solution, but some honest work is required to establish the estimates needed to show that the solution is nearstandard and S-continuous [see Lindstrøm (1983)].

However, granted that X is nearstandard and S-continuous, we see that $\tilde{T}_{t-\Delta t - s} \Xi(s, X(s))$ is a 2-lifting of $T_{t-s} \xi(s, {}^\circ X(s))$ on a set of measure one. Since $T_t x_0 = \tilde{T}_t X_0$, we see that by putting $x({}^\circ t) = {}^\circ X(t)$, we transform (15) into

$$(16) \qquad x(t) = T_t x_0 + \int_0^t T_{t-s} \xi(s, x(s)) \, db(s),$$

i.e., x is a continuous mild integral solution of (12). By induction, (15) is easily seen to be equivalent to

$$(17) \qquad X_t = X_0 + \sum_{s=\delta}^{t} \tilde{A}(X(s)) \, \Delta t + \sum_{s=\delta}^{t} \Xi(s, X(s)) \, \Delta \chi(s),$$

and this can be used to give a direct nonstandard proof of the fact that x is also a weakened solution of (12) [see Lindstrøm (1983)].

We round off with a few remarks on the full Navier–Stokes equation (4). We are interested in space–time statistical solutions of both the Navier–Stokes equation and a stochastic version of the equation in a bounded domain. There is a vast literature on this topic; we mention in particular the work of Bensoussan and Temam (1973) and the contribution of the Vishik school; for a survey and full references see the paper by Vishik *et al.* (1979). Here we indicate how this problem can be discussed within the present nonstandard framework.

The method is to replace the "infinite-dimensional" Navier–Stokes system by its "finite-dimensional" Galerkin approximations. In finite dimensions it is not too difficult to construct space–time statistical solutions. And provided we can prove the necessary uniform estimates we can use Prokhorov's theorem to produce a solution of the infinite-dimensional system.

In the nonstandard framework we can replace the given system by a hyperfinite-dimensional system; in the stochastic case the finite-dimensional approximations will be Itô-type stochastic equations, which we have studied in Section 4.5. We can then use the theory of Sections 3.4 and 3.5 to pass from the finite-dimensional approximations to the infinite system, provided—and this is, of course, the mathematical crux of the argument—that we establish the necessary uniform (i.e., dimension-independent) estimates for the finite-dimensional approximations.

We invite the reader to look at the theory from the present nonstandard perspective, which we believe offers a clean conceptual setting for this kind of problem. There are related questions concerning the Euler equation which also could be interpreted in the present framework; see Albeverio and Høegh-Krohn (1981) and Albeverio *et al.* (1979, 1985). For other interesting applications of infinite-dimensional stochastic differential equations, see Faris and Jona-Lasinio (1982), Jona-Lasinio and Mitter (1985) and Loges (1984).

4.8. WHITE NOISE AND LÉVY BROWNIAN MOTION

Let $(E, \mathscr{E}, m)$ be a measure space and let $\mathscr{E}_m$ be the sets in $\mathscr{E}$ with finite measure. Fix a probability space $(\Omega, \mathscr{B}, P)$. An n-dimensional *white noise* on $(E, \mathscr{E}, m)$ with respect to $(\Omega, \mathscr{B}, P)$ is a family

$$(1) \qquad X = \{X(A)\}_{A \in \mathscr{E}_m}$$

of random vectors $X(A): \Omega \to \mathbb{R}^n$ such that:

(i) $X(A)$ is Gaussian distributed with mean zero and covariance matrix $m(A) \cdot I$;
(ii) if $A \cap B = \varnothing$, then $X(A)$ and $X(B)$ are independent, and $X(A) + X(B) = X(A \cup B)$ and P-a.e.

4.8.1. EXAMPLE. Let $(E, \mathscr{E}, m)$ be the unit interval with the Borel σ-algebra and Lebesgue measure. Let $b: \Omega \times [0, 1] \to \mathbb{R}^n$ be an n-dimensional Brownian motion, and define $X = \{X(A)\}$ by

$$X(A) = \int_0^1 1_A \, db,$$

where 1_A is the indicator function of the set $A \subset E$. Then X is a white noise on $(E, \mathscr{E}, m)$.

By emphasizing the close relationship between white noise on $[0, 1]$ and Brownian motion, Example 4.8.1 introduces the theme of this section. Following Stoll (1982, 1985), we shall show that Anderson's construction of Brownian motion can easily be extended to a construction of white noise. Once we have white noise on $\mathbb{R}^d$, we shall reverse the procedure of Example 4.8.1 and obtain (Lévy) Brownian motion of d "time" parameters as a stochastic integral of white noise. Finally, we shall give a variant of a theorem of Stoll showing how this stochastic integral representation can be used to obtain a new invariance principle for Lévy Brownian motion.

A. Construction of White Noise

We shall first construct the nonstandard counterpart of white noise. Let $(Y, \mathscr{D}, \mu)$ be an *atomless* hyperfinite measure space; i.e., $\mathscr{D}$ is the set of all internal subsets of Y, and $\mu(\{x\}) \approx 0$ for all singletons $\{x\}$. If Ω is the set of all internal maps from Y to $\{-1, 1\}$, let $\mathscr{A}$ be the set of all internal subsets of Ω and P the uniform probability measure on Ω. For each $A \in \mathscr{D}$, define

$$(2) \qquad \chi(A) = \sum_{a \in A} \omega(a)\sqrt{\mu(a)};$$

the map $A \mapsto \chi(A)$ is called *S-white noise* on Y. Equation (2) is the obvious generalization of the white noise representation discussed in Section 3.3, and χ will play the same part in this section as Anderson's random walk did in Section 3.3.

Before we define the standard part of χ, we make a preliminary calculation. Let $A, B \in \mathscr{D}$, and define $\varepsilon : A \, \Delta \, B \to \{-1, 1\}$ to be 1 on $A - B$ and -1 on $B - A$. Then

$$(3) \qquad E((\chi(A) - \chi(B))^2) = E\left(\left(\sum_{a \in A \, \Delta \, B} \varepsilon(a)\omega(a)\sqrt{\mu(a)}\right)^2\right)$$

$$= E\left(\sum_{a, b \in A \, \Delta \, B} \varepsilon(a)\varepsilon(b)\omega(a)\omega(b)\sqrt{\mu(a)\mu(b)}\right)$$

$$= \sum_{a \in A \, \Delta \, B} \mu(a) = \mu(A \, \Delta \, B),$$

where the next to last step uses the independence of $\omega(a)$ and $\omega(b)$ when $a \neq b$. Thus if A and B are equal $L(\mu)$ almost everywhere, $\chi(A)$ and $\chi(B)$ are equal $L(P)$ almost everywhere.

For each Loeb measurable $A \subset Y$, choose an $A' \in \mathscr{D}$ such that $L(P)$-$(A \, \Delta \, A') = 0$. Define the standard part of χ to be the family

$$\{X(A)\}_{A \in L(\mathscr{D})_{L(\mu)}},$$

where

$$(4) \qquad X(A)(\omega) = {}^{\circ}\chi(A')(\omega).$$

It is not hard to check that $\{X(A)\}$ is a one-dimensional white noise on the Loeb space $(Y, L(\mathscr{D}), L(\mu))$; part (ii) of the definition follows from (3), and part (i) is proved by a calculation of the Fourier transform similar to the one in the second proof of 3.3.5(ii).

Let E be a Hausdorff space and m a σ-finite Radon measure on E. Assume that if x is an isolated point in E, then $m\{x\} = 0$. Fix a rich, hyperfinite subset Y of $*E$ [*rich* means that $\mathrm{st}(Y) = E$] and an atomless, internal measure μ on Y such that $m = L(\mu) \circ \mathrm{st}^{-1}$. If X is a white noise on $(Y, L(\mathscr{D}), L(\mu))$, the equation

(5) $$\tilde{X}(A) = X(\mathrm{st}^{-1}(A))$$

induces a white noise $\tilde{X}$ on $(E, \mathscr{E}, m)$. Thus on any σ-finite Radon space that does not charge isolated points, there is a white noise induced by an internal S-white noise.

Let us just remark that the condition given in the last sentence is stronger than what is strictly necessary, but since we shall be mainly concerned with white noise on $\mathbb{R}^d$, it is more than sufficient for our purposes.

Above we have only considered one-dimensional white noise, but it is trivial to extend the construction to n dimensions; just let n independent one-dimensional processes run along orthonormal axes.

B. Stochastic Integrals and the Continuity Theorem

We now turn to stochastic integration. Without filtrations and a concept of nonanticipation we can only expect to integrate deterministic functions, but it turns out that this is all we shall need in this section. Let χ be an S-white noise as in (2), and let $F: Y \to *\mathbb{R}$ be an internal function. The *stochastic integral* of F with respect to χ is defined as

(6) $$\left(\int F \, d\chi \right)(\omega) = \sum_{a \in Y} F(a)\chi(a)(\omega) = \sum_{a \in Y} F(a)\omega(a)\sqrt{\mu(a)}.$$

An easy calculation similar to (3) shows that

(7) $$E\left(\left(\int F \, d\chi \right)^2 \right) = \int F(a)^2 \, d\mu(a),$$

and thus a reasonable class of integrands is $SL^2(\mu)$.

Notice that if μ is atomless and $F \in SL^2(\mu)$, then the measure $\mu_F(A) = \int_A F^2 \, d\mu$ is also atomless, and the map $A \mapsto \int_0 1_A F \, d\chi$ is an S-white noise on $(Y, \mathscr{D}, \mu_F)$. In particular, $^{\circ}(\int 1_A F \, d\chi)$ is Gaussian distributed with mean zero and variance $^{\circ}(\int_A F^2 \, d\mu)$.

If $\tilde{X}$ is a white noise on a Radon space $(E, \mathscr{E}, m)$ induced by χ, we can for each $f \in L^2(m)$ define a stochastic integral by

(8) $$\int f \, d\tilde{X} = {}^{\circ}\left(\int F \, d\chi \right),$$

where $F \in SL^2(\mu)$ is a lifting of f. It follows from (7) that this integral is independent of the choice of lifting, and we leave it to the reader to check that it agrees with the one obtained from the standard approach. [The standard definition is the usual one; do the obvious thing for simple functions, and extend continuously using the L^2-isometry corresponding to (7).]

Again the extension to higher dimensions is straightforward; if χ is an n-dimensional S-white noise and $F: Y \to *\mathbb{R}^m \otimes *\mathbb{R}^n$ is internal [recall that $*\mathbb{R}^m \otimes *\mathbb{R}^n$ is the space of all $*\mathbb{R}$-valued $(m \times n)$ matrices], definition (5) extends to integrals of the form $\int F \, d\chi$ in the same way as before; put

$$\int F \, d\chi = \sum_{a \in Y} F(a) \cdot \chi(a),$$

where $\cdot$ is matrix multiplication.

The stochastic integrals we have defined in this section are just random variables; unlike integrals of martingales, they do not depend on an extra parameter. We have seen that we can remedy this somewhat by considering $(\int 1_A F(x) \, d\chi)_{A \in \mathcal{D}}$ as an S-white noise on $(Y, \mathcal{D}, \mu_F)$, but it is often more interesting to integrate a kernel $F(x, y)$ of two variables, and study the resulting integral $\int F(x, y) \, d\chi(y)$ as a stochastic process parametrized by Y. Stoll's construction of Lévy Brownian motion is in terms of such integrals, and as a first illustration we shall take a look at what happens in the one-dimensional case.

4.8.2. EXAMPLE. Let $T = \{0, \Delta t, 2\, \Delta t, \ldots, 1\}$ be a hyperfinite time line, and let μ be the normalized counting measure on T. If Ω is the set of all internal maps from T to $\{-1, 1\}$ and P is the uniform probability measure on Ω, then

$$\chi(A) = \sum_{a \in A} \omega(a)\sqrt{\mu(a)} = \sum_{a \in A} \omega(a)\sqrt{\Delta t}, \qquad \omega \in \Omega$$

is an S-white noise on T. Put

(9) $F(t, a) = \tfrac{1}{2} \operatorname{sgn}(t - a) + \tfrac{1}{2};$

then the process

(10) $\hat{\chi}(t) = \int_T F(t, a) \, d\chi(a)$

is Anderson's random walk. Hence we have reversed Example 4.8.1 and obtained Brownian motion as a stochastic integral of white noise on $[0, 1]$. Observe for future reference that

(11) $\hat{\chi}(t) - \hat{\chi}(s) = \frac{1}{2} \int (\operatorname{sgn}(t - a) - \operatorname{sgn}(s - a)) \, d\chi(a).$

Our plan is to construct Lévy Brownian motion on $\mathbb{R}^d$ in the spirit of this example, but before we turn to this, we must describe the nonstandard theory of white noise on Euclidean spaces in more detail.

An internal subset $\Gamma \subseteq {}^*\mathbb{R}^d$ is called a *hyperfinite lattice* if

$$\Gamma = \{k_1 \, \Delta t, \ldots, k_d \, \Delta t) \,|\, (k_1, \ldots, k_d) \in {}^*\mathbb{Z}^d, \max_{1 \leq i \leq d} |k_i| \leq N\}$$

for some $\Delta t \approx 0$ and some $N \in {}^*\mathbb{N}$ such that $N \cdot \Delta t$ is infinite. The uniform, internal measure μ on Γ defined by

$$\mu(\{(k_1 \, \Delta t, \ldots, k_d \, \Delta t)\}) = (\Delta t)^d$$

induces Lebesgue measure on $\mathbb{R}^d$. As usual Ω is the set of all internal maps from Γ to $\{-1, 1\}$, P is the uniform probability measure on Ω, and

$$\chi(A) = \sum_{a \in A} \omega(a)\sqrt{\mu(a)} = \sum \omega(a)(\Delta t)^{d/2}$$

is S-white noise on Γ.

We are interested in internal processes $I : \Omega \times \Gamma \to {}^*\mathbb{R}$ that can be obtained as stochastic integrals $I(x) = \int F(x, y) \, d\chi(y)$. To have standard parts processes of this kind must be reasonably regular; we shall use the following notion of S-continuity.

4.8.3. DEFINITION. An internal process $Z : \Omega \times \Gamma \to {}^*\mathbb{R}$ is *S-continuous* if for almost all ω, $Z(\omega, x) \approx Z(\omega, y)$ whenever x and y are nearstandard and $x \approx y$.

Notice that we are only concerned with S-continuity at nearstandard points.

A result we shall find useful in proving the S-continuity of various stochastic integrals is the following theorem.

4.8.4. THEOREM. Let Γ be a hyperfinite lattice in ${}^*\mathbb{R}^d$, and let $F : \Gamma \times \Gamma \to {}^*\mathbb{R}$ be an internal function. Assume that there are positive real numbers α and C such that

$$(12) \qquad \int |F(x, a) - F(y, a)|^2 \, d\mu(a) \leq C\|x - y\|^\alpha$$

for all $x, y \in \Gamma$. Then the stochastic integral $x \mapsto \int F(x, a) \, d\chi(a)$ is S-continuous.

The main probabilistic ingredient needed for the proof is a hyperfinite version of Kolmogorov's continuity theorem:

4.8.5. PROPOSITION. Let Γ be a hyperfinite lattice in ${}^*\mathbb{R}^d$, and let $Z : \Omega \times \Gamma \to {}^*\mathbb{R}$ be an internal process. If there exist positive real numbers p, r, K

such that

(13) $$E(|Z(x) - Z(y)|^p) \leq K\|x - y\|^{d+r}$$

for all $x, y \in \Gamma$, then Z is S-continuous.

Let us first show how 4.8.5 is used to prove 4.8.4.

PROOF OF THEOREM 4.8.4. Let $m \in \mathbb{N}$ be so large that $m\alpha > d$, say $m\alpha = d + r$. If $I(x) = \int F(x, a)\, d\chi(a)$, then by 4.8.5 it suffices to show that there is a $K \in \mathbb{R}$ such that

(14) $$E((I(x) - I(y))^{2m}) \leq K\|x - y\|^{m\alpha}$$

for all $x, y \in \Gamma$. We first observe that

(15) $$E((I(x) - I(y))^{2m}) = E\left(\left(\sum_{a \in \Gamma}(F(x, a) - F(y, a))\, d\chi(a)\right)^{2m}\right)$$

$$= \sum_{(a_1,\ldots,a_{2m}) \in \Gamma^{2m}} E\left(\prod_{i=1}^{2m}(F(x, a_i)\right.$$

$$\left. - F(y, a_i))\omega(a_i)\sqrt{\mu(a_i)}\right).$$

The next thing to notice is that since $\omega(a_i)$ and $\omega(a_j)$ are independent when $a_i \neq a_j$, most terms in the last sum are zero; e.g., if $a_i \neq a_j$ when $i \neq j$, then $E(\prod_{i=1}^{2m}\omega(a_i)) = 0$. More than that is true; if there is an i such that the set $\{j \leq 2m \,|\, a_j = a_i\}$ is of odd cardinality, then $E(\prod_{i=1}^{2m}\omega(a_i)) = 0$. To make this precise, we introduce an equivalence relation $\sim$ on Γ^m by putting $(a_1,\ldots,a_{2m}) \sim (b_1,\ldots,b_{2m})$ if $a_i = a_j$ iff $b_i = b_j$. Let Δ be the set of equivalence classes of $\sim$. An equivalence class $\delta \in \Delta$ is called *even* if whenever $(a_1,\ldots,a_{2m}) \in \delta$, then the sets $A_i = \{j \,|\, a_j = a_i\}$ all have an even number of elements. Split Δ into two parts; the even elements Δ_{even} and the rest Δ_{odd}. We have seen that Δ_{odd} makes no contribution to (15), and hence

$$E((I(x) - I(y))^{2m})$$

$$= \sum_{\delta \in \Delta_{\text{even}}} \sum_{(a_1,\ldots,a_{2m}) \in \delta} E\left(\prod_{i=1}^{m}(F(x, a_i) - F(y, a_i))\omega(a_i)\sqrt{\mu(a_i)}\right).$$

To prove (14), it thus suffices to show that if $\delta \in \Delta_{\text{even}}$, then

(16) $$\sum_{(a_1,\ldots,a_{2m}) \in \delta} E\left(\prod_{i=1}^{2m}(F(x, a_i) - F(y, a_i))\omega(a_i)\sqrt{\mu(a_i)}\right) \leq K_\delta\|x - y\|^{m\alpha}$$

for some constant $K_\delta \in \mathbb{R}$.

Fix $\delta \in \Delta_{\text{even}}$, and let $2p_1, \ldots, 2p_q$ be the number of elements in the various partition classes $A_i = \{j \mid a_j = a_i\}$; obviously $p_1 + p_2 + \cdots + p_q = m$. Since $\omega(a_i)^2 = 1$

$$\sum_{(a_1, \ldots, a_{2m}) \in \delta} E\left(\prod_{i=1}^{2m} (F(x, a_i) - F(y, a_i))\omega(a_i)\sqrt{\mu(a_i)} \right)$$

$$= \sum_{(b_1, \ldots, b_q) \in \Gamma^q} \prod_{i=1}^{q} (F(x, b_i) - F(y, b_i))^{2p_i} \mu(b_i)^{p_i}$$

$$= \prod_{i=1}^{q} \left(\int \mu(b)^{p_i - 1}(F(x, b) - F(y, b))^{2p_i} \, d\mu(b) \right).$$

Since $p_1 + p_2 + \cdots + p_q = m$, (16) follows if we can only prove that

$$(17) \qquad \int \mu(b)^{p_i - 1}(F(x, b) - F(y, b))^{2p_i} \, d\mu(b) \le K' \|x - y\|^{p_i \alpha}$$

for some $K' \in \mathbb{R}$.

Notice that since

$$|F(x, b) - F(y, b)|^2 \mu(b) \le \int |F(x, a) - F(y, a)|^2 \, d\mu(a) \le C \|x - y\|^{\alpha},$$

we have

$$(|F(x, b) - F(y, b)| / \|x - y\|^{\alpha/2})\mu(b)^{1/2} \le C^{1/2},$$

and thus

$$\int \mu(b)^{p_i - 1}(F(x, b) - F(y, b))^{2p_i} \, d\mu(b)$$

$$= \|x - y\|^{\alpha(p_i - 1)} \int \left(\frac{|F(x, b) - F(y, b)|}{\|x - y\|^{\alpha/2}} \mu(b)^{1/2} \right)^{2p_i - 2}$$

$$\times |F(x, b) - F(y, b)|^2 \, d\mu(b)$$

$$\le \|x - y\|^{\alpha(p_i - 1)} C^{p_i - 1} \int |F(x, b) - F(y, b)|^2 \, d\mu(b) \le C^{p_i} \|x - y\|^{p_i \alpha},$$

which is (17) with $K' = C^{p_i}$. The theorem is proved.

For the convenience of the reader we include a proof of 4.8.5, although it is almost identical to the standard argument given in, e.g., Simon (1979). To simplify the notation, we shall assume that $\Delta t = 2^{-\eta}$ for some $\eta \in {}^*\mathbb{N} - \mathbb{N}$.

PROOF OF PROPOSITION 4.8.5. Without any loss of generality we prove the continuity only at points $x = (x_1, \ldots, x_d)$ with $0 \le x_i \le 1$ for all i.

Fix a real number α, $0 < \alpha < r/p$, then

$$P\{\omega \,||\, Z(\omega, x) - Z(\omega, y)| \geq \|x - y\|^{\alpha}\}$$
$$\leq \|x - y\|^{-\alpha p} E(|Z(x) - Z(y)|^p) \leq K \|x - y\|^{d + \varepsilon},$$

where $\varepsilon = r - \alpha p$.

If $k = (k_1, \ldots, k_d) \in {}^*\mathbb{Z}^d$, let $k^{(i)} = (k_1, \ldots, k_i + 1, \ldots, k_d)$. For all $n \leq \eta$ and all $k \in {}^*\mathbb{Z}^d$, we have

$$P\left\{\omega \,\left|\, \left| Z\left(\omega, \frac{k^{(i)}}{2^n}\right) - Z\left(\omega, \frac{k}{2^n}\right)\right| \geq 2^{-n\alpha}\right.\right\} \leq K \cdot 2^{-nd} \cdot 2^{-n\varepsilon}.$$

Given $n \in {}^*\mathbb{N}$, let

$$H_n = \{(k_1, \ldots, k_d) \in {}^*\mathbb{Z}^d \,|\, 0 \leq k_i < 2^n \text{ for all } i\}.$$

For all $N \leq \eta$, we have

$$\sum_{n=N}^{\eta} \sum_{k \in H_n} \sum_{i=1}^{d} P\left(\left| Z\left(\frac{k^{(i)}}{2^n}\right) - Z\left(\frac{k}{2^n}\right)\right| \geq 2^{-n\alpha}\right) \leq K \cdot d \frac{e^{-N\varepsilon}}{1 - 2^{-\varepsilon}}.$$

Thus for almost all ω there is a number $\nu(\omega) \in \mathbb{N}$ such that

$$\left| Z\left(\omega, \frac{k^{(i)}}{2^n}\right) - Z\left(\omega, \frac{k}{2^n}\right)\right| \leq 2^{-n\alpha}$$

whenever $n \geq \nu(\omega)$.

Consider two elements $k/2^n = (k_1/2^n, \ldots, k_d/2^n)$ and $t = (k_1/2^n, \ldots, t_i, \ldots, k_d/2^n)$ in Γ such that $k_i/2^n \leq t_i \leq (k_i + 1)/2^n$. Then

$$t_i - \frac{k_i}{2^n} = \sum_{j=0}^{\eta-n} \gamma_j 2^{-(n+j)}$$

for some sequence $\{\gamma_j\}$ of zeros and ones. If $n \geq \nu(\omega)$, we thus have

$$\left| Z\left(\omega, \frac{k}{2^n}\right) - Z(\omega, t)\right| \leq \sum_{j=0}^{\eta-n} \gamma_j 2^{-\alpha(n+j)} \leq \frac{2^{-\alpha n}}{1 - 2^{-\alpha}}.$$

Similarly, we get

$$\left| Z\left(\omega, \frac{k^{(i)}}{2^n}\right) - Z(\omega, t)\right| \leq \frac{2^{-\alpha n}}{1 - 2^{-\alpha}}.$$

This implies that if $\nu(\omega)$ is finite, $Z(\omega, t)$ is S-continuous along all lines parallel to one of the axes. Hence Z is S-continuous.

REMARK. We were rather careless toward the end of this proof; with just a little more patience one could get that Z is Hölder continuous with index α.

C. Lévy Brownian Motion

At long last we can now turn to Lévy Brownian motion. Let us first explain what it is (Lévy, 1948).

4.8.6. DEFINITION. A *Lévy Brownian motion* on $\mathbb{R}^d$ is a stochastic process $L: \Omega \times \mathbb{R}^d \to \mathbb{R}$ such that:

   (i) $L(0) = 0$ a.e.;
   (ii) if $x_1, \ldots, x_n \in \mathbb{R}^d$, the random vector $(L_{x_1}, \ldots, L_{x_n})$ is Gaussian distributed with zero expectation;
   (iii) $E((L_x - L_y)^2) = \|x - y\|$ for all $x, y \in \mathbb{R}^d$;
   (iv) $x \mapsto L(\omega, x)$ is continuous for almost all ω.

Notice that the covariance of the process can be calculated from (i)–(iii);

$$E(L_x L_y) = \tfrac{1}{2}\{\|x\| + \|y\| - \|x - y\|\};$$

and that Lévy Brownian motion thus induces a unique probability measure on $C(\mathbb{R}^d, \mathbb{R})$.

Recall from Example 4.8.2 that one-dimensional Brownian motion can be obtained as a stochastic integral

$$\hat{\chi}(t) - \hat{\chi}(s) = \frac{1}{2} \int (\operatorname{sgn}(t - a) - \operatorname{sgn}(s - a))\, d\chi(a),$$

or, as we may also express it,

$$(18) \qquad \hat{\chi}(t) - \hat{\chi}(s) = \frac{1}{2} \int \left(\frac{t - a}{|t - a|} - \frac{s - a}{|s - a|} \right) d\chi(a).$$

Stoll's idea was that equation (18) may be generalized to dimension d by putting

$$(19) \quad \hat{\chi}(x) - \hat{\chi}(y) = C_d \int \left(\frac{x - a}{\|x - a\|^{(d+1)/2}} - \frac{y - a}{\|y - a\|^{(d+1)/2}} \right) d\chi(a),$$

and he showed that for the appropriate choice of the constant C_d, $\hat{\chi}$ induces a Lévy Brownian motion on $\mathbb{R}^d$. We shall use a slightly different representation (also mentioned by Stoll) when $d > 1$; we shall study a process λ satisfying

$$(20) \quad \lambda(x) - \lambda(y) = k_d \int \left(\frac{1}{\|x - a\|^{(d-1)/2}} - \frac{1}{\|y - a\|^{(d-1)/2}} \right) d\chi(a),$$

and prove that the standard part of λ is a Lévy Brownian motion for the right choice of k_d.

There is no substantial difference between the two approaches (19) and (20); all arguments that apply to one process apply almost automatically to the other. Stoll's original representation (19) has the advantage of working also in the one-dimensional case; on the other hand it needs an $\mathbb{R}^d$-valued white noise.

For the remainder of this section we assume $d > 1$, and use the convention $1/0 = 0$.

Let us assume that λ satisfies (20), and compute

$$(21) \quad E((\lambda(x) - \lambda(y))^2) = k_d^2 \int \left(\frac{1}{\|x - a\|^{(d-1)/2}} - \frac{1}{\|y - a\|^{(d-1)/2}} \right)^2 d\mu(a).$$

If $x, y \in \Gamma$, let $\Gamma_{x,y}$ be the lattice

$$\Gamma_{x,y} = \left\{ \frac{a - x}{\|x - y\|} \,\middle|\, a \in \Gamma \right\},$$

and let $\mu_{x,y}$ be the uniform measure on $\Gamma_{x,y}$ given by

$$\mu_{x,y}(\{b\}) = \left(\frac{\Delta t}{\|x - y\|} \right)^d$$

for all $b \in \Gamma_{x,y}$. Notice that if $\Delta t / \|x - y\| \approx 0$, then $\mu_{x,y}$ induces the Lebesgue measure on $\Gamma_{x,y}$. Defining a new variable k by

$$k = \frac{a - x}{\|x - y\|},$$

(21) turns into

$$(22) \quad E((\lambda(x) - \lambda(y))^2)$$

$$= k_d^2 \int \left(\frac{1}{\|k\|^{(d-1)/2} \|x - y\|^{(d-1)/2}} \right.$$

$$\left. - \frac{1}{\|w + k\|^{(d-1)/2} \|x - y\|^{(d-1)/2}} \right)^2 \|x - y\|^d \, d\mu_{x,y}(k)$$

$$= \|x - y\| k_d^2 \int \left(\frac{1}{\|k\|^{(d-1)/2}} - \frac{1}{\|w + k\|^{(d-1)/2}} \right)^2 d\mu_{x,y}(k),$$

where $w = (x - y)/\|x - y\|$ is a unit vector.

We shall use the following lemma.

4.8.7. LEMMA. For all $x, y \in \Gamma$, the function

$$J(k) = \left(\frac{1}{\|k\|^{(d-1)/2}} - \frac{1}{\|w + k\|^{(d-1)/2}} \right)$$

is in $SL^2(\mu_{x,y})$, and there exists a $K \in \mathbb{R}$ such that $\int J(k)^2 \, d\mu_{x,y}(k) < K$ for all $x, y \in \Gamma$. Moreover, if x and y are finite and not infinitesimally close, then

$$\int \left(\frac{1}{\|k\|^{(d-1)/2}} - \frac{1}{\|w+k\|^{(d-1)/2}} \right)^2 d\mu_{x,y}(k)$$

$$= \int \left(\frac{1}{\|z\|^{(d-1)/2}} - \frac{1}{\|z+e\|^{(d-1)/2}} \right)^2 dm(z),$$

where m is Lebesgue measure on $\mathbb{R}^d$, and e is any unit vector.

The proof is not hard but very tedious, and we leave it to the reader. One word of caution may be appropriate; the integrals

$$\int \frac{1}{\|k\|^{d-1}} d\mu_{x,y}(k) \qquad \text{and} \qquad \int \frac{1}{\|w+k\|^{d-1}} d\mu_{x,y}(k)$$

both diverge at infinity, and a little care is needed in showing that

$$\int \left(\frac{1}{\|k\|^{(d-1)/2}} - \frac{1}{\|w+k\|^{(d-1)/2}} \right)^2 d\mu_{x,y}(k)$$

converges.

One consequence of (22) and 4.8.7 is that if λ satisfies (20), then

$$E((\lambda(x) - \lambda(y))^2) \le k_d^2 K \|x - y\|$$

for all $x, y \in \Gamma$, and thus λ is S-continuous by 4.8.4. We can now define a process λ satisfying (20) and its standard part L; put

$$(23) \qquad \lambda(x) = k_d \int \left(\frac{1}{\|x-a\|^{(d-1)/2}} - \frac{1}{\|a\|^{(d-1)/2}} \right) d\chi(a),$$

where

$$(24) \qquad k_d = \left(\int \left(\frac{1}{\|z\|^{(d-1)/2}} - \frac{1}{\|z+e\|^{(d-1)/2}} \right)^2 dm(z) \right)^{-1/2},$$

and let $L : \Omega \times \mathbb{R}^d \to \mathbb{R}$ be defined by

$$(25) \qquad L(\omega, {}^\circ x) = {}^\circ \lambda(\omega, x)$$

for all $x \in \Gamma$.

4.8.8. THEOREM. L is a Lévy Brownian motion on $\mathbb{R}^d$.

PROOF. There is not much left to prove; 4.8.6(i) is immediate from (23); 4.8.6(ii) follows from the general theory for stochastic integration with respect to white noise; 4.8.6(iii) is a consequence of the choice of k_d [recall (22), (24), and 4.8.7]; and since we have already observed that λ is S-continuous, 4.8.6(iv) is also satisfied.

Notice that if $\tilde{X}$ is the white noise on $\mathbb{R}^d$ constructed from χ,

(26) $$L(x) = \int \left(\frac{1}{\|x - a\|^{(d-1)/2}} - \frac{1}{\|a\|^{(d-1)/2}} \right) d\tilde{X}(a)$$

is a standard representation of L as a stochastic integral of white noise.

REMARK. Stoll's stochastic integral representation of Lévy Brownian motion is not the first one; Chentsov (1957) [see also Gangolli (1967), McKean (1963), and Takenaka (1977)] gave a representation in terms of white noise on projective space, and Cartier (1971) discovered a representation in terms of two-dimensional white noise on $\mathbb{R}^d$, using a kernel that is a little more complicated than the one in (26). Stoll points out that his representation is invariant under translations T in a way Cartier's is not; if $\tilde{X}$ is the white noise in (26) and $\tilde{X}^T$ is the one obtained by $\tilde{X}^T(C) = \tilde{X}(T^{-1}(C))$, then $L_x - L_y = L^T_{T(x)} - L^T_{T(y)}$ almost everywhere, where L^T is the Lévy Brownian motion obtained from $\tilde{X}^T$. A more important result is a Donsker-type invariance principle for Lévy Brownian motion.

D. Invariance Principles

Let us first construct finite approximations to λ. Fix $N \in {}^*\mathbb{N} - \mathbb{N}$, and consider the lattices

$$\Gamma_n = \left\{ \left(\frac{k_1}{2^n}, \ldots, \frac{k_d}{2^n} \right) \, \middle| \, (k_1, \ldots, k_d) \in {}^*\mathbb{Z}^d, \, \max_{1 \le i \le d} k_i \le N \cdot 2^{-n} \right\}$$

for all n less than some hyperfinite integer η. Let Ω be the set of all internal maps from Γ_n into $\{-1, 1\}$, and let P be the uniform probability measure on Ω. The measure μ_n on Γ_n is uniform, and $\mu_n(\{a\}) = (1/2^n)^d$ for each $a \in \Gamma_n$. By analogy with (23), we define $\lambda_n : \Omega \times \Gamma_n \to {}^*\mathbb{R}$ by

(27) $$\lambda_n(\omega, x) = k_d \sum_{a \in \Gamma_n} \left(\frac{1}{\|x - a\|^{(d-1)/2}} - \frac{1}{\|a\|^{(d-1)/2}} \right) \omega(a) \sqrt{\mu_n(a)}.$$

Our goal is to show that the measures induced by the λ_n's on $C(\mathbb{R}^d, \mathbb{R})$ converge weakly to the measure induced by Lévy Brownian motion. Let us first explain how we get a measure on $C(\mathbb{R}^d, \mathbb{R})$ from λ_n. If $x \in {}^*\mathbb{R}^d$, let $[x]_n$ be the half-open cube centered at x with sides of length 2^{-n}, i.e.,

$$[x]_n = \{(y_1, \ldots, y_d) \in {}^*\mathbb{R}^d \, | \, x_i - 2^{-(n+1)} \le y_i < x_i + 2^{-(n+1)} \text{ for all } i\}.$$

If $F : \Gamma_n \to {}^*\mathbb{R}$ is an internal function, we extend it to a function $\bar{F} : {}^*\mathbb{R}^d \to {}^*\mathbb{R}$ by putting

$$\bar{F}(y) = F(x)$$

if $x \in \Gamma_n$ and $y \in [x_n]$. The function $\bar{F}$ is not S-continuous, but we can turn it into an S-continuous function $\tilde{F}$ by putting

(28) $$\tilde{F}(x) = {}^*m([x]_n)^{-1} \cdot \int_{[x]_n} \bar{F}(y) \, d^*m(y),$$

where m is Lebesgue measure on $\mathbb{R}^d$. Notice that $\tilde{F}$ agrees with F on Γ_n.

If $n \in \mathbb{N}$, let $L_n : \Omega \times \mathbb{R}^d \to \mathbb{R}$ be the standard part of $\tilde{\lambda}_n$, and define a measure ν_n on $C(\mathbb{R}^d, \mathbb{R})$ by

$$\nu_n(B) = L(P)\{\omega \,|\, L_n(\omega, \cdot) \in B\}$$

for all Borel sets B. If $\tilde{\nu}_n$ is the internal measure on ${}^*C(\mathbb{R}^d, \mathbb{R})$ given by

$$\tilde{\nu}_n(A) = P\{\omega \,|\, \tilde{\lambda}_n(\omega, \cdot) \in A\},$$

then

(29) $$\nu_n = L(\tilde{\nu}_n) \circ \mathrm{st}^{-1}.$$

On the other hand, if $n \in {}^*\mathbb{N} - \mathbb{N}$, then according to Theorem 4.8.8, $L(\tilde{\nu}_n) \circ \mathrm{st}^{-1}$ is the measure induced on $C(\mathbb{R}^d, \mathbb{R})$ by Lévy Brownian motion. If we call this measure ν, we see that for all $n \in {}^*\mathbb{N} - \mathbb{N}$ and all bounded, continuous functions $f : C(\mathbb{R}^d, \mathbb{R}) \mapsto \mathbb{R}$,

$$\int f \, d\nu = {}^\circ\!\int {}^*f \, d\tilde{\nu}_n.$$

Given an $\varepsilon \in \mathbb{R}_+$, there must be an $n_0 \in \mathbb{N}$, such that for all $n \in \mathbb{N}$, $n \geq n_0$,

$$\left| \int f \, d\nu - \int f \, d\nu_n \right| < \varepsilon,$$

and hence the sequence $\{\nu_n\}$ converges weakly to ν.

Before we formulate this result as a theorem, let us redefine L_n and ν_n in entirely standard terms. Let $\hat{\Gamma}_n$ be the standard lattice consisting of the finite points in Γ_n, and let $l_n : \Omega \times \hat{\Gamma}_n \to \mathbb{R}$ be defined by

$$l_n(\omega, x) = \sum_{a \in \hat{\Gamma}_n} \left(\frac{1}{\|x - a\|^{(d-1)/2}} - \frac{1}{\|a\|^{(d-1)/2}} \right) \omega(a) \cdot 2^{-nd/2}.$$

Obviously, $l_n(\omega, x) = {}^\circ\lambda_n(\omega, x)$ for almost all ω, and hence L_n equals the extension $\tilde{l}_n$ of l_n (notice that $\tilde{l}_n$ can be defined from l_n in standard terms; the cube $[x]_n$ is now a standard cube with sides of length 2^{-n}). We define ν_n from L_n as before.

4.8.9. THEOREM. For each $n \in \mathbb{N}$ let $\hat{\Gamma}_n$ be the lattice

$$\hat{\Gamma}_n = \{(k_1/2^n, \dots, k_d/2^n) \,|\, (k_1, \dots, k_d) \in Z^d\},$$

and define for $x \in \hat{\Gamma}_n$

$$l_n(\omega, x) = \sum_{a \in \hat{\Gamma}_n} \left(\frac{1}{\|x - a\|^{(d-1)/2}} - \frac{1}{\|a\|^{(d-1)/2}} \right) \omega(a) \cdot 2^{-nd/2},$$

where the $\omega(a)$'s are independent random variables taking the values -1 and 1 with probability $\frac{1}{2}$. Let $L_n : \Omega \times \mathbb{R}^d \to \mathbb{R}$ be the continuous extension of l_n defined above, and let ν_n be the measure on $C(\mathbb{R}^d, \mathbb{R})$ induced by L_n. Then $\{\nu_n\}_{n \in \mathbb{N}}$ converges weakly to Lévy Brownian motion.

This is our version of Stoll's invariance principle (Stoll, 1982, 1985); the original result is more complicated to formulate as it uses general interpolation functions to turn l_n into a continuous process. The conditions we have put on the random variables $\omega(a)$ are obviously much stronger than necessary; all we need is that they induce a white noise in the limit.

REMARK. Nonstandard techniques for proving invariance principles were first introduced by Müller (1969) before the invention of Loeb measure. Anderson (1976) combined Müller's ideas with the Loeb construction and his own hyperfinite random walk to give an almost trivial proof of Donsker's invariance principle for Brownian motion. Later Keisler (1984) and Kosciuk (1982) proved related results for solutions of stochastic differential equations.

To emphasize that Lévy Brownian motion is not the only process that can be obtained as a stochastic integral of white noise, we close this section with a construction of the Yeh-Wiener process.

4.8.10. DEFINITION. A *Yeh-Wiener process* on $\mathbb{R}^d$ is a stochastic process $W : \Omega \times \mathbb{R}^d \to \mathbb{R}$ such that

(i) for all $z_1, \dots, z_n \in \mathbb{R}^d$, the random vector $(W_{z_1}, \dots, W_{z_n})$ is Gaussian distributed with mean zero;

(ii) if $x = (x_1, \dots, x_d)$ and $y = (y_1, \dots, y_d)$ are two points in $\mathbb{R}^d$, then $E(W_x W_y) = \prod_{i=1}^d |x_i| \wedge |y_i|$;

(iii) for almost all ω, the path $x \mapsto W(\omega, x)$ is continuous.

Let Γ be a hyperfinite lattice in $*\mathbb{R}^d$ and define $F : \Gamma \times \Gamma \mapsto *\mathbb{R}$ by

$$F(x, a) = \begin{cases} 2^{-\frac{d}{2}} & \text{if } |a_i| < |x_i| \text{ for all } i, \\ 0 & \text{elsewhere}, \end{cases}$$

where $x = (x_1, \ldots, x_d)$ and $a = (a_1, \ldots, a_d)$. If χ is S-white noise on Γ, let Y be the process

$$Y(x) = \int F(x, a) \, d\chi(a).$$

For $x = (x_1, \ldots, x_d)$ and $y = (y_1, \ldots, y_d)$ in Γ, we get

$$E(Y(x)Y(y)) = E\left(\sum_{a,b \in \Gamma} F(x, a)F(y, a)\omega(a)\omega(b)\sqrt{\mu(a)\mu(b)} \right)$$

$$= \sum_{a \in \Gamma} F(x, a)F(y, a)\mu(a) \approx \prod_{i=1}^{d} |x_i| \wedge |y_i|,$$

which shows that Y has the right covariance. Also,

$$\int (F(x, a) - F(y, a))^2 \, d\mu(a) \le d \max_{1 \le i \le d} \left| |y_i| - |x_i| \right| (\|x\| + \|y\|)^{d-1},$$

which shows that for each $K \in \mathbb{N}$, there is a constant C_K such that for all $x, y \in \Gamma$ with $\|x\|, \|y\| \le K$,

$$\int (F(x, a) - F(y, a))^2 \, d\mu(a) \le C_K \|x - y\|.$$

By changing F outside $\Gamma_K = \{x \in \Gamma \,|\, \|x\| \le K\}$ if necessary, we can use Theorem 4.8.4 to conclude that Y is continuous on each Γ_K, and hence it is S-continuous.

The standard part W of Y is obviously a Yeh–Wiener process. Reasoning as in the proof of 4.8.9, one gets a simple proof of Kuelbs' (1968, 1973) and Wichura's (1969) invariance principle for this process [see Stoll (1982) for the details].

REMARK. All the main results in this section are from Stoll (1982), but we have changed the exposition somewhat. Stoll first proves the representation (26) by standard methods, and then uses this result to prove Theorem 4.8.8. By introducing the Continuity Theorem 4.8.4, we have obtained a way of proving that λ is S-continuous, and this makes it possible to reverse Stoll's construction. Knowing that λ is continuous also helps to simplify the proof of 4.8.9. Stoll's treatment of the Yeh–Wiener process is much like the one we have given, although he uses a different argument to show that Y is S-continuous. In the shorter and more recent exposition of his work Stoll (1985) is using an approach closer to ours.

NOTES

Even in a chapter of this length it is not possible to give a complete survey of the nonstandard theory of stochastic processes, and the reader should consult Stroyan and Bayod (1985) and Cutland (1983c) for further information and references. The most serious of our omissions are probably Perkins's theory of local time (recall the remarks and references at the end of Section 3.3); Keisler's study of Markov processes (1984); and the now well-developed field of probability logic (see the references in Section 4.4.D). As a matter of fact, we shall have much to say about both Markov processes (Chapter 5) and local time (Sections 6.4 and 7.5), but from quite different perspectives than Keisler and Perkins. Probability logic is further from our chosen topic, and we just refer the reader to Keisler (1985) (probably the best starting point for a logician) and Hoover and Keisler (1984) (the most natural point of departure for the probabilist). Let us finally mention that Lawler's important and well-known work on self-avoiding random walks is reviewed briefly in Section 6.4. A completely standard, but very stimulating discussion of the various branches of stochastic analysis and their interconnections is given in Föllmer (1984).

REFERENCES

S. Albeverio and R. Høegh-Krohn (1981). Stochastic methods in quantum field theory and hydrodynamics. *Phys. Rep.* **77**.

S. Albeverio, M. de Faria, and R. Høegh-Krohn (1979). Stationary measures for the Euler flow in two dimensions. *J. Statist. Phys.* **20**.

S. Albeverio, R. Høegh-Krohn, and D. Merlini (1985). Euler flows, associated generalized random fields and Coulomb systems. *In* (S. Albeverio, ed.) *Infinite Dimensional Analysis and Stochastic Processes*, pp. 216–244, Res. Notes Math. Pitman, London.

R. M. Anderson (1976). A nonstandard representation for Brownian motion and Itô integration. *Israel J. Math.* **25**.

R. M. Anderson (1982). Star-finite representations of measure spaces. *Trans. Amer. Math. Soc.* **271**.

L. Arkeryd (1984). Loeb-Sobolev spaces with applications to variational integrals and differential equations (preprint). Chalmers Inst. Technol., Gothenborg, Sweden.

L. Arkeryd and J. Bergh (1985). Some properties of Loeb-Sobolev spaces (preprint). Chalmers Inst. Technol., Gothenborg, Sweden.

M. T. Barlow (1981). Construction of a martingale with given absolute value. *Ann. Probab.* **9**.

M. T. Barlow (1982). One dimensional stochastic differential equation with no strong solution. *J. London Math. Soc.* **26**.

M. T. Barlow and E. Perkins (1984). One-dimensional stochastic differential equation involving a singular increasing process. *Stochastics* **12**.

A. Bensoussan and R. Temam (1972). Equations aux dérivées partielles stochastiques non linéaires I. *Israel J. Math.* **11**.

A. Bensoussan and R. Temam (1973). Equations stochastiques du type Navier-Stokes. *J. Funct. Anal.* **13**.

O. Berg and C. Blomberg (1976). Association kinetics with coupled diffusional flows. Special application to the Lac Repressor-Operator system. *Biophys. Chem.* **4**.

O. Berg and C. Blomberg (1977). Association kinetics with coupled diffusion. An extension to coiled-chain macro-molecules applied to Lac Repressor-Operator system. *Biophys. Chem.* **7**.

P. Billingsley (1968). *Convergence of Probability Measures*, Wiley, New York.

P. Cartier (1971). Introduction à l'étude de mouvement browniens à plusieurs paramètres. *Sem. Probab. V. Lecture Notes in Math.* **191**. Springer-Verlag, Berlin and New York.

N. N. Chentsov (1957). Lévy Brownian motion for several parameters and generalized white noise. *Theory Probab. Appl.* **2**.

A. Chojnowskaja-Michalik (1979). Stochastic differential equations in Hilbert Spaces. *In* (Z. Ciesielski, ed.) *Probability Theory*, Banach Center Publications 5, PWN, Warzaw.

P.-L. Chow (1978). Stochastic partial differential equations in turbulence related problems. *In* (A. T. Bharucha-Reid, ed.) *Probabilistic Analysis and Related Topics*, Vol. 1. Academic Press, New York.

N. Christopheit (1983). Discrete approximations of continuous time stochastic control systems. *SIAM J. Control Optim.* **21**.

K.-L. Chung and R. Williams (1983). *Introduction to Stochastic Integration*. Birkhäuser, Basel.

N. J. Cutland (1982). On the existence of solutions to stochastic differential equations on Loeb spaces. *Z. Wahrsch. Verw. Gebiete* **60**.

N. J. Cutland. (1983a). Internal controls and relaxed controls. *J. London Math. Soc.* **27**.

N. J. Cutland (1983b). Optimal controls for partially observed stochastic systems: an infinitesimal approach. *Stochastics* **8**.

N. J. Cutland (1983c). Nonstandard measure theory and its applications. *Bull. London Math. Soc.* **15**.

N. J. Cutland (1985a). Partially observed stochastic controls based on cumulative digital readouts of the observations. In *Proc. IFIP Work. Conf. Stochastic Differential Systems*, 4th, Marseille-Luminy, 1984. Springer-Verlag, Berlin and New York.

N. J. Cutland (1985b). Infinitesimal methods in control theory: Deterministic and stochastic. *Acta App. Math.* (to appear).

N. J. Cutland (1985c). Simplified existence for solutions to stochastic differential equations. *Stochastics* **14**.

M. H. A. Davis (1984). *Lectures on Stochastic Control and Nonlinear Filtering*. Springer-Verlag, Berlin and New York.

C. Doléans (1968). Existence du processus croissant naturel associé à un potentiel de classe (D). *Z. Wahrsch. Verw. Gebiete* **9**.

C. Doléans-Dade (1976). On the existence and unicity of solutions of stochastic integral equations. *Z. Wahrsch. Verw. Gebiete* **36**.

J. L. Doob (1953). *Stochastic Processes*. Wiley, New York.

J. L. Doob (1984). *Classical Potential Theory and its Probabilistic Counterpart.* Springer-Verlag, New York and Berlin.

R. J. Elliott (1982). *Stochastic Calculus and Applications.* Springer-Verlag, New York and Berlin.

R. J. Elliott and M. Kohlmann (1982). On the existence of optimal partially observed controls. *Appl. Math. Optim.* **9**.

S. Fajardo (1984). Completeness theorems for the general theory of stochastic processes. *Methods in Math. Logic, Lecture Notes in Math.*, **1030**, Springer-Verlag, Berlin and New York.

S. Fajardo (1985). Probability logic with conditional expectation. *Ann. Pure Appl. Logic* **28**.

W. G. Faris and G. Jona-Lasinio (1982). Large fluctuations for a non-linear heat equation with noise. *J. Phys. A* **15**.

A. F. Filippov (1962). On certain questions in the theory of optimal control. *J. Soc. Indust. Appl. Math.*, Ser. A, Control. 1.

H. Föllmer (1984). Von der Brownschen Bewegung zum Brownschen Blatt: einige neuere Richtungen in der Theorie der Stochastischen Prozesse. In (W. Jäger *et al.*, eds.) *Perspectives in Mathematics.* Birkhäuser, Basel.

R. Gangolli (1967). Positive definite kernels on homogeneous spaces and certain stochastic processes related to Lévy's Brownian motion of several parameters. *Ann. Inst. H. Poincaré Sect. B* **3**.

L. L. Helms and P. A. Loeb (1982). A nonstandard proof of the martingale convergence theorem. *Rocky Mountain J. Math.* **12**.

C. W. Henson (1986). Banach space model theory (in preparation).

D. N. Hoover (1978). Probability logic. *Ann. Math. Logic.* **14**.

D. N. Hoover (1982). A normal form theorem for $L_{\omega_1 P}$ with applications. *J. Symbolic Logic* **47**.

D. N. Hoover (1984) Synonymity, generalized martingales and subfiltrations. *Ann. Probab.* **12**.

D. N. Hoover (1985). A probabilistic interpolation theorem (to appear).

D. N. Hoover and H. J. Keisler (1984). Adapted probability distributions. *Trans. Amer. Math. Soc.* **286**.

D. N. Hoover and E. Perkins (1983a). Nonstandard construction of the stochastic integral and applications to stochastic differential equations I. *Trans. Amer. Math. Soc.* **275**.

D. N. Hoover and E. Perkins (1983b). Nonstandard construction of the stochastic integral and applications to stochastic differential equations II. *Trans. Amer. Math. Soc.* **275**.

N. Ikeda and S. Watanabe (1981). *Stochastic Differential Equations and Diffusion Processes.* North-Holland Publ., Amsterdam.

K. Itô (1944). Stochastic integral. *Proc. Imp. Acad. (Tokyo)* **20**.

J. Jacod and J. Memin (1981). Existence of weak solutions for stochastic differential equations with driving semimartingales. *Stochastics* **4**.

G. Jona-Lasinio and P. K. Mitter (1985). On the stochastic quantization of field theory. *Commun. Math. Phys.* **101**.

H. J. Keisler (1977). Hyperfinite model theory. *In* (R. O. Gandy and J. M. E. Hyland, eds.) *Logic Colloq. 1976.* North-Holland Publ., Amsterdam.

H. J. Keisler (1983). A non-tatónnement process with infinitesimal traders (preliminary version). Univ. of Wisconsin, Madison.

H. J. Keisler (1984). An infinitesimal approach to stochastic analysis. *Mem. Amer. Math. Soc.* **297**.

H. J. Keisler (1985). Probability quantifiers. *In* (J. Barwise and S. Feferman eds.) *Model Theoretical Logics.* Springer-Verlag, Berlin and New York.

S. A. Kosciuk (1982). Nonstandard stochastic methods in diffusion theory. Ph.D. thesis. Univ. of Wisconsin, Madison.

S. A. Kosciuk (1983). Stochastic solutions to partial differential equations. *In* (A. E. Hurd, ed.) *Nonstandard Analysis—Recent Contributions* pp. 113–119. Springer-Verlag, Berlin and New York.

N. V. Krylov (1972). On Itô's stochastic integral equation. *Theory Probab. Appl.* **14** (1969). [See also the correction in *Theory Probab. Appl.* **17**.]

N. V. Krylov (1974). Some estimates of the probability density of a stochastic integral. *Math. USSR-Izv.* **8**.

N. V. Krylov (1980). *Controlled Diffusion Processes.* Springer-Verlag, New York and Berlin.

J. Kuelbs (1968). The invariance principle for a lattice of random variables. *Ann. Math. Stat.* **39**.

J. Kuelbs (1973). The invariance principle for Banach space valued random variables. *J. Multivariate Anal.* **3**.

H. Kunita and S. Watanabe (1967). On square integrable martingales. *Nagoya Math. J.* **30**.

H.-H. Kuo (1975). *Gaussian Measures in Banach Spaces.* Springer-Verlag, Berlin and New York.

P. Lévy (1948). *Processus stochastiques et mouvement Brownien.* Gauthier-Villars, Paris.

T. Lindstrøm (1980a). Hyperfinite stochastic integration I: The nonstandard theory. *Math. Scand.* **46**.

T. Lindstrøm (1980b). Hyperfinite stochastic integration II: Comparison to the standard theory. *Math. Scand.* **46**.

T. Lindstrøm (1980c). Hyperfinite stochastic integration III: Hyperfinite representations of standard martingales. *Math. Scand.* **46**.

T. Lindstrøm (1983). Stochastic integration in hyperfinite dimensional linear spaces. *In* (A. E. Hurd, ed.) *Nonstandard Analysis—Recent Developments.* Springer-Verlag, New York and Berlin.

T. Lindstrøm (1985). The structure of hyperfinite stochastic integrals. *Z. Wahrsch. Verw. Gebiete* (to appear).

T. Lindstrøm (1986). A standard characterization of nonstandard stochastic integrals (in preparation).

V. Luges (1984), Girsanov's theorem in Hilbert space and an application to the statistics of Hilbert space-valued stochastic differential equations, *Stoch. Proc. and Appl.* **17**.

H. P. McKean (1963). Brownian motion with a several dimensional time. *Theory Probab. Appl.* **8**.

E. J. McShane (1967). Relaxed controls and variational problems. *SIAM J. Control Optim.* **5**.

M. Metivier and J. Pellaumail (1980). *Stochastic Integration.* Academic Press, New York.

P. A. Meyer (1976). Un cours sur les intégrales stochastiques. *In Sém. Probab. X,* Lecture Notes in Math. **511**, Springer-Verlag, Berlin and New York.

D. W. Müller (1969). Nonstandard proofs of invariance principles in probability theory. *In* (W. A. J. Luxemburg, ed.) *Applications of Model Theory to Algebra, Analysis, and Probability.* Holt, Rinehart, and Winston, New York.

J. Neveu (1975). *Discrete Parameter Martingales.* North-Holland Publ., Amsterdam.

H. Osswald (1984). On the existence of solutions to stochastic integral equations with respect to square integrable continuous martingales on Loeb spaces (preprint). Univ. of München.

H. Osswald (1985). Introduction to nonstandard measure theory I-II. Lecture notes. 1984-1985. Univ. of München.

R. L. Panetta (1978). Hyperreal probability spaces: some applications of the Loeb construction. Ph.D. thesis. Univ. of Wisconsin, Madison.

E. Perkins (1982). On the construction and distribution of a local martingale with a given absolute value. *Trans. Amer. Math. Soc.* **271**.

E. Perkins (1983). Stochastic processes and nonstandard analysis. *In* (A. E. Hurd, ed.) *Nonstandard Analysis—Recent Developments.* Springer-Verlag, Berlin and New York.

P. E. Protter (1977a). On the existence, uniqueness, convergence, and explosions of solutions of systems of stochastic integral equations. *Ann. Probab.* **5**.

P. E. Protter (1977b). Right continuous solutions of systems of stochastic integral equations. *J. Multivariate Anal.* **7**.

P. H. Richter and M. Eigen (1974). Diffusion control reaction rates in spheroidal geometry. Application to repressor-operator association and membrane bound enzymes. *Biophys. Chem.* **2**.

H. Rodenhausen (1982a). The completeness theorem for adapted probability logic. Doctoral dissertation. Univ. of Heidelberg.

H. Rodenhausen (1982b). A characterization of nonstandard liftings of measurable functions and stochastic processes. *Israel J. Math.* **43**.

D. Ross (1984). Completeness theorem for probability logic with function symbols (preprint). Univ. of Iowa.

B. Simon (1979). *Functional Integration and Quantum Physics.* Academic Press, New York.

A. Stoll (1982). A Nonstandard Construction of Lévy Brownian Motion with Applications to Invariance Principles. Diplomarbeit, Freiburg.

A. Stoll (1985). A nonstandard construction of Lévy Brownian motion. *Z. Wahrsch. Verw. Gebiete* (to appear).

D. W. Stroock and S. R. S. Varadhan (1979). *Multidimensional Diffusion Processes.* Springer-Verlag, Berlin and New York.

K. D. Stroyan (1985). Previsible sets for hyperfinite filtrations. *Z. Wahrsch. Verw. Gebiete* (to appear).

K. D. Stroyan and J. M. Bayod (1985). *Foundations of Infinitesimal Stochastic Analysis.* North-Holland Publ. (to appear).

S. Takenaka (1977). On projective invariance of multi-parameter Brownian motion. *Nagoya Math. J.* **67**.

M. J. Vishik, A. J. Komech, and A. V. Fursikov (1979). Some mathematical problems of statistical hydrodynamics. *Russian Math. Surveys* **34**.

J. Warga (1967). Functions of relaxed controls. *SIAM J. Control Optim.* **5**.

J. Warga (1972). *Optimal Control of Differential and Functional Equations.* Academic Press, New York and London.

M. J. Wichura (1969). Inequalities with applications to the weak convergence of random processes with multidimensional time parameters. *Ann. Math. Statist.* **40**.

CHAPTER 5

HYPERFINITE DIRICHLET
FORMS AND
MARKOV PROCESSES

The interplay between methods from functional analysis and stochastic processes is one of the most important and exciting aspects of mathematical physics today. It is a highly technical and sophisticated theory based on decades of research in both areas. Our purpose in this chapter is to develop a more elementary approach where hyperfinite linear algebra and Markov chains replace functional analysis and continuous time Markov processes. In the first two sections we describe a general theory for non-negative quadratic forms on hyperfinite-dimensional spaces, but from Section 5.3 on we restrict ourselves to Markov forms and begin the analysis of the associated Markov processes. While Section 5.3 is devoted to the hyperfinite processes themselves and contains, e.g., the Beurling–Deny formula, Fukushima's decomposition theorem, and the Feynman–Kac formula, Sections 5.4 and 5.5 contain a study of their standard parts. The theory in Section 5.5 is rather complicated and has probably not yet reached its final form; therefore, the reader may find it more rewarding first to take a look at the applications in Section 5.6.

Unless otherwise specified, all linear spaces in this chapter are over the reals or hyperreals.

5.1. HYPERFINITE QUADRATIC FORMS AND THEIR DOMAINS

We shall develop a hyperfinite theory of non-negative, symmetric, quadratic forms on infinite-dimensional spaces. It is well known that in the Hilbert space case the theory for closed forms of this kind is equivalent to the theory of non-negative, self-adjoint operators; in fact, there is a natural correspondence between forms E and operators A given by $E(u, v) = \langle A^{1/2}u, A^{1/2}v \rangle$. We have chosen to present the theory in terms of forms and not operators for two reasons: partly because forms are real-valued, and this makes it simpler to take standard parts, but also because in most of our applications, the form is what is naturally given.

A. The Domain

Let us begin by recalling some results from Section 2.2. If H is an internal, hyperfinite-dimensional linear space with an inner product $\langle \cdot, \cdot \rangle$ generating a norm $\| \cdot \|$, we let Fin(H) be the set of all elements in H with finite norm. By defining $x \approx y$ if $\|x - y\| \approx 0$, Proposition 2.2.1 tells us that the space

$$(1) \qquad {}^{\circ}H = \text{Fin}(H)/\approx$$

is a Hilbert space with respect to the inner product $({}^{\circ}x, {}^{\circ}y) = \text{st}\langle x, y \rangle$, where ${}^{\circ}x$ denotes the equivalence class of x. We call $({}^{\circ}H, (\cdot, \cdot))$ the *hull* of $(H, \langle \cdot, \cdot \rangle)$.

Given a non-negative, symmetric, bilinear form

$$\mathscr{E} : H \times H \to {}^*\mathbb{R},$$

we want to define its standard part E as a bilinear form on ${}^{\circ}H$. If $\mathscr{E}$ is *S-bounded*, i.e., there exists a $K \in \mathbb{R}$ such that

$$|\mathscr{E}(u, v)| \le K \|u\| \|v\|$$

for all $u, v \in H$, then we can simply define E by

$$E({}^{\circ}u, {}^{\circ}v) = {}^{\circ}\mathscr{E}(u, v).$$

If $\mathscr{E}$ is not S-bounded, we run into two difficulties; we no longer have that $\mathscr{E}(u, v) \approx \mathscr{E}(\tilde{u}, \tilde{v})$ whenever $u \approx \tilde{u}$ and $v \approx \tilde{v}$, and there may be elements $v \in \text{Fin}(H)$ such that $\mathscr{E}(\tilde{v}, \tilde{v})$ is infinite for all $\tilde{v} \approx v$. The last problem should not surprise us; it is an immediate consequence of the fact that unbounded forms on Hilbert spaces cannot be defined everywhere; we shall solve it by simply letting $E({}^{\circ}v, {}^{\circ}v)$ be undefined when $\mathscr{E}(\tilde{v}, \tilde{v})$ is infinite for all $\tilde{v} \in {}^{\circ}v$. The most natural solution to the first problem may be to define

$$(2) \qquad E({}^{\circ}u, {}^{\circ}u) = \inf\{{}^{\circ}\mathscr{E}(v, v) \mid v \in {}^{\circ}u\},$$

and then extend E to a bilinear form by the usual trick

$$E(°u, °v) = \tfrac{1}{2}\{E(°u + °v, °u + °v) - E(°u, °u) - E(°v, °v)\}.$$

The disadvantage of this approach is that it gives us very little understanding of how the infimum in (2) is obtained; for an easier access to the regularity properties of $\mathscr{E}$ and E, we prefer a more indirect way of attack. Our plan is to define a subset $\mathscr{D}[\mathscr{E}]$ of $\mathrm{Fin}(H)$—called the *domain* of $\mathscr{E}$—satisfying

(3)   if $°\mathscr{E}(u, u) < \infty$, there is a $v \in \mathscr{D}[\mathscr{E}]$ such that $v \approx u$;

(4)   if $u, v \in \mathscr{D}[\mathscr{E}]$ and $u \approx v$, then $°\mathscr{E}(u, u) = °\mathscr{E}(v, v) < \infty$.

We then define E by

(5) $$E(°u, °u) = °\mathscr{E}(v, v),$$

when $v \in \mathscr{D}[\mathscr{E}] \cap °u$. It turns out that the two definitions (2) and (5) agree (see 5.1.14).

Before giving the precise definition of $\mathscr{D}[\mathscr{E}]$, we shall introduce a few useful notions.

Since we shall often be concerned with the relationship between the form $\mathscr{E}$ and the inner product $\langle \cdot, \cdot \rangle$, it will be convenient to work with forms incorporating both; for $\alpha \in {}^*\mathbb{R}$, $\alpha \geq 0$, we define

(6) $$\mathscr{E}_\alpha(u, v) = \mathscr{E}(u, v) + \alpha\langle u, v \rangle.$$

Each of these forms generates a norm (possibly a seminorm in the case $\alpha = 0$):

(7) $$|u|_\alpha = (\mathscr{E}_\alpha(u, u))^{1/2}.$$

Remember that the original Hilbert space norm on H is denoted by $\|\cdot\|$.

Since $\mathscr{E}$ is a symmetric, non-negative form on the hyperfinite-dimensional space H, elementary linear algebra tells us that there is a unique, symmetric, non-negative definite operator $A : H \to H$ such that

(8) $$\mathscr{E}(u, v) = \langle Au, v \rangle$$

for all $u, v \in H$.

If $\|A\|$ is the operator norm of A, we fix an infinitesimal Δt such that

(9) $$0 < \Delta t \leq \frac{1}{\|A\|},$$

and define a new operator $Q^{\Delta t}$ by

(10) $$Q^{\Delta t} = I - \Delta t A.$$

Notice that by (9), the operator $Q^{\Delta t}$ is non-negative, and since A is non-negative, the operator norm of $Q^{\Delta t}$ is less than or equal to one.

Introduce a nonstandard time line T by

(11) $$T = \{k \, \Delta t \, | \, k \in {}^*\mathbb{N}\},$$

and for each element $t = k \, \Delta t$ in T, define Q^t to be the operator

(12) $$Q^t = (Q^{\Delta t})^k.$$

The family $\{Q^t\}_{t \in T}$ is obviously a semigroup, and we shall call it the *semigroup associated with $\mathscr{E}$ and Δt.* Whenever we refer to $\mathscr{E}$, A, T, and Q^t in the remainder of this section we shall assume that they are linked by (8)–(12). In applications, the primary object will often be the semigroup $\{Q^t\}$, and we can then define A (and hence $\mathscr{E}$) by

(13) $$A = (1/\Delta t)(I - Q^{\Delta t}).$$

The operator A is called the *infinitesimal generator* of $\{Q^t\}$.

Since A and Q^t are non-negative operators, they have unique non-negative square roots, which we denote by $A^{1/2}$ and $Q^{t/2}$, respectively.

For each $t \in T$, we may define an approximation $A^{(t)}$ of A by

(14) $$A^{(t)} = \frac{1}{t}(I - Q^t),$$

and from $A^{(t)}$ we get the form

(15) $$\mathscr{E}^{(t)}(u, v) = \langle A^{(t)}u, v \rangle.$$

Be careful not to confuse $\mathscr{E}^{(t)}$ with the form $\mathscr{E}_\alpha$ defined in (6).

Notice that even when $\mathscr{E}$ is not S-bounded, $\mathscr{E}^{(t)}$ is S-bounded for all noninfinitesimal t. One of the motivations behind our definition of the domain $\mathscr{D}[\mathscr{E}]$ is that we want to single out the elements where $\mathscr{E}$ really is approximated by the bounded forms $\mathscr{E}^{(t)}$, $t \neq 0$, i.e., those $u \in H$ such that

(16) $$°\mathscr{E}(u, u) = \lim_{\substack{t \downarrow 0 \\ t \neq 0}} °\mathscr{E}^{(t)}(u, u).$$

We could have taken this to be our definition of $\mathscr{D}[\mathscr{E}]$, but for technical and expository reasons we have chosen another one which we shall soon (see Proposition 5.1.5) show to be equivalent to (16).

5.1.1. DEFINITION. Let $\mathscr{E}$ be a non-negative, symmetric quadratic form on a hyperfinite-dimensional linear space H. The domain $\mathscr{D}[\mathscr{E}]$ of $\mathscr{E}$ is the set of all $u \in H$ satisfying:

(i) $°\mathscr{E}_1(u, u) < \infty$.
(ii) For all $t \approx 0$, $\mathscr{E}(Q^t u, Q^t u) \approx \mathscr{E}(u, u)$.

Let us try to convey the intuition behind this definition. Think of A as a differential operator; then the elements of $\mathscr{D}[\mathscr{E}]$ are "smooth" functions, and Q^t is a "smoothing" operator often given by an integral kernel. If an element u is already smooth, then an infinitesimal amount of smoothing Q^t, $t \approx 0$, should not change it noticeably, and hence $\mathscr{E}(Q^t u, Q^t u) \approx \mathscr{E}(u, u)$. We shall give a partial justification of this rather crude image later, when we show that if $°\mathscr{E}_1(u, u) < \infty$, then the "smoothed" elements $Q^t u$, $t \neq 0$, are all in $\mathscr{D}[\mathscr{E}]$ (Lemma 5.1.7; see also Corollary 5.1.9).

Our first task will be to establish a list of alternative definitions of $\mathscr{D}[\mathscr{E}]$, among them (16). We begin with the following simple identity giving the relationship between $\mathscr{E}$ and $\mathscr{E}^{(t)}$:

5.1.2. LEMMA. For all $u \in H$ and $t \in T$

$$\mathscr{E}^{(t)}(u, u) = \frac{\Delta t}{t} \sum_{0 \leq s < t} \mathscr{E}(Q^s u, u) = \frac{\Delta t}{t} \sum_{0 \leq s < t} \mathscr{E}(Q^{s/2} u, Q^{s/2} u).$$

PROOF. This is just an easy calculation:

$$\mathscr{E}^{(t)}(u, u) = \frac{1}{t} \langle (I - Q^t)u, u \rangle = \frac{1}{t} \sum_{0 \leq s < t} \langle (Q^s - Q^{s + \Delta t})u, u \rangle$$

$$= \frac{\Delta t}{t} \sum_{0 \leq s < t} \mathscr{E}(Q^s u, u) = \frac{\Delta t}{t} \sum_{0 \leq s < t} \mathscr{E}(Q^{s/2} u, Q^{s/2} u).$$

Among other things, Lemma 5.1.2 tells us that $\mathscr{E}^{(t)}$ is non-negative.

5.1.3. LEMMA. Let B, $C : H \to H$ be non-negative, symmetric operators commuting with A and each other. Then the functions

$$t \mapsto \langle Q^t Bu, Cu \rangle \quad \text{and} \quad t \mapsto \mathscr{E}^{(s)}(Q^t Bu, Cu)$$

are non-negative and decreasing for all $u \in H$ and set.

PROOF. We first notice that the $\mathscr{E}^{(s)}$ part follows from the other one since

$$\mathscr{E}^{(s)}(Q^t Bu, Cu) = (1/s)\langle Q^t(I - Q^s)Bu, Cu \rangle,$$

and the operator $B' = (I - Q^s)B$ is non-negative and commutes with A and C.

If $t > r$, then

$$\langle Q^r Bu, Cu \rangle - \langle Q^t Bu, Cu \rangle = \langle (I - Q^{t-r})Q^r Bu, Cu \rangle$$

$$= (t - r)\mathscr{E}^{(t-r)}(Q^{r/2}B^{1/2}C^{1/2}u, Q^{r/2}B^{1/2}C^{1/2}u)$$

$$\geq 0$$

since $\mathscr{E}^{(t-r)}$ is non-negative, and hence $t \mapsto \langle Q^t Bu, Cu \rangle$ decreases. For the positivity, observe that

$$\langle Q^t Bu, Cu \rangle = \langle Q^{t/2} B^{1/2} C^{1/2} u, Q^{t/2} B^{1/2} C^{1/2} u \rangle \geq 0.$$

From 5.1.3 we may now obtain our main inequalities.

5.1.4. PROPOSITION. For all $u \in H$, $t \in T$:

(i) $0 \leq \mathscr{E}(u, u - Q^t u) \leq \mathscr{E}(u, u) - \mathscr{E}(Q^t u, Q^t u) \leq 2\mathscr{E}(u, u - Q^t u)$.

(ii) $0 \leq \mathscr{E}(Q^{\Delta t} u, Q^{\Delta t} u) - \mathscr{E}(Q^{2\Delta t} u, Q^{2\Delta t} u) \leq \mathscr{E}(u, u) - \mathscr{E}(Q^{\Delta t} u, Q^{\Delta t} u)$.

PROOF. By trivial algebra

$$\mathscr{E}(u, u) - \mathscr{E}(Q^t u, Q^t u) = \mathscr{E}(u, u - Q^t u) + \mathscr{E}(Q^t u, u - Q^t u).$$

Applying Lemma 5.1.3 with $B = I$, $C = I - Q^t$, we see that

$$0 \leq \mathscr{E}(Q^t u, u - Q^t u) \leq \mathscr{E}(u, u - Q^t u),$$

and part (i) follows.

(ii) The non-negativity is immediate from (i), and as above we have

$$\mathscr{E}(u, u) - \mathscr{E}(Q^{\Delta t} u, Q^{\Delta t} u) = \mathscr{E}(u, u - Q^{\Delta t} u) + \mathscr{E}(Q^{\Delta t} u, u - Q^{\Delta t} u).$$

Applying 5.1.3 to each of the last two terms, using $B = I$, $C = I - Q^{\Delta t}$ in the first case, and $B = Q^{\Delta t}$, $C = I - Q^{\Delta t}$ in the second, we get

$$\begin{aligned}
\mathscr{E}(u, u) - \mathscr{E}(Q^{\Delta t} u, Q^{\Delta t} u) &\geq \mathscr{E}(Q^{2\Delta t} u, u - Q^{\Delta t} u) + \mathscr{E}(Q^{3\Delta t} u, u - Q^{\Delta t} u) \\
&= \mathscr{E}(Q^{2\Delta t} u, u) - \mathscr{E}(Q^{2\Delta t} u, Q^{\Delta t} u) \\
&\quad + \mathscr{E}(Q^{3\Delta t} u, u) - \mathscr{E}(Q^{3\Delta t} u, Q^{\Delta t} u) \\
&= \mathscr{E}(Q^{\Delta t} u, Q^{\Delta t} u) - \mathscr{E}(Q^{2\Delta t} u, Q^{2\Delta t} u).
\end{aligned}$$

The proposition is proved.

The inequalities above are what we need to establish a reasonable theory for $\mathscr{D}[\mathscr{E}]$. We first give our promised list of alternative definitions of the domain of $\mathscr{E}$:

5.1.5. PROPOSITION. The following are equivalent:

(i) u is in the domain of $\mathscr{E}$.

(ii) $°\mathscr{E}_1(u, u) < \infty$, and for all $t \approx 0$, we have $\mathscr{E}(u, u - Q^t u) \approx 0$.

(iii) $°\mathscr{E}_1(u, u) < \infty$, and for all $t \approx 0$, we have $\mathscr{E}(u - Q^t u, u - Q^t u) \approx 0$.

(iv) $°\mathscr{E}_1(u, u) < \infty$, and for all $t \approx 0$, we have $\mathscr{E}^{(t)}(u, u) \approx \mathscr{E}(u, u)$.

PROOF. (i) $\Leftrightarrow$ (ii). Follows immediately from 5.1.4(i).

(ii) $\Rightarrow$ (iii). We have

$$0 \leq \mathscr{E}(u - Q^t u, u - Q^t u) = \mathscr{E}(u, u - Q^t u) - \mathscr{E}(Q_u^t, u - Q^t u),$$

and by 5.1.3 the term $\mathscr{E}(Q^t u, u - Q^t u)$ is positive.

(iii) $\Rightarrow$ (i).   Recall that $|u|_0 = \mathcal{E}(u, u)^{1/2}$ is a seminorm. By 5.1.3 and the triangle inequality

$$0 \le |u|_0 - |Q^t u|_0 \le |u - Q^t u|_0.$$

Multiplying both sides by $|u|_0 + |Q^t u|_0$, we get

$$0 \le |u|_0^2 - |Q^t u|_0^2 \le |u - Q^t u|_0(|u|_0 + |Q^t u|_0) \le 2|u|_0|u - Q^t u|_0.$$

Hence if $^\circ\mathcal{E}_1(u, u) < \infty$ and $\mathcal{E}(u - Q^t u, u - Q^t u) \approx 0$, we have $\mathcal{E}(u, u) - \mathcal{E}(Q^t u, Q^t u) \approx 0$.

(ii) $\Rightarrow$ (iv).   Follows at once from 5.1.2.

(iv) $\Rightarrow$ (ii).   Follows from 5.1.2 and the fact that $s \mapsto \mathcal{E}(Q^s u, u)$ is decreasing.

The different characterizations of $\mathcal{D}[\mathcal{E}]$ are useful for different purposes; as an illustration we use 5.1.5(iii) to prove that the domain has the right linear structure.

5.1.6. COROLLARY.   Let $u, v \in \mathcal{D}[\mathcal{E}]$, and assume that $\alpha \in {}^*\mathbb{R}$ is nearstandard. Then αu and $u + v$ are elements of $\mathcal{D}[\mathcal{E}]$.

PROOF.   The αu part is trivial. For $u + v$ we use 5.1.5(iii) and the triangle inequality;

$$\|(u + v) - Q^t(u + v)\|_0 = \|u - Q^t u + v - Q^t v\|_0$$
$$\le \|u - Q^t u\|_0 + \|v - Q^t v\|_0$$

and the last two terms are infinitesimal when $t \approx 0$.

The second part of 5.1.4 informs us that $Q^t u$ is more likely to be in $\mathcal{D}[\mathcal{E}]$ than u is. The next lemma pins this down more precisely.

5.1.7. LEMMA.   Assume $^\circ\mathcal{E}_1(u, u) < \infty$. Then for all noninfinitesimal t, we have $Q^t u \in \mathcal{D}[\mathcal{E}]$.

PROOF.   By 5.1.3

$$^\circ\mathcal{E}_1(Q^t u, Q^t u) \le {}^\circ\mathcal{E}_1(u, u) < \infty.$$

To prove that 5.1.1(ii) is satisfied, notice that according to 5.1.4(ii), the function

$$t \mapsto \mathcal{E}(Q^t u, Q^t u)$$

is decreasing and convex, and hence

$$\frac{1}{s}[\mathcal{E}(Q^t u, Q^t u) - \mathcal{E}(Q^{t+s} u, Q^{t+s} u)] \le \frac{1}{t}[\mathcal{E}(u, u) - \mathcal{E}(Q^t u, Q^t u)]$$

for all $s > 0$. Multiplying through by s, we get

(17) $$0 \le \mathcal{E}(Q^t u, Q^t u) - \mathcal{E}(Q^{t+s} u, Q^{t+s} u)$$
$$\le (s/t)[\mathcal{E}(u, u) - \mathcal{E}(Q^s u, Q^s u)].$$

For $s \approx 0$ and $t \not\approx 0$, the expression on the right is infinitesimal, and the lemma follows.

We shall now strengthen the lemma above and show that if $\mathscr{E}_1(u, u) < \infty$, then there is an infinitesimal t such that $Q^t u \in \mathscr{D}[\mathscr{E}]$. This is a special case of our next result. First some terminology; a subset $\mathscr{F}$ of H is called $\mathscr{E}$-*closed* if for all sequences $\{u_n\}_{n \in \mathbb{N}}$ of elements from $\mathscr{F}$ such that $^\circ|u_n - u_m|_1 \to 0$ as $n, m \to \infty$, there exists an element u in $\mathscr{F}$ such that $^\circ|u_n - u|_1 \to 0$ as $n \to \infty$.

5.1.8. PROPOSITION. $\mathscr{D}[\mathscr{E}]$ is $\mathscr{E}$-closed. Moreover, if $\{u_n\}_{n \in \mathbb{N}}$ is a $|\cdot|_1$-Cauchy sequence from $\mathscr{D}[\mathscr{E}]$, and $\{u_n\}_{n \in {^*\mathbb{N}}}$ is an internal extension, then there is a $\gamma \in {^*\mathbb{N}} - \mathbb{N}$ such that $u_\eta \in \mathscr{D}[\mathscr{E}]$ for all $\eta \leq \gamma$.

PROOF. Let $\{u_n\}_{n \in \mathbb{N}}$ be a $|\cdot|_1$-Cauchy sequence from $\mathscr{D}[\mathscr{E}]$, and let $\{u_n\}_{n \in {^*\mathbb{N}}}$ be an internal extension of it. There is an element $\gamma \in {^*\mathbb{N}} - \mathbb{N}$ such that $|u_n - u_m|_1 \approx 0$ whenever n and m are infinite and less than γ. Let $\eta \in {^*\mathbb{N}} - \mathbb{N}$, $\eta \leq \gamma$. By choice of γ, $^\circ\mathscr{E}_1(u_\eta, u_\eta) < \infty$ and $^\circ|u_n - u_\eta|_1 \to 0$ as n approaches infinity in $\mathbb{N}$. All that remains is to prove that $u_\eta \in \mathscr{D}[\mathscr{E}]$.

Assume not, then by 5.1.5(iii) there is an $\varepsilon \in \mathbb{R}_+$ and a $t \approx 0$ such that

$$|u_\eta - Q^t u_\eta|_0 > \varepsilon.$$

Choose $m \in \mathbb{N}$ so large that

$$|u_\eta - u_m|_0 < \varepsilon/4;$$

then by 5.1.3

$$|Q^t u_\eta - Q^t u_m|_0 < \varepsilon/4.$$

Combining the three inequalities above,

$$\varepsilon < |u_\eta - Q^t u_\eta|_0 \leq |u_\eta - u_m|_0 + |u_m - Q^t u_m|_0 + |Q^t u_m - Q^t u_\eta|_0$$
$$\leq \varepsilon/2 + |u_m - Q^t u_m|_0,$$

but since $u_m \in \mathscr{D}[\mathscr{E}]$, the last term is infinitesimal by 5.1.5 (iii). We have the contradiction we wanted.

5.1.9. COROLLARY. If $\mathscr{E}_1(u, u) < \infty$, there is a $t_0 \approx 0$ such that $Q^t u \in \mathscr{D}[\mathscr{E}]$ for all $t \geq t_0$.

PROOF. First notice that if $Q^{t_0} u \in \mathscr{D}[\mathscr{E}]$, so is $Q^t u$ for all $t > t_0$. Put $u_n = Q^{1/n} u$. Then the sequence $\{|u_n|_1\}$ is increasing and bounded by $|u|_1$, and we can apply the proposition to it. The corollary follows.

REMARK. Proposition 5.1.8 is rather surprising since there exist standard forms that are not closed. In fact, there are numerous applications where the main difficulty is to show that the form constructed is closed, or at least can be extended to a closed form [see, e.g., Albeverio and Høegh-Krohn

(1976, 1977a,b; 1981a,b, 1982, 1984), Albeverio *et al.* (1977, 1980, 1981, 1984b, 1985), Carmona (1979), Fukushima (1980, 1985), Reed and Simon (1975), Röckner and Wielens (1985), and Wielens (1985)]. If we know that it comes from a hyperfinite form, this follows immediately from 5.1.8. In Chapter 6, we shall see various examples of how useful this observation is; for the time being we only remark that since we shall soon (Theorem 5.2.1) show that all standard, closed forms can be obtained from hyperfinite forms, the method is quite general.

Notice that if we can show that whenever $°\mathscr{E}_1(u, u) < \infty$, then $\|u - Q'u\| \approx 0$ for all $t \approx 0$, Corollary 5.1.9 will imply the first part of our program, i.e., (3) above.

5.1.10. LEMMA. If $°\mathscr{E}(u, u) < \infty$, then for all $t \approx 0$

$$\|u - Q'u\| \approx 0.$$

PROOF.

$$\|u - Q'u\|^2 = \langle u - Q'u, u - Q'u \rangle = t\mathscr{E}^{(t)}(u, u - Q'u)$$
$$= t[\mathscr{E}^{(t)}(u, u) - \mathscr{E}^{(t)}(u, Q'u)] \le t\mathscr{E}(u, u) \approx 0$$

for $t \approx 0$.

Let us turn our attention to our second main goal (4).

5.1.11. LEMMA. If $u, v \in \mathscr{D}[\mathscr{E}]$ and $u \approx v$, then

$$\mathscr{E}(u, u) \approx \mathscr{E}(v, v).$$

PROOF. It is obviously enough to show that if $u \in \mathscr{D}[\mathscr{E}]$ and $u \approx 0$, then $\mathscr{E}(u, u) \approx 0$. But if $u \in \mathscr{D}[\mathscr{E}]$, we know from 5.1.5(iv):

$$(18) \qquad °\mathscr{E}(u, u) = \lim_{\substack{t \downarrow 0 \\ t \ne 0}} °\mathscr{E}^{(t)}(u, u).$$

Also

$$\mathscr{E}^{(t)}(u, u) = (1/t)\langle (I - Q')u, u \rangle = (1/t)(\langle u, u \rangle - \langle Q'u, u \rangle) \le (1/t)\|u\|,$$

which is infinitesimal for $t \ne 0$. Combining this with (18), the lemma follows.

We may now sum up our results on $\mathscr{D}[\mathscr{E}]$ in one statement.

5.1.12. THEOREM. Let $\mathscr{E}$ be a symmetric, non-negative quadratic form on a hyperfinite-dimensional linear space H:

(i)　If $u, v \in \mathscr{D}[\mathscr{E}]$ and α is a finite element of $^*\mathbb{R}$, then $\alpha u, u + v \in \mathscr{D}[\mathscr{E}]$.

(ii)　$\mathscr{D}[\mathscr{E}]$ is $\mathscr{E}$-closed.

(iii)   If $°\mathscr{E}_1(u, u) < \infty$, then there exists a $v \in \mathscr{D}[\mathscr{E}]$ with $\|u - v\| \approx 0$. Moreover,

$$°\mathscr{E}(v, v) = \lim_{\substack{t \downarrow 0 \\ t \neq 0}} °\mathscr{E}(Q^t u, Q^t u) = \lim_{\substack{t \downarrow 0 \\ t \neq 0}} °\mathscr{E}^{(t)}(u, u).$$

(iv)   If $u, v \in \mathscr{D}[\mathscr{E}]$ and $u \approx v$, then $\mathscr{E}(u, u) \approx \mathscr{E}(v, v)$.

The following definition now makes sense.

5.1.13. DEFINITION.   The *standard part* of $\mathscr{E}$ is the quadratic form E on $°H$ defined by:

(i)   The domain $\mathscr{D}[E]$ of E is the set of all equivalence classes $°u \in °H$ such that $\inf\{\mathscr{E}_1(v, v) \mid v \in °u\} < \infty$.

(ii)   If $x, y \in °H$ are in the domain of E, let $E(x, y) = °\mathscr{E}(u, v)$, where $u \in x, v \in y$ are in $\mathscr{D}[\mathscr{E}]$.

An E_1-*Cauchy sequence* is a sequence $\{x_n\}$ of elements from $\mathscr{D}[E]$ such that $E_1(u_n - u_m, u_n - u_m) \to 0$ as $n, m \to \infty$. We say that E is *closed* if all E_1-Cauchy sequences converge in E_1-norm to an element in $\mathscr{D}[E]$. The next proposition follows immediately from 5.1.12 and the definition of E.

5.1.14. PROPOSITION.   Let E be the standard part of $\mathscr{E}$. Then E is closed, and for all $x \in °H$

$$(19) \qquad E(x, x) = \inf\{°\mathscr{E}(u, u) \mid u \in x\},$$

where we take the value ∞ on the right to mean that the expression on the left is undefined.

Notice that (19) is just our original suggestion (2) for the standard part of $\mathscr{E}$.

B. The Resolvent

Up to now we have only been interested in the relationship between the form $\mathscr{E}$ and the associated semigroup $\{Q^t\}$. For the remainder of the section, we turn our attention to the resolvent $\{G_\alpha\}$ of $\mathscr{E}$. The goal is to give a description of $\mathscr{E}$ and $\mathscr{D}[\mathscr{E}]$ in terms of $\{G_\alpha\}$, similar to the one we have just given using the semigroup. The main result (Theorem 5.1.19) will allow us to reconstruct a form from its resolvent, and it will play an essential part in our study of singular perturbations of operators in Chapter 6.

The operator G_α is defined to be $(A - \alpha)^{-1}$ whenever this exists, and the formal calculation

$$(20) \qquad (A - \alpha)^{-1} = (I - (I - \Delta t(A - \alpha)))^{-1} \Delta t = \sum_{k=0}^{\infty} (I - \Delta t(A - \alpha))^k \Delta t$$

$$= \sum_{k=0}^{\infty} (Q^{\Delta t} + \alpha \, \Delta t)^k \, \Delta t$$

tells us that G_α will exist if the series on the right-hand side converges. Since $Q^{\Delta t}$ is a non-negative, symmetric operator with norm at most one, all its eigenvalues must be between zero and one. Choosing α such that $|\alpha|\,\Delta t \leq \frac{1}{2}$, we get that the absolute value of all eigenvalues of $Q^{\Delta t} + \alpha\,\Delta t$ must be less than $1 + \alpha\,\Delta t$. Hence if $\alpha < 0$ and $|\alpha|\,\Delta t \leq \frac{1}{2}$, the series in (20) converges, and we get the following proposition.

5.1.15. PROPOSITION. Let $\mathscr{E}$ be a non-negative, symmetric quadratic form, and let $\alpha \in {}^*\mathbb{R}$, $-1/2\,\Delta t \leq \alpha < 0$. Then G_α exists and

$$(21) \qquad G_\alpha = \sum_{k=0}^{\infty} (Q^{\Delta t} + \alpha\,\Delta t)^k\,\Delta t.$$

Moreover, in operator norm $\|G_\alpha\| \leq 1/|\alpha|$.

In the standard theory, the formula corresponding to (21) is

$$G_\alpha = \int_0^\infty e^{-t(A-\alpha)}\,dt = \int_0^\infty e^{-tA}\,e^{\alpha t}\,dt,$$

giving G_α as a weighted sum of the elements e^{-tA} in the semigroup. Since $\int_0^\infty -\alpha\,e^{\alpha t}\,dt = 1$, it is convenient to multiply this equation by $-\alpha$ to obtain

$$(22) \qquad -\alpha G_\alpha = \int_0^\infty -\alpha\,e^{-tA}\,e^{\alpha t}\,dt.$$

It is not quite obvious that this result carries over to the hyperfinite theory, since the equation $(Q^{\Delta t} + \alpha\,\Delta t)^k = Q^{k\Delta t}(1 + \alpha\,\Delta t)^k$ (corresponding to $e^{-t(A-\alpha)} = e^{-tA} \cdot e^{\alpha t}$) is false. But the next result shows that the two operators are close enough for our purposes:

5.1.16. LEMMA. For $\alpha \in {}^*\mathbb{R}$, $-1/\sqrt{\Delta t} \leq \alpha < 0$, and all $u \in H$ with ${}^\circ\mathscr{E}_1(u, u) < \infty$.

$$(23) \qquad \left| \alpha G_\alpha u - \left(\alpha \sum_{k=0}^{\infty} Q^{k\Delta t}(1 + \alpha\,\Delta t)^k\,\Delta t \right) u \right|_1 \approx 0.$$

PROOF. Let $\{e_i\}_{i \leq N}$ be an orthonormal basis of eigenvectors for A, and let a_i be the ith eigenvalue. Defining $b_i = a_i + 1$, we notice that if $u = \sum_{i=1}^N u_i e_i$, then

$$(24) \qquad \mathscr{E}_1(u, u) = \sum_{i=1}^N b_i u_i^2.$$

Summing geometric series, we see that

$$\alpha G_\alpha(e_i) = \left(\alpha \sum_{k=0}^{\infty} (1 - \Delta t a_i + \Delta t \alpha)^k\,\Delta t \right) e_i = \frac{\alpha}{a_i - \alpha}\,e_i$$

and similarly

$$\left(\alpha \sum_{k=0}^{\infty} Q^{k\Delta t}(1 + \alpha \, \Delta t)^k \, \Delta t \right)(e_i) = \frac{\alpha}{a_i - \alpha + a_i \alpha \, \Delta t} \, e_i,$$

which yields

$$\alpha G_\alpha(u) - \left(\alpha \sum_{k=0}^{\infty} Q^{k\Delta t}(1 + \alpha \, \Delta t)^k \, \Delta t \right)(u)$$

$$= \sum_{i=1}^{N} \frac{a_i \, \Delta t u_i e_i}{(1 - (a_i/\alpha))(1 - (a_i/\alpha) - a_i \, \Delta t)}.$$

Taking the $|\cdot|_1$-norm of this, we get from (24)

$$\left| \alpha G_\alpha u - \left(\alpha \sum_{k=0}^{\infty} Q^{k\Delta t}(1 + \alpha \, \Delta t)^k \, \Delta t \right) u \right|_1^2$$

$$= \sum_{k=1}^{N} b_i u_i^2 \frac{a_i^2 \, \Delta t^2}{(1 - (a_i/\alpha))^2 (1 - (a_i/\alpha) - a_i \, \Delta t)^2} \leq \eta \mathscr{E}_1(u, u),$$

where

$$\eta = \max_{1 \leq i \leq N} \left(\frac{a_i^2 \, \Delta t^2}{(1 - (a_i/\alpha))^2 (1 - (a_i/\alpha) - a_i \, \Delta t)^2} \right).$$

All that remains is to show that η is infinitesimal. First observe that since $\|A\| \, \Delta t \leq 1$ [recall (9)], we have $a_i \, \Delta t \leq 1$ for all i. Hence

$$\eta \leq \max_{1 \leq i \leq N} \left\{ \frac{a_i^2 \, \Delta t^2}{(1 - (a_i/\alpha))^2 ((a_i/\alpha))^2} \right\} = \max_{1 \leq i \leq N} \left\{ \frac{\alpha^4 \, \Delta t^2}{(\alpha - a_i)^2} \right\}.$$

Since $-1/\sqrt{\Delta t} \leq \alpha < 0$, the last term is always infinitesimal, and the proof is finished.

Equation (23) is the nonstandard counterpart of (22). Notice that

$$(25) \qquad \sum_{k=0}^{\infty} -\alpha(1 + \alpha \, \Delta t)^k = 1,$$

and that if α is infinite, then there is $t_\alpha \approx 0$ such that

$$(26) \qquad \sum_{0 \leq k\Delta t \leq t_\alpha} -\alpha(1 + \alpha \, \Delta t)^k \approx 1.$$

On the other hand, if α is finite, then

$$(27) \qquad \sum_{t \leq k\Delta t} -\alpha(1 + \alpha \, \Delta t)^k \approx 1$$

for all infinitesimal t. We can now begin our description of $\mathscr{E}$ and $\mathscr{D}[\mathscr{E}]$ in terms of G_α.

5.1.17. LEMMA. If $-1/\sqrt{\Delta t} < \alpha < 0$ and $°\mathscr{E}(u, u) < \infty$, then

(i) If α is infinite, $\|-\alpha G_\alpha u - u\| \approx 0$.

(ii) If α is finite, $-\alpha G_\alpha u \in \mathscr{D}[\mathscr{E}]$.

(iii) There is an infinite α such that $-\alpha G_\alpha u \in \mathscr{D}[\mathscr{E}]$.

PROOF. According to 5.1.16 it suffices to prove the statements we get after replacing G_α by

$$\hat{G}_\alpha = \sum_{k=0}^{\infty} Q^{k\Delta t}(1 + \alpha \, \Delta t)^k \, \Delta t.$$

(i) Let $t_\alpha \approx 0$ be as in (26); then

$$\|-\alpha\hat{G}_\alpha u - u\| = \left\| \sum_{k=0}^{\infty} (Q^{k\Delta t}u - u)(-\alpha)(1 + \alpha \, \Delta t)^k \, \Delta t \right\|$$

$$\approx \left\| \sum_{0 < k\Delta t \le t_\alpha} (Q^{k\Delta t}u - u)(-\alpha)(1 + \alpha \, \Delta t)^k \, \Delta t \right\| \approx 0,$$

where the last step uses Lemma 5.1.10.

(ii) Choose $t \approx 0$ such that $Q^t u \in \mathscr{D}[\mathscr{E}]$. If $\{e_i\}_{i \le N}$ is an orthonormal basis of eigenvectors for A, and a_i is the ith eigenvalue, we have

$$\sum_{0 \le k\Delta t < t} Q^{k\Delta t}(1 + \alpha \, \Delta t)^k e_i = \sum_{0 \le k\Delta t < t} (1 - a_i \, \Delta t)^k (1 + \alpha \, \Delta t)^k e_i$$

$$= \frac{1 - (1 + \alpha \, \Delta t)^{t/\Delta t}(1 - a_i \, \Delta t)^{t/\Delta t}}{a_i - \alpha + a_i\alpha \, \Delta t} e_i.$$

If $u = \sum_{i=1}^{N} u_i e_i$, we get

$$\left| \sum_{0 \le k\Delta t < t} Q^{k\Delta t}u(1 + \alpha \, \Delta t)^k \, \Delta t \right|_1^2$$

$$= \sum_{i=1}^{N} b_i u_i^2 \left(\frac{1 - (1 + \alpha \, \Delta t)^{t/\Delta t}(1 - a_i \, \Delta t)^{t/\Delta t}}{a_i - \alpha + a_i\alpha \, \Delta t} \right)$$

where $b_i = a_i + 1$. Since α is finite, it is easy to check that

$$\frac{1 - (1 + \alpha \, \Delta t)^{t/\Delta t}(1 - a_i \, \Delta t)^{t/\Delta t}}{a_i - \alpha + a_i\alpha \, \Delta t} \approx 0$$

for all i, and hence

(28)
$$\left| \sum_{0 \le k\Delta t < t} Q^{k\Delta t}u(1 + \alpha \, \Delta t)^k \, \Delta t \right|_1 \approx 0.$$

Observe that since $Q^{k\Delta t}u \in \mathscr{D}[\mathscr{E}]$ for all $k \, \Delta t \ge t$, we must have

$$\sum_{t \le k\Delta t} Q^{k\Delta t}u(1 + \alpha \, \Delta t)^k \, \Delta t \in \mathscr{D}[\mathscr{E}].$$

But by (28), $|\hat{G}_\alpha u - \sum_{t \leq k\Delta t} Q^{k\Delta t} u (1 + \alpha \, \Delta t)^k \, \Delta t|_1 \approx 0$, and hence $\hat{G}_\alpha u \in \mathscr{D}[\mathscr{E}]$.

(iii)   Notice that

$$(29) \quad \mathscr{E}_1(-\alpha G_\alpha u, -\alpha G_\alpha u) = \sum_{i=1}^N b_i u_i^2 \frac{\alpha^2}{(a_i - \alpha)^2} = \sum_{i=1}^N b_i u_i^2 \frac{1}{(a_i/\alpha - 1)^2}$$

increases as $\alpha \to -\infty$, and is bounded by $\mathscr{E}_1(u, u)$. Applying 5.1.8 to the sequence $u_n = nG_{-n}u$, the lemma follows.

The next proposition adds two new characterizations of $\mathscr{D}[\mathscr{E}]$ to the list in 5.1.5.

5.1.18. PROPOSITION.   The following are equivalent:

   (i) $u \in \mathscr{D}[\mathscr{E}]$,
   (ii) $^\circ\mathscr{E}_1(u, u) < \infty$ and $\lim_{\circ \alpha \to -\infty} {}^\circ\mathscr{E}_1(u + \alpha G_\alpha u, u + \alpha G_\alpha u) = 0$,
   (iii) $^\circ\mathscr{E}_1(u, u) = \lim_{\circ \alpha \to -\infty} {}^\circ\mathscr{E}_1(-\alpha G_\alpha u, -\alpha G_\alpha u) < \infty$.

PROOF.   (i) $\Rightarrow$ (ii). Pick an infinite α such that $-\alpha G_\alpha u \in \mathscr{D}[\mathscr{E}]$. Then $u + \alpha G_\alpha u \in \mathscr{D}[\mathscr{E}]$, $u + \alpha G_\alpha u \approx 0$, and hence $\mathscr{E}_1(u + \alpha G_\alpha u, u + \alpha G_\alpha u) \approx 0$. Part (ii) follows.

(ii) $\Rightarrow$ (iii). By (29) and the triangle inequality $0 \leq |u|_1 - |-\alpha G_\alpha u|_1 \leq |u + \alpha G_\alpha u|_1$, and multiplying by $|u|_1 + |-\alpha G_\alpha u|_1 \leq 2\mathscr{E}_1(u, u)$, $0 \leq |u|_1^2 - |-\alpha G_\alpha u|_1^2 \leq 2\mathscr{E}_1(u, u)|u + \alpha G_\alpha u|_1$, which shows that (ii) $\Rightarrow$ (iii).

(iii) $\Rightarrow$ (i). Pick an infinite α such that $-\alpha G_\alpha u \in \mathscr{D}[\mathscr{E}]$. Then $\|u + \alpha G_\alpha u\| \approx 0$ and $\mathscr{E}_1(u, u) \approx \mathscr{E}_1(-\alpha G_\alpha u, -\alpha G_\alpha u)$, and hence $u \in \mathscr{D}[\mathscr{E}]$.

We have now reached the final theorem of this section. As we mentioned above, the result gives a way of reconstructing a form from its resolvent. In our study of singular perturbations in Chapter 6, we shall find it much easier to control the resolvent of the perturbed form than the form itself. Once we have a good grasp of the resolvent, Theorem 5.1.19 will give us the form.

5.1.19. THEOREM.   Let $\mathscr{E}$ be a symmetric, non-negative, hyperfinite form on H, and let E be its standard part. For all $x \in {}^\circ H$ and all $v \in x$

$$(30) \qquad E(x, x) = -\lim_{\circ \alpha \to -\infty} {}^\circ(\alpha^2 \langle G_\alpha v, v \rangle + \alpha \langle v, v \rangle).$$

PROOF.   Notice that since G_α is bounded, it does not matter which $v \in x$ we use. We split the proof into two cases.

(i)   *x is not in the domain of E:* Let $\{e_i\}_{i \leq N}$ be an orthonormal basis of eigenvectors for A, and assume that the corresponding eigenvalues $\{a_i\}_{i \leq N}$

are in decreasing order. Pick $v = \sum_{i \leq N} v_i e_i$ in x. An easy calculation shows that

$$-(\langle \alpha^2 G_\alpha v, v \rangle + \alpha \langle v, v \rangle) = \sum_{i=1}^{N} a_i v_i^2 \frac{-\alpha}{a_i - \alpha}.$$

Assume for contradiction that the limit in (30) is finite; then there is an infinite α such that

$$(31) \qquad \overset{\circ}{\sum_{i=1}^{N}} a_i v_i^2 \frac{-\alpha}{a_i - \alpha} < \infty.$$

If H is the largest integer such that $a_H > |\alpha|$,

$$\sum_{i=1}^{N} a_i v_i^2 \frac{-\alpha}{a_i - \alpha} = \sum_{i=1}^{H} a_i v_i^2 \frac{-\alpha}{a_i - \alpha} + \sum_{i=H+1}^{N} a_i v_i^2 \frac{-\alpha}{a_i - \alpha}$$

$$= -\alpha \sum_{i=1}^{H} v_i^2 \frac{1}{1 - \alpha/a_i} + \sum_{i=H+1}^{N} a_i v_i^2 \frac{1}{1 - a_i/\alpha}$$

$$\geq -\frac{\alpha}{2} \sum_{i=1}^{H} v_i^2 + \frac{1}{2} \sum_{i=H+1}^{N} a_i v_i^2,$$

and hence the last two terms are finite. But if $-(\alpha/2) \sum_{i=1}^{H} v_i^2$ is finite, v is infinitely close to

$$v' = \sum_{i=H+1}^{N} v_i e_i,$$

and if in addition $\frac{1}{2} \sum_{i=H+1}^{N} a_i v_i^2$ is finite, then $^\circ \mathscr{E}_1(v', v') < \infty$. This contradicts the assumption that $x \notin \mathscr{D}[E]$.

(ii)   *Assume that $x \in \mathscr{D}[E]$*: Let $v \in x$ be such that

$$(32) \qquad\qquad ^\circ \mathscr{E}_1(v, v) < \infty.$$

Since $\mathscr{E}_{-\alpha}(G_\alpha u, w) = \langle u, w \rangle$, we have

$$\mathscr{E}(-\alpha G_\alpha v, -\alpha G_\alpha v) = \mathscr{E}_{-\alpha}(\alpha G_\alpha v, \alpha G_\alpha v) + \alpha \langle \alpha G_\alpha v, \alpha G_\alpha v \rangle$$

$$= \alpha^2 \langle G_\alpha v, v \rangle + \alpha^3 \langle G_\alpha v, G_\alpha v \rangle.$$

The theorem will follow from 5.1.18(iii) if we can prove that for all v satisfying (32),

$$\lim_{^\circ \alpha \to -\infty} {}^\circ(\alpha^2 \langle G_\alpha v, v \rangle + \alpha^3 \langle G_\alpha v, G_\alpha v \rangle + \alpha^2 \langle G_\alpha v, v \rangle + \alpha \langle v, v \rangle) = 0.$$

By simple algebra, this is the same as

$$\lim_{°\alpha \to -\infty} °(\alpha \|\alpha G_\alpha v + v\|^2) = 0.$$

Pulling α inside the norm and reformulating the problem in nonstandard terms, we see that what we have to prove is

(33) $$\||\alpha|^{3/2} G_\alpha v - |\alpha|^{1/2} v\|^2 \approx 0$$

for all infinite, negative α of sufficiently small absolute value.

If $v = \sum_{i=1}^{N} v_i e_i$ is the eigenvector expansion of v, we see that

(34) $$\||\alpha|^{3/2} G_\alpha v - |\alpha|^{1/2} v\|^2 = \sum_{i=1}^{N} \left(\frac{|\alpha|^{3/2}}{a_i - \alpha} - |\alpha|^{1/2} \right)^2 v_i^2$$

$$= \sum_{i=1}^{N} \frac{-\alpha a_i^2}{(a_i - \alpha)^2} v_i^2 = \sum_{i=1}^{N} a_i v_i^2 \left(\frac{1}{\beta_i + \beta_i^{-1} + 2} \right),$$

where $\beta_i = -\alpha / a_i$.

Notice that if a_i is infinitesimal compared to α or α is infinitesimal compared to a_i, then $1/(\beta_i + \beta_i^{-1} + 2)$ is infinitesimal. To get the sum on the right-hand side of (34) to be infinitesimal, we only have to choose α such that the contributions from the terms satisfying neither of these requirements are infinitesimal.

Assuming that the eigenvalues $\{a_i\}$ are given in descending order, we define

$$\gamma = \sup \left\{ °\sum_{i=1}^{k} a_i v_i^2 \,\middle|\, a_k \text{ is infinite} \right\}.$$

Since $°\sum_{i=1}^{k} a_i v_i^2 \leq \mathscr{E}(v, v)$ is finite by (32), γ is a real number.

Using saturation on the sets

$$A_n = \left\{ j \in {}^*\mathbb{N} \,\middle|\, \sum_{i=1}^{j} a_i v_i^2 > \gamma - \frac{1}{n} \text{ and } a_j > n \right\},$$

we find a hyperinteger K such that a_K is infinite and

$$\sum_{i=1}^{K} a_i v_i^2 \approx \gamma.$$

We choose $|\alpha|$ to be infinitely large, but infinitesimal compared to a_K.

For each $\varepsilon \in \mathbb{R}_+$, let

$$M_\varepsilon = \inf \left\{ k \,\middle|\, \sum_{i=1}^{k} a_i v_i^2 \geq \gamma + \varepsilon \right\}.$$

By choice of γ, the term a_{M_ε} must be finite. But

$$\sum_{i=1}^n a_i v_i^2 \frac{1}{\beta_i + \beta_i^{-1} + 2} \le \sum_{i=1}^K a_i v_i^2 \frac{1}{\beta_i + \beta_i^{-1} + 2}$$
$$+ \sum_{i=K+1}^{M_\varepsilon - 1} a_i v_i^2 + \sum_{i=M_\varepsilon}^N a_i v_i^2 \frac{1}{\beta_i + \beta_i^{-1} + 2},$$

where the first term is infinitesimal since each β_i is; the second term is less than 2ε by choice of M_ε; and the last term is infinitesimal since each β_i is infinite. Since $\varepsilon \in \mathbb{R}_+$ is arbitrary, the sum on the left must be infinitesimal. This proves (33), and hence the theorem.

REMARK. The hyperfinite part of the theory in this section is new, but most of the results correspond to well-known standard theorems [see, e.g., Fukushima (1980) and Reed and Simon (1975)].

5.2. CONNECTIONS TO STANDARD THEORY

We shall interrupt our development of the hyperfinite theory for a moment, and relate the results we have obtained so far to the standard theory. In the last section we proved that a quadratic form on a hyperfinite-dimensional space H induces a closed form on the hull $^\circ H$ of H. For most applications the space $^\circ H$ is too large; what we really want is a form defined on a Hilbert space K given in advance of the nonstandard construction. If K can be identified with a subspace of $^\circ H$, we get the desired form by restricting the form on $^\circ H$ to K; notice that since K is a closed subspace of $^\circ H$, the restricted form is also closed. The result we shall prove in this section states that any closed, symmetric, non-negative form on any Hilbert space K can be obtained from a hyperfinite form in this way. There are two reasons for proving such a representation theorem; the first and most important is to obtain the "correct" relationship between already established standard theory on the one hand, and the new hyperfinite theory on the other. The second, closely related reason is to show that no generality is lost by working within the nonstandard framework.

Let K be a standard Hilbert space. A hyperfinite-dimensional subspace H of $^*K$ is called *S-dense* in $^*K$ if for all $x \in K$, there is a $y \in H$ such that $\|x - y\| \approx 0$. Recall that $\approx$ is the equivalence relation on H given by $u \approx v$ if and only if $\|u - v\| \approx 0$, and that $^\circ u$ denotes the equivalence class of u under $\approx$. If H is S-dense in $^*K$, we can identify K with a subspace of $^\circ H$ by identifying x and $^\circ u$ whenever $\|x - u\| \approx 0$.

If $\mathscr{E}$ is a quadratic, symmetric, non-negative form on H, we let E denote the standard part of $\mathscr{E}$ as defined in 5.1.13, and we let E_K be the restriction of E to K. As mentioned above, our goal is to show that all closed forms on K can be obtained in this way, but before we prove this, we shall recall a few results from the standard theory of quadratic forms [Fukushima (1980, Section 1.3) is a convenient reference].

To each closed, symmetric, non-negative, densely defined form F on K, there is associated a unique, strongly continuous semigroup $\{T_t\}_{t \in \mathbb{R}_+}$ of contraction operators. The form can be recovered from the semigroup by

(1) $$F(x, x) = \lim_{t \downarrow 0} \frac{1}{t} \langle (I - T_t)x, x \rangle,$$

(2) $$\mathscr{D}[F] = \left\{ x \in K \,\middle|\, \lim_{t \downarrow 0} \frac{1}{t} \langle (I - T_t)x, x \rangle < \infty \right\},$$

where the functions $t \mapsto \langle (I - T_t)x, x \rangle$ are positive and decreasing. Equations (1) and (2) are just the standard counterparts of 5.1.12(iii).

Let st_K be the standard part map from $^*K$ to K.

5.2.1. PROPOSITION. Let F be a closed, densely defined, nonnegative, symmetric form on a Hilbert space K. Let $\{T_t\}$ be the semigroup generated by F, and let H be an S-dense, hyperfinite-dimensional subspace of $^*K$. Then there exists a non-negative, symmetric form $\mathscr{E}$ on H—associated with an internal time line T—such that

(3) $$F = E_K.$$

Moreover, if $\{Q^s\}_{s \in T}$ is the semigroup generated by $\mathscr{E}$ and T, then for all $t \in \mathbb{R}_+$, $s \in T$, $u \in K$, $v \in H$ such that $t = {}^\circ s$, $u = \mathrm{st}_K(v)$, we have

(4) $$\mathrm{st}_K Q^s v = T_t u.$$

PROOF. Let P be the projection of $^*K$ on H, and write $\{^*T_t\}_{t \in {}^*\mathbb{R}_+}$ for $^*(\{T_t\}_{t \in \mathbb{R}_+})$. Our plan is first to define the internal semigroup by putting $Q^{\Delta t} = P^*T_{\Delta t}$ for a carefully chosen infinitesimal Δt, and then let

$$\mathscr{E}(u, v) = (1/\Delta t)\langle (I - Q^{\Delta t})u, v \rangle.$$

Notice that if $u \in H$, then

$$\langle Q^{\Delta t}u, u \rangle = \langle P^*T_{\Delta t}u, u \rangle = \langle ^*T_{\Delta t}u, u \rangle \geq 0,$$

which shows that $Q^{\Delta t}$ is positive on H. Also, since the operator norm of $^*T_{\Delta t}$ is less than or equal to one, so is the norm of $Q^{\Delta t}$, and hence conditions (5.1.8)–(5.1.12) are satisfied.

We shall now choose Δt such that (3) and (4) hold. If u is nearstandard and $^\circ t < \infty$, then

$$\| P^*T_t u - {}^*T_t u \| \approx 0$$

since $^*T_t$ takes nearstandard elements to nearstandard elements, and H is S-dense in $^*K$. By induction we get

(5)
$$\|(P^*T_t)^n u - {}^*T_{t\cdot n}u\| \approx 0$$

for all $n \in \mathbb{N}$. For each $u \in K$, let $v_u = Pu$. Then $v_u \in H$ and $\|u - v_u\| \approx 0$. Consider the sets

$$A_u = \{n \in {}^*\mathbb{N} | \forall k \le 2^{2n}(\|(P^*T_{2^{-n}})^k v_u - {}^*T_{k2^{-n}}v_u\| \le 1/n)\}.$$

By (5), this set contains $\mathbb{N}$, and since it is internal, it must contain an infinite internal segment $\{n \in {}^*\mathbb{N} | n \le n_u\}$. Using saturation, we find an infinite n smaller than all the n_u's.

Next we consider the set

$$B_u = \left\{ m \in {}^*\mathbb{N} | \forall k \le 2^{2m}(|(2^m/k)\langle(I - (P^*T_{2^{-m}})^k)v_u, v_u\rangle \right.$$
$$\left. - (2^m/k)\langle(I - {}^*T_{k\cdot 2^{-m}})v_u, v_u\rangle| \le \frac{1}{m} \right\}.$$

For each $u \in K$, this set contains $\mathbb{N}$, and hence an initial segment $\{m \in {}^*\mathbb{N} | m \le m_u\}$. By saturation there is an infinite m smaller than all the m_u's, $u \in K$.

We now take Δt to be the largest of the two infinitesimals 2^{-n} and 2^{-m}. Equation (4) follows immediately from the definition of A_u and the choice of n. By the definition of B_u and the choice of m, we see that

$$\lim_{t \to 0} \frac{1}{t}\langle(I - T_t)u, u\rangle = \lim_{\substack{t \to 0 \\ t \ne 0}} {}^\circ\mathscr{E}^{(t)}(v_u, v_u).$$

The proposition follows from (1), (2), and Theorem 5.1.12(iii).

REMARK. Let us make a few comments on 5.2.1. The assumption that F is densely defined is for convenience only; if it is not satisfied, we just apply the proposition to the closure of $\mathscr{D}[F]$. If F is not closed, we obviously cannot obtain F as E_K for any hyperfinite form $\mathscr{E}$ since 5.1.14 tells us that E_K is always closed However, if F is closable (i.e., there exists a closed form extending F), all closed extensions of F can be represented as standard parts of hyperfinite forms. A natural representation for a closable form F would be a representation of its smallest closed extension—the Friedrichs extension. If F is not closable, no hyperfinite representation (in our sense) is possible; any representation we try will change some F values, and restrict and extend $\mathscr{D}[F]$ in different directions in order to turn F into a closed form. As we commented in Section 5.1, the fact that nonclosable forms do not have hyperfinite representations is more a blessing than a curse; in standard theory a lot of effort goes into showing that the forms one constructs are closable; in the hyperfinite theory this is an immediate consequence of the construction.

Let us finally remark that there is an equation between the resolvents of F and $\mathscr{E}$ similar to (4); we leave the precise statement and the proof to the reader.

In Proposition 5.2.1, the space H was just any S-dense, hyperfinite-dimensional subspace of $^*K$, but in applications we often want to choose special kinds of subspaces appropriate for the problems we have in mind. We shall take a look at the case where $K = L^2(X, m)$ for some Hausdorff space X.

Let us make the following assumptions about the measure space $(X, \mathscr{B}, m)$: the measure m is a completed Borel measure such that $m(K) < \infty$ for all compact sets K, and it is a *Radon measure* in the sense that for all $B \in \mathscr{B}$,

$$(6) \qquad m(B) = \sup\{m(K) \mid K \subset B, K \text{ compact}\},$$

and for all $B \in \mathscr{B}$ with $m(B) < \infty$,

$$(7) \qquad m(B) = \inf\{m(O) \mid B \subset O, O \text{ open}\}.$$

A subset Y of $^*X$ is called *rich* if it is hyperfinite and $\operatorname{st}(Y) = X$. If Y is rich, it is easy to construct a hyperfinite measure μ on Y such that $m = L(\mu) \circ \operatorname{st}^{-1}$. Let H be the hyperfinite-dimensional space of all internal functions $f: Y \to {}^*\mathbb{R}$, given the inner product

$$\langle f, g \rangle = \int f \cdot g \, d\mu.$$

We want to show that all closed forms on $K = L^2(X, \mathscr{B}, m)$ can be represented as hyperfinite forms on H.

One way of constructing a rich set Y in $^*X$ and a measure μ on Y representing m is as follows (recall the proof of 3.4.10). If $O_1, \ldots, O_n$ are open sets in X, let $\mathscr{P}_{O_1,\ldots,O_n}$ be the collection of all hyperfinite partitions $\mathscr{P}$ of $^*X$ such that if P is a partition class in $\mathscr{P}$, then P is *Borel and for all $i \leq n$ either $P \subset {}^*O_i$ or $P \cap {}^*O_i = \varnothing$. Using saturation on the family $\{\mathscr{P}_{O_1,\ldots,O_n}\}$, we find a partition $\mathscr{P}$ which is in all these collections. Let Y be an internal set containing one point from each partition class of $\mathscr{P}$. By construction of $\mathscr{P}$, this set must be rich in $^*X$. For each $y \in Y$, let P_y be the corresponding equivalence class in $\mathscr{P}$; it is easy to see that if y is nearstandard, then P_y is contained in the monad of y. We define the internal measure μ on Y by

$$(8) \qquad \mu(\{y\}) = {}^*m(P_y)$$

the equality $m = L(\mu) \circ \operatorname{st}^{-1}$ follows from (6), (7), and 3.4.8.

Let $\tilde{H}$ be the subspace of $*K$ consisting of all functions constant on each class $P_y \in \mathscr{P}$. Since H and $\tilde{H}$ obviously are isomorphic, Proposition 5.2.1 will give us a representation of closed forms on K in terms of internal forms on H if we can only show that $\tilde{H}$ is S-dense in $*K$.

From Section 3.2 we know that each function f in K has an S-square-integrable lifting $\tilde{f}$ in $\tilde{H}$ such that ${}^\circ\tilde{f}(x) = f({}^\circ x)$ for almost all nearstandard x. If we can show that $\|*f - \tilde{f}\| \approx 0$, then $\tilde{H}$ is dense in $*K$. By (6), it suffices to show this when f is bounded and of compact support, and for such functions the statement is an immediate consequence of Anderson's non-standard version of Lusin's theorem (Corollary 3.4.9). We have proved:

5.2.2. COROLLARY. Let X be a Hausdorff space, m a Radon measure on X, and F a densely defined, closed, non-negative, symmetric form on $L^2(X, m)$. Then there exists a hyperfinite, rich subset Y of $*X$, an internal measure μ on Y, and a non-negative, symmetric form $\mathscr{E}$ on $L^2(Y, \mu)$ representing F in the following sense: $m = L(\mu) \circ \mathrm{st}^{-1}$ and for all $u \in L^2(X, m)$

$$F(u, u) = \inf\{{}^\circ\mathscr{E}(v, v) \,|\, v \text{ is a 2-lifting of } u\}.$$

Moreover, if $\{T_t\}$ and $\{Q^s\}$ are the semigroups generated by F and $\mathscr{E}$, respectively, then $Q^s v$ is a 2-lifting of $T_t u$ whenever v is a 2-lifting of u and $s \approx t$.

In the remaining sections of this chapter, we shall be mostly interested in quadratic forms generating Markov processes. A bounded operator S on $L^2(X, m)$ is called a *Markov operator* if it maps non-negative functions to non-negative functions, and

$$\|S(f)\|_\infty \leq \|f\|_\infty$$

for all bounded functions f. A quadratic form F on $L^2(X, m)$ is called a *Dirichlet form* if it is closed, densely defined, symmetric, and non-negative, and generates a semigroup $\{T_t\}$ of Markov operators. Hyperfinite Dirichlet forms are defined analogously. We shall see in the next section that the Dirichlet forms are exactly the forms generating Markov processes; for the time being we only make the following simple observation.

5.2.3. COROLLARY. If the form F in 5.2.2 is a Dirichlet form, we can also take $\mathscr{E}$ to be a Dirichlet form.

PROOF. We shall prove that if F is a Dirichlet form, then the $\mathscr{E}$ obtained from the proofs of 5.2.1 and 5.2.2 is a Dirichlet form. Observe first that it is enough to prove that $Q^{\Delta t}$ is a Markov operator. In the proof of 5.2.1, we defined $Q^{\Delta t}$ as $P*T_{\Delta t}$ for a suitable infinitesimal Δt, where P is the orthogonal projection of $*K$ onto H. With the choice of H made in the proof of 5.2.2, this projection is just the conditional expectation with respect to the algebra generated by the partition $\mathcal{P}$. Since conditional expectations preserve nonnegativity and decrease the supremum norm, the corollary follows.

When are the standard forms generated by two hyperfinite forms different? The last result we shall prove shows that to answer this question, it is enough to check whether the forms have the same resolvents. Recall that in Theorem 5.1.19 we found a way of reconstructing a form from its resolvent. This representation will be used later to study singular perturbations of operators. Lemma 5.2.4 will be useful in checking that certain perturbations are nontrivial, i.e., that the perturbed form is different from the original one.

5.2.4. LEMMA. Let K be a Hilbert space and H an S-dense, hyperfinite-dimensional subspace of $^*K$. Let $\mathscr{E}$ and $\hat{\mathscr{E}}$ be two nonnegative, symmetric forms on H inducing E_K and $\hat{E}_K$, respectively, on K. Let $\{G_\alpha\}$ and $\{\hat{G}_\alpha\}$ be the resolvents of $\mathscr{E}$ and $\hat{\mathscr{E}}$. Assume that for some finite, noninfinitesimal $\alpha \in {}^*\mathbb{R}_-$, there is a $u \in H$ with $^\circ \mathscr{E}_1(u, u) < \infty$ such that $v = G_\alpha u$, $w = \hat{G}_\alpha u$ are both nearstandard, but $^\circ \|v - w\| \neq 0$. Then $E_K \neq \hat{E}_K$.

PROOF. Assume for contradiction that $E_K = \hat{E}_K$. Pick $\tilde{v} \approx v$, $\tilde{w} \approx w$ such that $\tilde{v} \in \mathscr{D}[\hat{\mathscr{E}}]$, $\tilde{w} \in \mathscr{D}[\mathscr{E}]$, and notice that by 5.1.17, $v \in \mathscr{D}[\mathscr{E}]$, $w \in \mathscr{D}[\hat{\mathscr{E}}]$. We have

$$(9) \qquad \langle u, v - w \rangle \approx \langle u, v - \tilde{w} \rangle = \mathscr{E}_{-\alpha}(v, v - \tilde{w}),$$

and since v, w are nearstandard and $E_K = \hat{E}_K$,

$$(10) \qquad {}^\circ \mathscr{E}_{-\alpha}(v, v - \tilde{w}) = {}^\circ \hat{\mathscr{E}}_{-\alpha}(\tilde{v}, \tilde{v} - w).$$

On the other hand,

$$(11) \qquad \langle u, v - w \rangle \approx \langle u, \tilde{v} - w \rangle = \hat{\mathscr{E}}_{-\alpha}(w, \tilde{v} - w).$$

Combining (9), (10), and (11), we see that

$$0 = {}^\circ \hat{\mathscr{E}}_{-\alpha}(\tilde{v} - w, \tilde{v} - w) \geq {}^\circ |\alpha| {}^\circ \|v - w\|^2 > 0,$$

and the lemma is proved.

REMARK. The representation theorems in this section are intended for general theoretical purposes. When studying a particular problem concerning a specific operator, it is often more convenient to work with a hyperfinite form constructed directly from our intuitive insight into the problem, rather than with the one obtained by an appeal to Proposition 5.2.1. For example, if we are interested in the form F generated by $-\Delta$ (where Δ is the Laplace operator) on $\mathbb{R}^d$, i.e., the closure of

$$F(f, g) = -\int_{\mathbb{R}^d} \Delta f g \, dx,$$

a simple hyperfinite representation can be constructed as follows. Let

$$\Gamma = \left\{ (n_1 \varepsilon, \dots, n_d \varepsilon) \,\middle|\, (n_1, \dots, n_d) \in {}^*\mathbb{Z}^d, \max_{1 \le i \le d} n_i \le \frac{1}{\varepsilon} \right\},$$

where $\varepsilon \approx 0$, be a hyperfinite lattice in ${}^*\mathbb{R}^d$, and define

$$Af(x) = \sum_{|y-x|=\varepsilon} \frac{f(y) - f(x)}{\varepsilon^2}$$

for all internal functions $f : \Gamma \to {}^*\mathbb{R}$, using the convention that $f(y) = 0$ when $y \notin \Gamma$. The form

$$\mathscr{E}(f, g) = -\sum_{x \in \Gamma} Af(x)g(x)\varepsilon^d$$

is a simpler and more intuitive representation of F than the one obtained from Corollary 5.2.2. Hyperfinite representations of this kind will be important in Sections 5.6 and 7.5.

5.3. HYPERFINITE DIRICHLET FORMS

For the remainder of this chapter we shall restrict our attention to the hyperfinite forms generated by Markov processes, the Dirichlet forms. The aim is to give a reasonably detailed account of the relationship between properties of these forms and the behavior of the associated processes.

Consider a particle which can be in $N + 1$ different states $s_0, s_1, \dots, s_N$. Assume that if the particle is in the state s_i at some instant t, then— independently of what its past history may be—the probability that it will be in state s_j at the next instant $t + \Delta t$ is given by a fixed number q_{ij}. This is the familiar setting for the theory of stationary Markov chains with finite state spaces; see, e.g., Chung (1960) and Dynkin and Yushkevich (1969). We shall be interested in the case where $S = \{s_0, s_1, \dots, s_N\}$ is a hyperfinite set, and $T = \{k \, \Delta t \,|\, k \in {}^*\mathbb{N}\}$ is a hyperdiscrete time line with $\Delta t \approx 0$ (for technical reasons it is convenient to have a *-infinite time line to work with in this chapter). The idea is to use the hyperfinite setup to reduce the highly sophisticated theory of continuous parameter Markov processes taking values in topological spaces [see Fukushima (1980) and Silverstein (1974, 1976)] to the much simpler theory of finite Markov chains. Thus we shall have in mind the situation where S is an S-dense subset of $*Y$ for some Hausdorff space Y, and where a standard Markov process is to be defined as the standard part of our hyperfinite Markov chain.

A. Hyperfinite Markov Processes and the Definition of Dirichlet Forms

To make the assumptions a little more precise, let $Q = \{q_{ij}\}$ be an $(N + 1) \times (N + 1)$ matrix with non-negative entries, and assume that

$$(1) \qquad \sum_{j=0}^{N} q_{ij} = 1$$

for all i. Let m be a hyperfinite measure on $S = \{s_0, s_1, \ldots, s_N\}$; we shall write m_i for $m(\{s_i\})$, and—whenever it is convenient—$q_{s_i s_j}$ for q_{ij}. If (Ω, P) is an internal measure space, and $X : \Omega \times T \to S$ is an internal process, let

$$(2) \qquad [\omega]_t^X = \{\omega' \in \Omega \mid X(\omega', s) = X(\omega, s) \text{ for all } s \leq t\}.$$

If for all ω

$$(3) \qquad P([\omega]_0^X) = m\{X(\omega, 0)\},$$

and whenever $X(\omega, t) = s_i$,

$$(4) \qquad P\{\omega' \in [\omega]_t^X \mid X(t + \Delta t, \omega') = s_j\} = q_{ij} P([\omega]_t^X),$$

then we call X a *Markov process with initial distribution m and transition matrix Q*. Notice that we do not assume that m and P are probability measures; $P(\Omega)$ could be an infinite, hyperfinite number.

Given Q and m, it is easy to construct an associated Markov process X; just let Ω be the set of all internal functions $\omega : T \to S$, let X be the coordinate function $X(\omega, t) = \omega(t)$, and take P to be the measure generated by

$$(5) \qquad P([\omega]_{k\Delta t}^X) = m(\{\omega(0)\}) \prod_{n=0}^{k-1} q_{\omega(n\Delta t), \omega((n+1)\Delta t)}.$$

A special case is when the initial distribution is concentrated in one point s_i and has mass one, i.e., $m_j = \delta_{ij}$ (where δ_{ij} is the Kronecker symbol), then

$$(6) \qquad P_i([\omega]_{k\Delta t}^X) = \delta_{i\omega(0)} \prod_{n=0}^{k-1} q_{\omega(n\Delta t), \omega((n+1)\Delta t)}$$

generates a probability measure.

We next introduce a few regularity conditions. The state s_0 is a *trap*, i.e.,

$$(7) \qquad q_{0i} = 0 \qquad \text{for all} \quad i \neq 0.$$

The initial measure m and the transition matrix Q satisfy the symmetry conditions

(8) $m_i q_{ij} = m_j q_{ji}$ for all $i \neq 0, j \neq 0$.

Finally, we assume that

(9) $m_i \neq 0$ for at least one $i \neq 0$.

It is easy to find examples of transition matrices Q such that no m satisfies (8) and (9), and thus these assumptions may be regarded as conditions on Q.

Notice that for most i, the transition probability q_{i0} should be of order of magnitude Δt, since if not the process will die in infinitesimal time.

A process X satisfying (3), (4), (7), (8), and (9) is called a *symmetric Markov process associated with m and Q*, and it is this class of processes we now shall study in some detail. The main tool will be the theory of quadratic forms and semigroups developed in Section 5.1. We shall first obtain the form from the process.

If

(10) $S_0 = \{s_1, s_2, \ldots, s_N\}$

is the state space S without the trap s_0, we let H be the linear space of all internal functions $u: S_0 \to {}^*\mathbb{R}$ with the inner product

(11) $$\langle u, v \rangle = \sum_{i=1}^{N} u(s_i) v(s_i) m(s_i).$$

Just as we usually write m_i for $m(s_i)$, we shall write $u(i)$ or u_i for $u(s_i)$. From time to time we shall identify H with the set of all internal functions $u: S \to {}^*\mathbb{R}$ such that $u(s_0) = 0$.

Our convention of letting the trap s_0 be the zeroth element is notationally convenient, but it does create certain pitfalls for the careless reader; e.g., failure to distinguish between sums of the forms $\sum_{i=0}^{N}$ and $\sum_{i=1}^{N}$ may cause problems.

For $t \in T$ and $u \in H$ we define a new function $Q^t u \in H$ by

(12) $Q^t u(i) = E_i(u(X(t)))$,

where E_i is the expectation with respect to the measure P_i defined in (6). Intuitively, $Q^t u(i)$ is the expected value of $u(X(t))$ for a particle starting in state s_i. Notice that

(13) $Q^{\Delta t} u = Q \cdot u$

where $\cdot$ is matrix multiplication, and that since $q_{ij}^{(t+s)} = \sum q_{ik}^{(t)} q_{kj}^{(s)}$, we must have

(14) $Q^{t+s} = Q^t \cdot Q^s$

(here $q_{ij}^{(t)}$ is the transition probability given by Q^t). Hence the family $\{Q^t\}_{t \in T}$ is a semigroup of operators on H, and

$$(15) \qquad Q^{\Delta t} u(i) = \sum_{j=1}^{N} u(j) q_{ij}.$$

The infinitesimal generator A of this semigroup is given by

$$(16) \qquad Au(i) = \frac{1}{\Delta t} \left(u(i) - \sum_{j=1}^{N} u(j) q_{ij} \right),$$

and *the Dirichlet form associated with Q and m* is defined to be

$$(17) \qquad \mathscr{E}(u, v) = \langle Au, v \rangle = \sum_{i=1}^{N} Au(i) v(i) m_i.$$

Combining (16) and (17), we get

$$(18) \qquad \mathscr{E}(u, v) = \frac{1}{\Delta t} \sum_{i=1}^{N} \left[u(i) v(i) m_i - \sum_{j=1}^{N} u(j) v(i) q_{ij} m_i \right].$$

We have already given a different definition of Dirichlet forms in Section 5.2, but we shall prove in Proposition 5.3.3 that if a form is a Dirichlet form with respect to some pair m, Q, then it is a Dirichlet form in the sense of Section 5.2 and vice versa. Until this equivalence is established, we shall use the definition above.

B. Alternative Descriptions of Dirichlet Forms

Our first result gives an alternative way of expressing a Dirichlet form in terms of m_i and q_{ij}; it is a nonstandard version of the Beurling and Deny (1959) formula, and is often more useful than (18):

5.3.1. LEMMA. Let $\mathscr{E}$ be the Dirichlet form of Q and m. Then

$$(19) \qquad \mathscr{E}(u, v) = \frac{1}{\Delta t} \left[\sum_{1 \le i < j \le N} (u(i) - u(j))(v(i) - v(j)) q_{ij} m_i \right.$$
$$\left. + \sum_{i=1}^{N} u(i) v(i) q_{i0} m_i \right].$$

PROOF. Notice that

$$\mathscr{E}(u, v) = \frac{1}{\Delta t} \left[\sum_{i=1}^{N} u(i) v(i) m_i - \sum_{i=1}^{N} \sum_{j=1}^{N} u(j) v(i) q_{ij} m_i \right]$$
$$= \frac{1}{\Delta t} \left[\sum_{i=1}^{N} \sum_{j=0}^{N} u(i) v(i) q_{ij} m_i - \sum_{i=1}^{N} \sum_{j=1}^{N} u(j) v(i) q_{ij} m_i \right]$$
$$= \frac{1}{\Delta t} \left[\sum_{1 \le i, j \le N} (u(i) - u(j)) v(i) q_{ij} m_i + \sum_{i=1}^{N} u(i) v(i) q_{i0} m_i \right],$$

where the first line is a trivial modification of (18); the second line follows from the first since $\sum_{j=0}^{N} q_{ij} = 1$; and the last line is just a rearrangement of the second.

Fix a pair (i, j), and consider the terms in the last expression above involving both i and j. If $i = j$, there is only one such term, and that term is zero. If $i \neq j$, there are two terms to consider,

$$(u(i) - u(j))v(i)q_{ij}m_i \quad \text{and} \quad (u(j) - u(i))v(j)q_{ji}m_j,$$

and since $q_{ij}m_i = q_{ji}m_j$, their sum equals

$$(u(i) - u(j))(v(i) - v(j))q_{ij}m_i.$$

Summing over all pairs (i, j), the lemma follows.

As an immediate consequence we have

5.3.2. COROLLARY. A Dirichlet form is symmetric and non-negative.

The next result gives us three ways of deciding whether a given form is a Dirichlet form without actually constructing an associated Markov process. But first a few definitions.

If $u \in H$, the function $\tilde{u} = (0 \vee u) \wedge 1$ is called the *unit contraction* of u. A quadratic form $\mathscr{E}$ is said to have the *Markov property* if for all u

$$\mathscr{E}(\tilde{u}, \tilde{u}) \leq \mathscr{E}(u, u).$$

Recall that a *Markov operator* $T : H \to H$ is an operator which maps non-negative functions to non-negative functions, and which never increases the supremum norm, i.e.,

$$\|Tu\|_\infty \leq \|u\|_\infty$$

for all $u \in H$.

5.3.3. PROPOSITION. Let

$$\mathscr{E}(u, v) = \sum_{i, j=1}^{N} b_{ij}u(i)v(j)$$

be a nonzero, symmetric form on H, and let $\{Q^t\}$ be the associated semigroup. The following statements are equivalent:

(i) $\mathscr{E}$ is the Dirichlet form of some Q and m.
(ii) $Q^{\Delta t}$ is a Markov operator.
(iii) $\mathscr{E}$ has the Markov property.
(iv) Whenever $i \neq j$, $b_{ij} < 0$; but $b_{ii} \geq -\sum_{j \neq i} b_{ij}$ for all i.

PROOF. We shall prove $(i) \Rightarrow (ii) \Rightarrow (iv) \Rightarrow (i)$ and $(i) \Rightarrow (iii) \Rightarrow (iv)$.
$(i) \Rightarrow (ii)$. This follows immediately from (12).
$(ii) \Rightarrow (iv)$. Notice that

$$Q^{\Delta t}u(i) = u(i) - \Delta t \sum_{j=1}^{n} b_{ij}u(j).$$

For each $j \in S_0$, let u_j be given by $u_j(i) = \delta_{ij}$. Since $Q^{\Delta t}$ is Markov, we have for all $i \neq j$

$$0 \leq Q^{\Delta t}u_j(i) = -\Delta t b_{ij},$$

and thus $b_{ij} \leq 0$.

On the other hand, applying $Q^{\Delta t}$ to the function which is constant one and using that $Q^{\Delta t}$ cannot increase the supremum norm, we get

$$1 \geq Q^{\Delta t}1(i) = 1 - \Delta t \sum_{j=1}^{n} b_{ij}.$$

It follows that $b_{ii} \geq -\sum_{j \neq i} b_{ij}$.

$(iv) \Rightarrow (i)$. We first observe that since $\mathscr{E}$ is symmetric

$$\sum_{1 \leq i < j \leq N} b_{ij}(u(i) - u(j))^2$$

$$= -\sum_{i \neq j} b_{ij}u(i)u(j) + \sum_{1 \leq i < j \leq N} b_{ij}u(i)^2 + \sum_{1 \leq i < j \leq N} b_{ij}u(j)^2$$

$$= -\mathscr{E}(u, u) + \sum_{1 \leq i,j \leq N} b_{ij}u(i)^2,$$

and thus

$$(20) \qquad \mathscr{E}(u, u) = - \sum_{1 \leq i < j \leq N} b_{ij}(u(i) - u(j))^2 + \sum_{1 \leq i,j \leq N} b_{ij}u(i)^2.$$

The plan is to define Q and m by matching the two expressions (19) and (20) term by term. It turns out that we have one degree of freedom for each i; we may choose

$$(21) \qquad 0 \leq q_{ii} < 1.$$

Comparing (19) and (20), we notice that we must have

$$(22) \qquad (1/\Delta t)m_i q_{ij} = -b_{ij} \qquad \text{for} \quad j \neq 0, i$$

and

$$(23) \qquad \frac{1}{\Delta t} m_i q_{i0} = \sum_{j=1}^{N} b_{ij}.$$

In order to have $\sum_{j=0}^{N} q_{ij} = 1$ satisfied,

$$m_i = \sum_{j=0}^{N} m_i q_{ij} = m_i q_{ii} - \Delta t \sum_{j \neq i} b_{ij} + \Delta t \sum_{j=1}^{N} b_{ij} = m_i q_{ii} + b_{ii} \Delta t,$$

which gives us

(24) $$m_i = b_{ii} \Delta t / (1 - q_{ii}).$$

Combining this with (22), we get

(25) $$q_{ij} = -b_{ij}(1 - q_{ii})/b_{ii} \qquad \text{for} \quad j \neq 0, i,$$

and from (23), we have

(26) $$q_{i0} = \left(\sum_{j=1}^{N} b_{ij}(1 - q_{ii}) \right) \Big/ b_{ii}.$$

This construction breaks down when $b_{ii} = 0$, but in that case $b_{ij} = 0$ for all j, and we get $m_i = 0$ and can choose the q_{ij}'s arbitrarily.

We finally observe that since $\mathscr{E}$ is symmetric, (22) implies (8), and that the nontriviality of $\mathscr{E}$ implies (9).

(i) $\Rightarrow$ (iii). This follows immediately from Lemma 5.3.1.

(iii) $\Rightarrow$ (iv). Fix $k, l \in S_0$, $k \neq l$, let $\varepsilon \in {}^*\mathbb{R}_+$, and define u by

$$u(i) = \begin{cases} -\varepsilon & \text{if} \quad i = k, \\ 1 & \text{if} \quad i = l, \\ 0 & \text{otherwise.} \end{cases}$$

If $\tilde{u}$ is the unit contraction of u, we get from (20)

$$
\begin{aligned}
0 \leq \mathscr{E}(u, u) - \mathscr{E}(\tilde{u}, \tilde{u}) &= \left(-\sum_{i \neq k, l} b_{ik}\varepsilon^2 - \sum_{i \neq k, l} b_{il}1^2 - b_{kl}(1 + \varepsilon)^2 \right. \\
&\qquad \left. + \sum_{j=1}^{N} b_{jk}\varepsilon^2 + \sum_{j=1}^{N} b_{jl}1^2 \right) \\
&\quad - \left(-\sum_{i \neq k, l} b_{il}1^2 - b_{kl}1^2 + \sum_{j=1}^{N} b_{jl}1^2 \right) \\
&= -\varepsilon^2 \sum_{i \neq k, l} b_{ik} - 2\varepsilon b_{kl} - b_{kl}\varepsilon^2 + \varepsilon^2 \sum_{j=1}^{N} b_{jk}.
\end{aligned}
$$

Choosing ε small enough, the term $-2\varepsilon b_{kl}$ dominates the rest, and it follows that $b_{kl} \leq 0$.

To get the second half of (iv), we fix a $k \in S_0$ and consider the function

$$v(i) = \begin{cases} 1 + \varepsilon & \text{if} \quad i = k, \\ 1 & \text{otherwise.} \end{cases}$$

If $\tilde{v}$ is the unit contraction of v,

$$0 \le \mathscr{E}(v, v) - \mathscr{E}(\tilde{v}, \tilde{v})$$

$$= \left(-\sum_{i \ne k} b_{ik}\varepsilon^2 + \sum_{i \ne k}\sum_{j=1}^{N} b_{ij} + \sum_{j=1}^{N} b_{kj}(1 + \varepsilon)^2 \right) - \sum_{i,j} b_{ij}1^2$$

$$= -\varepsilon^2 \sum_{i \ne k} b_{ik} + 2\varepsilon \sum_{j=1}^{N} b_{kj} + \varepsilon^2 \sum_{j=1}^{N} b_{kj},$$

and choosing ε small enough, we see that $\sum_{j=1}^{N} b_{kj}$ must be nonnegative. The proposition is proved.

Notice that 5.3.3(ii) is the definition of Dirichlet forms we used in Section 5.2.

Let us try to illustrate the theory by a simple example.

5.3.4. EXAMPLE. (Brownian motion on a circle). Let $N = (\Delta t)^{-1/2}$ be an even hyperfinite integer, and let $S_0 = \{s_1, \ldots, s_N\}$ be uniformly distributed on a circle of circumference one. If $i, j = \{1, 2, \ldots, N\}$, let the transition probability q_{ij} be $\frac{1}{2}$ if s_i and s_j are neighbors, and 0 otherwise. The semigroup $\{Q^t\}$ is given by

$$Q^{\Delta t}u(i) = \tfrac{1}{2}u(i + 1) + \tfrac{1}{2}u(i - 1),$$

where the addition is modulo N inside the u's. The infinitesimal generator is

$$(27) \qquad Au(i) = -\frac{u(i + 1) - 2u(i) + u(i - 1)}{2\,\Delta t},$$

and if $m_i = 1/N$ for all i, the associated Dirichlet form $\mathscr{E}$ is given by

$$\mathscr{E}(u, v) = -\frac{N}{2} \sum_{i=1}^{N} [u(i + 1) - 2u(i) + u(i - 1)]v(i),$$

or—in the Beurling–Deny formulation—

$$(28) \qquad \mathscr{E}(u, v) = \frac{N}{2} \sum_{i=1}^{N} [u(i + 1) - u(i)][v(i + 1) - v(i)].$$

Notice that by (27), the infinitesimal generator A is a nonstandard version of the operator

$$\tilde{A}f = -\tfrac{1}{2}f'',$$

and by (28), the form $\mathscr{E}$ is a representation of

$$E(f, g) = \frac{1}{2} \int_C f'g' \, dm,$$

where m is the Lebesgue measure on the circle C, and all derivatives are taken along the circle. Passing from the original expression for $\mathscr{E}$ to the Beurling–Deny version amounts to an integration by parts.

C. Equilibrium Potentials

We now turn to a closer study of the relationship between a Dirichlet form $\mathscr{E}$ and the associated process X. Recall that for $\alpha \in {}^*\mathbb{R}_+$, the form $\mathscr{E}_\alpha$ is defined by

$$\mathscr{E}_\alpha(u, v) = \mathscr{E}(u, v) + \alpha \int uv \, dm.$$

The first problem we consider is the following. Let $f: D_f \to {}^*\mathbb{R}$ be an internal function defined on an internal subset D_f of S_0. We want to find the function $e_\alpha(f)$ agreeing with f on D_f and minimizing $\mathscr{E}_\alpha(e_\alpha(f), e_\alpha(f))$.

We first observe that it suffices to find an extension $e_\alpha(f)$ of f such that

$$(29) \qquad \mathscr{E}_\alpha(e_\alpha(f), u) = 0$$

for all u which are zero on D_f. To see this, let w be another extension of f; then $u = w - e_\alpha(f)$ is zero on D_f, and by (29)

$$(30) \qquad \mathscr{E}_\alpha(w, w) = \mathscr{E}_\alpha(e_\alpha(f) + u, e_\alpha(f) + u)$$

$$= \mathscr{E}_\alpha(e_\alpha(f), e_\alpha(f)) + 2\mathscr{E}_\alpha(e_\alpha(f), u) + \mathscr{E}_\alpha(u, u)$$

$$= \mathscr{E}_\alpha(e_\alpha(f), e_\alpha(f)) + \mathscr{E}_\alpha(u, u).$$

It still remains to find a function $e_\alpha(f)$ satisfying (29). Let σ_f be the stopping time

$$(31) \qquad \sigma_f(\omega) = \min\{t \in T \,|\, X(\omega, t) \in D_f\},$$

and define

$$(32) \qquad e_\alpha(f)(i) = E_i((1 + \alpha \, \Delta t)^{-\sigma_f/\Delta t} f(X(\sigma_f))),$$

where E_i is the expectation with respect to the measure P_i. We let

$$(1 + \alpha \, \Delta t)^{-\sigma_f/\Delta t} = 0$$

if $X(\omega, \cdot)$ never hits D_f. Note that if $i \notin D_f$, then

$$(33) \qquad (1 + \alpha \, \Delta t)e_\alpha(f)(i) = \sum_{j=1}^{N} e_\alpha(f)(j)q_{ij}.$$

If u is zero on D_f, then

$$\mathscr{E}_\alpha(e_\alpha(f), u) = \frac{1}{\Delta t} \sum_{i=1}^{N} \left[e_\alpha(f)(i) - \sum_{j=1}^{N} e_\alpha(f)(j) q_{ij} \right] u(i) m_i$$

(34)
$$+ \alpha \sum_{i=1}^{N} e_\alpha(f)(i) u(i) m_i = \frac{1}{\Delta t}$$

$$\times \sum_{i=1}^{N} \left[(1 + \alpha \, \Delta t) e_\alpha(f)(i) - \sum_{j=1}^{N} e_\alpha(f)(j) q_{ij} \right] u(i) m_i = 0,$$

since the expression in the bracket is zero when $i \notin D_f$, and $u(i) = 0$ when $i \in D_f$. We have proved

5.3.5. PROPOSITION. Let $\mathscr{E}$ be a hyperfinite Dirichlet form, and let $f: D_f \to$ *$\mathbb{R}$ be an internal function defined on a subset of S_0. The extension of f having the smallest $\mathscr{E}_\alpha$ value is

$$e_\alpha(f)(i) = E_i((1 + \alpha \, \Delta t)^{-\sigma_f/\Delta t} f(X(\sigma_f))).$$

If f is constant one on its domain $A = D_f$, we write $e_\alpha(A)$ for $e_\alpha(f)$. Observe that by (29)

$$\mathscr{E}_\alpha(e_\alpha(A), e_\alpha(A)) = \mathscr{E}_\alpha(e_\alpha(A), 1),$$

and applying (19) to $\mathscr{E}_\alpha(e_\alpha(A), 1)$, we get

$$(35) \quad \mathscr{E}_\alpha(e_\alpha(A), e_\alpha(A)) = \alpha \sum_{i=1}^{N} e_\alpha(A)(i) m_i + \sum_{i=1}^{N} e_\alpha(A)(i) \frac{q_{i0}}{\Delta t} m_i.$$

The functions $e_\alpha(f)$ are called *equilibrium potentials*, and they will serve as bridgeheads in our campaign to unify the analytic theory of Dirichlet forms and the probabilistic theory of Markov processes. The key to their importance is the observation that they have natural interpretations both in analytic and in probabilistic terms; on the one hand they minimize the forms $\mathscr{E}_\alpha$ under suitable side conditions; on the other they describe how the process X hits subsets of S_0. For a fuller understanding of the probabilistic description, note that for finite α

$$°e_\alpha(A)(i) = E_i(°e^{-\alpha \sigma_A}),$$

and that the function $\alpha \mapsto E_i(°e^{-\alpha \sigma_A})$ is the Laplace transform of σ_A. Hence the distribution of σ_A can be completely recovered from the functions $e_\alpha(A)$. This fact will be of great importance in Section 5.5.

Often it is of no consequence which α we work with, and we shall then choose $\alpha = 1$. To simplify notation, we write e_A and e_f for $e_1(A)$ and $e_1(f)$, respectively.

As a first example of the use of equilibrium potentials, we prove the following proposition.

5.3.6. PROPOSITION. Let X be a hyperfinite Markov process and $\mathscr{E}$ its Dirichlet form. Then for all $u \in H$ and $t \in T$

$$(36) \qquad P\{\omega \mid \exists s \le t(|u(X(\omega, s))| \ge \varepsilon)\} \le \frac{2(1 + \Delta t)^{t/\Delta t}}{\varepsilon^2} \mathscr{E}_1(u, u).$$

PROOF. Let $A = \{i \in S_0 \mid u(i) \ge \varepsilon\}$, then

$$P\{\omega \mid \exists s \le t(u(X(\omega, s)) \ge \varepsilon)\}$$

$$= \int P_i\{\omega \mid \exists s \le t(u(X(\omega, s)) \ge \varepsilon)\} \, dm(i)$$

$$= \int P_i\{\omega \mid (1 + \Delta t)^{-\sigma_A/\Delta t} \ge (1 + \Delta t)^{-t/\Delta t}\} \, dm(i)$$

$$\le \int E_i\left(\frac{(1 + \Delta t)^{-\sigma_A/\Delta t}}{(1 + \Delta t)^{-t/\Delta t}}\right) dm(i)$$

$$= (1 + \Delta t)^{t/\Delta t} \int e_A(i) \, dm(i)$$

$$\le (1 + \Delta t)^{t/\Delta t} \mathscr{E}_1(e_A, e_A),$$

where the last step uses (35). Since $\mathscr{E}$ has the Markov property, we get from 5.3.5 that

$$\mathscr{E}_1(e_A, e_A) \le \mathscr{E}_1\left(\frac{u}{\varepsilon}, \frac{u}{\varepsilon}\right),$$

and hence

$$P\{\omega \mid \exists s \le t(u(X(\omega, s)) \ge \varepsilon)\} \le \frac{(1 + \Delta t)^{t/\Delta t}}{\varepsilon^2} \mathscr{E}_1(u, u).$$

Applying the same argument to $-u$, the proposition follows.

5.3.7. COROLLARY. Let $u, u_n, n \in \mathbb{N}$ be elements in H, and assume that $^\circ\mathscr{E}_1(u - u_n, u - u_n) \to 0$ as $n \to \infty$. There is a subsequence $\{u_{n_k}\}$ such that for a.a. ω, $u_{n_k}(X)$ converges uniformly to $u(X)$ on all S-bounded subsets of T.

PROOF. By the proposition and basic measure theory.

D. Fukushima's Decomposition Theorem

So far we have not used the theory of domains from Section 5.1, but for the next theorem it will be needed. The condition (5.1.9) requiring that

$$(37) \qquad \|A\| \, \Delta t \le 1,$$

or—equivalently—that $Q^{\Delta t}$ is positive, played a not unimportant technical role in Section 5.1, and by choosing Δt sufficiently small, we can always assume that it is satisfied:

5.3.8. DEFINITION. A Dirichlet form $\mathscr{E}$ is *normal* (with respect to Δt) if for all $u \in H$

$$(38) \qquad \langle Q^{\Delta t} u, u \rangle \ge 0.$$

With this definition we can apply the machinery of Section 5.1 to normal Dirichlet forms. As a first example of the power of the theory, we shall prove a hyperfinite version of Fukushima's (1979, 1980) decomposition theorem. For each $t \in T$, we let $\mathscr{A}_t$ be the internal algebra generated by the sets $[\omega]_t^X$ defined in (2), and we ask the reader to recall the notions of quadratic variation (Definition 4.2.4) and λ^2-martingale (Definition 4.2.6).

5.3.9. THEOREM. Let $\mathscr{E}$ be a normal, hyperfinite Dirichlet form associated to a Markov process $X : \Omega \times T \to S$. For each $u \in \mathscr{D}[\mathscr{E}]$ there exist two processes N^u, $M^u : \Omega \times T \to {}^*\mathbb{R}$ such that:

   (i) $u(X(\omega, t)) = N^u(\omega, t) + M^u(\omega, t)$ for all ω, t.

   (ii) M^u is a λ^2-martingale with respect to $(\Omega, \{\mathscr{A}_t\}, P)$.

   (iii) N^u is S-continuous, and $E([N^u](t)) \approx 0$ for all finite $t \in T$.

PROOF. Define N^u by

$$(39) \qquad N^u(\omega, 0) = u(X(\omega, 0)),$$

$$(40) \qquad \Delta N^u(\omega, t) = Q^{\Delta t} u(X(\omega, t)) - u(X(\omega, t)).$$

By definition of $Q^{\Delta t}$, the process

$$(41) \qquad M^u = u(X) - N^u$$

is an $\{\mathscr{A}_t\}$-martingale. To prove that M^u is a λ^2-martingale, we observe that if $X(\omega, t) = s_i$ and $X(\omega, t + \Delta t) = s_j$, then

$$(42) \qquad \Delta M^u(\omega, t) = u(j) - Q^{\Delta t} u(i).$$

Since $P\{\omega \mid X(\omega, t) = s_i\} \le m_i$, we get from (42):

$$(43) \qquad E([M_t^u])^{1/2} \le \left(\sum_0^t \sum_{i,j} (u(j) - Q^{\Delta t}u(i))^2 q_{ij}m_i \right)^{1/2}$$

$$\le \left(\sum_0^t \sum_{i,j} (u(j) - u(i))^2 q_{ij}m_i \right)^{1/2}$$

$$+ \left(\sum_0^t \sum_{i,j} (u(i) - Q^{\Delta t}u(i))^2 q_{ij}m_i \right)^{1/2}$$

$$\le (2t\mathscr{E}(u, u))^{1/2} + (t\mathscr{E}(u, u - Q^{\Delta t}u))^{1/2} < \infty,$$

and hence M^u is a λ^2-martingale.

It remains to prove (iii). The quadratic variation part is easy:

$$\frac{1}{\Delta t} E(\Delta N^u(t)^2) \le \frac{1}{\Delta t} \sum_{i=1}^N (u(i) - Q^{\Delta t}u(i))^2 m_i \le \mathscr{E}(u, u - Q^{\Delta t}u),$$

which is infinitesimal by 5.1.5(ii). Hence $E([N^u](t)) \le t\mathscr{E}(u, u - Q^{\Delta t}u)$ is infinitesimal for all finite t.

For each $n \in \mathbb{N}$, let $u_n = Q^{1/n}u$. Our plan is first to show that N^{u_n} is S-continuous for each n, and then use 5.3.7 to deduce that N^u must also be S-continuous.

We first recall that by (5.1.17)

$$\mathscr{E}(u_n, u_n) - \mathscr{E}(Q^{\Delta t}u_n, Q^{\Delta t}u_n) \le n\mathscr{E}(u, u) \Delta t,$$

and since according to 5.1.4(i)

$$0 \le \mathscr{E}(u_n, u_n - Q^{\Delta t}u_n) \le \mathscr{E}(u_n, u_n) - \mathscr{E}(Q^{\Delta t}u_n, Q^{\Delta t}u_n),$$

we get

$$\frac{1}{\Delta t} E(\Delta N^{u_n}(t)^2) \le \mathscr{E}(u_n, u_n - Q^{\Delta t}u_n) \le n\mathscr{E}(u, u) \Delta t.$$

It follows that

$$E((N^{u_n}(t) - N^{u_n}(s))^2) \le n\mathscr{E}(u, u)(t - s)^2$$

for all $t, s \in T$. By Kolmogorov's continuity theorem (this is Proposition 4.8.5 with $d = 1$; the proof can be read independently of the rest of Section 4.8), each N^{u_n} is S-continuous.

If we can find a subsequence of $\{u_n\}$ such that $N^{u_{n_k}}$ converges uniformly to N^u on compacts, N^u is S-continuous. By 5.3.7 there is a subsequence such that $u_{n_k}(X) \to u(X)$ uniformly on compacts, and since

$$N^v = v(X) - M^v,$$

it suffices to establish the corresponding convergence for a subsequence $\{M^{u_{n_k}}\}$. If $v_n = u - u_n$, we get from (43) and Doob's inequality

$$E(\max_{s \le t}(M^u(s) - M^{u_n}(s))^2)^{1/2} \le 2E((M^u(t) - M^{u_n}(t))^2)^{1/2}$$

$$\le 2(2t\mathscr{E}(v_n, v_n))^{1/2}$$

$$+ 2(t\mathscr{E}(v_n, v_n - Q^{\Delta t}v_n))^{1/2},$$

which gives us the convergence we need. The theorem is proved.

We know from Chapter 4 that λ^2-martingales have left and right limits. Fukushima's decomposition theorem tells us that so has $u(X)$, and that the discontinuities of $u(X)$ are those of the martingale M^u. If our state space S_0 is embedded in the *-version of a topological space Y, and the domain $\mathscr{D}[\mathscr{E}]$ contains enough S-continuous functions, this will imply that X itself has left and right limits. In the next section we shall use information about $u(X)$ and $\mathscr{D}[\mathscr{E}]$ to carry out a detailed analysis of the process X. Another application of Theorem 5.3.9 is to stochastic differential equations with singular drift coefficients; see Section 5.6.

Let us finally mention an alternative way of proving 5.3.9; instead of approximating with the functions $u_n = Q^{1/n}u$, we could have used

$$u_n = nG_{-n}u,$$

where $\{G_\alpha\}$ is the resolvent of $\mathscr{E}$. This approach avoids Kolmogorov's theorem, but uses instead the theory of resolvents from Section 5.1. The reader is invited to carry out the proof, using Fukushima (1980, Theorem 5.2.2) as a reference if needed.

E. The Hyperfinite Feynman–Kac Formula

The last result we shall prove in this section is a hyperfinite version of the Feynman–Kac formula—the cornerstone of the functional integration approach to quantum physics. Given the infinitesimal generator A of a Markov process X, and a function V on the state space, the Feynman–Kac formula gives a description of the semigroup $e^{-t(A+V)}$ in terms of V and X. Much of the formula's importance in quantum mechanics stems from the fact that the Hamiltonian H_0 of a free particle is just the infinitesimal generator $-(\hbar^2/2m)\,\Delta$ of a Brownian motion, while the Hamiltonian H of a particle moving in a potential V is given by $H = H_0 + V$. The Feynman–Kac formula thus gives us a probabilistic method of approaching the Schrödinger operators $H = H_0 + V$. For systematic accounts of the remarkable success of this approach, the reader should consult the books by Simon (1979) and Glimm and Jaffe (1981); in the present work we shall only make use of it on two occasions—during the discussion of polymer measures in Section 6.4 and when explaining the connection between polymer models and quantum fields in Section 7.5.

In the nonstandard setting, A will be a hyperfinite Markov operator and V an internal function on the state space. The idea of the proof is to show that the semigroup $\{T^t\}_{t \in T}$ generated by A is infinitely close to the semigroup $\{S^t\}_{t \in T}$ given by

$$(44) \qquad S^{\Delta t} = (1 - V \Delta t)(1 - A \Delta t).$$

Once this relation is established, the following easy lemma will give the probabilistic interpretation of T^t:

5.3.10. LEMMA. Let A be a hyperfinite Markov operator and let $X : \Omega \times T \to S$ be the associated process. If $V : S_0 \to {}^*\mathbb{R}$ is an internal function, and S^t is the semigroup given by (44), then

$$S^t u(i) = e^{-t\eta(t)} E_i \left\{ u(X(t)) \exp\left(-\int_0^t V(X(s)) \, ds \right) \right\},$$

where $0 \le \eta(t) \le \frac{1}{2} \| V \|_\infty^2 \Delta t$.

PROOF. If $\{Q^t\}$ is the semigroup generated by A,

$$S^t = (1 - V \Delta t) Q^{\Delta t} (1 - V \Delta t) \cdots Q^{\Delta t} (1 - V \Delta t) Q^{\Delta t}.$$

Using the probabilistic interpretation of $Q^{\Delta t}$ repeatedly, we see that

$$S^t u(i) = E_i \left[u(X(t)) \prod_{0 \le s < t} (1 - V(X(s)) \Delta t) \right].$$

But

$$\prod_{0 \le s < t} (1 - V(X(\omega, s)) \Delta t) = \exp\left(\sum_0^t \ln(1 - V(X(\omega, s)) \Delta t) \right)$$

$$= \exp\left(-\sum_0^t V(X(\omega, s)) \Delta t - \frac{1}{2} \sum_0^t \Theta(\omega, s)^2 \Delta t^2 \right)$$

where $\Theta(\omega, s)$ lies between 0 and $V(X(\omega, s))$. Hence

$$\exp(-(t/2) \| V \|_\infty^2 \Delta t) \cdot \exp\left(-\int_0^t V(X(s)) \, ds \right) \le \prod_0^t (1 - V(X(s)) \Delta t)$$

$$\le \exp\left[-\int_0^t V(X(s)) \, ds \right],$$

and the lemma follows.

Note that if $\|V\|_\infty^2 \Delta t \approx 0$, then the lemma says that $S^t u$ and $E_{(\cdot)}(u(X(t)) \exp[-\int_0^t V(X(s)) \, ds])$ are infinitely close in any reasonable sense.

We are ready to prove a hyperfinite version of the Feynman–Kac formula:

5.3.11. THEOREM. Let A be a hyperfinite Markov operator and let $X : \Omega \times T \to S$ be the associated process. Assume that $V : S_0 \to {}^*\mathbb{R}$ is an internal function such that

(a)   there is a $\beta \in \mathbb{R}$ such that $\langle (A + V)u, u \rangle \ge \beta \|u\|^2$ for all u,
(b) $\|V\|_\infty / \ln(\Delta t) \approx 0$.

If $\{T^t\}$ denotes the semigroup generated by $A + V$, then

$$\left\| T^t u - E_{(\cdot)}\left(u(X(t)) \cdot \exp\left[-\int_0^t V(X(s)) \, ds \right] \right) \right\| \approx 0$$

for all finite t and all u with finite norm.

PROOF.   By the lemma it suffices to prove that $T^t u \approx S^t u$. Observe first that

$$S^t = S^{t-\Delta t}(1 - (A + V)\,\Delta t + VA\,\Delta t^2) = S^{t-\Delta t}T^{\Delta t} + S^{t-\Delta t}VA\,\Delta t^2$$

$$= S^{t-2\Delta t}(1 - (A + V)\,\Delta t + VA\,\Delta t^2)T^{\Delta t} + S^{t-\Delta t}VA\,\Delta t^2$$

$$= S^{t-2\Delta t}T^{2\Delta t} + S^{t-2\Delta t}VAT^{\Delta t}\,\Delta t^2 + S^{t-\Delta t}VA\,\Delta t^2.$$

Continuing in this manner, we finally get

$$S^t = T^t + \sum_{s=0}^t S^{t-s-\Delta t}VAT^s\,\Delta t^2.$$

To show that the sum $\sum S^{t-s-\Delta t}VAT^s\,\Delta t^2$ is infinitesimal, we write it as

(45) $$\sum_{s=0}^t S^{t-s-\Delta t}V(A + V)T^s\,\Delta t^2 - \sum_{s=0}^t S^{t-s-\Delta t}V^2 T^s\,\Delta t^2.$$

Considering the second term first, we note that since

$$\|S^{t-s-\Delta t}\| \le e^{\|V\|_\infty(t-s)} \le e^{\|V\|_\infty t} \quad \text{and} \quad \|T^s\| \le e^{|\beta|s} \le e^{|\beta|t},$$

we have

$$\left\| \sum_{s=0}^t S^{t-s-\Delta t}V^2 T^s\,\Delta t^2 \right\| \le e^{\|V\|_\infty t}\|V\|_\infty^2 e^{|\beta|t}\,\Delta t,$$

which is infinitesimal by (b).

Turning our attention to the first term in (45), we note that

$$\left\| \sum_{s=0}^{t} S^{t-s-\Delta t} V(A + V) T^s \, \Delta t^2 \right\| \le e^{\|V\|_\infty t} \|V\|_\infty \sum_{s=0}^{t} \|(A + V) T^s\| \, \Delta t^2.$$

In a moment we shall prove that

(46) $$\|(A + V) T^s\| \le \frac{M}{s + \Delta t}$$

for a finite constant M, but let us first show that this is sufficient to finish the proof: since

$$\sum_{s=0}^{t} \frac{\Delta t}{s + \Delta t} \le 2 \ln\left(\frac{1}{\Delta t}\right),$$

we get

$$\left\| \sum_{s=0}^{t} S^{t-s-\Delta t} V(A + V) T^s \, \Delta t^2 \right\| \le M e^{\|V\|_\infty t} \|V\|_\infty \sum_{s=0}^{t} \frac{\Delta t}{s + \Delta t} \Delta t$$

$$\le 2 M e^{\|V\|_\infty t} \|V\|_\infty \ln\left(\frac{1}{\Delta t}\right) \Delta t,$$

which is infinitesimal for all finite t by (b).

It only remains to prove (46). Obviously, $\|(A + V) T^s\|$ equals the largest of the values $|\lambda(1 - \lambda \, \Delta t)^{s/\Delta t}|$ when λ ranges over the eigenvalues of $A + V$. If λ is negative, it is bounded from below by the β in (a), and thus $|\lambda(1 - \lambda \, \Delta t)|^{s/\Delta t} \le |\beta| \cdot e^{|\beta| t}$. If λ is positive, $\|(A + V) T^s\|$ cannot exceed the maximal value

$$\frac{1}{s + \Delta t}\left(1 - \frac{\Delta t}{s + \Delta t}\right)^{s/\Delta t}$$

of the function $\lambda \mapsto \lambda(1 - \lambda \, \Delta t)^{s/\Delta t}$. Since

$$\frac{1}{s + \Delta t}\left(1 - \frac{\Delta t}{s + \Delta t}\right)^{s/\Delta t} \le \frac{1}{s + \Delta t},$$

we see that we can choose M equal to the maximum of $|\beta| t \, e^{|\beta| t}$ and 1.

REMARK. If the proof above seems rather complicated, the reason is that the conditions are quite weak; for instance, the boundedness condition (b) only requires that V has a certain infinite bound. This makes it possible to apply the theorem to potentials which are much too singular to be described by standard functions—a fact which will be important in Section 6.4.

The following integrated version of the Feynman-Kac formula will be needed in Section 7.5.

5.3.12. COROLLARY. Assume that A and V are as in Theorem 5.3.11, and that there is a positive real number c such that $V(s) \geq c$ for all $s \in S_0$. Then

$$\left\| \int_{\Delta t}^{\infty} \frac{1}{t} \left(T^t u - E_{(\cdot)} \left(\int_0^t u(X(t)) \exp\left[-\int_0^t V(X(s))\, ds \right] \right) \right) dt \right\| \approx 0$$

for all u with finite norm.

There are two difficulties to overcome in order to prove this corollary: the behavior of the integrand at zero and at infinity. Since the integrand decays as e^{-ct}/t for large t, the behavior at infinity does not really create any problem. At the origin, we have a singularity of order $\ln(1/\Delta t)$, but since the error estimates in the proof of 5.3.11 allow an extra factor $\ln(1/\Delta t)$, the corollary follows. We leave the details to the reader.

REMARK. The systematic theory of Dirichlet forms was first studied by Beurling and Deny (1958, 1959), and the theory has since been developed by a number of authors; the reader should consult the books by Silverstein (1974, 1976) and Fukushima (1980) for further references. It is interesting to note that in their first papers [but see Deny (1970)] Beurling and Deny only considered finite Markov chains; in a certain sense the hyperfinite theory developed above is a return to the origins of the theory. The Feynman-Kac formula grew out of Kac's attempt to give a mathematical foundation for Feynman's ideas; for the probabilistic interpretation of the formula, see, e.g., Williams (1979).

5.4. STANDARD PARTS AND MARKOV PROCESSES

In this section we shall study the standard parts of hyperfinite Markov processes. In order to take standard parts we need a topology; we shall assume that with the exception of the trap s_0, the state space S is embedded in the nonstandard version $^*Y$ of some Hausdorff space Y. If X is a hyperfinite Markov process taking values in S, we want to find conditions that guarantee that the standard part of X exists and is a Y-valued Markov process. It turns out that there are two difficulties we shall have to overcome; the first is that the paths of X may be so irregular that no natural standard part process exists; the second is that even when a standard part does exist, there is no reason why it should automatically be a Markov process—taking standard parts we may lump together states that should be kept apart. We shall see that the theory of right standard parts that we developed in Chapter 4 is sufficient to solve the first of these problems. Thus most of the work in this section will be directed to the second problem under discussion. Before we delve into the technicalities, we shall discuss the problem informally in somewhat more detail.

Assume that $x : \Omega \times \mathbb{R}_+ \to Y$ is the standard part of X and that we want to prove that x is a Markov process with respect to the filtration it generates. Given that $x(t) = y$, there will in general be several states $s \in S_0$ such that $y = \text{st}(s)$, and X may be in any one of them. From the nonstandard point of view, these states are totally unrelated, and hence the past and the future of the process may differ widely from one state to the next. Observation of the past may indicate which states are the more likely, and thus influence our prediction of the future. This explains why in general x is not a Markov process.

Note, however, that if the process started at s_i and the process started at s_j have the "same" future whenever $s_i \approx s_j$, the above argument breaks down, and it is reasonable to expect that x is Markov. One way of formulating this condition is to demand that

(1) $$L(P_i)\{\omega \,|\, x(\omega, t) \in B\} = L(P_j)\{\omega \,|\, x(\omega, t) \in B\}$$

for all $t \in \mathbb{R}_+$ and all Borel sets B. But (1) as it stands turns out to be too strict for the applications we have in mind; instead of demanding that it holds for *all* infinitely close s_i, s_j, we shall only demand that it holds for all such s_i, s_j outside an *exceptional set* (i.e., a set which the process hits with probability zero). It may at first seem that little is achieved by allowing a condition to fail on an exceptional set, but in fact the extra freedom and flexibility we gain will turn out to be very useful. The situation is reminiscent of measure theory; by Anderson's Lusin theorem 3.4.9, a measurable function is one that fails to be S-continuous on a set of measure zero—a condition similar to the weaker version of (1).

This section falls naturally into two halves. In the first part we develop the necessary theory for exceptional sets. In the second we show that if (1) plus some additional conditions are satisfied, the standard part is a strong Markov process. In the next section we shall translate these probabilistic conditions into the language of Dirichlet forms, and find conditions on forms which guarantee that the standard parts of the associated processes are Markov.

A. Exceptional Sets

By a *hyperfinite Markov process* we shall in this section understand a stationary hyperfinite Markov chain as described in Section 5.3. However, the symmetry condition (5.3.8)—that $m_i q_{ij} = m_j q_{ji}$ for all $i, j \neq 0$—will not be needed. We shall instead assume that for each $i \in S_0$, the function

(2) $$t \mapsto P\{\omega \,|\, X(\omega, t) = s_i\}$$

is decreasing.

We shall further assume that $S_0 \subset {}^*Y$ for some topological space Y, but that the trap s_0 is not an element of ${}^*Y$. To keep the measure theoretical complications to a minimum, we introduce the following condition on Y:

5.4.1. DEFINITION. A Hausdorff space Y is *almost σ-compact* if for each hyperfinite probability measure P on a subset S_0 of ${}^*Y$, the set $\bar{S}_0 = S_0 \cap Ns({}^*Y)$ is Loeb measurable, and

$$L(P)(\bar{S}_0) = \sup\{L(P)(S_0 \cap \mathrm{st}^{-1}(K)) \,|\, K \text{ compact}\}.$$

Note that if Y is almost σ-compact, then any internal probability measure P on S_0 has a standard part $L(P) \circ \mathrm{st}^{-1}$ (by Theorem 3.4.6); this standard part is Radon, but need not be a probability measure. In the new terminology Proposition 3.4.7 can be recast as follows:

5.4.2. PROPOSITION. All locally compact spaces and all complete metric spaces are almost σ-compact.

Throughout this section we shall assume that Y is almost σ-compact. We have already defined

$$(3) \qquad\qquad \bar{S}_0 = S_0 \cap Ns({}^*Y),$$

and we now introduce

$$(4) \qquad\qquad \bar{\Omega} = \{\omega \in \Omega \,|\, X(\omega, 0) \in \bar{S}_0\}.$$

Assuming that $\bar{S}_0$ and $\bar{\Omega}$ are $L(m)$- and $L(P)$-measurable, respectively, we define new measures $\bar{m}$ and $\bar{P}$ by

$$(5) \qquad\qquad \bar{m}(A) = L(m)(A \cap \bar{S}_0),$$

$$(6) \qquad\qquad \bar{P}(B) = L(P)(B \cap \bar{\Omega}).$$

If $\delta \in T$, let T_δ be the subline

$$T_\delta = \{0, \delta, 2\delta, \ldots\}$$

and set

$$T_\delta^r = \{t \in T_\delta \,|\, t \le r\}, \qquad T_\delta^{\mathrm{fin}} = \{t \in T_\delta \,|\, t \text{ is finite}\}.$$

We shall write $X^{(\delta)}$ for the restriction $X \upharpoonright T_\delta$.

5.4.3. DEFINITION. A subset A of S_0 is called *δ-exceptional* if for all $\varepsilon \in \mathbb{R}_+$, there is an internal set $B \supset A$ such that

$$(7) \qquad\qquad \bar{P}\{\omega \,|\, \exists t \in T_\delta^1(X(\omega, t) \in B)\} < \varepsilon.$$

A set is *exceptional* if it is δ-exceptional for some infinitesimal δ.

Note that if A is δ-exceptional, then because of (2),

$$(8) \qquad\qquad \bar{P}\{\omega \,|\, \exists t \in T_\delta^{\mathrm{fin}}(X(\omega, t) \in A)\} = 0.$$

The larger δ is, the more sets are δ-exceptional. It is therefore convenient to be able to restrict to coarser time lines, and as long as we are only interested in the right standard part of X, we may do so without any loss of generality.

Our first lemma contains two extremely simple but also very useful observations. The proof is left to the reader.

5.4.4. LEMMA. (i)   All internal sets $B \subset S_0$ with $m(B) \approx 0$ are exceptional.

(ii)   The families of exceptional and δ-exceptional sets are closed under countable unions.

Why did we use (7) and not (8) to define exceptional sets? Simply because there is no reason to believe that the set in (8) is measurable when A is fairly complicated. Obviously, the set is measurable when A is internal, but although

$$(9) \quad \left\{ \omega \,\Big|\, \exists t \in T_\delta^r \Big(X(\omega, t) \in \bigcup_{n \in \mathbb{N}} A_n \Big) \right\} = \bigcup_{n \in \mathbb{N}} \{\omega \,|\, \exists t \in T_\delta^r (X(\omega, t) \in A_n)\}$$

for all sequences $\{A_n\}_{n \in \mathbb{N}}$ of subsets of S, the corresponding formula for intersections is false in general. It does hold, however, if the sequence is decreasing and consists of internal sets. This is the observation behind the next lemma.

5.4.5. LEMMA.   Let $A \subset S$, and assume that there is a family $\{B_{m,n}\}_{m,n \in \mathbb{N}}$ of internal sets such that

$$(10) \qquad\qquad A = \bigcup_{m \in \mathbb{N}} \bigcap_{n \in \mathbb{N}} B_{m,n}$$

and for each m, the sequence $\{B_{m,n}\}_{n \in \mathbb{N}}$ is decreasing. Then

$$(11) \quad \{\omega \,|\, \exists t \in T_\delta^r (X(\omega, t) \in A)\} = \bigcup_{m \in \mathbb{N}} \bigcap_{n \in \mathbb{N}} \{\omega \,|\, \exists t \in T_\delta^r (X(\omega, t) \in B_{m,n})\}.$$

PROOF.   It suffices to show that

$$(12) \left\{ \omega \,\Big|\, \exists t \in T_\delta^r \Big(X(\omega, t) \in \bigcap_{n \in \mathbb{N}} B_{m,n} \Big) \right\} = \bigcap_{n \in \mathbb{N}} \{\omega \,|\, \exists t \in T_\delta^r (X(\omega, t) \in B_{m,n})\}$$

for all m, since (11) then follows from (9). Also, it is immediately clear that the left-hand side of (12) is included in the right-hand side. To prove the opposite inclusion, choose

$$\omega_0 \in \bigcap_{n \in \mathbb{N}} \{\omega \,|\, \exists t \in T_\delta^r (X(\omega, t) \in B_{m,n})\},$$

and consider the set

$$\{n \in {}^*\mathbb{N} \,|\, \exists t \in T_\delta^r (X(\omega_0, t) \in \tilde{B}_{m,n})\},$$

where $\{\tilde{B}_{m,n}\}_{n\in {}^*\mathbb{N}}$ is some internal, decreasing extension of $\{B_{m,n}\}_{n\in\mathbb{N}}$. By choice of ω_0, this set contains $\mathbb{N}$, and since it is internal, it must have an infinite member η. Thus

$$\omega_0 \in \{\omega\,|\,\exists t \in T_\delta^r(X(\omega, t) \in \tilde{B}_{m,\eta})\} \subset \left\{\omega\,\Big|\,\exists t \in T_\delta^r\Big(X(\omega, t) \in \bigcap_{n\in\mathbb{N}} B_{m,n}\Big)\right\},$$

and the lemma is proved.

An exceptional set A is hit by a set of paths of $\bar{P}$-measure zero, but there may still be states $s_i \notin A$ such that a particle starting at s_i hits A with positive $L(P_i)$ probability.

5.4.6. DEFINITION. A δ-exceptional subset A of S_0 is called *properly δ-exceptional* if there is a family $\{B_{m,n}\}_{m,n\in\mathbb{N}}$ of internal sets such that

$$(13) \qquad\qquad A = \bigcup_{m\in\mathbb{N}} \bigcap_{n\in\mathbb{N}} B_{m,n},$$

and for all $s_i \notin A$

$$(14) \qquad\qquad L(P_i)\{\omega\,|\,\exists t \in T_\delta^{\text{fin}}(X(\omega, t) \in A)\} = 0.$$

A set is *properly exceptional* if it is properly δ-exceptional for some $\delta \approx 0$.

By the lemma above, the set in (14) is measurable. Also note that the classes of properly δ-exceptional and properly exceptional sets are closed under countable unions.

The properly exceptional sets are the "nice" exceptional sets in two ways; they are impenetrable from the outside, and they are constructed from the internal sets in a simple and uniform fashion. The next lemma shows that the class is also large enough.

5.4.7. LEMMA. If $A \subset S$ is δ-exceptional, there is a properly δ-exceptional set $B \supset A$.

PROOF. Since A is δ-exceptional, there is for each pair (m, n) of natural numbers an internal set $B_{m,n}$ containing A such that

$$\bar{P}\{\exists t \in T_\delta^{m+1}(X(\omega, t) \in B_{m,n})\} \le 1/n^2 m.$$

Define

$$C_{m,n} = \{i \in S\,|\,P_i\{\exists t \in T_\delta^m(X(t) \in B_{m,n})\} \ge 1/nm\},$$

and set

$$\bar{A} = \bigcup_{m\in\mathbb{N}} \bigcap_{n\in\mathbb{N}} C_{m,n}.$$

We first show that $\bar{A}$ is δ-exceptional. By definition of $C_{m,n}$

$$\bar{P}\{\omega \,|\, \exists t \in T_\delta^1(X(\omega, t) \in C_{m,n})\} \cdot 1/nm$$

$$\leq \bar{P}\{\omega \,|\, \exists t \in T_\delta^{m+1}(X(\omega, t) \in B_{m,n})\} \leq 1/n^2 m,$$

and thus

$$\bar{P}\{\omega \,|\, \exists t \in T_\delta^1(X(\omega, t) \in C_{m,n})\} \leq 1/n.$$

This shows that for each m the set $\bigcap_{n\in\mathbb{N}} C_{m,n}$ is δ-exceptional, and since the class of δ-exceptional sets is closed under countable unions, $\bar{A}$ is also δ-exceptional.

From the definition of the family $\{C_{m,n}\}$, we see that if $s_i \notin \bar{A}$, then

(15) $$L(P_i)\{\omega \,|\, \exists t \in T_\delta^{\text{fin}}(X(\omega, t) \in A)\} = 0.$$

We now iterate the construction above countably many times to get an increasing sequence

$$A_0 \subset A_1 \subset A_2 \subset \cdots$$

of δ-exceptional sets, where $A_0 = A$ and $A_{i+1} = \bar{A}_i$ for each i. Let

$$B = \bigcup_{i\in\mathbb{N}} A_i;$$

then B is a δ-exceptional set of the form

$$B = \bigcup_{i\in\mathbb{N}} \bigcup_{m\in\mathbb{N}} \bigcap_{n\in\mathbb{N}} C_{m,n}^{(i)}$$

for internal sets $C_{m,n}^{(i)}$, and it follows from (15) that if $s_i \notin B$, then

$$L(P_i)\{\omega \,|\, \exists t \in T_\delta^{\text{fin}}(X(\omega, t) \in B)\} = 0.$$

The lemma is proved.

Most of the exceptional sets we encounter in the theory of Markov processes are sets we would like the process to avoid. From the standard point of view, this means that a point y in Y should be avoided if all its nonstandard representations $s_i \in \text{st}^{-1}(y) \cap S_0$ are in the exceptional set. To study this relationship more closely, we define the *inner standard part* A° of a subset A of S by

(16) $$A^\circ = \{y \in Y \,|\, \text{st}^{-1}(y) \cap S_0 \subset A\}.$$

We have already defined the *standard part* of A by

(17) $$^\circ A = \text{st}(A) = \{y \in Y \,|\, \exists s \in A(\text{st}(s) = x)\}.$$

The inner standard part can also be defined in terms of the standard part operation:

(18) $$A^\circ = \complement(\text{st}(\complement A)),$$

where the outer complement is with respect to Y, the inner with respect to S_0.

It is trivial to check that standard parts commute with arbitrary unions. Using (18), we get that inner standard parts commute with arbitrary intersections. The other way around is less nice; standard parts and intersections do not commute, and neither do inner standard parts and unions. All we can say is the following:

5.4.8. LEMMA. If $\{A_n\}_{n \in \mathbb{N}}$ is a decreasing family of internal sets,

(19) $$\left(\bigcap_{n \in \mathbb{N}} A_n \right)^\circ = \bigcap_{n \in \mathbb{N}} {}^\circ A_n.$$

If $\{B_n\}_{n \in \mathbb{N}}$ is an increasing family of internal sets,

(20) $$\left(\bigcup_{n \in \mathbb{N}} B_n \right)^\circ = \bigcup_{n \in \mathbb{N}} B_n^\circ.$$

PROOF. Obviously, the left-hand side of (19) is contained in the set on the right. To prove the converse, let

$$x \in \bigcap_{n \in \mathbb{N}} {}^\circ A_n.$$

For each $n \in \mathbb{N}$, pick $y_n \in A_n$ such that $x = \text{st}(y_n)$. Extend $\{A_n\}_{n \in \mathbb{N}}$ to a decreasing internal family $\{A_n\}_{n \in {}^*\mathbb{N}}$ and $\{y_n\}_{n \in \mathbb{N}}$ to an internal sequence $\{y_n\}_{n \in {}^*\mathbb{N}}$ such that $y_n \in A_n$ for all $n \in {}^*\mathbb{N}$.

For each neighborhood O of x, consider the set

$$N_O = \{n \in {}^*\mathbb{N} \mid y_n \in {}^*O\}.$$

All these sets are internal and contain $\mathbb{N}$, and hence we can find an infinite integer η that is in all of them. But then

$$x = \text{st}(y_\eta),$$

and since the family $\{A_n\}_{n \in {}^*\mathbb{N}}$ is decreasing, $y_\eta \in A_\eta \subset \bigcap_{n \in \mathbb{N}} A_n$. This proves (19).

We now get (20) from (19) by using (18):

$$\left(\bigcup_{n \in \mathbb{N}} B_n \right)^\circ = \complement \, \text{st} \, \complement \bigcup B_n = \complement \, \text{st} \bigcap_{n \in \mathbb{N}} \complement B_n$$

$$= \complement \bigcap_{n \in \mathbb{N}} \text{st} \, \complement B_n = \bigcup_{n \in \mathbb{N}} \complement \, \text{st} \, \complement B_n = \bigcup_{n \in \mathbb{N}} B_n^\circ.$$

Since in general

$$\left(\bigcup_{m \in \mathbb{N}} \bigcap_{n \in \mathbb{N}} B_{m,n} \right)^{\circ} \neq \bigcup_{m \in \mathbb{N}} \bigcap_{n \in \mathbb{N}} B_{m,n}^{\circ},$$

there is no obvious reason to believe that the inner standard part of a properly exceptional set is always Borel, but we shall now prove that it must at least be universally measurable.

5.4.9. LEMMA. Assume $A = \bigcup_{m \in \mathbb{N}} \bigcap_{n \in \mathbb{N}} B_{m,n}$ for a family $\{B_{m,n}\}_{m,n \in \mathbb{N}}$ of internal sets. For any completed Borel probability measure μ on Y, the inner standard part A° is μ-measurable, and there exists a family $\{D_{m,n}\}_{m,n \in \mathbb{N}}$ of internal sets such that

$$(21) \qquad A \subset \bigcap_{n \in \mathbb{N}} \bigcup_{m \in \mathbb{N}} D_{m,n}$$

and

$$(22) \qquad \mu \left(\bigcap_{n \in \mathbb{N}} \bigcup_{m \in \mathbb{N}} D_{m,n}^{\circ} - A^{\circ} \right) = 0.$$

PROOF. We may assume that the family $\{B_{m,n}\}$ is increasing in m and decreasing in n. Note that

$$A = \bigcup_{m \in \mathbb{N}} \bigcap_{n \in \mathbb{N}} B_{m,n} = \bigcap_{f \in \mathbb{N}^{\mathbb{N}}} \bigcup_{m \in \mathbb{N}} B_{m,f(m)}.$$

Let $\bar{f}(m)$ be the sequence $\langle f(0), f(1), \ldots, f(m) \rangle$, and define

$$C_{\bar{f}(m)} = \bigcup_{k \leq m} B_{k,f(k)}.$$

For fixed f, the sequence $\{C_{\bar{f}(m)}\}_{m \in \mathbb{N}}$ is increasing, and hence by 5.4.8 and the fact that inner standard parts and arbitrary intersections commute,

$$(23) \qquad A^{\circ} = \bigcap_{f \in \mathbb{N}^{\mathbb{N}}} \bigcup_{m \in \mathbb{N}} C_{\bar{f}(m)}^{\circ}.$$

Since $\complement C_{\bar{f}(m)}^{\circ} = \operatorname{st} \complement C_{\bar{f}(m)}$ is closed, the complement

$$\complement A^{\circ} = \bigcup_{f \in \mathbb{N}^{\mathbb{N}}} \bigcap_{m \in \mathbb{N}} \complement C_{\bar{f}(m)}^{\circ}$$

of A can be derived from the closed sets by the Souslin operation, and hence A° is measurable with respect to any completed Borel measure [see Saks (1937), p. 50, for any easy proof].

It only remains to find the family $\{D_{m,n}\}$. From (23) it follows that for each $\varepsilon \in \mathbb{R}_{+}$, there is a function $f_{\varepsilon} : \mathbb{N} \to \mathbb{N}$ such that

$$\mu \left(\bigcap_{g \leq f_{\varepsilon}} \bigcup_{m \in \mathbb{N}} C_{\bar{g}(m)}^{\circ} - A^{\circ} \right) < \varepsilon,$$

where $g \leq f_\varepsilon$ means that $g(n) \leq f_\varepsilon(n)$ for all $n \in \mathbb{N}$. Since our original sequences $\{B_{m,n}\}_{n \in \mathbb{N}}$ are decreasing,

$$\bigcap_{g \leq f_\varepsilon} \bigcup_{m \in \mathbb{N}} C^\circ_{\bar{g}(m)} = \bigcup_{m \in \mathbb{N}} C^\circ_{\bar{f}_\varepsilon(m)}.$$

Putting $D_{m,n} = C_{\bar{f}_{1/n}(m)}$, the lemma follows.

REMARK. The argument above shows that a subset of a Hausdorff space can be derived from the closed sets by using the Souslin operation if and only if it is the standard part of a set derived from the internal sets by the same operation. This result—and the proof we have given—is due to Henson (1979).

We have now reached the last lemma we shall need before we can return to our Markov process. It will be used to pick hyperfinite representations of measures avoiding properly exceptional sets.

5.4.10. LEMMA. Let D be a subset of S_0 of the form $D = \bigcap_{n \in \mathbb{N}} \bigcup_{m \in \mathbb{N}} D_{m,n}$ for a family $\{D_{m,n}\}$ of internal sets. If μ is a Radon probability measure on Y with $\mu(D^\circ) = 0$, there exists an internal probability measure ν on S_0 such that $\mu = L(\nu) \circ \mathrm{st}^{-1}$ and $L(\nu)(D) = 0$.

PROOF. We may obviously assume that the family $\{D_{m,n}\}$ is increasing in m and decreasing in n. Defining new measures $\mu_{m,n}$ by

$$(24) \qquad \mu_{m,n}(B) = \mu(B - D^\circ_{m,n}),$$

we have

$$(25) \qquad \mu(B) = \sup_{n \in \mathbb{N}} \inf_{m \in \mathbb{N}} \mu_{m,n}(B).$$

The set $S_0 - D_{m,n}$ is S-dense in $Y - D^\circ_{m,n}$, and hence there is an internal measure $\nu_{m,n}$ concentrated on $S_0 - D_{m,n}$ such that

$$(26) \qquad \mu_{m,n} = L(\nu_{m,n}) \circ \mathrm{st}^{-1}.$$

We may choose these measures such that the family $\{\nu_{m,n}\}$ is decreasing in m and increasing in n. Extending to an internal sequence $\{\nu_{m,n}\}_{m,n \in {}^*\mathbb{N}}$, we first pick a $\gamma \in {}^*\mathbb{N} - \mathbb{N}$ such that

$$^\circ\nu_{\gamma,n}(S_0) = \lim_{m \to \infty} {}^\circ\nu_{m,n}(S_0)$$

for all $n \in \mathbb{N}$, and then an $\eta \in {}^*\mathbb{N} - \mathbb{N}$ such that

$$^\circ\nu_{\gamma,\eta}(S_0) = \lim_{n \to \infty} {}^\circ\nu_{\gamma,n}(S_0).$$

By using (25), (26), and the definitions of γ and η, we see that $L(\nu_{\gamma,\eta})(D) = 0$ and $\mu = L(\nu_{\gamma,\eta}) \circ \mathrm{st}^{-1}$. The measure $\nu_{\gamma,\eta}$ has all the properties of the desired measure, except that it need not be a probability measure.

However, it is clear that $\nu_{\gamma,\eta}(S_0) = 1 + \varepsilon$ for some $\varepsilon \approx 0$, and putting $\nu = (1 + \varepsilon)^{-1}\nu_{\gamma,\eta}$, the lemma is proved.

By combining 5.4.9 and 5.4.10 we get:

5.4.11. COROLLARY. Let $A \subset S_0$ be a properly exceptional set, and let μ be a completed Borel probability measure on Y such that $\mu(A^\circ) = 0$. Then there is an internal probability measure ν on S_0 such that $\mu = L(\nu) \circ \mathrm{st}^{-1}$ and $L(\nu)(A) = 0$.

B. Strong Markov Processes and Modified Standard Parts

Having completed our study of exceptional sets, we now turn to the real subject matter of this section, an investigation of the standard parts of hyperfinite Markov processes. Let us first describe what kind of processes we would like to obtain as standard parts.

Given a topological space Y, we let Y_Δ be the set $Y \cup \{\Delta\}$ obtained by adding a new element Δ to Y, and we let Y_Δ have the σ-algebra $\mathscr{B}_\Delta$ generated by the Borel sets on Y and the singleton $\{\Delta\}$. Our standard processes will be Y_Δ-valued, and the new element Δ will serve as a trap.

Recall that a filtration $\{\mathscr{F}_t\}_{t \in \mathbb{R}_+}$ of σ-algebras is *right-continuous* if $\mathscr{F}_t = \bigcap_{s>t} \mathscr{F}_s$ for all $t \in \mathbb{R}_+$, and that a mapping $\sigma : \Omega \to [0, \infty]$ is a *stopping time* with respect to $\{\mathscr{F}_t\}$ if

$$\{\omega \,|\, \sigma(\omega) \le t\} \in \mathscr{F}_t$$

for all t. Each stopping time σ introduces a σ-algebra $\mathscr{F}_\sigma$ by

$$(27) \qquad \mathscr{F}_\sigma = \{A \in \mathscr{F}_\infty \,|\, \forall t(A \cap \{\sigma \le t\} \in \mathscr{F}_t)\},$$

where $\mathscr{F}_\infty$ is the σ-algebra generated by the union of the $\mathscr{F}_t$'s.

Let $\mathscr{M}(Y)$ be the set of all Radon measures on Y with finite mass. A set or a function is called *universally measurable* if it is measurable with respect to all $\mu \in \mathscr{M}(Y)$. If for each $y \in Y$ we are given a probability measure $\bar{P}_y$ on a set Ω, then for each $\mu \in \mathscr{M}(Y)$, we let $\bar{P}_\mu$ be the measure defined by

$$(28) \qquad \bar{P}_\mu(A) = \int \bar{P}_y(A) \, d\mu(y)$$

[provided, of course, that this makes sense; i.e., $y \mapsto \bar{P}_y(A)$ is μ-measurable].

Finally, if x is a Y_Δ-valued process, we let the *lifetime* ζ of x be the stopping time defined by

$$\zeta(\omega) = \inf\{t \in \mathbb{R}_+ \,|\, x(\omega, t) = \Delta\}.$$

5.4.12. DEFINITION. A *strong Markov process* is a quadruple $\langle \Omega, \{\mathscr{F}_t\}_{t \in \mathbb{R}_+}, \{\bar{P}_y\}_{y \in Y_\Delta}, x \rangle$ where $\{\mathscr{F}_t\}$ is a right-continuous filtration on Ω, each $\bar{P}_y$ is a probability measure on $\mathscr{F}_\infty$, and $x : \Omega \times \mathbb{R}_+ \to Y_\Delta$ is a stochastic process on $\langle \Omega, \mathscr{F}_\infty, \bar{P}_y \rangle$ for each y. Moreover, the following conditions must be satisfied:

   (i)   For all $t \geq 0$ and all measurable $E \subset Y \cup \{\Delta\}$, the map $y \mapsto \bar{P}_y\{x_t \in E\}$ is universally measurable.
   (ii)   For all $t > \zeta(\omega)$, we have $x(\omega, t) = \Delta$.
   (iii)   For each $y \in Y_\Delta$, the process x is adapted to $(\Omega, \{\mathscr{F}_t\}, \bar{P}_y)$ and is right-continuous with left limits at all $t < \zeta(\omega) \, \bar{P}_y$ a.e.
   (iv)   For all $\{\mathscr{F}_t\}$ stopping times σ, all measurable $E \subset Y_\Delta$, all $\mu \in \mathscr{M}(Y)$, and all $s \in \mathbb{R}_+$:

$$\bar{P}_\mu\{x_{\sigma+s} \in E \,|\, \mathscr{F}_\sigma\} = \bar{P}_{x_\sigma}\{x_s \in E\}, \qquad \bar{P}_\mu \text{ a.e.}$$

In order to prove that a class of hyperfinite Markov chains have standard parts that are strong Markov processes, we shall have to overcome two main difficulties: the construction of the family of measures $\{\bar{P}_y\}$, and the proof of the strong Markov property 5.4.12(iv). But first we must introduce the necessary regularity conditions on our nonstandard processes.
   Recall that $\bar{S}_0 = S_0 \cap \mathrm{Ns}(^*Y)$ and that $X^{(\delta)} = X \restriction T_\delta$. The *lifetime* ζ_δ of $X^{(\delta)}$ is defined by

$$\zeta_\delta(\omega) = \inf\{^\circ t \,|\, X^{(\delta)}(\omega, t) \notin \bar{S}_0\},$$

and we define the *right standard part* $^\circ X^{(\delta)+}$ as follows. If $t < \zeta_\delta(\omega)$, let

$$^\circ X^{(\delta)+}(\omega, t) = S - \lim_{\substack{s \downarrow t \\ s \approx t}} X^{(\delta)}(\omega, s)$$

if this limit exists, and

$$^\circ X^{(\delta)+}(\omega, t) = \Delta$$

else. If $t \geq \zeta_\delta(\omega)$, we always put $^\circ X^{(\delta)+}(\omega, t) = \Delta$.

   5.4.13. DEFINITION. A subset A of S_0 is called a *set of irregularities* of X if there is a (positive) infinitesimal δ_0 such that for $\delta \approx 0$, $\delta \geq \delta_0$:

   (i)   For all $s_i \in \bar{S}_0 - A$ and $L(P_i)$ a.a. ω, the path $X^{(\delta)}(\omega, \cdot)$ has S-right and S-left limits at all $t < \zeta_\delta(\omega)$.
   (ii)   For all $s_i \in \bar{S}_0 - A$, the set

$$\{\omega \,|\, \exists t \in T_\delta^{\text{fin}}(^\circ t > \zeta_\delta(\omega) \wedge X(\omega, t) \in \bar{S}_0)\}$$

has $L(P_i)$ measure zero.
   (iii)   For all infinitely close $s_i, s_j \in \bar{S}_0 - A$,

$$L(P_i)\{^\circ X^{(\delta)+}(\omega, t) \in B\} = L(P_j)\{^\circ X^{(\delta)+}(\omega, t) \in B\}$$

for all finite $t \in T_\delta$ and all Borel sets B.

X has *exceptional irregularities* if it has an exceptional set of irregularities.

The first condition above guarantees that X has a reasonable standard part; the second says that "infinity" is a trap; the third is a version of (1). Putting $st(s_i) = \Delta$ when $s_i \in S - \bar{S}_0$, we have the following definition of a modified standard part.

5.4.14. DEFINITION. Assume that X has exceptional irregularities, and let A be a properly δ_0-exceptional set of irregularities (where δ_0 is as in 5.4.13). Let $x : \Omega \times \mathbb{R}_+ \to Y_\Delta$ be defined by:

(i) if $X(\omega, 0) \notin A$, then $x(\omega) = {}^\circ X^{(\delta)+}(\omega)$;
(ii) if $X(\omega, 0) \in A$, then $x(\omega, t) = st(X(\omega, 0))$ for all $t \in \mathbb{R}_+$.

Then x is called a *modified standard part* of X.

Our aim is to prove that if X has exceptional irregularities, then with the appropriate definition of the family $\{\bar{P}_y\}$ of measures, all modified standard parts of X are strong Markov processes. The first step toward the definition of $\{\bar{P}_y\}$ is the following lemma—a "smeared out" version of condition 5.4.13(iii).

If ν is an internal probability measure on S, let P_ν be the measure on Ω defined [in analogy with (27)] by

$$(28') \qquad P_\nu(C) = \int P_i(C)\, d\nu(s_i).$$

Recall that the space Y is assumed to be almost σ-compact (Definition 5.4.1).

5.4.15. LEMMA. Let A be a properly exceptional set of irregularities of X, and let x be the corresponding modified standard part of X. Let ν_1, ν_2 be two internal probability measures on S_0 such that $L(\nu_1)(A) = L(\nu_2)(A) = 0$ and $L(\nu_1) \circ st^{-1} = L(\nu_2) \circ st^{-1}$. Then for all $t \in \mathbb{R}_+$ and all Borel sets B

$$L(P_{\nu_1})\{x(\omega, t) \in B\} = L(P_{\nu_2})\{x(\omega, t) \in B\}.$$

PROOF. Let $\mu = L(\nu_1) \circ st^{-1} = L(\nu_2) \circ st^{-1}$. Choose $\tilde{t} \approx t$ so large that

$${}^\circ X(\omega, \tilde{t}) = x(\omega, t), \quad L(P_{\nu_1}), \quad \text{and} \quad L(P_{\nu_2}) \quad \text{a.e.}$$

and pick an internal set $\tilde{B}$ such that

$$L(P_{\nu_i})(\{X(\omega, \tilde{t}) \in \tilde{B}\} \triangle \{{}^\circ X(\omega, \tilde{t}) \in B\}) = 0$$

for $i = 1, 2$.

Define a function $f : Y \to \mathbb{R}$ as follows: if $y \notin A^\circ$, let $f(y) = L(P_i)\{x(\omega, t) \in B\}$ for some (i.e., all) $s_i \in st^{-1}(y) \cap S_0 - A$; if $y \in A^\circ$, define

$f(y)$ arbitrarily. The function $s_i \mapsto P_i\{X(\omega, \tilde{t}) \in \tilde{B}\}$ is a lifting of f with respect to both ν_1 and ν_2. Hence

$$L(P_{\nu_1})\{x(\omega, t) \in B\} = {}^\circ P_{\nu_1}\{X(\omega, \tilde{t}) \in \tilde{B}\}$$

$$= {}^\circ\! \int P_i\{X(\omega, \tilde{t}) \in \tilde{B}\} \, d\nu_1(s_i)$$

$$= \int f(y) \, d\mu(y)$$

$$= {}^\circ\! \int P_i\{X(\omega, \tilde{t}) \in \tilde{B}\} \, d\nu_2(s_i)$$

$$= {}^\circ P_{\nu_2}\{X(\omega, \tilde{t}) \in \tilde{B}\}$$

$$= L(P_{\nu_2})\{x(\omega, t) \in B\},$$

and the lemma is proved.

5.4.16. LEMMA. Assume that X has a properly exceptional set A of irregularities, and let x be the corresponding modified standard part. For all infinitely close $s_i, s_j \in \bar{S}_0 - A$; all finite sequences $t_1 < t_2 < t_3 < \cdots < t_n$ from $\mathbb{R}_+$; and all Borel sets $B_1, B_2, \ldots, B_n$ we have

$$L(P_i)\{x(\omega, t_1) \in B_1 \wedge x(\omega, t_2) \in B_2 \wedge \cdots \wedge x(\omega, t_n) \in B_n\}$$

$$= L(P_j)\{x(\omega, t_1) \in B_1 \wedge x(\omega, t_2) \in B_2 \wedge \cdots \wedge x(\omega, t_n) \in B_n\}.$$

PROOF. We shall prove this by induction on the length n of the sequences $t_1 < t_2 < \cdots < t_n, B_1, B_2, \ldots, B_n$. The case $n = 1$ is part of Definition 5.4.13. Assume that the lemma holds for all sequences of length $n - 1$, and pick $\tilde{t}_1, \tilde{t}_2, \ldots, \tilde{t}_n$ such that $\tilde{t}_1 \approx t_1, \ldots, \tilde{t}_n \approx t_n$ and

$$ {}^\circ X(\omega, \tilde{t}_l) = x(\omega, t_l) \qquad L(P_k) \qquad \text{a.e.}$$

for $l = 1, 2, \ldots, n$ and $k = i, j$. Choose internal sets $\tilde{B}_1, \tilde{B}_2, \ldots, \tilde{B}_n$ such that

$$L(P_k)(\{X(\omega, \tilde{t}_l) \in \tilde{B}_l\} \triangle \{{}^\circ X(\omega, \tilde{t}_l) \in B_l\})$$

for $l = 1, 2, \ldots, n$ and $k = i, j$.

We define two measures ν_i, ν_j on S by putting

$$\nu_k(s) = P_k\{\omega \,|\, X(\omega, \tilde{t}_{n-1}) = s \text{ and for all } l < n - 1, X(\omega, \tilde{t}_l) \in \tilde{B}_l\}$$

for all $s \in S$ and $k = i, j$. By the induction hypothesis

$$L(\nu_i) \circ \mathrm{st}^{-1} = L(\nu_j) \circ \mathrm{st}^{-1},$$

and since A is properly exceptional, $L(\nu_i)(A) = L(\nu_j)(A) = 0$. Applying Lemma 5.4.15 with $t = t_n - t_{n-1}$, we get

$$L(P_{\nu_i})\{x(t_n - t_{n-1}) \in B_n\} = L(P_{\nu_j})\{x(t_n - t_{n-1}) \in B_n\}.$$

Since X is Markov and time invariant, the lemma follows.

We now have what we need in order to define the missing ingredients $\{\mathscr{F}_t\}$ and $\{\bar{P}_y\}$ of our Markov process. For each $t \in \mathbb{R}_+$, let $\mathscr{F}_t^\circ$ be the σ-algebra generated by the sets

$$\{\omega \,|\, x(\omega, t_1) \in B_1 \wedge \cdots \wedge x(\omega, t_n) \in B_n\},$$

where $0 \le t_1 < t_2 < \cdots < t_n \le t$ and $B_1, \ldots, B_n$ are Borel sets in Y. Define

$$\mathscr{F}_t = \bigcap_{s>t} \mathscr{F}_t^\circ;$$

the filtration $\{\mathscr{F}_t\}_{t\in\mathbb{R}_+}$ is obviously right-continuous. We define $\mathscr{F}_\infty$ to be the σ-algebra generated by all $\mathscr{F}_t$, $t \in \mathbb{R}_+$.

It follows from Lemma 5.4.16 that for all $C \in \mathscr{F}_\infty$ and all infinitely close s_i, s_j in $\bar{S}_0 - A$

$$(29) \qquad\qquad L(P_i)(C) = L(P_j)(C).$$

This observation makes possible the following definition of a family $\{\bar{P}_y\}_{y \in Y_\Delta}$ of measures on $\mathscr{F}_\infty$.

If $y \notin A^\circ$, let for all $C \in \mathscr{F}_\infty$

$$(30) \qquad\qquad \bar{P}_y(C) = L(P_i)(C) \text{ for all } s_i \in \mathrm{st}^{-1}(y) - A.$$

If $y \in A^\circ \cup \{\Delta\}$, let

$$(31) \qquad \bar{P}_y(C) = \begin{cases} 1 & \text{if } C \text{ contains all constant paths } x(t) = y, \\ 0 & \text{else.} \end{cases}$$

Observe that since a set $C \in \mathscr{F}_\infty$ contains either all or none of the constant paths $x(\omega, t) = y$, the set function $\bar{P}_y$ is a measure.

We have reached our goal:

5.4.17. THEOREM. Assume that S_0 is a hyperfinite subset of $^*Y$ for some almost σ-compact space Y, and let $X: \Omega \times T \to S$ be a hyperfinite Markov process with exceptional irregularities. If x is a modified standard part of X, then $\langle \Omega, \{\mathscr{F}_t\}_{t\in\mathbb{R}_+}, \{\bar{P}_y\}_{y\in Y_\Delta}, x \rangle$ is a strong Markov process.

PROOF. Since we already know that $\{\mathscr{F}_t\}$ is right continuous, all we have to do is to check conditions 5.4.12(i)–(iv). The second and the third of these conditions are immediate from the construction, and we can concentrate on (i) and (iv).

Given a Radon probability measure μ on Y, we define two new measures μ_0 and μ_1 on Y by

(32) $$\mu_0(B) = \mu(B - A^\circ),$$

(33) $$\mu_1(B) = \mu(B \cap A^\circ),$$

where A is the properly exceptional set of irregularities used in the construction of x. By Corollary 5.4.11, we can find an internal measure ν on S_0 such that

(34) $$\mu_0 = L(\nu) \circ \mathrm{st}^{-1}$$

and

(35) $$L(\nu)(A) = 0.$$

We first prove 5.4.12(i):

(i)   Given $\alpha \in [0, 1]$, a Borel set E, and $t \in \mathbb{R}_+$, we must show that

$$\{y \in Y \,|\, \bar{P}_y\{x(t) \in E\} > \alpha\}$$

is μ-measurable for all finite Radon measures μ. Since

$$\{y \in A^\circ \,|\, \bar{P}_y\{x(t) \in E\} > \alpha\} = A^\circ \cap E$$

is universally measurable by 5.4.9, it suffices to prove that

(36) $$\{y \notin A^\circ \,|\, \bar{P}_y\{x(t) \in E\} > \alpha\}$$

is μ-measurable. In fact, by the definition of μ_0 we only have to prove that (36) is μ_0-measurable.

Since X has S-right limits a.e. with respect to the probability measure P_ν constructed from the ν in (34) by using (28'), we can find a $\tilde{t} \approx t$ such that

$$L(P_\nu)(\{x(t) \in E\} \,\triangle\, \{^\circ X(\tilde{t}) \in E\}) = 0.$$

There must be an internal set $\tilde{E}$ such that

$$L(P_\nu)\{X(\tilde{t}) \in (\mathrm{st}^{-1}(E) \,\triangle\, \tilde{E})\} = 0,$$

and hence

$$L(P_i)\{x(t) \in E\} = {}^\circ P_i\{X(\tilde{t}) \in \tilde{E}\}$$

for $L(\nu)$ a.a. s_i.

Combining this with the definition of $\bar{P}_y$, we see that $y \mapsto \bar{P}_y\{x(t) \in E\}$ has $i \mapsto P_i\{X(\tilde{t}) \in \tilde{E}\}$ as a ν-lifting, and hence it is a μ_0-measurable function. This proves that the set in (36) is μ_0-measurable, and 5.4.12(i) follows.

(iv) We must show that for all $\{\mathscr{F}_t\}$-stopping times σ, all sets $B \in \mathscr{F}_\sigma$, and all $s \in \mathbb{R}_+$, the equation

$$(37) \qquad \bar{P}_\mu\{\omega \in B \,|\, x_{\sigma+s} \in E\} = \int_B \bar{P}_{x_\sigma}\{x_s \in E\} \, d\bar{P}_\mu$$

holds for all Radon probability measure μ on Y and all Borel sets E.

First notice that since the paths of x are constant $\bar{P}_{\mu_1}$ a.e., we have

$$\bar{P}_{\mu_1}\{\omega \in B \,|\, x_{\sigma+s} \in E\} = \int_B \bar{P}_{x_\sigma}\{x_s \in E\} \, d\bar{P}_{\mu_1},$$

and it suffices to prove

$$(38) \qquad \bar{P}_{\mu_0}\{\omega \in B \,|\, x_{\sigma+s} \in E\} = \int_B \bar{P}_{x_\sigma}\{x_s \in E\} \, d\bar{P}_{\mu_0}.$$

If ν is the nonstandard representation of μ_0 given in (34)–(35), the hyperfinite counterpart of (38) is

$$(39) \qquad P_\nu\{\omega \in C \,|\, X_{\tau+s} \in F\} = \int_C P_{X_\tau}\{X_s \in F\} \, dP_\nu,$$

where C and F are internal sets and C is measurable in the *-algebra generated by the internal stopping time τ. Since X is a time-invariant Markov process, (39) holds. Our plan is to reduce (38) to a version of (39).

We first pick an internal stopping time τ such that $^\circ\tau = \sigma \, L(P_\nu)$ a.e., and such that there is a τ-measurable set C satisfying

$$(40) \qquad L(P_\nu)(B \bigtriangleup C) = 0.$$

If P^τ is the internal measure on Ω given by

$$P^\tau(D) = \int P_{X_\tau}(D) \, dP_\nu,$$

we observe that since A is properly exceptional,

$$(41) \qquad L(P^\tau)\{X_t \in A\} = 0$$

for all $t \in T_{\delta_0}^{\text{fin}}$ (where δ_0 is the infinitesimal used in the construction of x; see Definition 5.4.14). Hence we can choose an $\tilde{s} \in T_{\delta_0}^{\text{fin}}$, $\tilde{s} = s$, such that

$$(42) \qquad {}^\circ X_{\tilde{s}} = x_s \qquad L(P^\tau) \quad \text{a.e.}$$

$$(43) \qquad {}^\circ X_{\tau+\tilde{s}} = x_{\sigma+s} \qquad L(P_\nu) \quad \text{a.e.}$$

Finally, we pick an internal set F such that

$$(44) \qquad L(P_\nu)\{X_{\tau+\tilde{s}} \in \mathrm{st}^{-1}(E) \bigtriangleup F\} = 0,$$

$$(45) \qquad L(P^\tau)\{X_{\tilde{s}} \in \mathrm{st}^{-1}(E) \bigtriangleup F\} = 0.$$

We now have

$$(46) \quad \bar{P}_{\mu_0}\{\omega \in B \,|\, x_{\sigma+s} \in E\}$$

$$= \int \bar{P}_y\{\omega \in B \,|\, x_{\sigma+s} \in E\} \, d\mu_0(y) \qquad \text{(by def. of } \bar{P}_{\mu_0})$$

$$= \int L(P_i)\{\omega \in B \,|\, x_{\sigma+s} \in E\} \, dL(\nu)(s_i) \qquad \text{(by def. of } \bar{P}_y)$$

$$= L(P_\nu)\{\omega \in B \,|\, x_{\sigma+s} \in E\} \qquad \text{(by def. of } P_\nu)$$

$$= {}^\circ P_\nu\{\omega \in C \,|\, X_{\tau+\bar{s}} \in F\} \qquad \text{(by (40), (43), (44))}$$

$$= {}^\circ\!\!\int_C P_{X_\tau}\{X_{\bar{s}} \in F\} \, dP_\nu \qquad \text{(by (39))}$$

$$= \int_B L(P_{X_\tau})\{x_s \in E\} \, dL(P_\nu) \qquad \text{(by (40), (42), (45))}$$

$$= \int_B \bar{P}_{x_\sigma}\{x_s \in E\} \, d\bar{P}_{\mu_0} \qquad \text{(by def. of } \nu, \tau, \text{ and } \{\bar{P}_y\}).$$

This proves (38) and the theorem.

We illustrate the use of 5.4.17 by the following—rather trivial—example. More worthwhile applications will be given in Section 5.6.

5.4.18. EXAMPLE. Let $\eta \in {}^*\mathbb{N} - \mathbb{N}$, and set

$$S_0 = \left\{ \frac{k}{\sqrt{\eta}} \,\middle|\, k \in {}^*\mathbb{Z}, |k| \leq \eta \right\}.$$

Define the transition probabilities $q_{s,s'}$ by

$$q_{s,s'} = \begin{cases} \frac{1}{2} & \text{if } |s - s'| = 1/\sqrt{\eta}, \\ 0 & \text{otherwise,} \end{cases}$$

if $s, s' \in S_0$, and if s_0 is the trap, let

$$q_{s,s_0} = \begin{cases} \frac{1}{2} & \text{if } s = \pm\sqrt{\eta}, \\ 0 & \text{otherwise.} \end{cases}$$

Let m be the internal measure on S_0 given by

$$m(s) = 1/\sqrt{\eta}$$

for all $s \in S_0$, and let the time line be

$$T = \left\{ \frac{k}{\eta} \,\middle|\, k \in {}^*\mathbb{N} \right\}.$$

The process X is Anderson's random walk with a uniform initial distribution corresponding to the Lebesgue measure. To prove that the standard part of X is a strong Markov process, we show that the empty set is a set of irregularities of X. Since X is S-continuous $L(P_i)$ a.e. for all nearstandard $s_i \in S_0$, the two first conditions in 5.4.13 are obviously satisfied. If $s_i \approx s_j$, the paths starting at s_i look exactly like the paths starting at s_j except for an infinitesimal translation. Hence they induce the same standard paths, and 5.4.13(iii) follows.

We have focused on the standard notion of a strong Markov process, but there is nothing canonical about this choice; in fact, from the point of view of the standard theory of Markov processes as presented, e.g., in Blumenthal and Getoor (1968) and Fukushima (1980), it would have been more natural to introduce conditions on X guaranteeing that the modified standard part x is a Hunt process. However, to be a Hunt process a strong Markov process has to satisfy several rather technical requirements, and since we feel that this section already has its fair share of technicalities, we leave these further developments to the reader.

The material in this section is new, but the hyperfinite theory for exceptional sets developed here follows the standard theory (Blumenthal and Getoor, 1968; Fukushima, 1980) closely. Keisler (1984) approached the nonstandard theory of Markov processes from a slightly different point of view, characterizing different kinds of standard Markov processes in terms of what kinds of liftings they allow, and giving applications to solutions of stochastic differential equations. Keisler's processes can be time dependent, but the conditions corresponding to (5.4.1) are on the sample space Ω and not on the state space S_0.

5.5. REGULAR FORMS AND MARKOV PROCESSES

Combining the results of the two previous sections, we shall now obtain conditions on hyperfinite Dirichlet forms which guarantee that all modified standard parts of the associated Markov chains are strong Markov processes. The method we shall apply is simple; we just use the relationship between forms and processes established in 5.3.5–5.3.9 to translate the conditions of Theorem 5.4.17 into the language of Dirichlet forms.

We need slightly stronger assumptions in this section than in the last. First we reintroduce the symmetry condition $m_i q_{ij} = m_j q_{ji}$ ($i, j \neq 0$) of formula 5.3.8, which in Section 5.4 was temporarily replaced by the weaker assumption in formula 5.4.2. We are thus back in the setting of Section 5.3. Next we need stricter topological requirements. Recall that a Hausdorff space is *regular*—or T_3 if you like—if for all closed sets F and all $x \notin F$, there are disjoint open sets O_1, O_2 such that $x \in O_1$, $F \subset O_2$. A space is *second countable* when there is a countable base for the topology. In this section we shall assume that the state space Y is second countable, regular, and almost σ-compact. These conditions are satisfied by the spaces we are most interested in—locally compact spaces with countable bases, and complete, separable metric spaces. As in Section 5.4 we assume that the hyperfinite state space S_0 is a subset of $*Y$.

A. Separation of Compacts

One by one we shall reformulate the conditions of 5.4.13 in terms of Dirichlet forms. The following assumption will take care of 5.4.13(i).

5.5.1. DEFINITION. A Dirichlet form $\mathscr{E}$ *separates compacts* if there is a countable family π of open sets such that for all disjoint compacts K_1, K_2, there are sets O_1, $O_2 \in \pi$ and an internal function $u : S_0 \to *\mathbb{R}$ satisfying $K_1 \subset O_1$, $K_2 \subset O_2$, $u \upharpoonright S_0 \cap *O_1 = 1$, $u \upharpoonright S_0 \cap *O_2 = 0$, and

$$°\mathscr{E}_1(u, u) < \infty.$$

Note that since there are only countably many pairs O_1, O_2, we can choose all the functions above from a countable family. In fact, if $e_{*O_2^{'}}^{*O_1^{'}}$ is the function which has the smallest $\mathscr{E}_1$-value among those which are one on $*O_1$ and zero on $*O_2$, we can always choose u from the family

$$\{e_{*O_2^{'}}^{*O_1^{'}} \mid O_1, O_2 \in \pi \text{ are disjoint}\}.$$

Such a collection of u's is called a *separating family* for $\mathscr{E}$ and π. Note that by Proposition 5.3.5, each $e_{*O_2^{'}}^{*O_1^{'}}$ will only take values between zero and one.

Let $\zeta = \zeta_{\Delta t}$ be the lifetime of X, i.e.,

$$\zeta(\omega) = \inf\{°t \mid X(\omega, t) \in \bar{S}_0\},$$

where $\bar{S}_0 = S_0 \cap \mathrm{Ns}(*Y)$.

5.5.2. LEMMA. Assume that $\mathscr{E}$ separates compacts and that $\mathscr{F}$ is a separating family. If the path $X(\omega, \cdot)$ fails to have an S-left or S-right limit at some $t < \zeta(\omega)$, then so does $u(X(\omega, \cdot))$ for some $u \in \mathscr{F}$.

PROOF. Fix an $\omega \in \Omega$ and a $t \in \mathbb{R}_+$, $t < \zeta(\omega)$. Given a sequence $\{t_n\}_{n \in \mathbb{N}}$ from T such that the standard part $°t_n$ increases strictly to t, we shall first show that the sequence

$$\{°X(\omega, t_n)\}_{n \in \mathbb{N}}$$

has a cluster point.

Assume not; then for each $y \in Y$ there is a neighborhood O_y and an integer $n_y \in \mathbb{N}$ such that

$$°X(\omega, t_n) \notin O_y$$

when $n \geq n_y$. Since Y is regular, we can find a neighborhood G_y of y such that the closure $\bar{G}_y$ is contained in O_y, and hence

$$X(\omega, t_n) \notin {}^* G_y$$

when $n \geq n_y$. Extend $\{t_n\}_{n \in \mathbb{N}}$ to an internal sequence $\{t_n\}_{n \in {}^*\mathbb{N}}$ of elements of T less than t, and consider the set

$$A_y = \{n \in {}^*\mathbb{N} \,|\, n \leq n_y \text{ or } X(\omega, t_n) \notin {}^* G_y\}.$$

Since A_y is internal and contains $\mathbb{N}$, there is an $\eta_y \in {}^*\mathbb{N} - \mathbb{N}$ such that all $\eta \leq \eta_y$ are elements of A_y. By saturation there is an infinite η less than all η_y. But then

$$X(\omega, t_\eta) \notin {}^* G_y$$

for all y. This implies that $X(\omega, t_\eta)$ is not nearstandard, contradicting our assumption that $t < \zeta(\omega)$.

Let x be a cluster point of $\{{}^\circ X(\omega, t)\}$. If x is not the S-left limit of $X(\omega, \cdot)$ at t, there must be another sequence $\{s_n\}_{n \in \mathbb{N}}$ increasing to t such that x is not a cluster point for $\{{}^\circ X(\omega, s_n)\}$. Repeating the argument above, we see that $\{{}^\circ X(\omega, s_n)\}$ must have a cluster point y. Let $u \in \mathscr{F}$ be one on a neighborhood of x and zero on a neighborhood of y. Obviously, $u(X(\omega, \cdot))$ does not have an S-left limit at t. This proves the S-left limit case of the lemma; the S-right limit case is similar.

We shall now use 5.5.2 and Fukushima's decomposition theorem 5.3.9 to show that if $\mathscr{E}$ separates compacts, then the associated Markov process X satisfies 5.4.13(i). Recall that if $\delta \in T$, the subline T_δ is defined by

$$T_\delta = \{0, \delta, 2\delta, \ldots\};$$

$X^{(\delta)}$ is the restriction $X \restriction T_\delta$; and ζ_δ is the lifetime of $X^{(\delta)}$,

$$\zeta_\delta(\omega) = \inf\{{}^\circ t \,|\, X^{(\delta)}(\omega, t) \notin \bar{S}_0\}.$$

5.5.3. PROPOSITION. Let $\mathscr{E}$ be a normal Dirichlet form which separates compacts. There exist a $\delta_0 \approx 0$ and an exceptional set A_0 such that for all $s_i \in \bar{S}_0 - A_0$ and all infinitesimal $\delta \geq \delta_0$, the restricted process $X^{(\delta)}$ has S-left and S-right limits at all $t < \zeta_\delta$ $L(P_i)$-a.e.

PROOF. To find δ_0, we fix a separating family $\mathscr{F}$ for $\mathscr{E}$. Since for any $u \in \mathscr{F}$,

$$°\mathscr{E}(u, u) < \infty,$$

we can find a $\delta_u \approx 0$ such that $u \in \mathscr{D}[\mathscr{E}^{(\delta)}]$ for all infinitesimal $\delta \geq \delta_u$. By saturation there is a $\delta_0 \approx 0$ larger than all δ_u.

Turning to A_0, we first observe that by Lemma 5.5.2 and the countability of $\mathscr{F}$, it suffices to show that for each $u \in \mathscr{F}$, there is an exceptional set A such that for all $s_i \in S_0 - A$, the process $u(X^{(\delta)})$ has S-right and S-left limits $L(P_i)$ a.e. We can then take A_0 to be the union of these A's. Since $u(X^{(\delta_0)})$ takes values between zero and one, the only way it can fail to have one-sided S limits is by oscillating too wildly. We shall use stopping times to keep track of the oscillations.

Given two rationals p, q, $0 \le p < q \le 1$, we define a sequence $\{\tau^n_{(p,q)}\}$ of stopping times as follows:

$$\tau^0_{(p,q)} = \min\{t \in T_{\delta_0} \mid u(X^{(\delta_0)}(\omega, t)) \le p\}$$

$$\tau^{2n}_{(p,q)} = \min\{t \in T_{\delta_0} \mid t > \tau^{2n-1}_{(p,q)}(\omega) \wedge u(X^{(\delta_0)}(\omega, t)) \le p\}$$

$$\tau^{2n+1}_{(p,q)} = \min\{t \in T_{\delta_0} \mid t > \tau^{2n}_{(p,q)}(\omega) \wedge u(X^{(\delta_0)}(\omega, t)) \ge q\}.$$

Let $A \subset S_0$ be defined by

$$A = \bigcup_{(p,q)} \bigcup_{m \in \mathbb{N}} \bigcup_{k \in \mathbb{N}} \bigcap_{n \in \mathbb{N}} \{i \mid P_i\{\tau^n_{(p,q)} \le m\} \ge 1/k\}.$$

If $X^{(\delta_0)}$ fails to have S-right or S-left limits with positive $L(P_i)$ probability, then $s_i \in A$. We must show that A is exceptional.

By Lemma 5.4.5,

$$\{\omega \mid \exists t \in T^1_{\delta_0}(X(\omega, t) \in A)\}$$

$$= \bigcup_{(p,q)} \bigcup_{m \in \mathbb{N}} \bigcup_{k \in \mathbb{N}} \bigcap_{n \in \mathbb{N}} \left\{ \omega \mid \exists t \in T^1_{\delta_0}\left(P_{X^{(\delta_0)}(\omega,t)}\{\tau^n_{(p,q)}\} \le m\} \ge \frac{1}{k}\right)\right\}.$$

If A is not δ_0-exceptional, there must be a pair (p, q) of rationals, integers m, $k \in \mathbb{N}$, and an infinite number $\eta \in {}^*\mathbb{N}$ such that

$$L(P)\left\{ \omega \mid \exists t \in T^1_\delta\left(P_{X^{(\delta_0)}(\omega,t)}\{\tau^\eta_{(p,q)} \le m\} \ge \frac{1}{k}\right)\right\} > 0.$$

This implies that with positive $L(P)$ probability, $u(X^{(\delta_0)})$ jumps back and forth between p and q more than η times before time $m + 1$.

Since $u \in \mathscr{D}[\mathscr{E}^{(\delta_0)}]$, Fukushima's decomposition theorem 5.3.9 tells us that

$$u(X^{(\delta_0)}) = N + M,$$

where N is S-continuous $L(P)$ a.e., and M is a λ^2-martingale. If $u(X^{(\delta_0)})$ jumps η times between p and q before $t = m + 1$, there must be an infinitesimal interval where it jumps back and forth infinitely many times. Since N is S-continuous—and hence almost constant on infinitesimal intervals—most of this jumping is done by M. Hence the quadratic variation of M is infinite on a set of positive measure, contradicting the fact that it is a λ^2-martingale.

We can conclude that A must be δ_0-exceptional, and the proposition is thus proved.

B. Nearstandardly Concentrated Forms

Let A and B be two disjoint, internal subsets of S_0, and let $f : A \cup B \to {}^*\mathbb{R}$ be the function which is constant one on A and constant zero on B. We shall write

(1) $$e_A^B = e_1(f)$$

for the equilibrium potential defined in (5.3.32). The notation e_A for $e_A^\varnothing$ has already been introduced in Section 5.3. Recall that $|\cdot|_1$ denotes the norm generated by $\mathscr{E}_1$, i.e.,

$$|u|_1 = \mathscr{E}_1(u, u)^{1/2}.$$

The following condition on $\mathscr{E}$ implies 5.4.13(ii).

5.5.4. DEFINITION. A Dirichlet form $\mathscr{E}$ is *nearstandardly concentrated* if there exist a countable family $\mathscr{C}$ and an increasing sequence $\{B_n\}_{n\in\mathbb{N}}$ both consisting of internal subsets of S_0, such that:

(i)   The sets $\bigcup_{n\in\mathbb{N}} B_n - \bar{S}_0$ and $\bar{S}_0 - \bigcup\{C \,|\, C \in \mathscr{C}\}$ are both exceptional.
(ii) For all $C \in \mathscr{C}$, $\lim_{n\to\infty} \lim_{m\to\infty} {}^\circ|e_{C\cap B_n}^{\complement B_m} - e_C|_1 = 0$.

REMARK. The first condition of 5.5.4 more or less asserts that

$$\bigcup_{n\in\mathbb{N}} B_n \subset \bar{S}_0 \subset \bigcup\{C \,|\, C \in \mathscr{C}\},$$

but for technical reasons it is more convenient to use the slightly weaker version we have given. Condition 5.5.4(ii) is often difficult to verify, and later in this section we shall introduce modifications of it that are easier to handle.

5.5.5. LEMMA. Let X be the Markov process generated by a nearstandardly concentrated Dirichlet form $\mathscr{E}$. If $B = \bigcup_{n\in\mathbb{N}} B_n$, the set

$$D = \{s_i \in \complement B \,|\, \overline{L(P_i)}\{\exists t \in T^{\mathrm{fin}}(X(t) \in \bar{S}_0)\} > 0\}$$

is exceptional.

PROOF. We have used the outer measure $\overline{L(P_i)}$ in the definition of D since there is no obvious reason why

$$\{\exists t \in T^{\mathrm{fin}}(X(t) \in \bar{S}_0)\}$$

should be $L(P_i)$-measurable. However, since $\bar{S}_0 - \bigcup \{C \mid C \in \mathscr{C}\}$ is exceptional, it suffices to show that

$$D^C = \{s_i \in \complement B \mid L(P_i)\{\exists t \in T^{\mathrm{fin}}(X(t) \in C)\} > 0\}$$

is exceptional for each $C \in \mathscr{C}$.

Pick two increasing sequences $\{n_k\}_{k\in\mathbb{N}}$, $\{m_k\}_{k\in\mathbb{N}}$ such that

$$\delta_k \equiv \big| e^{\complement B_{m_k}}_{B_{n_k} \cap C} - e_C \big|_1 < 2^{-k},$$

and define

$$D_k^C = \{s_i \in \complement B_{m_k} \mid e_C(s_i) \geq \sqrt{\delta_k}\}.$$

Note that if $s_i \in D_k^C$, then

$$e_C(s_i) - e^{\complement B_{m_k}}_{C \cap B_{n_k}}(s_i) \geq \sqrt{\delta_k},$$

and hence by Proposition 5.3.6

$$\begin{aligned}
(2) \qquad P\{\omega \,|\, &\exists t \leq 1(X(\omega, t) \in D_k^C)\} \\
&\leq P\{\omega \,|\, \exists t \leq 1(|e_C - e^{\complement B_{m_k}}_{C \cap B_{n_k}}|(X(\omega, t)) \geq \sqrt{\delta_k})\} \\
&\leq \frac{2(1 + \Delta t)^{1/\Delta t}}{\delta_k} \mathscr{E}_1(e_C - e^{\complement B_{m_k}}_{C \cap B_{n_k}}, e_C - e^{\complement B_{m_k}}_{C \cap B_{n_k}}) \\
&= 2(1 + \Delta t)^{1/\Delta t} \delta_k \approx 2 \cdot e \cdot \delta_k.
\end{aligned}$$

For each $K \in \mathbb{N}$, we have $D^C \subset \bigcup_{k=K}^{\infty} D_k^C$. By (2) this implies that for all $K \in \mathbb{N}$

$$P\{\omega \,|\, \exists t \leq 1(X(\omega, t) \in D^C)\} \leq 2e \sum_{k=K}^{\infty} \delta_k \leq e2^{-K+2}$$

since $\delta_k < 2^{-k}$, and hence D^C is exceptional.

5.5.6. PROPOSITION. If $\mathscr{E}$ is nearstandardly concentrated, there is an exceptional set A_1 such that for all $s_i \in \bar{S}_0 - A_1$,

$$L(P_i)\{\omega \,|\, \exists t \in T^{\mathrm{fin}}(^\circ t > \zeta(\omega) \wedge X(\omega, t) \in \bar{S}_0)\} = 0.$$

PROOF. Let $\{B_n\}_{n\in\mathbb{N}}$ be as in 5.5.4, and set $B = \bigcup_{n\in\mathbb{N}} B_n$. We let A_1 be any properly exceptional set containing $B - S_0$ and the set D in 5.5.5.

Choose an internal, increasing sequence $\{B_n\}_{n\in{}^*\mathbb{N}}$ extending $\{B_n\}_{n\in\mathbb{N}}$, and define

$$\sigma_n(\omega) = \min\{t \in T \mid X(\omega, t) \notin B_n\}.$$

If

$$\sigma(\omega) = \inf\{^\circ t \mid X(\omega, t) \notin B\},$$

it is not hard to check that

$$\sigma(\omega) = \sup\{{}^\circ\sigma_n(\omega) \,|\, n \in \mathbb{N}\}.$$

Given an $s_i \in \bar{S}_0 - A_1$, we can find an $\eta \in {}^*\mathbb{N} - \mathbb{N}$ such that

(3) $$\sigma(\omega) = {}^\circ\sigma_\eta(\omega), \qquad L(P_i) \quad \text{a.e.}$$

It follows from Lemma 5.5.5 that $\bar{S}_0 - B \subset D \subset A_1$, and by definition of A_1, we have $B - \bar{S}_0 \subset A_1$. Since A_1 is properly exceptional and $s_i \notin A_1$, this implies that

(4) $$\sigma(\omega) = \zeta(\omega), \qquad L(P_i) \quad \text{a.e.}$$

Combining (3) and (4), we get

(5) $${}^\circ\sigma_\eta(\omega) = \zeta(\omega), \qquad L(P_i) \quad \text{a.e.}$$

Since $s_i \notin A_1$, and A_1 contains D and is properly exceptional,

$$L(P_i)(X(\omega, \sigma_\eta(\omega)) \in D) = 0,$$

and hence by definition of D

$$L(P_i)(\exists t \in T^{\text{fin}}(t > \sigma_\eta(\omega) \wedge X(\omega, t) \in \bar{S}_0)) = 0.$$

The proposition follows from (5).

C. Quasi-Continuous Extensions

We shall begin our study of condition 5.4.13(iii) by introducing the notion of quasi-continuity. An internal function $f: D_f \to {}^*\mathbb{R}$ defined on a subset of S_0 is called *quasi-continuous* if there is an exceptional set A such that for all infinitely close $s_i, s_j \in \bar{S}_0 \cap D_f \cap \complement A$, we have $f(s_i) \approx f(s_j)$.

REMARK. Recall that for an internal function $f: S_0 \to {}^*\mathbb{R}$ the following two definitions of S-continuity at a point $x \in Y$ agree:

   (i)   For all $y, z \in S_0$, $y \approx x$, $z \approx x$, we have $f(y) \approx f(z)$.
   (ii) For all $\varepsilon \in \mathbb{R}_+$, there is a neighborhood O of x such that $|f(y) - f(z)| < \varepsilon$ for all $y, z \in {}^*O$.

When as above we restrict the domain of continuity to an external set $\complement A$, this equivalence breaks down. Clearly (ii) still implies (i), but it is not hard to find examples which show that the converse is false.

If A is an internal subset of S_0 and $\delta \in T$, let

$$\sigma_A^\delta(\omega) = \min\{t \in T_\delta \,|\, X^{(\delta)}(\omega, t) \in A\}.$$

We shall write $e_\alpha^{(\delta)}(f)$ for the equilibrium potential of f with respect to the form $\mathscr{E}_\alpha^{(\delta)}$, i.e.,

$$e_\alpha^{(\delta)}(f)(s_i) = E_i((1 + \alpha\delta)^{-\sigma_{D_f}^\delta / \delta} f(X^{(\delta)}(\sigma_{D_f}^\delta))).$$

Recall that we abbreviate $e_\alpha^{(\delta)}(f)$ by $e_\alpha(f)$ when $\delta = \Delta t$. An internal function $f: D_f \to {}^*\mathbb{R}$ has *finite energy* if

$$(6) \qquad\qquad {}^\circ\mathscr{E}_1(e_1(f), e_1(f)) < \infty.$$

Note that if (6) is satisfied,

$$ {}^\circ\mathscr{E}_\alpha^{(\delta)}(e_\alpha^{(\delta)}(f), e_\alpha^{(\delta)}(f)) < \infty $$

for all finite, positive α and δ. An internal set $A \subset S_0$ has *finite energy* if

$$ {}^\circ\mathscr{E}_1(e_1(A), e_1(A)) < \infty. $$

5.5.7. DEFINITION. A Dirichlet form $\mathscr{E}$ *generates quasi-continuous extensions* if there is a $\delta_1 \approx 0$ such that for all infinitesimal $\delta \geq \delta_1$ and all quasi-continuous, S-bounded internal functions $f: D_f \to {}^*\mathbb{R}$, of finite energy $e_\alpha^{(\delta)}(f)$ is quasi-continuous for all non-negative real numbers α.

We observed in Section 5.3 that since

$$ e_\alpha(A)(s_i) = E_i((1 + \alpha\,\Delta t)^{-\sigma_A/\Delta t}), $$

the function

$$ \alpha \mapsto {}^\circ e_\alpha(A)(s_i) $$

is the Laplace transform of the measure λ_i defined on $\mathbb{R}$ by

$$ \lambda_i(C) = L(P_i)\{\omega \mid {}^\circ\sigma_A(\omega) \in C\}. $$

If the functions $s_i \mapsto e_\alpha(A)(s_i)$ are quasi-continuous, then for all infinitely close s_i, s_j outside an exceptional set, the particles starting at s_i hit A with the same time distribution as the ones starting at s_j. Hence quasi-continuity of equilibrium potentials implies a certain uniformity in the future behavior of the process. To prove condition 5.4.13(iii), we must extend this observation to also take into account what happens after the first time X hits A.

Given a finite sequence $A = \langle A_1, A_2, \ldots, A_n \rangle$ of internal sets and an integer $j \leq n$, let

$$ A_{\leq j} = \langle A_1, A_2, \ldots, A_j \rangle $$

and

$$ A_{\geq j} = \langle A_j, A_{j+1}, \ldots, A_n \rangle. $$

We define an internal stopping time σ_A by induction on the length n of A as follows: if $A = \langle A_1 \rangle$ has length one, σ_A is just the first hitting time σ_{A_1}, i.e.,

$$ \sigma_A(\omega) = \inf\{t \in T \mid X(\omega, t) \in A\}. $$

If σ_B has been defined for all sequences B of length less than n, let

$$ \sigma_A(\omega) = \inf\{t \in T \mid t \geq \sigma_{A_{\leq n-1}}(\omega) \wedge X(\omega, t) \in A_n\}. $$

When we apply this definition to a restriction $X^{(\delta)}$ of X, we write σ_A^δ for the resulting stopping time.

For each finite sequence A of internal sets, we define "generalized equilibrium potentials" $e_\alpha^{(\delta)}(A)$ by

(7) $$e_\alpha^{(\delta)}(A)(s_i) = E_i((1 + \alpha\delta)^{-\sigma_A^\delta/\delta}).$$

5.5.8. LEMMA. For all finite sequences A of internal subsets of S_0, all positive α and δ, and all j less than the length of A, we have

(8) $$e_\alpha^{(\delta)}(A_{\geq j}) = e_\alpha^{(\delta)}(e_\alpha^{(\delta)}(A_{\geq j+1}) \restriction A_j),$$

i.e., $e_\alpha^{(\delta)}(A_{\geq j})$ is the equilibrium potential with respect to $\mathscr{E}_\alpha^{(\delta)}$ of the restriction of $e_\alpha^{(\delta)}(A_{\geq j+1})$ to A_j. In particular, the function

(9) $$j \mapsto \mathscr{E}_\alpha^{(\delta)}(e_\alpha^{(\delta)}(A_{\geq j}), e_\alpha^{(\delta)}(A_{\geq j}))$$

is increasing.

PROOF. That the function in (9) is increasing follows from (8) and 5.3.5. To prove (8), just observe that since X is Markov

$$
\begin{aligned}
e_\alpha^{(\delta)}(A_{\geq j})(s_i) &= E_i[(1 + \alpha\delta)^{-\sigma_{A_{\geq j}}^\delta/\delta}] \\
&= E_i[(1 + \alpha\delta)^{-\sigma_{A_j}^\delta/\delta}(1 + \alpha\delta)^{-(\sigma_{A_{\geq j}}^\delta - \sigma_{A_j}^\delta)/\delta}] \\
&= E_i[(1 + \alpha\delta)^{-\sigma_{A_j}^\delta/\delta} e_\alpha^{(\delta)}(A_{\geq j+1})(X(\sigma_{A_j}^\delta))] \\
&= e_\alpha^{(\delta)}(e_\alpha^{(\delta)}(A_{\geq j+1}) \restriction A_j)(s_i).
\end{aligned}
$$

The lemma is proved.

5.5.9. COROLLARY. Assume that $\mathscr{E}$ generates quasi-continuous extensions, and that $A = \langle A_1, \ldots, A_n \rangle$ is a finite sequence of internal sets where A_n has finite energy. Then $e_\alpha^{(\delta)}(A)$ is quasi-continuous and has finite energy for all finite, positive α and all δ.

PROOF. Use induction on k and the lemma to prove the statement for all $A_{\geq n-k}$, $0 \leq k < n$.

We are now ready for 5.4.13(iii).

5.5.10. PROPOSITION. Let $\mathscr{E}$ be a nearstandardly concentrated, normal Dirichlet form which separates compacts and allows quasi-continuous extensions. There is a properly exceptional set A of irregularities of X such that if x is the modified standard part with respect to A, then for all infinitely close $s_i, s_j \in \bar{S}_0 - A$, all $t \in \mathbb{R}_+$, and all Borel sets $B \subset Y$

$$L(P_i)\{\omega \,|\, x(\omega, t) \in B\} = L(P_j)\{\omega \,|\, x(\omega, t) \in B\}.$$

PROOF. We shall first construct the exceptional set A. Let A_0 and A_1 be as in 5.5.3 and 5.5.6, respectively. Let π be the countable family in Definition 5.5.1, and observe that all $O \in \pi$ have finite energy. Let Σ be the set of all finite, alternating sequences

$$\mathscr{S} = \langle {}^*O_1, {}^*O_2, {}^*O_1, {}^*O_2, \ldots, {}^*O_1, {}^*O_2 \rangle$$

where $O_1, O_2 \in \pi$. Fix an infinitesimal δ larger than the δ_0 of 5.5.3 and the δ_1 of 5.5.7, and let

$$\mathscr{F} = \{ e_q^{(\delta)}(\mathscr{S}) \,|\, q \in \mathbb{Q}_+, \mathscr{S} \in \Sigma \}.$$

Since $\mathscr{F}$ is countable, Corollary 5.5.9 tells us that there is an exceptional set A_2 such that all functions in $\mathscr{F}$ are continuous off A_2. Let A be a properly exceptional set containing $A_0 \cup A_1 \cup A_2$.

The functions $\alpha \mapsto e_\alpha^{(\delta)}(\mathscr{S})$ are obviously S-continuous, and thus

(10) $$^{\circ}E_i((1 + \alpha\delta)^{-\sigma_\mathscr{S}^\delta/\delta}) = {}^{\circ}E_j((1 + \alpha\delta)^{-\sigma_\mathscr{S}^\delta/\delta})$$

for all infinitely close $s_i, s_j \in \bar{S}_0 - A$. Since

$$^{\circ}E_i((1 + \alpha\delta)^{-\sigma_\mathscr{S}^\delta/\delta}) = \int e^{-\alpha {}^{\circ}\sigma_\mathscr{S}^\delta} \, dL(P_i)$$

for all $\alpha \in \mathbb{R}_+$, the function

$$\alpha \mapsto {}^{\circ}E_i((1 + \alpha\delta)^{-\sigma_\mathscr{S}^\delta/\delta})$$

is the Laplace transform of the distribution of $^{\circ}\sigma_\mathscr{S}^\delta$ with respect to $L(P_i)$. Hence it follows from (10) that for all $s_i, s_j \in \bar{S}_0 - A$, $s_i \approx s_j$,

(11) $$L(P_i)\{\omega \,|\, {}^{\circ}\sigma_{\mathscr{S}_{\leq m}}^\delta(\omega) \leq t < {}^{\circ}\sigma_{\mathscr{S}_{\leq m+1}}^\delta\}$$
$$= L(P_j)\{\omega \,|\, {}^{\circ}\sigma_{\mathscr{S}_{\leq m}}^\delta(\omega) \leq t < {}^{\circ}\sigma_{\mathscr{S}_{\leq m+1}}^\delta\}$$

for all $t \in \mathbb{R}_+$, and all $\mathscr{S} \in \Sigma$, $m < $ length $\mathscr{S}$.

Assume that the proposition is false; then we can find $s_i, s_j \in \bar{S}_0 - A$ and a Borel set B such that

(12) $$L(P_i)\{\omega \,|\, x(\omega, t) \in B\} - L(P_j)\{\omega \,|\, x(\omega, t) \in B\} = \varepsilon > 0$$

for some $t \in \mathbb{R}_+$. Pick $\tilde{t} \approx t$ so large that $^{\circ}X(\omega, \tilde{t}) = x(\omega, t)$ $L(P_i)$- and $L(P_j)$- a.e.; this is possible since $A_0, A_1 \subset A$. The measures μ_i and μ_j defined by

$$\mu_i(C) = L(P_i)\{\omega \,|\, {}^{\circ}X(\omega, \tilde{t}) \in C\}, \qquad \mu_j(C) = L(P_j)\{\omega \,|\, {}^{\circ}X(\omega, \tilde{t}) \in C\}$$

are Radon measures on Y, and hence there is a compact set $K \subset B$ such that

(13) $$\mu_i(K) > \mu_j(K) + 3\varepsilon/4.$$

We can also find a compact set $K' \subset Y - K$ such that

$$\mu_i((Y - K) - K') < \varepsilon/4, \qquad \mu_j((Y - K) - K') < \varepsilon/4.$$

Since $\mathscr{E}$ separates compacts, we can find two disjoint open sets $O, O' \in \pi$ such that $K \subset O$, $K' \subset O'$. Let m be so large that the $L(P_i)$ and $L(P_j)$ probabilities that a path should jump back and forth between $*O$ and $*O'$ more than m times before time $t + 1$ are both less than $\varepsilon/4$. Let

$$\mathscr{S} = \langle *O, *O', *O, *O', \ldots, *O, *O' \rangle$$

have length $2m$. Then

$$L(P_j)\{\omega \,|\, x(\omega, t) \in K\} > L(P_j) \bigcup_{n < m} \{\omega \,|\, {}^{\circ}\sigma^{\delta}_{\mathscr{S}_{\leq 2n+1}}(\omega) \leq t < {}^{\circ}\sigma^{\delta}_{\mathscr{S}_{\leq 2n+2}}(\omega)\} - \frac{2\varepsilon}{4}$$

$$= L(P_i) \bigcup_{n < m} \{\omega \,|\, {}^{\circ}\sigma^{\delta}_{\mathscr{S}_{\leq 2n+1}}(\omega) \leq t < {}^{\circ}\sigma^{\delta}_{\mathscr{S}_{\leq 2n+2}}(\omega)\} - \frac{2\varepsilon}{4}$$

$$> L(P_i)\{\omega \,|\, x(\omega, t) \in K\} - \frac{3\varepsilon}{4},$$

contradicting (13). The proposition is proved.

REMARK. Note that we do not really have to know that $\mathscr{E}$ generates quasi-continuous extensions to carry through the proof of 5.5.10; all that is needed is that the functions $e_q^{(\delta)}(\mathscr{S})$ are quasi-continuous.

We can now combine the results of this section with Theorem 5.4.17.

5.5.11. THEOREM. Assume that S_0 is a hyperfinite subset of $*Y$ for some second countable, regular, almost σ-compact space Y. Let $\mathscr{E}$ be a nearstandardly concentrated, normal Dirichlet form on S_0 that separates compacts and generates quasi-continuous extensions. Then the associated hyperfinite Markov process has a modified standard part which is a strong Markov process.

PROOF. By 5.5.3, 5.5.6, and 5.5.10, the hyperfinite Markov process has exceptional irregularities, and thus the result follows from 5.4.17.

D. Regular Forms

The conditions in 5.5.11 are not in any sense canonical; there are other possibilities which are just as reasonable. So far we have tended to choose assumptions that would simplify the theory rather than those most suitable for applications; we shall now reformulate our hypotheses so that they will be easier to verify.

First, a small observation whose proof we leave to the reader. We say that $\mathscr{E}$ *separates points* if Definition 5.5.1 only applies to singletons $K_1 = \{x\}$, $K_2 = \{y\}$. It turns out that if $\mathscr{E}$ separates points it also separates compacts. Hence it suffices to prove that our forms separate points.

As we have already mentioned, it is not always easy to verify that a form is nearstandardly concentrated; the problem is to prove that

$$(14) \qquad \lim_{n \to \infty} \lim_{m \to \infty} {}^{\circ}|e^{CB_m}_{C \cap B_n} - e_C|_1 = 0.$$

Recall that this condition is needed in order to prove that the process does not return from "infinity." We shall indicate how these difficulties can be circumvented by turning more of the non-nearstandard points into traps.

Let $\tilde{S}_0$ be an internal subset of S_0, and let τ be the first hitting time,

$$\tau(\omega) = \min\{t \in T \,|\, X(\omega, t) \in \tilde{S}_0\}.$$

We define a process $\tilde{X}$ taking values in $\tilde{S} = \tilde{S}_0 \cup \{s_0\}$ by

$$(15) \qquad \tilde{X}(\omega, t) = \begin{cases} X(\omega, t) & \text{when} \quad t < \tau(\omega), \\ s_0 & \text{when} \quad t \geq \tau(\omega). \end{cases}$$

According to the Beurling–Deny formula, 5.3.1, the associated Dirichlet form $\tilde{\mathscr{E}}$ is given by

$$\tilde{\mathscr{E}}(u, u) = \frac{1}{\Delta t} \left[\sum_{\substack{i,j \in \tilde{S}_0 \\ i < j}} (u(i) - u(j))^2 q_{ij} m_i + \sum_{i \in \tilde{S}_0} u(i)^2 m_i \sum_{j \in S - \tilde{S}_0} q_{ij} \right].$$

Let $\langle \cdot, \cdot \rangle^{\sim}$ be the L^2-inner product on $(\tilde{S}_0, m)$;

$$\langle u, v \rangle^{\sim} = \sum_{i \in \tilde{S}_0} u(i) v(i) m(i),$$

and let

$$\tilde{\mathscr{E}}_\alpha(u, v) = \tilde{\mathscr{E}}(u, v) + \alpha \langle u, v \rangle^{\sim}.$$

If $f : D_f \to {}^*\mathbb{R}$ is an internal function defined on a subset of $\tilde{S}_0$, let $\tilde{f} : D_f \cup (S_0 - \tilde{S}_0) \to {}^*\mathbb{R}$ be the internal function defined by

$$\tilde{f}(s_i) = \begin{cases} f(s_i) & \text{if} \quad s_i \in D_f, \\ 0 & \text{if} \quad s_i \in S_0 - \tilde{S}_0. \end{cases}$$

Observe that if $\tilde{e}_\alpha(f)$ denotes the equilibrium potential of f with respect to $\tilde{\mathscr{E}}_\alpha$,

$$(16) \qquad \widetilde{e_\alpha(f)} = e_\alpha(\tilde{f}).$$

For all internal $u : \tilde{S}_0 \to {}^*\mathbb{R}$, we have

$$(17) \qquad \tilde{\mathscr{E}}_\alpha(u, u) = \mathscr{E}_\alpha(\tilde{u}, \tilde{u}).$$

The next lemma is a slight variation of the definition of a nearstandardly concentrated form.

5.5.12. LEMMA. Let $\mathscr{E}$ be a hyperfinite Dirichlet form, and assume that there exist a countable family $\mathscr{C}$ and an increasing sequence $\{B_n\}_{n \in \mathbb{N}}$, both consisting of internal subsets of S_0, such that the sets $\bigcup_{n \in \mathbb{N}} B_n - \bar{S}_0$ and $\bar{S}_0 - \bigcup\{C \colon C \in \mathscr{C}\}$ are exceptional, and for all $C \in \mathscr{C}$

$$(18) \qquad \lim_{n \to \infty} |e_{C \cap B_n}|_1 = |e_C|_1 < \infty,$$

$$(19) \qquad \lim_{m \to \infty} |e_{C \cap B_n}^{\complement B_m}|_1 = |e_{C \cap B_n}|_1 \qquad \text{for all} \quad n \in \mathbb{N}.$$

Then $\mathscr{E}$ is nearstandardly concentrated.

PROOF. Recall that if $w \colon S_0 \to {}^*\mathbb{R}$ agrees with f on D_f, then by (5.3.30)

$$\mathscr{E}_1(e_1(f) - w, e_1(f) - w) = \mathscr{E}_1(w, w) - \mathscr{E}_1(e_1(f), e_1(f)).$$

Applying this twice, we have

$$|e_{C \cap B_n}^{\complement B_m} - e_C|_1 \le |e_{C \cap B_n}^{\complement B_m} - e_{C \cap B_n}|_1 + |e_{C \cap B_n} - e_C|_1$$

$$= (|e_{C \cap B_n}^{\complement B_m}|_1^2 - |e_{C \cap B_n}|_1^2)^{1/2} + (|e_C|_1^2 - |e_{C \cap B_n}|_1^2)^{1/2},$$

and hence 5.5.4(ii) follows from (18) and (19).

The sequences $\{|e_{C \cap B_n}|_1\}_{n \in \mathbb{N}}$, $\{|e_{C \cap B_n}^{\complement B_m}|_1\}_{m \in \mathbb{N}}$ are monotone, and thus the limits

$$\lim_{n \to \infty} {}^\circ|e_{C \cap B_n}|_1, \qquad \lim_{m \to \infty} {}^\circ|e_{C \cap B_n}^{\complement B_m}|_1$$

always exist. The problem is that they may be infinite, or be finite but fail to agree with (18) and (19). Granted the finiteness, we shall show how the second problem can be solved by restricting the form to a suitable subset $\tilde{S}_0$ of S_0 as above.

5.5.13. DEFINITION. A Dirichlet form $\mathscr{E}$ is called *locally finite* if there is a countable family $\{C \mid C \in \mathscr{C}\}$ and an increasing sequence $\{B_n\}_{n \in \mathbb{N}}$ both consisting of internal subsets of S_0, such that

(i) the sets $\bar{S}_0 \triangle \bigcup_{n \in \mathbb{N}} B_n$ and $\bar{S}_0 - \bigcup\{C \mid C \in \mathscr{C}\}$ are exceptional;
(ii) for each $C \in \mathscr{C}$, we have $\lim_{n \to \infty} \lim_{m \to \infty} {}^\circ|e_{C \cap B_n}^{\complement B_m}|_1 < \infty$.

5.5.14. LEMMA. If $\mathscr{E}$ is a locally finite Dirichlet form, there is an internal subset $\tilde{S}_0$ of S_0 such that $\bar{S}_0 - \tilde{S}_0$ is exceptional, and $\tilde{\mathscr{E}}$ is nearstandardly concentrated.

PROOF. Extend $\{B_n\}_{n\in\mathbb{N}}$ to an increasing, internal sequence $\{B_n\}_{n\in{}^*\mathbb{N}}$. There is an infinite $\mu \in {}^*\mathbb{N}$ such that for all $n \in \mathbb{N}$,

$$(20) \qquad {}^\circ|e^{\mathsf{C}B_\mu}_{C\cap B_n}|_1 = \lim_{m\to\infty} {}^\circ|e^{\mathsf{C}B_m}_{C\cap B_n}|_1.$$

Moreover, there is a $\nu \in {}^*\mathbb{N} - \mathbb{N}$ such that

$$(21) \qquad {}^\circ|e^{\mathsf{C}B_\mu}_{C\cap B_\nu}|_1 = \lim_{n\to\infty} {}^\circ|e^{\mathsf{C}B_\mu}_{C\cap B_n}|_1 = \lim_{n\to\infty}\lim_{m\to\infty} {}^\circ|e^{\mathsf{C}B_m}_{C\cap B_n}|_1.$$

Let $\tilde{S}_0 = B_\mu$; we shall show that $\tilde{\mathscr{E}}$ is nearstandardly concentrated with respect to the families $\{B_n\}_{n\in\mathbb{N}}$ and $\mathscr{C}' = \{C \cap B_\nu \,|\, C \in \mathscr{C}\}$. Note that since $\bar{S}_0 \,\triangle\, \bigcup_{n\in\mathbb{N}} B_n$ is exceptional, it suffices to prove the equivalents of (18) and (19). Let $|\cdot|_1^{\tilde{}}$ be the norm generated by $\tilde{\mathscr{E}}_1$. Using (16), (17), (20), and (21), we get for all $C \cap B_\nu \in \mathscr{C}'$:

$$\lim_{n\to\infty} {}^\circ|\tilde{e}_{(C\cap B_\nu)\cap B_n}|_1^{\tilde{}} = \lim_{n\to\infty} {}^\circ|e^{\mathsf{C}B_\mu}_{C\cap B_n}|_1 = {}^\circ|e^{\mathsf{C}B_\mu}_{C\cap B_\nu}|_1 = {}^\circ|\tilde{e}_{C\cap B_\nu}|_1^{\tilde{}} < \infty$$

and

$$\lim_{m\to\infty} {}^\circ|\tilde{e}^{\mathsf{C}B_m}_{(C\cap B_\nu)\cap B_n}|_1^{\tilde{}} = \lim_{m\to\infty} {}^\circ|e^{\mathsf{C}B_m}_{C\cap B_n}|_1 = {}^\circ|e^{\mathsf{C}B_\mu}_{C\cap B_n}|_1 = {}^\circ|\tilde{e}_{(C\cap B_\nu)\cap B_n}|_1^{\tilde{}},$$

which correspond to (18) and (19), respectively.

Since $\bar{S}_0 - \tilde{S}_0$ is exceptional, a modified standard part of the process $\tilde{X}$ generated by the form $\tilde{\mathscr{E}}$ in 5.5.14 is a perfectly adequate version of the standard part of X.

There are still a few difficulties in connection with Definition 5.5.13. Although it should not be too hard to check that

$$\lim_{n\to\infty}\lim_{m\to\infty} {}^\circ|e^{\mathsf{C}B_m}_{C\cap B_n}|_1 < \infty$$

once the families $\{B_n\}$ and $\mathscr{C}$ are chosen, it may be more problematic to prove that the sets

$$\bar{S}_0 \,\triangle\, \bigcup_{n\in\mathbb{N}} B_n \quad \text{and} \quad \bar{S}_0 - \bigcup\{C \,|\, C \in \mathscr{C}\}$$

are exceptional. However, when Y is locally compact, we can usually pick the B's and C's such that both sets are empty.

5.5.15. PROPOSITION. Let Y be a second countable, locally compact Hausdorff space, and let $\mathscr{E}$ be a hyperfinite Dirichlet form defined on a subset S_0 of ${}^*Y$. Assume that there is a covering $\{O_n\}_{n\in\mathbb{N}}$ of Y consisting of open sets with compact closures, such that for each $n \in \mathbb{N}$ there is an $m \in \mathbb{N}$ with

$$ {}^\circ|e^{\mathsf{C}\bigcup_{i=1}^{m}{}^*O_i}_{{}^*O_n}|_1 < \infty.$$

Then there is an internal set $\tilde{S}_0$, such that $\bar{S}_0 \subset \tilde{S}_0 \subset S_0$ and $\tilde{\mathscr{E}}$ is nearstandardly concentrated.

PROOF. Put $B_n = \bigcup_{i=1}^{n} {}^*O_i \cap S_0$, and let

$$\mathscr{C} = \{{}^*O_n \cap S_0 \mid n \in \mathbb{N}\}.$$

Since

$$\bigcup_{n \in \mathbb{N}} B_n = \bar{S}_0 = \bigcup \{C \mid C \in \mathscr{C}\},$$

the proposition follows from Lemma 5.5.14.

We shall finally take a look at quasi-continuous extensions.

5.5.16. LEMMA. Assume that for each S-bounded $u \in \mathscr{D}[\mathscr{E}]$ there is a sequence $\{u_n\}_{n \in \mathbb{N}}$ of quasi-continuous functions such that ${}^{\circ}|u - u_n|_1 \to 0$ as $n \to \infty$. Then all internal S-bounded functions v with ${}^{\circ}\mathscr{E}_1(v, v) < \infty$ are quasi-continuous, and $\mathscr{E}$ generates quasi-continuous extensions.

PROOF. If v is S-bounded and ${}^{\circ}\mathscr{E}_1(v, v) < \infty$, there is an S-bounded $u \in \mathscr{D}[\mathscr{E}]$ such that $\|u - v\| \approx 0$. For all $k \in \mathbb{N}$

$$m\left\{s_i \mid |u(s_i) - v(s_i)| > \frac{1}{k}\right\} < \frac{1}{k},$$

and thus this inequality also holds for some infinite k. This implies that the set

$$D = \{s_i \mid u(s_i) \not\approx v(s_i)\}$$

is contained in an internal set of infinitesimal measure. By 5.4.4(i), D is exceptional, and hence v is quasi-continuous if and only if u is.

To prove that u is quasi-continuous, we first observe that by passing to a subsequence if necessary, we may assume that

$$(22) \qquad \mathscr{E}_1(u_{k+1} - u_k, u_{k+1} - u_k) < 2^{-4k}.$$

Extend $\{u_n\}_{n \in \mathbb{N}}$ to an internal sequence $\{u_n\}_{n \leq \eta}$, $\eta \in {}^*\mathbb{N} - \mathbb{N}$, such that (22) holds for all $k < \eta$. Define

$$G_k = \{s_i \in S_0 \mid |u_{k+1}(s_i) - u_k(s_i)| > 2^{-k}\},$$

and note that

$$\mathscr{E}_1(e_{G_k}, e_{G_k}) \leq 2^{2k}\mathscr{E}_1(u_{k+1} - u_k, u_{k+1} - u_k) \leq 2^{-2k}.$$

Let

$$F_k = \bigcup_{k \leq l \leq \eta} G_l.$$

Since $|e_{F_k}|_1 \leq |\sum_{k \leq l < \eta} e_{G_l}|_1 \leq \sum_{k \leq l < \eta} |e_{G_l}|_1 \leq 2^{-k+1}$,

$$\mathscr{E}(e_{F_k}, e_{F_k}) \leq 2^{-2k+2}.$$

By Proposition 5.3.6

$$P\{\omega \mid \exists t \le 1(X(\omega, t) \in F_k)\} = P\{\omega \mid \exists t \le 1(e_{F_k}(X(\omega, t)) \ge 1)\}$$
$$\le (1 + \Delta t)^{1/\Delta t} \mathscr{E}_1(e_{F_k}, e_{F_k}) \le e \cdot 2^{-2k+2}.$$

Thus

$$F = \bigcap_{k \in \mathbb{N}} F_k$$

is exceptional.

Let A be an exceptional set such that for all $n \in \mathbb{N}$, u_n is S-continuous on $S_0 - A$. Since $\|u - u_n\| \approx 0$, we are done if we can only show that u_n is S-continuous on the complement of $F \cup A$. Assume not, and pick $\varepsilon \in \mathbb{R}_+$, $s_i, s_j \in S_0 - (F \cup A)$ such that

(23) $$|u_n(s_i) - u_n(s_j)| \ge \varepsilon,$$

but $s_i \approx s_j$. Choose $k \in \mathbb{N}$ so large that $s_i, s_j \notin F_k$ and $2^{-k+2} < \varepsilon$. Since s_i, $s_j \notin F_k$

$$|u_k(s_i) - u_n(s_i)| \le 2^{-k+1}, \qquad |u_k(s_j) - u_n(s_j)| \le 2^{-k+1},$$

and since u_k is S-continuous on $S_0 - A$

$$|u_k(s_i) - u_k(s_j)| \approx 0.$$

Hence

$$|u_n(s_i) - u_n(s_j)| \le |u_n(s_i) - u_k(s_i)| + |u_k(s_i) - u_k(s_j)| + |u_k(s_j) - u_n(s_j)|$$
$$\le 2^{-k+1} + |u_k(s_i) - u_k(s_j)| + 2^{-k+1} < \varepsilon,$$

contradicting (23). This completes the proof.

We shall say that an internal function $f : S_0 \to {}^*\mathbb{R}$ has *compact support* if there is a compact set K such that $f(s_i) \approx 0$ for all $s_i \in S_0 - {}^*K$.

5.5.17. DEFINITION. A hyperfinite Dirichlet form $\mathscr{E}$ is called *regular* if it is normal and

(i)   for all S-continuous functions u of compact support there is a sequence $\{u_n\}_{n \in \mathbb{N}}$ of internal functions of finite energy such that $\sup\{{}^\circ|u(s_i) - u_n(s_i)| \mid s_i \in S_0\} \to 0$ as $n \to \infty$;

(ii)   for all S-bounded $u \in \mathscr{D}[\mathscr{E}]$, there is a sequence $\{u_n\}_{n \in \mathbb{N}}$ of quasi-continuous functions such that ${}^\circ|u - u_n| \to 0$ as $n \to \infty$.

The last result we shall prove in this section is a nonstandard counterpart of a theorem of Fukushima stating that all regular (in a similar, standard sense) Dirichlet forms on second countable, locally compact spaces generate Hunt processes [see Fukushima (1971, 1980) and Silverstein (1974)].

5.5.18. THEOREM. Let S_0 be a hyperfinite subset of $*Y$ for some second countable, locally compact Hausdorff space Y. If $\mathscr{E}$ is a regular Dirichlet form on S_0, there is an internal set $\tilde{S}_0$, $\bar{S}_0 \subset \tilde{S}_0 \subset S_0$, such that a modified standard part of the associated process $\hat{X}$ is a strong Markov process.

PROOF. It follows easily from 5.5.17(i) that $\mathscr{E}$ separates compacts and satisfies the conditions of 5.5.15. By 5.5.17(ii) and 5.5.16, $\mathscr{E}$ must generate quasi-continuous extensions. The theorem follows from 5.5.11.

REMARK. The theory in this chapter is new but several of the fundamental ideas are taken from the standard theory, as described in Fukushima (1980) and Silverstein (1974, 1976). In particular, our proof of Proposition 5.5.10 is closely modeled on Silverstein's proof of the standard version of 5.5.18 [see Silverstein (1974) and Fukushima (1980)]. The existing standard theory flows smoothly for second countable, locally compact spaces, but gets more complicated when the space is not assumed to be locally compact [see Albeverio and Høegh-Krohn (1976, 1977a,b), Kusuoka (1982), Paclet (1977/78, 1979), and Takeda (1984)]. One of the advantages of the hyperfinite approach is that it allows a unified treatment of locally compact spaces and most other interesting spaces (e.g., separable Hilbert and Banach spaces). However, it cannot be denied that the nonstandard theory is also easier to apply in locally compact spaces (compare Proposition 5.5.15).

In the next section we shall take a look at some typical applications of the theory developed in the last two sections.

5.6. APPLICATIONS TO QUANTUM MECHANICS AND STOCHASTIC DIFFERENTIAL EQUATIONS

We shall end this chapter by taking a brief look at some of the areas in which the theory of Dirichlet forms has been applied. The results we are going to present are not new [most of them are taken from Albeverio *et al.* (1977)], but they illustrate how Dirichlet forms provide a natural unification of ideas from analysis, probability, and mathematical physics. There are also strong reasons to believe that combined with the theory developed in the previous sections, these examples may serve as the starting point of important generalizations, especially in the infinite-dimensional case. As this section is primarily expository in nature, we do not strive for the greatest possible generality and shall feel free to omit proofs whenever we find it convenient.

A. Hamiltonians and Energy Forms

Let us begin by explaining the relationship between Dirichlet forms and quantum mechanics. Recall that a quantum mechanical particle moving in a potential V is governed by its Hamiltonian

(1) $H = -\Delta + V,$

where $\Delta = \sum_{i=1}^{3} \partial^2/\partial x_i^2$ is the three-dimensional Laplace operator. If $0 = \inf \text{spec}(H)$ is a simple eigenvalue corresponding to a strictly positive eigenfunction φ, it is well known that under reasonable conditions on V, the Hamiltonian H is the infinitesimal generator of a symmetric Markov semigroup e^{-tH}; see, e.g., Simon (1979). The interactions we shall be interested in here are too singular for this approach, and we shall study the relationship between Markov processes and quantum physics from a slightly different point of view.

Recall that to fit into the general framework of quantum physics, H must be interpreted as a self-adjoint operator on $L^2(\mathbb{R}^d, dx)$, where dx denotes the Lebesgue measure. Another way of formulating this is to say that the form F defined on a reasonable class of functions (e.g., $C_0^2(\mathbb{R}^d)$) by

(2) $$F(f, g) = \int (-\Delta f + Vf)g\, dx$$

is closable. Through a sequence of formal calculations we shall now find a function φ and a Dirichlet form E such that

(3) $E(f, g) = F(\varphi f, \varphi g).$

The calculations will be formal in the sense that we shall assume without justification that the operations we perform are legitimate. Later in this section we shall give several examples illustrating the importance of (3). Before we begin our computations, let us mention that the reader who is totally unfamiliar with quantum mechanics may find it helpful first to take a look at the introduction to Section 6.2.

We shall have to assume that there is a function φ such that

(4) $(-\Delta + V)\varphi = 0.$

This is a rather innocent assumption; if φ is a positive generalized [in the sense that it need not belong to $L^2(\mathbb{R}^3, dx)$] eigenfunction of $-\Delta + V$ corresponding to an eigenvalue E_0, then

(5) $(-\Delta + \tilde{V})\varphi = 0,$

where $\tilde{V} = V - E_0$. Since potentials are only determined up to an additive constant, V and $\tilde{V}$ describe the same physical situation, and we might just as well work with $\tilde{V}$ as with V.

Assuming (4), we can rewrite (1) as

(6) $H = -\Delta + (\Delta\varphi/\varphi).$

Applying H to a function of the form φf, we see that

(7) $\quad H(\varphi f) = -\Delta(\varphi f) + (\Delta\varphi)f = -(\Delta\varphi)f - 2\nabla\varphi\,\nabla f - \varphi\,\Delta f + (\Delta\varphi)f$

$$= \varphi(-\Delta f - 2(\nabla\varphi/\varphi)\nabla f),$$

and if A is the operator

(8) $$Af = -\Delta f - 2(\nabla\varphi/\varphi)\nabla f,$$

equation (7) can be written as

(9) $$H\varphi = \varphi A.$$

In terms of the form F, we get

(10) $$F(\varphi f, \varphi g) = \int H(\varphi f)\varphi g\,dx = \int (Af)g\varphi^2\,dx.$$

Introducing a measure μ by

$$d\mu = \varphi^2\,dx,$$

it is natural to define a form on $L^2(\mathbb{R}^3, d\mu)$ by

(11) $$E(f, g) = \int (Af)g\,d\mu = \int (Af)g\varphi^2\,dx = F(\varphi f, \varphi g).$$

Comparing (11) and (3) we see that E is the natural candidate for the Dirichlet form associated with F.

To get E on a simpler form we integrate by parts in the first term of

$$E(f, g) = \int (-\Delta f)g\varphi^2\,dx - 2\int \nabla\varphi\,\nabla fg\varphi\,dx.$$

Assuming that f and g and their partial derivatives vanish at infinity, we get

$$E(f, g) = \int \nabla f\nabla(g\varphi^2)\,dx - 2\int \nabla\varphi\,\nabla fg\varphi\,dx$$

$$= \int \nabla f\nabla g\varphi^2\,dx + \int \nabla f\nabla\varphi^2 g\,dx - 2\int \nabla\varphi\,\nabla fg\varphi\,dx$$

$$= \int \nabla f\nabla g\,d\mu.$$

We shall refer to E as the *energy form* of the measure μ.

To summarize, we have shown that

(12) $$E(f, g) = F(\varphi f, \varphi g),$$

where

(13)
$$E(f, g) = \int \nabla f \, \nabla g \, d\mu$$

and

(14)
$$F(f, g) = \int (-\Delta f + Vf) g \, dx.$$

The connection between μ and V is that $V = \Delta\varphi/\varphi$ and $d\mu = \varphi^2 \, dx$.

There are several reasons for passing from F to E. The first concerns the important technical question of closability; recall that to be accepted as a Hamiltonian form, F must have a closed extension. As we shall now see, closability of E implies closability of F, and we shall later show that it is quite easy to find conditions which guarantee that E can be closed.

If $\mathcal{D}_0$ is the domain of E, and F is defined on its domain

$$\varphi\mathcal{D}_0 = \{\varphi f : f \in \mathcal{D}_0\}$$

by

$$F(\varphi f, \varphi g) = E(f, g),$$

then F is closable in $L^2(\mathbb{R}^d, dx)$ if E is closable in $L^2(\mathbb{R}^d, d\mu)$. The reason is quite simply that if E_1 and F_1 are the forms

(15)
$$E_1(f, g) = E(f, g) + \int fg \, d\mu,$$

(16)
$$F_1(f, g) = F(f, g) + \int fg \, dx,$$

then

(17)
$$F_1(\varphi f, \varphi g) = E_1(f, g).$$

This means that if $\{\varphi f_n\}$ is a Cauchy sequence with respect to the F_1 norm, then $\{f_n\}$ is a Cauchy sequence with respect to the E_1 norm, and from this the claim follows immediately.

The second reason for passing from F to E is perhaps more exciting. There are interactions which are so irregular that they cannot be conveniently modeled by operators of the form

(18)
$$H = -\Delta + V;$$

e.g., the potential may be too singular to be represented by a function V. But using the heuristic calculations above, it is often possible to reformulate the problem in terms of energy forms in a way that makes perfectly good sense. The point is that the measure $d\mu = \varphi^2 \, dx$ is usually a much less singular object than the function $V = \Delta\varphi/\varphi$. An example of this phenomenon is provided by the theory of point interactions, which we shall study both below (Example 5.6.4) and in Chapter 6.

There is a third reason for studying energy forms which should appeal to probabilists as well as physicists. We have seen that the quadratic form E on $L^2(\mathbb{R}^d, d\mu)$ is generated by the operator

$$(19) \qquad A = -\Delta - 2(\nabla\varphi/\varphi)\nabla,$$

which is the infinitesimal generator of the stochastic differential equation

$$(20) \qquad dx(t) = f(x(t)) \, dt + db(t),$$

where $f = 2(\nabla\varphi/\varphi)$ and b is a Brownian motion with variance parameter 2. If it exists, the Markov process associated with E must be a solution to (20). Using the theory developed in the last section [or the corresponding standard theory of Fukushima (1980) and Silverstein (1974)], we shall see that E generates a Markov process in situations where $\nabla\varphi/\varphi$ is too singular to exist as a function or even as a distribution. Hence energy forms can be used to study stochastic differential equations with generalized drifts. Note that since $f = 2(\nabla\varphi/\varphi) = \nabla \ln \varphi^2$, we must assume that f is a gradient.

The stochastic differential equation (20) is worth a study in its own right, but it gains additional relevance from the part it plays in quantum mechanics. Not only is it a useful technical tool (Albeverio et al., 1977, 1980, 1981, 1984a,b) but it has attained conceptual importance through the attempts that been made to provide quantum theory with a purely probabilistic foundation [see e.g., Nelson (1967, 1985), Guerra (1981), Guerra and Morato (1983), Carlen (1984), Yasue (1981), Zheng and Meyer (1984), and contributions to S. Albeverio et al. (1986a)]. For applications to other fields of mathematical physics, see Albeverio and Høegh-Krohn (1981a,b, 1982, 1984), Albeverio et al. (1983, 1984, 1985), Fukushima (1985), and their references. In infinite dimensions Dirichlet forms and their associated processes have found applications in quantum field theory and hydrodynamics [see, e.g., Albeverio and Høegh-Krohn (1981a,b, 1982, 1984), Albeverio et al. (1985c), and Takeda (1984)]; the mathematical framework was developed in Albeverio and Høegh-Krohn (1977a,b), Kusuoka (1982), Paclet (1978, 1979), and Takeda (1984). Although we shall not treat it in any detail here, it is in the infinite-dimensional theory that we have the highest hopes for the nonstandard approach; the reason is the ease with which we have been able to handle infinite-dimensional problems not only in Sections 5.4 and 5.5, but also in connection with Gross's theorem (Section 3.5) and stochastic integration (Section 4.7). We shall return briefly to these ideas at the end of the section.

B. Standard and Nonstandard Energy Forms

Before we can understand the applications above in detail, we need to know more about energy forms. It should come as no surprise to the reader that we are going to use a hyperfinite approach, studying hyperfinite forms $\mathscr{E}$ on a hyperfinite lattice Y. Let

$$(21) \qquad Y = \left\{ (k_1 \, \Delta x, \ldots, k_d \, \Delta x) \colon \forall i \le d \left(k_i \in {}^*\mathbb{Z} \text{ and } |k_i| \le \frac{1}{\Delta x^2} \right) \right\}$$

for some $\Delta x \approx 0$. If $e \in \mathbb{R}^d$ is a unit vector of the form

$$(22) \qquad\qquad e = (0, \ldots, 0, \pm 1, 0, \ldots, 0),$$

let sgn(e) be the sign of the nonzero component of e. We shall call the set of all such unit vectors U; note that U splits naturally into a positive part U^+ and a negative part U^-.

Given an internal function $f \colon Y \to {}^*\mathbb{R}$ and an element $e \in U$, we define

$$(23) \qquad\qquad D_e f(y) = \frac{f(y + e \, \Delta x) - f(y)}{\text{sgn}(e) \, \Delta x}$$

when both y and $y + \Delta x e$ belong to Y. If $y \in Y$ and $y + \Delta x e \notin Y$, we let $D_e f(y) = 0$. The resulting function $D_e f \colon Y \to {}^*\mathbb{R}$ is an internal version of the partial derivative in e's direction. We shall study *hyperfinite energy forms* given by

$$(24) \qquad\qquad \mathscr{E}(f, g) = \frac{1}{2} \sum_{y \in Y} \sum_{e \in U} D_e f(y) D_e g(y) \nu(y),$$

where $\nu(y)$ is an internal measure on Y. The factor $\frac{1}{2}$ is included to compensate for the fact that each direction is counted twice. Using that $D_{-e} f(y + e \, \Delta x) = D_e f(y)$, we can rewrite (24) as

$$(25) \qquad \mathscr{E}(f, g) = \sum_{y \in Y} \sum_{e \in U^+} D_e f(y) D_e g(y) \frac{\nu(y) + \nu(y + e \, \Delta x)}{2}.$$

The form $\mathscr{E}$ obviously has the Markov property, and hence it is a Dirichlet form by Proposition 5.3.3. To construct the transition matrix Q and the invariant measure m of $\mathscr{E}$ we could have used the procedure in the proof of 5.3.3, but it is just as easy to do this directly. First we fix the time line

$$(26) \qquad\qquad T = \{0, \Delta t, 2 \, \Delta t, \ldots\},$$

where

$$(27) \qquad\qquad \Delta t = \Delta x^2 / 2d.$$

Observe that this is the usual relation $\Delta x = \sqrt{\Delta t}$ between the space and the time scale, with the factor $1/2d$ added for technical convenience. If the Markov process has no trap, the hyperfinite Beurling–Deny formula 5.3.1 and relation (27) tell us that

$$(28) \quad \mathscr{E}(f, g) = \frac{d}{\Delta x^2} \sum_{y \in Y} \sum_{y' \in Y} [f(y) - f(y')][g(y) - g(y')] q_{yy'} m(y).$$

Let us assume that

$$(29) \qquad\qquad q_{yy'} = 0 \quad \text{if } |y - y'| \neq \Delta x,$$

i.e., the process can only jump to a neighboring site. We can then write (28) as

$$(30) \qquad \mathscr{E}(f, g) = d \sum_{y \in Y} \sum_{e \in U} D_e f(y) D_e g(y) q_{y, y + e\Delta x} m(y).$$

By the symmetry condition $q_{y, y + e\Delta x} m(y) = q_{y + e\Delta x, y} m(y + e\,\Delta x)$ and the fact that $D_{-e}f(y + e\,\Delta x) = D_e f(y)$, the last equation can be rewritten as

$$(31) \qquad \mathscr{E}(f, g) = 2d \sum_{y \in Y} \sum_{e \in U^+} D_e f(y) D_e g(y) q_{y, y + e\Delta x} m(y).$$

Comparing (25) and (31) we see that

$$(32) \qquad\qquad 2dm(y) q_{y, y + e\Delta x} = \frac{\nu(y) + \nu(y + e\,\Delta x)}{2}.$$

By (29) we must have $\sum_{e \in U} q_{y, y + e\Delta x} = 1$, and thus

$$(33) \qquad\qquad 2dm(y) = \sum_{e \in U} \frac{\nu(y) + \nu(y + e\,\Delta x)}{2}$$

or

$$(34) \qquad\qquad m(y) = \frac{1}{2} \nu(y) + \frac{1}{4d} \sum_{e \in U} \nu(y + e\,\Delta x).$$

Putting this into (32), we get

$$(35) \quad q_{y, y + e\Delta x} = \frac{\nu(y) + \nu(y + e\,\Delta x)}{4dm(y)} = \frac{\nu(y) + \nu(y + e\,\Delta x)}{2d\nu(y) + \sum_{\tilde{e} \in U} \nu(y + \tilde{e}\,\Delta x)}.$$

The last two equations show how to construct m and Q from ν. Observe that although in general m is different from ν, we always have $L(m) \circ \mathrm{st}^{-1} = L(\nu) \circ \mathrm{st}^{-1}$, and thus m and ν induce the same standard measure on $\mathbb{R}^d$.

Let us just remark that the procedure above cannot always be reversed; given m and Q such that both (29) and the symmetry condition

$$(36) \qquad m_y q_{yy'} = m_{y'} q_{y'y}$$

hold, it may be impossible to find a measure ν satisfying (32). What is needed is the following generalization of (36): if $y_1, y_2, \ldots, y_n$ is a closed path on Y (i.e., $y_1 = y_n$ and $|y_i - y_{i+1}| = \Delta x$ for all i), then

$$(37) \qquad \sum_{i=1}^{n-1} (-1)^i m_{y_i} q_{y_i y_{i+1}} = 0.$$

Note that (36) is (37) applied to the path y, y', y. We shall not make use of (37) here, but would like to comment that it corresponds to the standard condition that f in (20) must be a gradient.

The average value of the transition probabilities $q_{y,y+e\Delta x}$ is $1/2d$ and the order of magnitude of the deviation is Δx. It is therefore convenient to write $q_{y,y+e\Delta x}$ in the form

$$(38) \qquad q_{y,y+e\Delta x} = (1/2d) + \mathrm{sgn}(e)\beta_e(y)\,\Delta x,$$

where the factor $\mathrm{sgn}(e)$ is included to facilitate comparison to the standard theory. Rewriting (35) as

$$(39) \quad q_{y,y+e\Delta x} = \frac{\nu(y) + \nu(y + e\,\Delta x)}{2d\nu(y) + \Sigma_{\tilde{e} \in U}\, \nu(y + \tilde{e}\,\Delta x)}$$

$$= \frac{1}{2d} + \frac{[\nu(y + e\,\Delta x) - \nu(y)] + [\nu(y) - (1/2d)\,\Sigma_{\tilde{e} \in U}\, \nu(y + \tilde{e}\,\Delta x)]}{2d\nu(y) + \Sigma_{\tilde{e} \in U}\, \nu(y + \tilde{e}\,\Delta x)}$$

$$= \frac{1}{2d} + \mathrm{sgn}(e)\frac{D_e \nu(y)}{4dm(y)}\,\Delta x - \frac{\tilde{\Delta}\nu(y)}{8d^2 m(y)}\,\Delta x^2,$$

where $\tilde{\Delta}$ is the discrete Laplacian

$$(40) \qquad \tilde{\Delta}f(y) = \frac{1}{\Delta x^2}\left(\sum_{e \in U} f(y + e\,\Delta x) - 2df(y) \right),$$

we see that

$$(41) \qquad \beta_e(y) = \frac{D_e \nu(y)}{4dm(y)} - \mathrm{sgn}(e)\frac{\tilde{\Delta}\nu(y)}{8d^2 m(y)}\,\Delta x.$$

By equation (5.3.16) the infinitesimal generator of $\mathscr{E}$ is

$$(42) \qquad \tilde{A}f(y) = \frac{1}{\Delta t}\left(f(y) - \sum_{e \in U} f(y + e\,\Delta x)q_{y,y+\Delta xe} \right).$$

Using (27) and (38) we can rewrite this as

(43) $\tilde{A}f(y) = \dfrac{2d}{\Delta x^2}\left(f(y) - \sum\limits_{e \in U} f(y + e\,\Delta x)\left(\dfrac{1}{2d} + \text{sgn}(e)\beta_e(y)\,\Delta x\right)\right)$

$= \dfrac{2df(y) - \Sigma_{e \in U} f(y + e\,\Delta x)}{\Delta x^2}$

$-2d \sum\limits_{e \in U} \dfrac{f(y + e\,\Delta x) - f(y)}{\text{sgn}(e)\,\Delta x}\,\beta_e(y)$

$= -\tilde{\Delta}f(y) - 2d \sum\limits_{e \in U} D_e f(y) \cdot \beta_e(y),$

and combining (41) and (43) we see that

(44) $\tilde{A}f(y) = -\tilde{\Delta}f(y) - \dfrac{1}{2m(y)} \sum\limits_{e \in U} D_e f(y) D_e \nu(y)$

$+ \dfrac{\tilde{\Delta}\nu(y)}{4dm(y)}\,\Delta x \sum\limits_{e \in U} \text{sgn}(e) D_e f(y).$

Under any reasonable conditions on f and ν, the last term in (44) is infinitesimal. Observe also that (44) is the nonstandard counterpart of (19).

We can now begin to explore the connections between the standard and the hyperfinite theory of energy forms. Let μ be a completed Borel measure on $\mathbb{R}^d$ such that $\mu(K) < \infty$ when K is compact. Pick a hyperfinite representation ν of μ supported on the lattice Y. In order to make the last term in (44) vanish, we choose ν such that $[\tilde{\Delta}\nu(y)/m(y)]\,\Delta x \approx 0$. Let E_0 be the form defined on the set $\mathscr{C}_0^1(\mathbb{R}^d)$ of continuously differentiable functions of compact support by

(45) $E_0(f, g) = \displaystyle\int \nabla f \cdot \nabla g\, d\mu.$

The corresponding hyperfinite form is

(46) $\mathscr{E}(f, g) = \dfrac{1}{2} \sum\limits_{y \in Y} \sum\limits_{e \in U} D_e f(y) D_e g(y) \nu(y).$

We know from Sections 5.1 and 5.2 that $\mathscr{E}$ induces a closed standard form E on $L^2(\mathbb{R}^d, \mu)$. A natural question to ask is when E is an extension of E_0; this would imply that E_0 is closable.

Note first that if $f \in \mathscr{C}_0^1(\mathbb{R}^d)$, then

(47) $^\circ\mathscr{E}(^*f, {}^*f) = E(f, f).$

But since E can be defined by

(48) $$E(f,f) = \inf\left\{{}^{\circ}\mathscr{E}(\tilde{f},\tilde{f})\ \bigg|\ \int |\tilde{f} - {}^*f|^2\, dm \approx 0\right\},$$

all we can conclude from (47) is the inequality

(49) $$E(f,f) \le E_0(f,f).$$

To get equality in (49) we need a condition on ν. It turns out that the "natural" one to use is

(50) $${}^{\circ}\!\int \sum_{e \in U} \left(\frac{D_e\nu(y)}{m(y)}\right)^2 dm(y) < \infty,$$

but that by working a little harder, we can get away with a slightly weaker, "local" version of (50). The extra generality will be helpful in our applications.

5.6.1. LEMMA. Assume that there is a closed set $N \subset \mathbb{R}^d$ of μ measure zero such that

(51) $${}^{\circ}\!\int_{(Y\cap{}^*K)-{}^*O} \sum_{e \in U} \left(\frac{D_e\nu(y)}{m(y)}\right)^2 dm(y) < \infty$$

whenever O is an open set containing N and K is compact. Then ${}^*f \in \mathscr{D}[\mathscr{E}]$ for all $f \in \mathscr{C}_0^1(\mathbb{R}^d)$ and hence E is an extension of E_0.

PROOF. If ${}^*f \notin \mathscr{D}[\mathscr{E}]$, there must be an internal function $\tilde{f}$ such that

(52) $$\int |{}^*f - \tilde{f}|^2\, dm \approx 0,$$

but

$${}^{\circ}\mathscr{E}({}^*f - \tilde{f}, {}^*f) > 0.$$

We can choose $\tilde{f} = Q^t {}^*f$ for a suitable infinitesimal t. By definition

$$\mathscr{E}({}^*f - \tilde{f}, {}^*f) = \int \frac{1}{2} \sum_{e \in U} D_e^*({}^*f - \tilde{f}) D_e^* f\, d\nu(y),$$

and since N has μ-measure zero and f has compact support, there is a compact set K_1 and an open set $O_1 \supset N$ such that

$${}^{\circ}\!\int_{(Y\cap{}^*K_1)-{}^*O_1} \sum_{e \in U} D_e({}^*f - \tilde{f}) D_e^* f\, d\nu(y) > 0.$$

Approximating D^*f by differentiable functions, we can find a $g \in \mathcal{C}_0^2(\mathbb{R}^d - N)$ (the set of all functions which vanish outside a compact subset of $\mathbb{R}^d - N$ and have continuous second derivatives) such that

$$^{\circ}\!\!\int_{(Y \cap {}^*K_1) - {}^*O_1} \sum_{e \in U} D_e({}^*f - \tilde{f}) D_e{}^*g \, d\nu(y) > 0.$$

If we are a little careful how we choose g outside $K_1 - O_1$, we can ensure that

$$^{\circ}\!\!\int_Y \sum_{e \in U} D_e({}^*f - \tilde{f}) D_e{}^*g \, d\nu(y) > 0.$$

Hence

(53) $\quad 0 < \displaystyle ^{\circ}\!\!\int ({}^*f - \tilde{f}) \tilde{A}^* g \, d\nu \leq \left(\int |{}^*f - \tilde{f}|^2 \, d\nu \right)^{1/2} \left(\int (\tilde{A}^* g)^2 \, d\nu \right)^{1/2}.$

Comparing (52) and (53), we see that we get a contradiction if we can only prove that

$$^{\circ}\!\!\int (\tilde{A}^* g)^2 \, d\nu < \infty.$$

By (44), $\tilde{A}^* g$ consists of three terms, the first of which is square-integrable since $g \in \mathcal{C}_0^2(\mathbb{R}^d)$, the second is square-integrable by condition (51) and the fact that g vanishes outside a compact subset of $\mathbb{R}^d - N$, and the third is infinitesimal by our standing condition $\tilde{\Delta}\nu(y) \, \Delta x / m(y) \approx 0$. Thus the assumption that ${}^*f \notin \mathscr{D}[\mathscr{E}]$ leads to a contradiction.

That E is an extension of E_0 now follows from (47).

To translate 5.6.1 into standard terms, we assume that $d\mu = \varphi^2 \, dx$ as above. What corresponds to the hyperfinite quantity $D_e\nu(y)/m(y)$ is then

$$\frac{\partial \varphi^2 / \partial x_e}{\varphi^2} = \frac{2}{\varphi} \frac{\partial \varphi}{\partial x_e},$$

and condition (51) becomes

$$\int_{K - O} \left(\frac{\nabla \varphi}{\varphi} \right)^2 \varphi^2 \, dx = \int_{K - O} (\nabla \varphi)^2 \, dx < \infty.$$

Hence we get:

5.6.2. COROLLARY. Let $d\mu = \varphi^2 \, dx$, where φ is locally square integrable. Assume that there is a closed set N of Lebesgue measure zero such that the distribution $\nabla \varphi$ is in $L_2^{\text{loc}}(\mathbb{R}^d - N)$. Then E_0 is closable.

By saying that a distribution T is in $L_2^{loc}(\mathbb{R}^d - N)$, we mean that for any $f \in \mathscr{C}_0^\infty(\mathbb{R}^d - N)$ the distribution $f(x)T(x)$ is actually a function in $L^2(\mathbb{R}^d)$.

Let us take a look at two applications of 5.6.2, both due to Albeverio *et al.* (1977). The first example is very simple, but it illustrates vividly the importance of allowing an "exceptional" set N.

5.6.3. EXAMPLE. Let D be an open subset of $\mathbb{R}^d$, and let $\varphi(x) = 1$ for $x \in D$ and $\varphi(x) = 0$ for $x \notin D$. The boundary $N = \partial D$ of D has Lebesgue measure zero, and $\nabla \varphi \in L^2(\mathbb{R}^d - N)$. Hence the form

$$E_0(f, g) = \int_D \nabla f \nabla g \, dx$$

is closable by Corollary 5.6.2, and the associated operator is the Laplacian with Neumann boundary conditions.

In the next example we shall finally exploit the connection between quantum mechanics and Dirichlet forms developed at the beginning of this section.

5.6.4. EXAMPLE. Let $d = 3$, and for each $m \in \mathbb{R}$, let

$$\varphi_m(x) = \frac{e^{-m|x|}}{|x|}.$$

It is easy to check that

$$(54) \qquad \nabla \varphi_m(x) = -(x + mx|x|) \frac{e^{-m|x|}}{|x|^3}$$

and

$$(55) \qquad \Delta \varphi_m(x) = m^2 \varphi_m(x)$$

when $x \neq 0$. From (54) we see immediately that $\nabla \varphi_m \in L_2^{loc}(\mathbb{R}^3 - \{0\})$, and hence the form

$$E_m^0(f, g) = \int \nabla f \nabla g \varphi_m^2 \, dx$$

can be extended to a closed form E_m. Let F_m be the closed form on $L^2(\mathbb{R}^3, dx)$ defined by

$$(56) \qquad F_m(f, g) = E_m(\varphi_m^{-1} f, \varphi_m^{-1} g).$$

From (55) and the calculations we did at the beginning of this section, one would expect that

$$(57) \qquad F_m(f, g) = \int (-\Delta f + m^2 f) g \, dx.$$

Going through the calculations once more, checking that each step is legitimate when $\varphi = \varphi_m$, we get that (57) does indeed hold as long as $f(0) = g(0) = 0$.

Defining

$$(58) \qquad \tilde{F}_m(f, g) = F_m(f, g) - m^2 \int fg \, dx,$$

we thus get a family of forms agreeing on functions vanishing at the origin. But since for $m > 0$

$$F_m(\varphi_m, g) = E_m(1, \varphi_m^{-1} g) = 0,$$

φ_m is an eigenfunction for $\tilde{F}_m$ with eigenvalue $-m^2$. This implies that $\tilde{F}_m \neq \tilde{F}_k$ when $m \neq k$; if not, φ_m and φ_k would be two positive eigenfunctions corresponding to different eigenvalues and this is clearly impossible. Letting $-\Delta_m$ be the operator generating $\tilde{F}_m$, we have a family of self-adjoint operators agreeing with the free Hamiltonian $-\Delta$ on functions vanishing at the origin.

We shall try to explain briefly why the family $\{-\Delta_m\}$ is of interest to physicists. Assume that we are trying to model forces of short range. If the source is at the origin and the range of the interaction is less than ε, the Hamiltonian is of the form

$$(59) \qquad H_\varepsilon = -\Delta + V_\varepsilon,$$

where V_ε vanishes outside the ball $B_\varepsilon = \{x : |x| < \varepsilon\}$. Another way of saying that V_ε vanishes outside B_ε is to demand that

$$(60) \qquad H_\varepsilon f = -\Delta f,$$

whenever f is zero on B_ε.

If the range of the interaction is extremely short (as, e.g., with nuclear forces), it is convenient to put $\varepsilon = 0$. In this case it is not at all obvious how to interpret (59), but with (60) there is no problem; what we want is a self-adjoint operator H such that

$$(61) \qquad Hf = -\Delta f,$$

if f vanishes at the origin. But this is exactly the kind of operators we have constructed above, and hence $-\Delta_m$ is a model of zero-range interactions (or point interactions, as they are also called).

When we return to point interactions at greater length in the next chapter, we shall see how to interpret (59) for $\varepsilon = 0$ as

$$(62) \qquad\qquad H = -\Delta - \lambda\delta$$

for suitable infinitesimals λ. Although it may not be totally clear from our exposition, the same idea is behind the construction above. To realize this one has to take a closer look at how the forms F_m are defined. Using Lemma 5.6.1, we first constructed E_m as the standard part of

$$(63) \qquad \mathscr{E}_m(f, g) = \frac{1}{2} \sum_{y \in Y} \sum_{e \in U} D_e f(y) D_e g(y) \tilde{\varphi}_m^2 \, \Delta x^d,$$

where $\tilde{\varphi}_m$ is a suitable nonstandard approximation to φ_m. It follows that F_m is the standard part of the form

$$(64) \qquad \mathscr{F}_m(f, g) = \mathscr{E}_m(\tilde{\varphi}_m^{-1} f, \tilde{\varphi}_m^{-1} g).$$

Carrying out the calculations from the beginning of this section in the hyperfinite setting, we get that the operator generating $\mathscr{F}_m$ is of the form

$$(65) \qquad\qquad -\Delta_1 + B,$$

where Δ_1 is a hyperfinite version of the Laplacian and $Bg(x) \approx m^2 g(x)$ when $\mathrm{st}(x) \neq 0$. But when $x \approx 0$, the singularity $\tilde{\varphi}_m$ has at the origin causes the relation $Bg(x) \approx m^2 g(x)$ to fail; in fact, B has a singularity which turns (65) into a version of (62) for a suitable infinitesimal λ depending on m. The calculations are rather messy, and we leave them to the reader.

For more information about point interactions, see Chapter 6 and the forthcoming monograph by Albeverio *et al.* (1986).

There are applications of Dirichlet forms to other kinds of singular potentials as well, but we shall not discuss them here. Instead we turn our attention to the Markov processes generated by energy forms.

C. Energy Forms and Markov Processes

Recall that by Theorem 5.5.18, a hyperfinite Dirichlet form on a locally compact space generates a standard strong Markov process if it is regular in the sense of Definition 5.5.17.

5.6.5. PROPOSITION. Let ν be an internal measure on Y such that $^\circ\nu(^*K) < \infty$ for all compact sets K and $0 < {}^\circ\nu(^*O)$ for all open, nonempty sets O. Put $\varphi(x) = \nu(x)/\Delta x^d$, and assume that there is a closed set $N \subset \mathbb{R}^d$ of Lebesgue measure zero such that for all $x \in \mathbb{R}^d - N$, there is an open neighborhood $O \subset \mathbb{R}^d$ with $\varphi^{-1} \in SL^2(*O \cap Y, \Delta x^d)$. Then the form

$$\mathscr{E}(f, g) = \frac{1}{2} \sum_{y \in Y} \sum_{e \in U} D_e f(y) D_e g(y) \nu(y)$$

is regular.

SKETCH OF PROOF. The definition of regularity consists of two parts. To show that the first part is satisfied, we must prove that all S-continuous functions of compact support can be approximated by internal functions of finite energy in the supremum norm. This follows immediately from the fact that $\mathscr{C}_0^1$ is dense in $\mathscr{C}_0$.

To prove that the second condition in the definition of regularity is satisfied, it suffices to show that there is no function $f \in \mathscr{D}[\mathscr{E}] \cap SL^2(\nu)$ such that $^\circ\mathscr{E}(f, f) \neq 0$ but $\mathscr{E}(f, g) \approx 0$ for all quasi-continuous, S-bounded $g \in \mathscr{D}[\mathscr{E}]$. Assume that such an f exists; we shall first show that it cannot be nearstandard [in the sense that it is a 2-lifting of a function in $L^2(\mathbb{R}^d, L(\nu) \circ \mathrm{st}^{-1})$].

If f is nearstandard, there is a sequence $\{f_n\}$ of elements of $\mathscr{D}[\mathscr{E}]$ such that $\|f - f_n\| \to 0$ and each $D_e f_n$ is nearstandard. Since both f and f_n belong to $\mathscr{D}[\mathscr{E}]$, it follows that $\mathscr{E}(f - f_n, f - f_n) \to 0$. Hence for each $e \in U$

$$^\circ\!\int |D_e f - D_e f_n|^2 \, d\nu \to 0,$$

implying that $D_e f$ is nearstandard. But if $D_e f$ is nearstandard, we cannot have $\mathscr{E}(f, g) \approx 0$ for all quasi-continuous, S-bounded g, unless $\mathscr{E}(f, f) \approx 0$.

Assume now that f is not nearstandard. Then there is an internal set A of infinitesimal measure such that

$$\alpha = \int_A |D_e f| \, \Delta x^d$$

is noninfinitesimal for some $e \in U$. We shall leave this estimate to the reader; it is trivial for $d = 1$ but takes work in higher dimensions. We can choose $A \subset {}^*O$ for one of the open neighborhoods O mentioned in the proposition.

By Hölder's inequality

$$\alpha = \int_A |D_e f| \, \Delta x^d = \int_A |D_e f| \varphi(x) \varphi^{-1}(x) \, \Delta x^d \leq \mathscr{E}(f, f)^{1/2} \left(\int_A \varphi^{-2} \, \Delta x^d \right)^{1/2}.$$

Since $\int_A \varphi^{-2} \, \Delta x^d$ is infinitesimal and α is not, $\mathscr{E}(f, f)$ must be infinite. But this is impossible since we have assumed $f \in \mathscr{D}[\mathscr{E}]$.

The conditions of Proposition 5.6.5 guarantee that the nonstandard energy form $\mathscr{E}$ generates a standard Markov process, but unless the conditions of Lemma 5.6.1 are also satisfied, we do not have a standard characterization of this process. To get a better grasp of the relationship between the form and the process, let us return to Example 5.6.4 and compute the infinitesimal generator of the process in this case.

5.6.6. EXAMPLE. Let

$$\varphi_m(x) = \frac{e^{-m|x|}}{|x|}$$

as in Example 5.6.4. It is easy to check that

(66)
$$\nabla(\varphi_m^2)(x) = -2\left[m + \frac{1}{|x|}\right]\varphi_m^2(x)\mathbf{u}_r,$$

where $\mathbf{u}_r = x/|x|$ is the unit radial vector. Integrating by parts we see that if $f, g \in \mathscr{C}_0^2$, then

$$E_m(f, g) = \int \nabla f \nabla g \varphi_m^2 \, dx = \int f(-\Delta g \varphi_m^2 - \nabla g \cdot \nabla \varphi_m^2) \, dx$$

$$= \int f\left(-\Delta g + 2\left[m + \frac{1}{|x|}\right]\nabla g \cdot \mathbf{u}_r\right) \varphi_m^2 \, dx.$$

Since

(67)
$$A = -\Delta + 2\left[m + \frac{1}{|x|}\right]\mathbf{u}_r \cdot \nabla$$

is the infinitesimal generator of the stochastic differential equation

(68)
$$dx(t) = -2\left(m + \frac{1}{|x(t)|}\right)\mathbf{u}_r \, \Delta t + db(t),$$

the Markov process generated by E_m is a solution of (68).

As we have already indicated, the theory of Dirichlet forms is capable of treating equations with more severe singularities than the one occurring in (68); we refer the reader to Albeverio *et al.* (1977, 1980) and Ōshima (1982) [see also Portenko (1979a,b) for an alternative approach]. In what sense the Markov process generated by an energy form solves the associated stochastic differential equation was studied in Albeverio *et al.* (1980, 1984b) where singular forms and processes were approximated by smooth ones.

We have promised to say a few words about the infinite-dimensional theory. On the nonstandard side everything is as before; we are working with a hyperfinite lattice

$$Y = \left\{ (k_1 \Delta x, \ldots, k_d \Delta x) \middle| \forall i \le d \left(k_i \in {}^*\mathbb{Z} \text{ and } |k_i| \le \frac{1}{\Delta x^2} \right) \right\},$$

an internal measure ν on Y, and an energy form

$$\mathscr{E}(f, g) = \frac{1}{2} \sum_{y \in Y} \sum_{e \in U} D_e f(y) D_e g(y) \nu(y).$$

The only difference is that we now assume that $d \in {}^*\mathbb{N} - \mathbb{N}$.

Let H be the standard Hilbert space generated by the orthonormal basis $\{e_i\}_{i \in \mathbb{N}}$, where

$$e_i = (0, 0, \ldots, 1, \ldots, 0) \in Y$$

with the one occurring in the ith coordinate. With the proper identifications, we obviously have

$$H \subset Y \subset {}^*H.$$

From the examples we considered in Sections 3.5 and 4.7, we know that it is unnatural to assume that ν is supported on the nearstandard sets in the Hilbert space topology. Instead we introduce a Banach space B in which H is densely and continuously embedded. Letting st : $Y \to H$ and st$_B$: $Y \to B$ denote the standard part maps in the H and B topologies, we shall assume that

$$\mu = L(\nu) \circ \text{st}_B^{-1}$$

is a nonzero Radon measure on B. By the theory of Section 5.1, the hyperfinite form $\mathscr{E}$ induces a closed form E on $L^2(B, \mu)$. The question is what we can say about this form from reasonable conditions on B and μ.

It is not our purpose to develop an extensive nonstandard theory here, but to get a feeling for what to expect, we shall take a brief look at the results which have been obtained by standard methods. For simplicity we shall concentrate on Kusuoka's (1982) work; the contributions by Albeverio and Høegh–Krohn (1977a,b) and Paclet (1978, 1979) are of a similar nature.

For sufficiently smooth functions $u \in L^2(B, \mu)$, let $Du : B \to H$ denote the Gâteaux derivative of u. Kusuoka defines a quadratic form on $L^2(B, \mu)$

by

$$E_0(f, g) = \int_B \langle Df, Dg \rangle \, d\mu$$

and shows that E_0 is closable under reasonable conditions when μ is a probability measure. Adding more conditions he then shows that E_0 generates a strong Markov process on B. These additional conditions basically say that the embedding of H into B is compact, and that the dual B^* of B is a sufficiently large subspace of H.

The basic idea of the proof is as follows. Fukushima's existence theorem (the standard version of 5.5.18) only applies to forms defined on locally compact spaces, and does not cover E_0. Kusuoka uses Gelfand's representation theorem to find a locally compact, separable metric space M in which B is densely and continuously embedded. He interprets E_0 as a form on $L^2(M, \tilde{\mu})$, where $\tilde{\mu}$ is the probability measure induced by μ through the embedding from B to M, and applies Fukushima's existence theorem to find an associated process on M. The argument is completed by showing that $M - B$ has capacity zero and that the process thus lives on B.

Our hope is that by using the hyperfinite theory developed in this chapter, it should be possible to refine and extend the results mentioned above. Instead of applying the trick of replacing B by a locally compact space M, the results of Sections 5.4 and 5.5 should enable us to work directly with B-valued processes. We have not checked whether the theory of Section 5.5 is strong enough to prove extensions of Kusuoka's theorems, or whether further refinements are needed, but this is certainly a promising area for future research.

Before we take our leave of the infinite-dimensional theory, there is one more question we would like to discuss. Assume that ν is a uniform measure on Y; then the process associated with $\mathscr{E}$ is a hyperfinite-dimensional random walk. If B is the completion of H with respect to a measurable norm, we know from Gross's theorem and Section 4.7 that the standard part of this random walk is a Brownian motion x living on B. The problem is that ν does not have a standard part on B, and that, in fact, there is no Dirichlet form on B associated with x. A natural question is whether it is possible to use well-defined hyperfinite forms such as $\mathscr{E}$ to study processes such as x which are not generated by standard forms. This would require an extension of the theory developed in this chapter, as we have always assumed that the underlying measure has a standard part.

The contents of this chapter can be characterized as probabilistic potential theory. We have, however, spent most of the time discussing very general existence problems of a probabilistic nature, and no time at all on potential theoretic topics of central importance such as boundary problems. It is too late to redress the balance now, but we would like to make the reader aware of Loeb's (1976, 1980, 1982) work.

Let W be a locally compact Hausdorff space which is connected and locally connected, but not compact. Let $\mathcal{H}$ be a family of harmonic functions such that (W, $\mathcal{H}$) is a Brelot space where 1 is superharmonic. These assumptions are satisfied if $\mathcal{H}$ consists of $\mathscr{C}^2$ solutions of an elliptic differential equation of the form

$$\sum_{i,j} a_{ij} \frac{\partial^2 u}{\partial x_i \, \partial x_j} + \sum_i b_i \frac{\partial u}{\partial x_i} + cu = 0$$

on a region of $\mathbb{R}^d$, where (a_{ij}) is positive definite, $c \geq 0$, and a_{ij}, b_i, and c are locally Lipschitz. Another example is given by the solutions to

$$\Delta u + cu = 0$$

on an open Riemann surface, where $c \geq 0$ is a smooth density.

Working in a nonstandard model, Loeb extends Wiener's generalized solution of the Dirichlet problem to arbitrary compactifications of the harmonic space W, and gives a Martin–Choquet integral representation of positive harmonic functions. Under the additional assumption that there exist a positive potential and a bounded, nonzero harmonic function on W, Loeb constructs an ideal boundary Δ such that the points of Δ correspond to nonnegative harmonic functions. This boundary supports the maximal representing measures (with respect to the Choquet ordering) for positive bounded and quasi-bounded harmonic functions, and, as opposed to Martin's boundary, almost all the points of Δ are regular for the Dirichlet problem. These results are all contained in Loeb (1976). A modification of the methods of that paper also gives a simple construction of the maximal representing measures for positive harmonic functions on W as the weak* limits of finite sums of point masses on $[0, +\infty]$. In this case W can also be a Bauer space; see Loeb (1980, 1982).

REFERENCES

S. Albeverio and R. Høegh-Krohn (1976). Quasi-invariant measures, symmetric diffusion processes and quantum fields. In *Le Méthodes Mathématiques de la Théorie Quantique des Champs*, Ed. CNRS, Paris.

S. Albeverio and R. Høegh-Krohn (1977a). Dirichlet forms and diffusion processes on rigged Hilbert spaces. *Z. Wahrsch. Verw. Gebiete* **40**.

S. Albeverio and R. Høegh-Krohn (1977b). Hunt processes and analytic potential theory on rigged Hilbert spaces. *Ann. Inst. H. Poincaré Sect. B* **13**.

S. Albeverio and R. Høegh-Krohn (1981a). Stochastic methods in quantum field theory and hydrodynamics. *Phys. Rep.* **77**.

S. Albeverio and R. Høegh-Krohn (1981b). Some Markov processes and Markov fields in quantum theory, group theory, hydrodynamics and C^*-algebras. *In* (D. Williams, ed.), *Stochastic Integrals.* Springer Verlag, Berlin and New York.

S. Albeverio and R. Høegh-Krohn (1982). Some remarks on Dirichlet forms and their applications to quantum mechanics and statistical mechanics. *In* (M. Fukushima, ed.), *Functional Analysis in Markov Processes.* Springer-Verlag, Berlin and New York.

S. Albeverio and R. Høegh-Krohn (1984). Diffusion fields, quantum fields, fields with values in Lie groups. *In* (M. Pinsky, ed.), *Stochastic Analysis and Applications*, pp. 1–97. Dekker, New York.

S. Albeverio, R. Høegh-Krohn, and L. Streit (1977). Energy forms, Hamiltonians, and distorted Brownian paths. *J. Math. Phys.* **18**.

S. Albeverio, R. Høegh-Krohn, and L. Streit (1980). Regularization of Hamiltonians and processes. *J. Math. Phys.* **21**.

S. Albeverio, M. Fukushima, W. Karwowski, and L. Streit (1981). Capacity and quantum mechanical tunneling. *Comm. Math. Phys.* **81**.

S. Albeverio, Ph. Blanchard, and R. Høegh-Krohn (1983). A stochastic model for the orbits of planets and satellites: An interpretation of Titius–Bode law. *Expo. Math.* **4**.

S. Albeverio, F. Gesztesy, W. Karwowski, and L. Streit (1984a). On the connection between Schrödinger and Dirichlet forms *J. Math. Phys.* **2C**.

S. Albeverio, S. Kusuoka, and L. Streit (1984b). Convergence of Dirichlet forms and associated Schrodinger operators (preprint). Math. Inst., Ruhr Univ., Bochum. (To appear in *J. Funct. Anal.*, 1986.)

S. Albeverio, Ph. Blanchard, and R. Høegh-Krohn (1984c). Newtonian diffusions and planets, with remark on nonstandard Dirichlet forms and polymers. *In* (A. Truman and D. Williams, eds.) *Proceedings of the LMS Symposium on Stochastic Analysis and Applications* (*Swansea, 1983*), pp 1–24. Lect. Notes Math. 1094, Springer-Verlag, Berlin and New York.

S. Albeverio, Ph. Blanchard, F. Gesztesy, and L. Streit (1985a). Quantum mechanical low energy scattering in terms of diffusion processes. *In* (S. Albeverio, Ph. Combe, and M. Sirugue-Collin; eds.) *Stochastic Aspects of Classical and Quantum Systems*, pp 207–227. Lect. Notes Math. 1109. Springer-Verlag, New York and Berlin.

S. Albeverio, Ph. Blanchard, and R. Høegh-Krohn (1985b). Diffusions sur une variéte riemanienne: Barrières infranchissable et applications. In *Colloque en l'honneur de Laurent Schwartz.* Astérique, **132**.

S. Albeverio, R. Høegh-Krohn, and H. Holden (1985c). Markov processes on infinite dimensional spaces, Markov fields and Markov cosurfaces. *In* (L. Arnold and P. Kotelenez, eds.) *Stochastic Space–Time Models, Limit Theorems*, pp 11–40. Reidel Publ., Dordrecht.

S. Albeverio, G. Casati, and D. Merlini, eds. (1986a). Stochastic Processes in classical and quantum systems. *Proceedings First International Ascona–Como Meeting.* Lect. Notes Phys., Springer-Verlag, Berlin and New York.

S. Albeverio, F. Gesztesy, R. Høegh-Krohn, and H. Holden (1986b). *Solvable Models in Quantum Mechanics* (in preparation).

A. Beurling and J. Deny (1958). Espaces de Dirichlet. *Acta Math.* **99**.

A. Beurling and J. Deny (1959). Dirichlet spaces. *Proc. Nat. Acad. Sci.* **45**.

R. M. Blumenthal and R. K. Getoor (1968). *Markov Processes and Potential Theory.* Academic Press, New York and London.

E. Carlen (1984). Conservative diffusions. *Comm. Math. Phys.* **94**.

R. Carmona (1979). Regularity properties of Schrödinger and Dirichlet semigroups. *J. Funct. Anal.* **33**.

K. L. Chung (1960). *Markov Chains with Stationary Transition Probabilities.* Springer-Verlag, Berlin and New York.

J. Deny (1970). Methodes Hilbertienns et théorie du potentiel. *In* (M. Brelot, ed.), *Potential Theory.* Edizioni Cremonese, Roma.

E. B. Dynkin and A. A. Yushkevich (1969). *Markov Processes.* Plenum Press, New York.

M. Fukushima (1971). Dirichlet spaces and strong Markov processes. *Trans. Amer. Math. Soc.* **162**.

M. Fukushima (1979). A decomposition of additive functionals of finite energy. *Nagoya Math. J.* **19**.

M. Fukushima (1980). *Dirichlet forms and Markov processes.* North-Holland Publ., Amsterdam, 1980.

M. Fukushima (1985). Energy forms and diffusion processes. *In* (L. Streit, ed.), *Mathematics + Physics; Lectures on Recent Results.* World Scientific Publ. Co., Singapore.

J. Glimm and A. Jaffe (1981). *Quantum Physics—a Functional Integral Point of View.* Springer-Verlag, New York and Berlin.

F. Guerra (1981). Structural aspects of stochastic mechanics and stochastic field theory. *In* (C. De Witt-Morette and K. D. Elworthy eds.), New stochastic methods in physics. *Phys. Rep.* **77**.

F. Guerra and L. Morato (1983). Quantization of dynamical systems and stochastic control theory. *Phys. Rev. D* **27**.

C. W. Henson (1979). Analytic sets, Baire sets and the standard part map. *Can. J. Math.* **31**.

J. G. Hooton (1979). Dirichlet forms associated with hypercontractive semigroups. *Trans. Amer. Math. Soc.* **253**.

H. J. Keisler (1984). An infinitesimal approach to stochastic analysis. *Mem. Amer. Math. Soc.* **297**.

S. Kusuoka (1982). Dirichlet forms and diffusion processes on Banach spaces. *J. Fac. Sci. Univ. Tokyo Sect. 1A Math* **29**.

T. Lindstrøm (1986). Nonstandard energy forms and diffusion on manifolds and fractals. *In* (S. Albeverio, G. Casati, and D. Merlini, eds.) *Stochastic Processes in Classical and Quantum Systems.* Proc. 1st Int. Ascona-Coma Meet., Lect. Notes Phys., Springer-Verlag, Berlin and New York.

P. A. Loeb (1976). Applications of nonstandard analysis to ideal boundaries in potential theory. *Israel J. Math.* **25**.

P. A. Loeb (1980). A regular metrizable boundary for solutions of elliptic and parabolic differential equations. *Math. Ann.* **251**.

P. A. Loeb (1982). A construction of representing measures for elliptic and parabolic differential equations. *Math. Ann.* **260**.

E. Nelson (1967). *Dynamical Theories of Brownian Motion.* Princeton Univ. Press, Princeton, New Jersey.

E. Nelson (1985). *Quantum Fluctuations.* Princeton Univ. Press, Princeton, New Jersey.

Y. Ōshima (1982). Some singular diffusion processes and their associated stochastic differential equations. *Z. Wahrsch. Verw. Gebiete* **59**.

P. Paclet (1977/1978). Espaces de Dirichlet et capacités fonctionelles sur triplets de Hilbert-Schmidt. *In Sem. Krée,* Exp. 5, 4e année, 1977–78, Université Pierre et Marie Curie, Paris.

Ph. Paclet (1979). Espaces de Dirichlet en dimension infinie. *C.R. Ac. Sci., Paris,* Ser. A **288**.

N. I. Portenko (1979a). Diffusion processes with generalized drift coefficients, *Theory Probab. Its Appl.* (Eng. Transl.) **24**.

N. I. Portenko (1979b). Stochastic differential equations with generalized drift vector. *Theory Prob. Its Appl.* (Eng. Transl.) **24**.

M. Reed and B. Simon (1975). *Methods of Modern Mathematical Physics II: Fourier Analysis, Self-Adjointness.* Academic Press, New York.

M. Röckner and N. Wielens (1985). Dirichlet forms—closability and change of speed measure. *In* (S. Albeverio, ed.) *Infinite Dimensional Analysis and Stochastic Processes.* Res. Notes Math., Pitman, London.

S. Saks (1937). *Theory of the integral,* Monograf. Mat. **7**, Polska Akademia Nauk, Warszawa.

M. L. Silverstein (1974). *Symmetric Markov Processes.* Springer-Verlag, Berlin and New York.

M. L. Silverstein (1976). *Boundary Theory for Symmetric Markov Processes.* Springer-Verlag, Berlin and New York.

B. Simon (1979). *Functional Integration and Quantum Physics.* Academic Press, New York.

M. Takeda (1984). On the uniqueness of Markovian extensions of diffusion operators on infinite dimensional spaces, Bi Bo S (preprint). Bielefeld.

M. Tomisaki (1982). Dirichlet forms associated with direct product diffusion processes. *In* (M. Fukushima, ed.), *Functional Analysis in Markov Processes,* pp 76–119. Springer-Verlag, Berlin and New York.

N. Wielens (1985). The essential self-adjointness of generalized Schrödinger operators. *J. Funct. Anal.* **61**.

D. Williams (1979). *Diffusions, Markov Processes, and Martingales,* Vol. I, Foundations. Wiley, New York.

K. Yasue (1981). Stochastic calculus of variations. *J. Funct. Anal.* **41**.

W. Zheng and P. A. Meyer (1984). Quelques resultats de "Mécanique Stochastique". *In* (J. Azéma, ed.) *Sem. Probab. XVII.* Lect. Notes Math., Springer-Verlag, Berlin and New York.

CHAPTER 6

TOPICS IN DIFFERENTIAL
OPERATORS

This chapter contains a mixed bag of applications. In Section 6.1 we discuss a singular Sturm–Liouville problem. In Section 6.2 we develop a general theory for singular perturbations of non-negative operators, which we apply in Section 6.3 to point interactions and in Section 6.4 to perturbations by local time functionals, in particular to polymer models. In Section 6.5 we present a nonstandard approach to the Boltzmann equation, proving among other things an existence result in the space-inhomogeneous case. In Section 6.6 we make a few remarks on a hyperfinite version of the Feynman path integral; we explain that there is in general no associated Loeb measure, but point out how the *internal* path space measure can be used to solve the Schrödinger equation and to discuss the classical limit.

Unless otherwise specified, all linear spaces in this chapter are over the reals or hyperreals. Where the physical examples require a complex setting, the results extend easily from the real case.

6.1. A SINGULAR STURM–LIOUVILLE PROBLEM

As an introduction to a nonstandard treatment of singular perturbation theory we shall discuss a singular Sturm–Liouville problem.

Let μ be a finite non-negative Borel measure on $[0, 1]$. We shall consider the eigenvalue problem

(1) $-Y''(x) + \mu Y(x) = \lambda Y(x), \qquad 0 \le x \le 1,$

where λ is a real parameter and with boundary conditions

(2) $Y(0) = Y(1) = 0.$

Equation (1) as it stands does not make classical sense. But there are several ways to give a standard interpretation of the equation.

First one can consider the associated quadratic form defined by

(3) $$A\varphi = \int_0^1 (\varphi')^2 \, dm + \int_0^1 \varphi^2 \, d\mu,$$

where φ is a continuously differentiable function on $[0, 1]$ satisfying (2) and dm denotes the Lebesgue measure on $[0, 1]$. We shall see that A has a countable family of "generalized eigenfunctions" $\{Y_n\}$ which behave in a way similar to the eigenfunctions of the classical Sturm–Liouville problem.

A second way of interpreting (1) is to pass to the associated integral equation

(4) $$Y(x) = xY'(0) + \int_0^x (x - s) Y(s) \, d\mu - \lambda \int_0^x (x - s) Y(s) \, dm.$$

Standard methods tell us that (4) has a continuous solution Y_λ on $[0, 1]$ for every λ. But it seems difficult to get precise information, e.g., to decide for which λ we have $Y_\lambda(1) = 0$. We shall see that the sequence Y_n, λ_n "solving" (3) also answers our questions in connection with (4).

We shall obtain the solution via an excursion into nonstandard territory. In our original work [Albeverio *et al.* (1979b)] we converted the measure μ into a nonstandard smooth function by replacing it with $\delta_\varepsilon * \mu$, where δ_ε is a $*C^\infty$ delta function with support in $[-\varepsilon, \varepsilon]$, ε being a positive infinitesimal. By transfer, 1.2.4, we could apply classical Sturm–Liouville theory to the nonstandard equation and in this way also obtain new information in the standard case.

There is a more radical nonstandard approach to the problem, namely to replace the standard differential equation (1) by a nonstandard, hyperfinite difference equation. This was done for the classical Sturm–Liouville problem by MacDonald (1976) and extended to the more general context by Birkeland (1980). We shall give a brief exposition of Birkeland's treatment.

The key idea is to replace the system (1) and (2) by the difference equation

(5) $N^2 \Delta^2 y(k) + (\lambda - q(k))y(k) = 0, \qquad 0 < k < N,$

with the boundary conditions

(6) $y(0) = y(N) = 0,$

where N is some large—but still finite—integer and $q = \langle q(0), \ldots, q(N) \rangle$ is a given vector such that $q(k) \geq 0$ for all $0 \leq k \leq N$. The difference operators are defined by

$$\Delta y(k) = y(k+1) - y(k),$$

$$\Delta^2 y(k) = \Delta(\Delta y(k-1)) = y(k+1) - 2y(k) + y(k-1).$$

The system (5) and (6) is nothing but a system of $N - 1$ linear equations for the $N - 1$ unknowns $y(1), \ldots, y(N-1)$. The corresponding matrix is of the form

$$A - \lambda I,$$

where I is the identity matrix and where

$$A_{ij} = 0, \qquad \text{if} \quad |i - j| > 1,$$

$$A_{ij} = N^2, \qquad \text{if} \quad |i - j| = 1,$$

$$A_{ii} = -2N^2 - q(i).$$

Since A is a symmetric matrix it follows from elementary linear algebra that (5), (6) has $N - 1$ pairwise orthogonal real eigenvectors $y_1, \ldots, y_{N-1}$ and that the corresponding eigenvalues $\lambda_1, \ldots, \lambda_{N-1}$ are real and simple.

This is standard and elementary. What is still elementary, but no longer so trivial, is to get sharp and uniform bounds on the eigenvectors and eigenvalues in order to obtain useful information about the system (1), (2) from the discrete problem (5), (6).

We have the following result from Birkeland (1980).

6.1.1. PROPOSITION. Let $\lambda_1 < \lambda_2 < \cdots < \lambda_{N-1}$ be the eigenvalues for the problem (5), (6) and let $y_1, \ldots, y_{N-1}$ be the corresponding eigenvectors normalized by setting $\|y_j\|_2^2 = N$. Define

$$m = \|q\|_1 / N$$

and

$$b = 2(1 + 12m).$$

Then for $\lambda_j \leq 3N^2$, in particular if $j \leq N/2$, we have the bounds

(7) $\|y_j\|_\infty \leq b,$

(8) $N \cdot \|\Delta y_j\|_\infty \leq 2 \cdot \lambda_j^{1/2} \cdot b.$

If $3N^2 < \lambda_j$ we have an inequality

(7') $\|y_j\|_\infty \leq b'$,

where $b' = 2(1 + 12m')$ and $m' = \|q\|_\infty - m$.

We should, perhaps, add the remark that the vector norms we use are the standard ones, $\|v\|_1 = \Sigma|v(k)|$, $\|v\|_2 = (\Sigma|v(k)|^2)^{1/2}$, and $\|v\|_\infty = \max|v(k)|$.

We give some brief remarks on the proof; the reader may consult Birkeland (1980) for full details on the various calculations involved.

The first trick is to use the following identity, well known from the theory of ordinary differential equations. Let y and z be the solutions of the difference equations

(9) $N^2 \Delta^2 y(k) + (\lambda - q(k))y(k) = 0$

and

(10) $N^2 \Delta^2 z(k) + \lambda \cdot z(k) = 0,$

respectively, and suppose that $y(0) = z(0) = 0$. Then we have the identity

$$y(k)z(1) = y(1)z(k) + N^{-2} \sum_{i=1}^{k-1} q(i)y(i)z(k-i).$$

To prove 6.1.1 we must obtain sufficiently strong estimates for the sum above; i.e., we want to obtain an inequality of the form

(11) $\left| \sum_{i=1}^{k} q(i)y(i)z(k-i) \right| \leq P \cdot \|y\|_2 \cdot \|z\|_\infty \cdot N^{1/2},$

where P is some real number. This inequality can be combined with explicit information about the eigenvectors and eigenvalues of (10) to obtain an estimate of the form

(12) $\|y\|_\infty \leq 2(1 + 3P\lambda^{-1/2})\|y\|_2 N^{-1/2}.$

It remains to choose a suitable P. If we were only interested in the classical Sturm-Liouville problem, we could get away with something less than (7) and (8). In fact, we could use the Cauchy-Schwarz inequality,

$$\left| \sum_{i=1}^{k} q(i)y(i)z(k-i) \right| \leq \|q\|_2\|y\|_2\|z\|_\infty,$$

to choose $P = \|q\|_2 N^{-1/2}$, or even weaker $P = \|q\|_\infty$, to satisfy (11) and to obtain a sufficiently strong bound via (12). In the singular case we need better inequalities, hence a more delicate choice of P.

A somewhat lengthy analysis (Birkeland, 1980) shows that (11) is true with the choice $P = 4m\lambda^{1/2}$. This value of P in (12) at once implies (7), recalling the normalization $\|y_j\|_2^2 = N$. The bound in (8) requires some further calculations; see Birkeland (1980) for details.

To complete the nonstandard program we need an analog of the classical Sturm–Liouville theory for difference equations, which is elementary but somewhat tedious to supply. Let us touch on one point: we need to estimate the number of zeros of a vector $y = \langle y(k) \rangle$. This requires the right definition of zero point; it may happen that for some k, both $y(k-1)$ and $y(k) \neq 0$ but $y(k-1) \cdot y(k) \le 0$. We use linear interpolation: if $y(k-1)y(k) \le 0$ and $y(k) \neq 0$, then the real number

$$\xi = k - \frac{y(k)}{\Delta y(k-1)}$$

is called a *zero point* or *node* for the vector y. With this notion we have the following result:

6.1.2. PROPOSITION. Let y_j and λ_j be the jth eigenvector and eigenvalue for the system (5), (6).

(i) y_j has exactly $j + 1$ zero points in the closed interval $[0, N]$, and between two zero points for y_j there is a zero point for y_{j+1}.

(ii) λ_j satisfy the following inequality:

$$(\pi^2 j^2)/6 \le \sigma_j \le \lambda_j \le (\sigma_j^{1/2} + m)^2,$$

where $\sigma_j = 2N^2(1 - \cos(j\pi/N))$ is the jth eigenvalue of equation (10) with boundary conditions $z(0) = z(N) = 0$.

The proof is given in detail in Birkeland (1980). We remark that Birkeland (1980) also contains more explicit information about the system (5), (6). We have touched upon a few facts needed to make the transition from (5) and (6) to the system (1) and (2).

The time has come to choose a hyperfinite integer $N \in {}^*\mathbb{N} - \mathbb{N}$. We want to represent the measure μ by a vector $q \in {}^*\mathbb{R}^{N+1}$. We need a little bit of care, but it is possible to show that given any P, $N \in {}^*\mathbb{N} - \mathbb{N}$ such that $P/N \approx 0$, there is a vector q such that for every continuous function φ on $[0, 1]$

$$\int_0^1 \varphi \, d\mu \approx N^{-1} \sum {}^*\varphi\left(\frac{k}{N}\right) q(k),$$

and such that

$$0 \le q(k) \le P.$$

With this choice of q consider the problems (5) and (6). By transfer, 1.2.4, we still have Propositions 6.1.1 and 6.1.2. In particular, it follows from the bound in (ii) and 6.1.2 that if j is finite, then $\text{st}(\lambda_j)$ exists and satisfies

(13) $\pi^2 j^2 \leq \text{st}(\lambda_j) \leq (\pi j + \mu([0, 1]))^2,$

since $\mu([0, 1]) = \text{st}(N^{-1} \sum q(k)) = \text{st}(N^{-1} \cdot \|q\|_1).$

Next, from the inequalities in 4.1.1 we obtain

(14) $|y_j(m) - y_j(l)| = \left| \sum_{k=l}^{m} \Delta y_j(k) \right| \leq 2 \cdot \lambda_j^{1/2} b \cdot \dfrac{m - l}{N}.$

If j is finite this implies that $\text{st}(y_j(m)) = \text{st}(y_j(l))$ whenever $m - l/N$ is infinitesimal. This means that the functions $Y_j(x) = \text{st}(y_j(k))$, where $x = \text{st}(k/N)$, are well defined. Y_j is continuous, but not necessarily differentiable. But if $Y_j''(x)$ existed, then

$$Y_j''(x) = \text{st}(N^{-2} \Delta^2 y_j(k)),$$

where $x = \text{st}(k/N)$. This is the intuition connecting (5) and (6) and (1) and (2). In the general case we must settle for something less. By a suitable double summation we derive from (5) the equation

(15) $y_j(n) - y_j(l) = (n - l) \Delta y_j(l)$

$$+ N^{-2} \sum_{k=l+1}^{n} (n - k) y_j(k) q(k)$$

$$- \lambda N^{-2} \sum_{k=l+1}^{m} (n - k) y_j(k).$$

If j is finite, we may take standard parts. This leads to the integral equation

(16) $Y_j(x) - Y_j(x_0) = (x - x_0) \cdot \kappa + \displaystyle\int_{x_0}^{x} (x - t) Y_j(t) \, d\mu$

$$- \lambda \int_{x_0}^{x} (x - t) Y_j(t) \, dt,$$

where $x = \text{st}(n/N)$, $x_0 = \text{st}(l/N)$, $t = \text{st}(k/N)$, and $\kappa = \text{st}(N \Delta y_j(l))$. We are thus back to (4). It is also not difficult to extract (3) and to prove the following theorem from the corresponding results about the *hyperfinite* system (5), (6).

6.1.3. THEOREM. Let μ be a finite Borel measure on $[0, 1]$ and define a quadratic form A on the space $C_0^1[0, 1]$ of continuously differentiable functions φ on $[0, 1]$ that satisfy $\varphi(0) = \varphi(1) = 0$ by

$$A\varphi = \int_0^1 (\varphi')^2 \, dm + \int_0^1 \varphi^2 \, d\mu.$$

Let $M = \mu([0, 1])$ and $B = 2(1 + 12M)$.

There exists a sequence $\{\tau_j\}$ of real numbers and a sequence $\{Y_j\}$ in $C_0^1[0, 1]$ such that

(a) the following inequalities hold, $0 \le x \le 1$, $1 \le j < \infty$:

$$\pi^2 j^2 \le \tau_j \le (\pi j + M)^2, \qquad |Y_j(x)| \le B;$$

(b) $\{Y_j\}$ is an orthonormal and complete sequence in $L^2[0, 1]$;

(c) if φ is twice continuously differentiable on $[0, 1]$ and $\varphi(0) = \varphi(1) = 0$, then its orthogonal expansion in terms of $\{Y_j\}$ converges uniformly to φ;

(d) if φ is continuously differentiable on $[0, 1]$, $\varphi(0) = \varphi(1) = 0$, and $\sum_{j=1}^{\infty} d_j Y_j$ is its expansion, then

$$A\varphi = \sum_{j=1}^{\infty} \tau_j d_j^2;$$

(e) the Y_j are solutions of the integral equation (4) with $\lambda = \tau_i$;

(f) Y_j has exactly $j + 1$ zeros in $[0, 1]$, and between two zeros for Y_j there is a zero for Y_{j+1}; and

(g) if $\mu(\{x\}) = 0$ or $Y_j(x) = 0$, $Y_j'(x)$ exists and is continuous at x; further, $|Y_j'(x)| \le 2B\tau_j^{1/2}$. If $\mu(\{x\})Y_j(x) \ne 0$, then Y_j' has a jump discontinuity at x of size $\mu(\{x\})Y_j(x)$. On intervals $I \subset [0, 1]$ where μ is of the form $d\mu = g\,dm$, $g \in L^1(I)$, Y_j' is absolutely continuous, and Y_j satisfies $Y'' + (g - \tau_j)Y = 0$ almost everywhere.

We have indicated a part of the proof of the theorem and round off our discussion by proving (b) and (c). First we remark that the orthogonality properties of the eigenvectors y_j at once imply that the functions $\{Y_j\}$ are orthonormal over $[0, 1]$.

To prove the completeness part of (b) as well as property (c) we consider a two times continuously differentiable function Φ on $[0, 1]$, with $\Phi(0) = \Phi(1) = 0$. Define for $0 \le k \le N$

$$\varphi(k) = {}^*\Phi(k/N).$$

Standard linear algebra and transfer tell us that

$$(17) \qquad \varphi(k) = N^{-1} \sum_{j=1}^{N-1} c_j y_j(k),$$

where

$$c_j = \langle \varphi, y_j \rangle = \sum_{k=0}^{N} \varphi(k) y_j(k).$$

For finite j we have

$$d_j = \mathrm{st}(c_j/N) = \mathrm{st}(N^{-1} \cdot \sum \varphi(k) y_j(k)) = \int_0^1 \Phi Y_j \, dm.$$

It follows that for any positive integer $M < \infty$

$$\operatorname{st}\left(N^{-1} \sum_{j=1}^{M} c_j y_j(k) \right) = \sum_{j=1}^{M} d_j Y_j(\operatorname{st}(k/N)).$$

To prove completeness it will thus be sufficient to control the tail part of the expansion (17); i.e., we must show that for any positive standard ε there is an integer $M_\varepsilon < \infty$ such that

(18) $N^{-1} \sum_{j=M_\varepsilon}^{N-1} |c_j y_j(k)| < \varepsilon, \qquad 0 \le k \le N.$

For y_j we have inequalities from Proposition 4.1.1. We need to prove a "good" inequality for c_j. We calculate

(19) $c_j = \langle \varphi, y_j \rangle = \langle \varphi, \lambda_j^{-1}(y_j q - N^2 \Delta^2 y_j) \rangle$

$$= \lambda_j^{-1}\left[\sum_{k=0}^{N} \varphi(k) y_j(k) q(k) - N^2 \sum_{k=0}^{N} \varphi(k) \Delta^2 y_j(k) \right].$$

Here we have used the fact that y_j, λ_j is a solution of the system (5), (6). From our assumptions on Φ we have a finite real p such that

$$|\varphi(k)| \le p \quad \text{and} \quad N^2 |\Delta^2 \varphi(k)| \le p, \qquad 0 \le k \le N.$$

This gives the following bounds in (19):

$$\left| \sum_{k=0}^{N} \varphi(k) y_j(k) q(k) \right| \le p \|q\|_1 \|y_j\|_\infty$$

and, using summation by parts two times,

$$\left| \sum_{k=0}^{N} \varphi(k) \Delta^2 y_j(k) \right| = \left| \sum_{k=0}^{N} y_j(k) \Delta^2 \varphi(k) \right| \le p \|y_j\|_\infty / N.$$

These estimates combined with (19) give the following inequality for c_j;

$$|c_j| \le \lambda_j^{-1} p \|y_j\|_\infty (\|q\|_1 + N).$$

This implies the following bound for the sum in (18):

(20) $N^{-1} \sum_{j=M_\varepsilon}^{N-1} |c_j y_j(k)| \le p\left(1 + \frac{\|q\|_1}{N}\right) \sum_{j=M_\varepsilon}^{N-1} \lambda_j^{-1} \|y_j\|_\infty^2.$

From Proposition 6.1.2 we know that $\pi^2 j^2 / 6 \le \lambda_j$ for all j, and from Proposition 6.1.1 we have bounds for $\|y_j\|_\infty$. This applied to (20) gives

(21) $\sum_{j=M_\varepsilon}^{N-1} \lambda_j^{-1} \|y_j\|_\infty^2 \le \sum_{j=M_\varepsilon}^{N/2} \frac{6 b^2}{\pi^2 j^2} + \sum_{j=N/2}^{N-1} \frac{6(b')^2}{\pi^2 j^2}.$

Since b is finite, $°b = 2(1 + 12\mu([0, 1]))$, the first sum can be made small by choosing $M_\varepsilon < \infty$ large enough. To complete the proof we need to show that the last sum is infinitesimal. We have the following bound:

$$\sum_{j=N/2}^{N-1} \frac{6(b')^2}{\pi^2 j^2} \le \frac{6}{\pi^2}(b')^2 \frac{N}{2}\left(\frac{2}{N}\right)^2 = \frac{12}{\pi^2} \frac{(b')^2}{N}$$

From 6.1.1 we easily see that $b' \le 24(\|q\|_\infty + 1)$. We can now use our freedom to choose a $P \in {}^*\mathbb{N} - \mathbb{N}$ satisfying $P/N \approx 0$ and $\|q\|_\infty \le P$ such that $(b')^2/N$ will be infinitesimal. And, indeed, any hyperinteger P less than $N^{1/4}$ will do.

This ends our remarks on the proof of Theorem 6.1.3. We have focused attention on those aspects that are novel and interesting from the point of view of nonstandard methodology. The proofs of the remaining parts follow standard patterns [see Birkeland (1980) and MacDonald (1976)].

6.1.4. REMARK. The results of this section have been discussed from a standard point of view, i.e., by using distribution theory and approximation techniques, in Persson (1981, 1984).

Is there a lesson to be learned from the nonstandard proof? We think so: replace the continuous by a hyperfinite discrete problem. Study the genuinely finite version of this problem and prove sharp and uniform inequalities. Then by transfer and standard parts you have a solution to the continuous problem. This is a theme which we shall meet more than once in applicable nonstandard analysis. And it does give nonstandard methods a certain concreteness and constructivity.

6.2. SINGULAR PERTURBATIONS OF NON-NEGATIVE OPERATORS

Let us consider a particle of mass m moving in a potential V. In classical mechanics the trajectory of the particle is determined by Newton's equation

(1)
$$\frac{d^2\mathbf{x}}{dt^2} = -\frac{1}{m} \nabla V(\mathbf{x})$$

and the initial conditions

(2)
$$\mathbf{x}(0) = \mathbf{x}_0,$$

(3)
$$d\mathbf{x}/dt(0) = \mathbf{v}_0,$$

where $\mathbf{x}(t)$ is the position at time t. In quantum mechanics this deterministic description is abandoned and all that is possible is a formulation in probabilistic terms. The probabilistic information is coded up in a complex-valued function $\psi(\mathbf{x}, t)$ of space and time called the wave function. Knowledge of the particle's physical behavior can be obtained by certain mathematical operations on the wave function; e.g., the probability that the particle is in a subset A of $\mathbb{R}^3$ at time t is given by

$$\int_A |\psi(\mathbf{x}, t)|^2 \, d\mathbf{x}.$$

Although the exact way the particle behaves cannot be predicted, the evolution of the probabilistic law governing it is deterministic and given by the Schrödinger equation

$$(4) \qquad i\hbar \frac{\partial \psi}{\partial t} = -\frac{\hbar^2}{2m} \Delta \psi + V\psi,$$

where $\hbar$ is Planck's constant divided by 2π, and $\Delta = \sum_{i=1}^{3} (\partial^2/\partial x_i^2)$ is the Laplace operator. This equation plays the same central role in quantum mechanics as (1) does in classical mechanics.

To understand the Schrödinger equation, we must first get a good grasp on the operators occurring on its right-hand side. Since the constant $\hbar^2/2m$ is of no mathematical significance, we shall study operators of the form

$$(5) \qquad\qquad H = -\Delta + V,$$

where V acts by multiplication. In order to get a reasonable mathematical theory we must interpret H as a self-adjoint operator on $L^2(\mathbb{R}^3)$; how this can be done and what consequences follow are well known for a large class of functions V [see, e.g., Kato (1976) and Simon (1982)]. The purpose of this and the next two sections is to extend the theory to some cases of considerable physical interest where the potential is no longer given by a function.

The examples we have in mind have in common that the potential vanishes outside a set of measure zero. One example is point interactions, where the force is concentrated at a single point; for a number of purposes this is a good and convenient model for extremely short-range interactions, e.g., nuclear forces [see Albeverio et al. (1984a, 1986b) and Sections 5.6 and 6.3]. Lattices of point potentials have been used successfully to simulate crystals [Grossmann et al. (1980a, b); Høegh-Krohn et al. (1985); Albeverio et al. (1986b)]. In another example the potential is concentrated on a Brownian path modeling a polymer molecule; as we shall explain in Sections 6.4 and 7.5, the four-dimensional version of this model is important in quantum field theory. Since the potentials vanish outside sets of measure zero, the corresponding Schrödinger operators cannot be of the form $-\Delta + V$ where V is a function; we must perturb by more singular objects.

Such perturbations are most easily described by quadratic forms. Recall that there is a one-to-one correspondence between non-negative self-adjoint operators A and non-negative closed forms E given by

$$(6) \qquad\qquad E(f, g) = \langle A^{1/2}f, A^{1/2}g \rangle.$$

A perturbation of A concentrated on a set C induces a perturbed form

$$(7) \qquad \tilde{E}(f, g) = \langle A^{1/2}f, A^{1/2}g \rangle + \int_C Vfg \, dx.$$

If, as above, C has measure zero, the last integral is zero and there is no perturbation. However, it is easy now to see how this can be fixed; just replace the function V by a measure μ supported on C:

$$(8) \qquad \tilde{E}(f, g) = \langle A^{1/2}f, A^{1/2}g \rangle + \int_C fg \, d\mu.$$

For technical reasons it is convenient to split μ into two parts—a probability measure $\tilde{\rho}$ on C and a function $(-\tilde{\lambda})$. Hence

$$(9) \qquad \tilde{E}(f, g) = \langle A^{1/2}f, A^{1/2}g \rangle - \int_C \tilde{\lambda}fg \, d\tilde{\rho}.$$

The idea is to let the self-adjoint operator generated by $\tilde{E}$ be the desired perturbation of A. The problem, of course, is that $\tilde{E}$ need not generate a self-adjoint operator. Since C has measure zero, the last term in (9) only makes sense for a subset of the domain of $A^{1/2}$—those functions that are continuous in a neighborhood of C. The question is whether $\tilde{E}$ can be extended to a closed form.

Closability questions of this kind are often very intricate, but in Section 5.1 we found a general method for dealing with them; since the standard part of a hyperfinite form is always closed, all we have to do is find a hyperfinite representation of $\tilde{E}$. The strategy is as follows: we shall assume that the underlying space X is a Hausdorff space, that C is a closed subset of X, and that m and $\tilde{\rho}$ are Radon measures on X and C, respectively, with $m(C) = 0$. We also assume that $m(K) < \infty$ for all compact sets K. Given a closed form E on $L^2(X, m)$, let $\mathscr{E}$ be a hyperfinite representation of it defined on $L^2(Y, \mu)$ as in Corollary 5.2.2. Let B be an internal subset of Y with $C = \mathrm{st}(B)$, and let ρ be an internal probability measure on B such that $\tilde{\rho} = L(\rho) \circ \mathrm{st}^{-1}$. If $\lambda : B \to {}^*\mathbb{R}$ is internal, the nonstandard counterpart of $\tilde{E}$ is

$$(10) \qquad \tilde{\mathscr{E}}(f, g) = \mathscr{E}(f, g) - \int_B \lambda fg \, d\rho.$$

If L is the operator defining $\mathscr{E}$, the operator generated by $\tilde{\mathscr{E}}$ is given by

$$(11) \qquad Hf(i) = Lf(i) - \lambda(i)f(i)\frac{\rho(i)}{\mu(i)}.$$

There are three things to be done: we must show that under reasonable conditions and a suitable choice of λ the form $\tilde{\mathscr{E}}$ is S-bounded from below (such that the theory of Section 5.1 applies); that it differs from $\mathscr{E}$ in standard part; and that the perturbation is carried on C. In this section we shall study these problems in a general setting, while in the next two sections we shall take a look at what happens in the examples we mentioned above—potentials concentrated in points and along Brownian paths. The most surprising discovery we shall make concerns the choice of λ; we shall find examples where it is necessary to choose λ infinitesimal, but where the standard form induced by $\tilde{\mathscr{E}}$ still is different from E. Hence there are perturbations of E which cannot standardly be expressed by (9), but which nevertheless are induced by hyperfinite forms as in (10). This happens for point interactions in $\mathbb{R}^2$ and $\mathbb{R}^3$ and perturbations along Brownian paths in $\mathbb{R}^4$ and $\mathbb{R}^5$.

A. The Computation

We can now begin the work. Assume that H is given by (11), and that L, λ, ρ, and μ all are non-negative. Our main tool will be the relationship between forms and resolvents established in Section 5.1, and we set out by computing the resolvent of H. If $G_\alpha = (L - \alpha)^{-1}$, we get

$$(12) \qquad (H - \alpha)^{-1} = G_\alpha\left(I - \frac{\lambda\rho}{\mu}G_\alpha\right)^{-1} = G_\alpha \sum_{l=0}^{\infty}\left(\frac{\lambda\rho}{\mu}G_\alpha\right)^l,$$

provided, of course, that the series on the right converges. Since $\|G_\alpha\| \le -1/\alpha$, there is a $z_1 \in {}^*\mathbb{R}_-$ such that it does converge when $\alpha < z_1$. There is no *a priori* reason why this z_1 should be finite, but we are not going to let this stop us; we just assume $\alpha < z_1$ and continue our calculations.

Applying (12) to a function f, we get

$$(13) \qquad (H - \alpha)^{-1}f(x) = \int (H - \alpha)^{-1}(x, y)f(y)\, d\mu(y)$$

$$= \int_Y G_\alpha(x, y)f(y)\, d\mu(y)$$

$$+ \sum_{l=1}^{\infty}\int_{Y^{l+1}}\cdots\int G_\alpha(x, x_1)\frac{\lambda(x_1)\rho(x_1)}{\mu(x_1)}$$

$$\times G_\alpha(x_1, x_2)\cdots\frac{\lambda(x_{l-1})\rho(x_{l-1})}{\mu(x_{l-1})}G_\alpha(x_{l-1}, x_l)$$

$$\times \frac{\lambda(x_l)\rho(x_l)}{\mu(x_l)}G_\alpha(x_l, y)f(y)\, d\mu(x_l)\cdots d\mu(x_1)\, d\mu(y),$$

where $G_\alpha(x, y)$ and $(H - \alpha)^{-1}(x, y)$ are the kernels (or matrices, if you want) of $(L - \alpha)^{-1}$ and $(H - \alpha)^{-1}$ with respect to μ. Since an integral $\int_Y \cdots d\mu(x_i)$ is in fact a hyperfinite sum $\sum_Y \cdots \mu(x_i)$ (and similarly for integrals with respect to ρ), we get from (13):

$$(14) \quad (H - \alpha)^{-1}f(x) = \int G_\alpha(x, y)f(y) \, d\mu(y)$$

$$+ \sum_{l=1}^{\infty} \int_{B^l} \cdots \iint_Y G_\alpha(x, x_1)\lambda(x_1)G_\alpha(x_1, x_2) \cdots \lambda(x_{l-1})$$

$$\times G_\alpha(x_{l-1}, x_l)\lambda(x_l)G_\alpha(x_l, y)f(y) \, d\mu(y) \, d\rho(x_l) \cdots d\rho(x_1).$$

To get (14) on a more transparent form, we introduce three new operators. First, let $\hat{G}_\alpha$ be the internal operator from $L^2(Y, \mu)$ to $L^2(B, \rho)$ given by

$$(15) \qquad\qquad (\hat{G}_\alpha g)(x) = \int_Y G_\alpha(x, y)g(y) \, d\mu(y),$$

and let $\hat{G}_\alpha^*$ be its adjoint:

$$(16) \qquad\qquad (\hat{G}_\alpha^* g)(x) = \int_B G_\alpha(x, y)g(y) \, d\rho(y).$$

Finally, let G'_α be the operator mapping $L^2(B, \rho)$ to itself according to the formula:

$$(17) \qquad\qquad (G'_\alpha g)(x) = \int_B G_\alpha(x, y)g(y) \, d\rho(y).$$

If we also let

$$(18) \qquad\qquad v(x) = \lambda(x)^{1/2},$$

we can write (14) as

$$(19) \qquad (H - \alpha)^{-1}f(x) = G_\alpha f(x) + \sum_{l=0}^{\infty} \hat{G}_\alpha^* v(vG'_\alpha v)^l v\hat{G}_\alpha f(x),$$

where v acts by multiplication.

There exists a $z_2 \in {}^*\mathbb{R}$ such that $\alpha < z_2$ implies

$$(20) \qquad\qquad \|vG'_\alpha v\| < 1,$$

and for such α

$$(21) \qquad (1 - vG'_\alpha v)^{-1} = \sum_{l=0}^{\infty} (vG'_\alpha v)^l.$$

Inserting this in (19), we have for $\alpha < \min(z_1, z_2)$

$$(22) \qquad (H - \alpha)^{-1} f(x) = G_\alpha f(x) + \hat{G}^*_\alpha \left(\frac{1}{\lambda} - G'_\alpha\right)^{-1} \hat{G}_\alpha f(x).$$

Happily calculating within the nonstandard universe, we have been able to express the resolvent of H as the sum of the resolvent of L and a perturbation term. The calculations leading us to (22) are only valid for $\alpha < \min(z_1, z_2)$, but since all operators involved are analytic functions of α, the final result must hold whenever the expression on the right is defined. This is the case when $(1/\lambda - G'_\alpha)$ is a strictly positive operator (recall 5.1.15), and according to the calculation

$$(23) \qquad \left\langle \left(\frac{1}{\lambda} - G'_\alpha\right) f, f \right\rangle_{L^2(B,\rho)}$$

$$= \int_B \left(\frac{f(x)}{\lambda(x)} - \int G_\alpha(x, y) f(y) \, d\rho(y)\right) f(x) \, d\rho(x)$$

$$= \int_B f(x)^2 \left(\frac{1}{\lambda(x)} - \int_B G_\alpha(x, y) \, d\rho(y)\right) d\rho(x)$$

$$+ \frac{1}{2} \iint_{BB} G_\alpha(x, y)(f(x) - f(y))^2 \, d\rho(x) \, d\rho(y)$$

it suffices to choose λ such that

$$(24) \qquad \frac{1}{\lambda(x)} \geq \int G_\alpha(x, y) \, d\rho(y) + \varepsilon$$

for some $\varepsilon \in {}^*\mathbb{R}_+$ and all $x \in B$. Notice also that for fixed f

$$\left\langle \left(\frac{1}{\lambda} - G'_\alpha\right) f, f \right\rangle_{L^2(B,\rho)} = \int_B \frac{f^2}{\lambda} \, d\rho - \left\langle (L - \alpha)^{-1} \frac{f\rho}{\mu}, \frac{f\rho}{\mu} \right\rangle_{L^2(Y,\mu)}$$

increases as α decreases, and thus if (24) holds for $\alpha = \alpha_0$, the operator $(1/\lambda - G'_\alpha)$ is positive for all $\alpha < \alpha_0$.

From (22) we get

$$(25) \quad \langle (H - \alpha)^{-1} f, f \rangle = \langle (L - \alpha)^{-1} f, f \rangle + \left\langle \left(\frac{1}{\lambda} - G'_\alpha\right)^{-1} (\hat{G}_\alpha f), (\hat{G}_\alpha f) \right\rangle_{L^2(B,\rho)}$$

and thus $(H - \alpha)$ is positive if $((1/\lambda) - G'_\alpha)$ is. This implies that the form $\tilde{\mathscr{E}}$ generated by H is bounded from below, and that its standard part $\tilde{E}$ can be defined by using the theory in Sections 5.1 and 5.2. Notice that since λ and ρ are positive and $\tilde{\mathscr{E}}$ is bounded from below, ${}^\circ|\tilde{\mathscr{E}}_1(u, u)| = \infty$ implies ${}^\circ|\mathscr{E}_1(u, u)| = \infty$, and hence the domain of $\tilde{E}$ is at least as large as that of the original form E. According to Theorem 5.1.19.

$$
(26) \qquad \tilde{E}(f, g) = - \lim_{{}^\circ\alpha \to -\infty} {}^\circ\Big\{ \alpha^2 \langle G_\alpha \hat{f}, \hat{g} \rangle + \alpha \langle \hat{f}, \hat{g} \rangle
$$
$$
+ \alpha^2 \Big\langle \Big(\frac{1}{\lambda} - G'_\alpha \Big)^{-1} \hat{G}_\alpha \hat{f}, \hat{G}_\alpha \hat{g} \Big\rangle_{L^2(B,\rho)} \Big\},
$$

where $\hat{f}, \hat{g} \in SL^2(Y, \mu)$ are liftings of f and g.

We can summarize our results as follows:

6.2.1. LEMMA. Assume that there is a finite $\alpha_0 \in \mathbb{R}_-$ such that

$$
(27) \qquad \frac{1}{\lambda(x)} \geq \int_B G_{\alpha_0}(x, y)\, d\rho(y) + \varepsilon
$$

for some positive $\varepsilon \in \mathbb{R}$ and all $x \in B$. Then $\tilde{\mathscr{E}}$ is bounded from below by α_0, and its standard part $\tilde{E}$ is a closed form given by (26). Moreover, $\mathscr{D}(\tilde{E}) \supset \mathscr{D}(E)$.

B. Nontriviality

The lemma above is just the starting point for our more serious investigations. Although it tells us when the perturbed form $\tilde{E}$ exists, it leaves the more important question of when the perturbation is nontrivial, i.e., $\tilde{E} \neq E$, unanswered, and it is to this problem that we now turn.

Recall that according to Lemma 5.2.4, all we have to do is find $\alpha \in {}^*\mathbb{R}_-$ and an internal function f satisfying the following four conditions:

$$
(28) \qquad -\infty < {}^\circ\alpha < {}^\circ\alpha_0,
$$

$$
(29) \qquad {}^\circ E_1(f, f) < \infty,
$$

$$
(30) \quad v = (L - \alpha)^{-1} f \quad \text{and} \quad w = (H - \alpha)^{-1} f \quad \text{are nearstandard,}
$$

$$
(31) \qquad {}^\circ\| v - w \| \neq 0.
$$

Since L is constructed from the standard operator A, the S-bounded operator $(L - \alpha)^{-1}$ maps nearstandard elements to nearstandard elements. By (22)

$$
(32) \qquad (H - \alpha)^{-1} f - (L - \alpha)^{-1} f = \hat{G}^*_\alpha (1/\lambda - G'_\alpha)^{-1} \hat{G}_\alpha f,
$$

and thus it suffices to find a nearstandard f satisfying (29) such that the right-hand side of (32) is nearstandard and noninfinitesimal.

We introduce the appropriate conditions in stages. First we shall give a quite abstract formulation in terms of the operators G'_α and $\hat{G}_\alpha$ and then, in the next subsection, a less lofty but much more useful one in terms of the integral kernel $G_\alpha(x, y)$ and the measures ρ and μ. But before the technical work begins, there is one more remark we would like to make. In the calculations above, we have used the L^2 norm on the space of internal functions on B. From now on we shall mainly use the L^1 norm. The reason for this is purely technical; although the two approaches yield almost exactly the same results, the L^1 case leads to simpler calculations both here and, especially, in Section 6.4.

We shall say that $\hat{G}_\alpha$ is ρ-*dense* if for all nearstandard $g \in L^1(B, \rho)$ and all $\varepsilon \in \mathbb{R}_+$, there is a nearstandard $u \in L^2(Y, \mu)$ such that $\hat{G}_\alpha u$ is S-bounded, ${}^\circ\mathscr{E}(u, u) < \infty$, and

(33) $$\|g - \hat{G}_\alpha u\|_{L^1(B, \rho)} < \varepsilon.$$

An internal operator $T: L^\infty(B, \rho) \to L^2(Y, \mu)$ is called *compact* if it maps all S-bounded functions to nearstandard ones (recall Propositions 2.1.15 and 2.2.4). If we instead consider T as an operator from $L^1(B, \rho)$ to $L^2(Y, \mu)$, we say that T is S-*bounded off* B if whenever K is a compact set such that

$$K \cap \text{st}(B) = \varnothing,$$

then

$$T: L^1(B, \rho) \to L^2(Y, \mu_K)$$

is S-bounded, where μ_K is the internal measure

$$\mu_K(A) = \mu(A \cap {}^*K).$$

The operator formulation of the nontriviality result now reads as follows.

6.2.2. LEMMA. Assume that $G_\beta(\cdot, \cdot) \geq 0$ for all $\beta \leq \alpha_0$ and that there is a finite α, ${}^\circ\alpha < {}^\circ\alpha_0$, such that $\hat{G}_\alpha$ is ρ-dense and $\hat{G}_\alpha^*$ is compact as an operator from $L^\infty(B, \rho)$ to $L^2(Y, \mu)$ and S-bounded off B as an operator from $L^1(B, \rho)$ to $L^2(Y, \mu)$. Assume further that there is an S-bounded function h such that

(34) $$g = (1/\lambda - G'_\alpha)h$$

is nearstandard in $L^1(B, \rho)$ and

(35) $${}^\circ\|\hat{G}_\alpha^* h\|_{L^2(Y, \mu)} \neq 0,$$

and, finally, that there is a positive $\varepsilon \in \mathbb{R}$ such that

(36) $$\frac{1}{\lambda(x)} - \int G_{\alpha_0}(x, y) \, d\rho(y) \geq \varepsilon$$

for all $x \in B$. Then the standard form $\tilde{E}$ which $\tilde{\mathscr{E}}$ induces on $L^2(X, m)$ is different from E.

PROOF. As we have already observed, it suffices to find a nearstandard function f satisfying (29) which makes the right-hand side of (32) nearstandard and noninfinitesimal.

Note that by the resolvent equation

$$G_{\alpha_0} - G_\alpha = (\alpha_0 - \alpha) G_\alpha G_{\alpha_0}$$

and the non-negativity of the kernels $G_\alpha(x, y)$ and $G_{\alpha_0}(x, y)$, the function $\alpha \mapsto G_\alpha(x, y)$ decreases as α goes to $-\infty$ for all x and y. Hence (36) holds with α_0 replaced by α, and consequently $(1/\lambda - G'_\alpha)$ is invertible and (34) can be rewritten as

$$(37) \qquad h = ((1/\lambda) - G'_\alpha)^{-1} g.$$

Since h is S-bounded and $\hat{G}^*_\alpha$ is compact, the function

$$u = \hat{G}^*_\alpha h = \hat{G}^*_\alpha ((1/\lambda) - G'_\alpha)^{-1} g$$

is nearstandard and noninfinitesimal in $L^2(Y, \mu)$.

Although we do not know if $g = \hat{G}_\alpha f$ for some nearstandard f in $L^2(Y, \mu)$, the ρ-denseness of G_α implies that there is a sequence $\{f_n\}_{n\in\mathbb{N}}$ of nearstandard functions such that each $\hat{G}_\alpha f_n$ is S-bounded, $°\mathscr{E}(f_n, f_n) < \infty$, and

$$\|\hat{G}_\alpha f_n - g\|_{L^1(B,\rho)} \leq 1/n.$$

It is easy to check that $(1/\lambda - G'_\alpha)^{-1}$ has norm less than $1/\varepsilon$ both as an operator on $L^1(B, \rho)$ and as an operator on $L^\infty(B, \rho)$. To see this for the L^∞ case, let

$$\varepsilon_\alpha(x) = \frac{1}{\lambda(x)} - \int G_\alpha(x, y)\, d\rho(y) \geq \varepsilon$$

and assume that f achieves its maximum at x. Then

$$\left\|\left(\frac{1}{\lambda} - G'_\alpha\right)f\right\|_\infty \geq \left|\frac{f(x)}{\lambda(x)} - \int G_\alpha(x, y)f(y)\, d\rho(y)\right|$$

$$= \left|\varepsilon_\alpha(x)f(x) + \int G_\alpha(x, y)(f(x) - f(y))\, d\rho(y)\right|$$

$$\geq |\varepsilon_\alpha(x)f(x)| \geq \varepsilon \|f\|_\infty,$$

where we have used the positivity of $G_\alpha(x, y)$. The L^1 case is left to the reader.

Applying $(1/\lambda - G'_\alpha)^{-1}$ to the sequence $\{\hat{G}_\alpha f_n\}$, we see that

$$\left\{\left(\frac{1}{\lambda} - G'_\alpha\right)^{-1} \hat{G}_\alpha f_n\right\}_{n\in\mathbb{N}}$$

is a sequence of S-bounded functions converging to h in $L^1(B, \rho)$. By the conditions on $\hat{G}_\alpha^*$, the functions

$$u_n = \hat{G}_\alpha^* \left(\frac{1}{\lambda} - G_\alpha' \right)^{-1} \hat{G}_\alpha f_n$$

are nearstandard in $L^2(Y, \mu)$ and converge to u in all the spaces $L^2(Y, \mu_K)$. It only remains to show that $^\circ \| u_n \|_{L^2(Y, \mu)} \neq 0$ for some n, as we can then take f to be the corresponding f_n.

But this is easy: m is a Radon measure with $m(\mathrm{st}(B)) = 0$ and u is nearstandard; hence there must be a compact K such that $K \cap \mathrm{st}(B) = \varnothing$ and $^\circ \| u \|_{L^2(\mu_K)} \neq 0$. Since $\{u_n\}$ converges to u in $L^2(\mu_K)$, there must be an n such that $0 < \, ^\circ \| u_n \|_{L^2(\mu_K)} \leq \, ^\circ \| u_n \|_{L^2(\mu)}$.

REMARK. The condition $G_\beta(\cdot, \cdot) \geq 0$ for all $\beta \leq \alpha_0$ is satisfied if, for instance,

$$\mathscr{E}(f, g) = \mathscr{E}_D(f, g) + \int Vfg \, d\mu,$$

where $\mathscr{E}_D$ is a Dirichlet form and V is a function bounded from below by α_0.

C. The Main Result

The conditions of Lemma 6.2.2 are complicated and numerous, and the reader may have the impression that they have been chosen for the sole purpose of making the proof work. To show that this is not really the case, we shall now translate them into simple and natural conditions on the integral kernels $G_\alpha(x, y)$.

Let us first fix the following standing conditions that are satisfied for the resolvent kernels of most interesting operators [see, e.g., Section B7 of Simon (1982) for the Schrödinger case]:

6.2.3. CONDITIONS. For all finite α less than α_0, we have

(i)   If x, y, x', y' are nearstandard elements of Y and $x \approx x'$, $y \approx y'$, but $x \not\approx y$, then $G_\alpha(x, y)$ is nearstandard and

$$G_\alpha(x, y) \approx G_\alpha(x', y').$$

(ii)   If $x \in \mathrm{Ns}(Y)$ and $y \in Y - \mathrm{Ns}(Y)$, then

$$G_\alpha(x, y) \approx 0.$$

An easy but important consequence of (i) and (ii) is:

(iii)   If K is compact, B is internal, and $K \cap \mathrm{st}(B) = \varnothing$, then the restriction of $G_\alpha(x, y)$ to $^*K \times B$ is S-bounded.

In addition to 6.2.3 we need integrability conditions. As a first example we have:

6.2.4. LEMMA. Assume that $G_\alpha(\cdot, \cdot)$ satisfies Conditions 6.2.3 and that

$$G_\alpha(x, y)G_\alpha(x, z) \in SL^1(\mu \times \rho \times \rho).$$

Then $\hat{G}_\alpha^*$ is compact as an operator from $L^\infty(B, \rho)$ to $L^2(Y, \mu)$ and S-bounded off B as an operator from $L^1(B, \rho)$ to $L^2(Y, \mu)$.

PROOF. We first check the S-boundedness. Let K be a compact set such that $K \cap \mathrm{st}(B) = \varnothing$. By (iii) above,

$$|G_\alpha(x, y)| \le N$$

for some $N \in \mathbb{N}$ and all $x \in {}^*K$, $y \in B$. Hence

$$\|\hat{G}_\alpha^* f\|_{L^2(\mu_K)}^2 = \int_{*K} \left(\int_B G_\alpha(x, y) f(y) \, d\rho(y) \right)^2 d\mu_K(x)$$

$$\le N^2 \int_{*K} \left(\int f(y) \, d\rho(y) \right)^2 d\mu_K(x)$$

$$\le N^2 \mu({}^*K)^2 \|f\|_{L^1(\rho)}^2,$$

and since $\mu({}^*K)$ is finite, this proves that $\hat{G}_\alpha^*$ is S-bounded off B.
For the compactness part, we first note that since

$$\int_A \left(\int G_\alpha(x, y) f(y) \, d\rho(y) \right)^2 d\mu(x)$$

$$\le \|f\|_\infty \int_{A \times B \times B} |G_\alpha(x, y) G_\alpha(x, z)| \, d(\mu \times \rho^2)(x, y, z)$$

for all internal f and A, the function $\hat{G}_\alpha^* f$ is in $SL^2(\mu)$ for all S-bounded f.
Observe next that by Keisler's Fubini theorem, 3.2.14, (suitably extended to cover infinite measures), the function

(38) $(y, z) \mapsto G_\alpha(x, y) G_\alpha(x, z)$

is in $SL^1(\rho^2)$ for almost all x. If x is such an element which is not nearstandard, the right-hand side of

$$|\hat{G}_\alpha^* f(x)|^2 = \left(\int G_\alpha(x, y) f(y) \, d\rho(y) \right)^2$$

$$\le \|f\|_\infty^2 \int |G_\alpha(x, y) G_\alpha(x, z)| \, d\rho^2(y, z)$$

is infinitesimal since $G_\alpha(x, y) \approx 0$ when y is nearstandard, and $L(\rho)(Y - \mathrm{Ns}(Y)) = 0$.

Finally, we pick two infinitely close, nearstandard points x_1, x_2 such that $\text{st}(x_1) \notin \text{st}(B)$ and the function (38) is in $SL^1(\rho^2)$ when $x = x_1$ or $x = x_2$. Then

$$|\hat{G}_\alpha^* f(x_1)^2 - \hat{G}_\alpha^* f(x_2)^2|$$

$$\leq \|f\|_\infty^2 \int |G_\alpha(x_1, y)G_\alpha(x_1, z) - G_\alpha(x_2, y)G_\alpha(x_2, z)| \, d\rho^2(y, z),$$

which is infinitesimal by 6.2.3(ii).

We have shown that $\hat{G}_\alpha^* f$ is in $SL^2(\mu)$, that it vanishes at almost all non-nearstandard points, and that it is S-continuous outside a set of measure zero. Hence it is nearstandard in $L^2(Y, \mu)$, and the lemma is proved.

What we have just proved takes care of the requirements 6.2.2 puts on $\hat{G}_\alpha^*$, but we still have to consider the conditions on $(1/\lambda) - G_\alpha'$. The main difficulty is that we have to operate with two different parameters α_0 and α; while we are allowed to *choose* λ such that

$$(1/\lambda) - G_{\alpha_0}'$$

satisfies the conditions of 6.2.1, we must *prove* that

$$(1/\lambda) - G_\alpha'$$

satisfies the conditions of 6.2.2. Since

$$(1/\lambda) - G_\alpha' = ((1/\lambda) - G_{\alpha_0}') + (G_{\alpha_0}' - G_\alpha'),$$

and the resolvent equation tells us that

$$G_{\alpha_0} - G_\alpha = (\alpha_0 - \alpha)G_\alpha G_{\alpha_0},$$

the operator we need to study is given by

$$(39) \quad (G_{\alpha_0}' - G_\alpha')f(y) = (\alpha_0 - \alpha) \int\int G_\alpha(x, y)G_{\alpha_0}(x, z)f(y) \, d\mu(x) \, d\rho(z).$$

6.2.5. LEMMA. Assume that G_α satisfies 6.2.3 and that

$$(40) \qquad G_\alpha(x, y)G_{\alpha_0}(x, z) \in SL^1(\mu \times \rho^2).$$

Then $(G_{\alpha_0}' - G_\alpha')h$ is nearstandard in $L^1(B, \rho)$ for all S-bounded, internal functions h.

PROOF. By (39), (40), and Keisler's Fubini theorem, 3.2.14, we get

$$(G_{\alpha_0}' - G_\alpha')h \in SL^1(\rho).$$

By the same theorem there is a set $A \subset B$ of $L(\rho)$ measure one such that

$$(41) \qquad (x, z) \mapsto G_\alpha(x, y)G_{\alpha_0}(x, z) \in SL^1(\mu \times \rho)$$

for all $y \in A$. Let y_1, y_2 be infinitely close, nearstandard elements of A. Then

$$|(G'_{\alpha_0} - G'_\alpha)h(y_1) - (G'_{\alpha_0} - G'_\alpha)h(y_2)|$$

$$\leq (\alpha_0 - \alpha)\|h\|_\infty \int_B \int_Y |G_\alpha(x, y_1)$$

$$- G_\alpha(x, y_2)||G_{\alpha_0}(x, z)|\, d\mu(x)\, d\rho(z),$$

which is infinitesimal by (41) and Conditions 6.2.3. This completes the proof.

We are now ready for the main theorem.

6.2.6. THEOREM. Assume that there are two finite numbers $\alpha_0, \alpha \in {}^*\mathbb{R}$, ${}^\circ\alpha < {}^\circ\alpha_0 < 0$, such that G_α is ρ-dense and satisfies Conditions 6.2.3, and $G_\beta(\cdot, \cdot) \geq 0$ for all $\beta \leq \alpha_0$. Assume further that

$$(42) \qquad G_{\alpha_0}(x, y)G_{\alpha_0}(x, z) \in SL^1(\mu \times \rho^2),$$

and that λ is chosen such that

$$(43) \qquad \frac{1}{\lambda(x)} - \int G_{\alpha_0}(x, y)\, d\rho(y)$$

is nearstandard in $L^1(B, \rho)$ and bounded from below by a positive real number. Then the standard part of $\tilde{\mathcal{E}}$ is a nontrivial perturbation of E.

PROOF. We shall use Lemma 6.2.2 with $h = 1$. Observe that since G_α is positive and ρ-dense, ${}^\circ\|\hat{G}^*_\alpha 1\|_{L^2(\mu)} \neq 0$. Note also that since $G_\beta(\cdot, \cdot) \geq 0$, the function $\alpha \mapsto G_\alpha(x, y)$ decreases with α for all x and y, and consequently (42) implies the integrability conditions in 6.2.4 and 6.2.5. By the first of these lemmas, the conditions 6.2.2 puts on $\hat{G}^*_\alpha$ are satisfied, and we only have to check that

$$(44) \qquad g = ((1/\lambda) - G'_\alpha)(1) = ((1/\lambda) - G'_{\alpha_0})(1) + (G'_{\alpha_0} - G'_\alpha)(1)$$

is nearstandard in $L^1(B, \rho)$. But $(G'_{\alpha_0} - G'_\alpha)(1)$ is nearstandard by 6.2.5 and

$$(45) \qquad \left(\frac{1}{\lambda} - G'_{\alpha_0}\right)(1)(x) = \frac{1}{\lambda(x)} - \int G_{\alpha_0}(x, y)\, d\rho(y)$$

by choice of λ, and the theorem is proved.

In the introduction to this section we mentioned as the most interesting case the situation where λ is infinitesimal, but where we still get a nontrivial perturbation. To see how this occurs, observe that we are forced to choose λ infinitesimal when

$$(46) \qquad \int G_{\alpha_0}(x, y)\, d\rho(y)$$

is infinite. As G_{α_0} usually has a singularity on the diagonal, this is likely to happen when ρ is concentrated on a "small" set. By our arguments above, it is not really the finiteness of (46) which determines when $\tilde{\mathscr{E}}$ is a nontrivial perturbation, but rather the finiteness of the difference

(47) $$\int G_{\alpha_0}(x, y)\, d\rho(y) - \int G_\alpha(x, y)\, d\rho(y).$$

At first glance it may seem improbable that this difference between infinite objects should be finite, but it is here that the resolvent equation enters the scene and reminds us that

$$\int G_{\alpha_0}(x, y)\, d\rho(y) - \int G_\alpha(x, y)\, d\rho(y) = (\alpha_0 - \alpha) \int G_\alpha G_{\alpha_0}(x, y)\, d\rho(y),$$

where

$$G_\alpha G_{\alpha_0}(x, y) = \int G_\alpha(x, z) G_{\alpha_0}(z, y)\, d\mu(z).$$

The key point is that thanks to the integration with respect to μ, the new kernel $G_\alpha G_{\alpha_0}(\cdot, \cdot)$ is usually much less singular than the old one $G_{\alpha_0}(\cdot, \cdot)$, and hence there is a fair chance that (47) will be finite even when (46) is not. It is in such cases that we get nontrivial perturbations from infinitesimal λ's.

To illustrate the difference between $G_{\alpha_0}(\cdot, \cdot)$ and $G_\alpha G_{\alpha_0}(\cdot, \cdot)$, let us just quote the following results about the resolvent R_α of the Laplacian $-\Delta$ in $\mathbb{R}^d$ (see Lemma 6.3.1 for more detailed information): for $d = 1$, $R_{\alpha_0}(x, y)$ is continuous and bounded; for $d = 2$, it has a singularity on the diagonal proportional to $-\ln\|x - y\|$; and for $d > 2$, the singularity on the diagonal is proportional to $\|x - y\|^{2-d}$. The integral kernel $R_{\alpha_0} R_{\alpha_0}(x, y)$, on the other hand, is bounded and continuous for $d \leq 3$; has a singularity of type $-\ln\|x - y\|$ for $d = 4$; and has a singularity of type $\|x - y\|^{4-d}$ for $d > 4$. As we shall see examplified in the next two sections, a consequence of this behavior is that perturbations of the Laplacian induced by infinitesimal λ's usually exist two dimensions higher than those coming from finite values of λ.

D. The Case of Standard λ's

But let us take a closer look at what happens in situations where we can choose λ noninfinitesimal. From the nonstandard point of view this case may seem less exciting than the other one, but the hyperfinite theory still has a significant contribution to make; it gives an efficient method for showing that a form

$$(48) \qquad E(f, g) - \int \tilde{\lambda}(x)\tilde{f}(x)\tilde{g}(x) \, d\tilde{\rho}(x)$$

is closable.

To set the stage for the theorem, let us recall the basic ingredients of the problem: E is a densely defined, closed, non-negative form on $L^2(X, m)$; the probability measure $\tilde{\rho}$ is supported on a closed set C of m measure zero, and $\tilde{\lambda}: C \to \mathbb{R}$ is a bounded Borel function. Let E_0 be the form defined on the bounded and continuous elements of $\mathscr{D}[E]$ by

$$(49) \qquad E_0(\tilde{f}, \tilde{g}) = E(\tilde{f}, \tilde{g}) - \int \tilde{\lambda}(x)\tilde{f}(x)\tilde{g}(x) \, d\tilde{\rho}(x).$$

We want to know when E_0 can be extended to a closed form.

For purely technical reasons we shall restrict ourselves to the case where E is a Dirichlet form. We can then choose a hyperfinite representation $\mathscr{E}$ that is itself a Dirichlet form, and if G_α is the resolvent of $\mathscr{E}$, then the kernel $G_\alpha(x, y)$ is non-negative for all x, y, and α, and

$$(50) \qquad \|-\alpha G_\alpha(f)\|_\infty \le \|f\|_\infty$$

for all α and f. In the remark which follows the proof of the theorem, we comment briefly on what happens when E is not a Dirichlet form.

6.2.7. THEOREM. Let $\mathscr{E}$ be a hyperfinite representation of a Dirichlet form E and ρ a hyperfinite representation of $\tilde{\rho}$. Let λ be an S-bounded lifting of $\tilde{\lambda}$ with respect to ρ. If G_α denotes the resolvent of $\mathscr{E}$, assume that

$$(51) \qquad \lim_{{}^\circ\alpha \to -\infty} {}^\circ\left\| \int G_\alpha(\cdot, y) \, d\rho(y) \right\|_{L^1(\rho)} = 0,$$

and that each $\tilde{f} \in \mathscr{D}[E_0]$ has an S-bounded $L^2(\mu + \rho)$-lifting f such that ${}^\circ\mathscr{E}(f, f) < \infty$ and

$$(52) \qquad \lim_{{}^\circ\alpha \to -\infty} {}^\circ\|f + \alpha \hat{G}_\alpha f\|_{L^2(\rho)} = 0.$$

If there exist an $\varepsilon \in \mathbb{R}_+$ and a finite $\alpha_0 \in {}^*\mathbb{R}_-$ such that

$$(53) \qquad \lambda(x) \le \left(\int G_\alpha(x, y) \, d\rho(y) + \varepsilon \right)^{-1}$$

for all x, then the standard part of $\tilde{\mathscr{E}}(f, g) = \mathscr{E}(f, g) - \int \lambda fg \, d\rho$ is a closed extension of E_0.

PROOF. That the standard part of $\tilde{E}$ is closed follows immediately from (53) and Lemma 6.2.1. To prove the extension part, let $\tilde{f}, \tilde{g} \in \mathscr{D}[E_0]$ and pick liftings f and g according to (52). By (26) it suffices to prove that

$$(54) \qquad E_0(\tilde{f}, \tilde{g}) = -\lim_{{}^\circ \alpha \to -\infty} {}^\circ\left(\alpha^2 \langle G_\alpha f, g \rangle + \alpha \langle f, g \rangle \right.$$
$$\left. + \alpha^2 \left\langle \left(\frac{1}{\lambda} - G'_\alpha \right)^{-1} \hat{G}_\alpha f, \hat{G}_\alpha g \right\rangle_{L^2(\rho)} \right),$$

and since by 5.1.19

$$(55) \qquad E(\tilde{f}, \tilde{g}) = -\lim_{{}^\circ \alpha \to -\infty} {}^\circ(\alpha^2 \langle G_\alpha f, g \rangle + \alpha \langle f, g \rangle),$$

this reduces to showing that

$$(56) \qquad \lim_{{}^\circ \alpha \to -\infty} {}^\circ\left\langle \left(\frac{1}{\lambda} - G'_\alpha \right)^{-1} (-\alpha \hat{G}_\alpha) f, (-\alpha \hat{G}_\alpha) g \right\rangle_{L^2(\rho)} = {}^\circ\!\int \lambda f g \, d\rho.$$

Since by (52)

$$\lim_{{}^\circ \alpha \to -\infty} {}^\circ\langle \lambda(-\alpha \hat{G}_\alpha) f, (-\alpha \hat{G}_\alpha) g \rangle_{L^2(\rho)} = {}^\circ\!\int \lambda f g \, d\rho,$$

(56) will follow if we can prove that

$$(57) \qquad \lim_{{}^\circ \alpha \to -\infty} {}^\circ\left\{ \left\langle \left(\frac{1}{\lambda} - G'_\alpha \right)^{-1} (-\alpha \hat{G}_\alpha) f, (-\alpha \hat{G}_\alpha) g \right\rangle_{L^2(\rho)} \right.$$
$$\left. - \langle \lambda(-\alpha \hat{G}_\alpha) f, (-\alpha \hat{G}_\alpha) g \rangle_{L^2(\rho)} \right\} = 0.$$

Now

$$(58) \qquad \left\langle \left(\frac{1}{\lambda} - G'_\alpha \right)^{-1} (-\alpha \hat{G}_\alpha) f, (-\alpha \hat{G}_\alpha) g \right\rangle_{L^2(\rho)}$$
$$- \langle \lambda(-\alpha \hat{G}_\alpha) f, (-\alpha \hat{G}_\alpha) g \rangle_{L^2(\rho)}$$
$$= \left\langle \left(\frac{1}{\lambda} - G'_\alpha \right)^{-1} (-\alpha \hat{G}_\alpha) f, (-\alpha \hat{G}_\alpha) g \right\rangle_{L^2(\rho)}$$
$$- \left\langle \lambda \left(\frac{1}{\lambda} - G'_\alpha \right) \left(\frac{1}{\lambda} - G'_\alpha \right)^{-1} (-\alpha \hat{G}_\alpha) f, (-\alpha \hat{G}_\alpha) g \right\rangle_{L_2(\rho)}$$
$$= \left\langle \lambda G'_\alpha \left(\frac{1}{\lambda} - G'_\alpha \right)^{-1} (-\alpha \hat{G}_\alpha) f, (-\alpha \hat{G}_\alpha) g \right\rangle_{L^2(\rho)},$$

and since $G_\alpha(x, y) \geq 0$, we know that $(1/\lambda - G'_\alpha)^{-1}$ has norm less than $1/\varepsilon$ as an operator on $L^\infty(B)$. By (50), the functions $(-\alpha \hat{G}_\alpha) f$ and $(-\alpha \hat{G}_\alpha) g$ are S-bounded, and hence (51) implies that the last term in (58) goes to zero.

REMARK. The only use we have made of the assumption that $\mathscr{E}$ is a Dirichlet form is in showing that $((1/\lambda) - G'_\alpha)^{-1}(-\alpha\hat{G}_\alpha)f$ and $(-\alpha\hat{G}_\alpha)g$ are S-bounded; in the general case, all that is immediately clear is that they are in $L^2(B, \rho)$. This suffices if we replace (51) by the stronger assumption

$$(59) \qquad \lim_{{}^\circ\alpha \to -\infty} {}^\circ\left\| \int G_\alpha(x, y)^2 \, d\rho(y) \right\|_{L^1(\rho)} = 0,$$

but since the singularities of $G_\alpha(x, y)$ blow up when we square them, (59) is much too restrictive. Hence a satisfactory treatment of general forms requires finer estimates, and these we leave to the reader.

E. Translation into Standard Terms

The results above are phrased in terms of the resolvent G_α of the hyperfinite form $\mathscr{E}$. From a nonstandard point of view this is only natural, but in a standard setting a formulation using the resolvent R_α of the standard form E is often more convenient. Since translating our results into standard terms is not quite as trivial as it may look at first glance, we shall give a brief sketch of what seems to be the best approach.

The idea is simple; given a standard form E satisfying standard counterparts of the conditions of Theorem 6.2.6 or 6.2.7, we shall show that it is the standard part of a hyperfinite form satisfying the original conditions. In Section 5.2, we used the associated semigroup to construct hyperfinite representations, but in the present context, where semigroups are less important than resolvents, an alternative line of attack is preferable.

We return to the setting of Section 5.2. Thus K is a standard Hilbert space, and H is a hyperfinite-dimensional, S-dense subset of $^*K$. Assume that E is a symmetric, non-negative, densely defined form on K, and let R_α be its resolvent. Fix an $\alpha_0 \in \mathbb{R}_-$, and define

$$(60) \qquad G_{\alpha_0} = P^* R_{\alpha_0} P,$$

where $P: {}^*K \to H$ is the orthogonal projection. Clearly, G_{α_0} is invertible as a linear map from H to H, and

$$(61) \qquad A = G_{\alpha_0}^{-1} + \alpha_0$$

is a symmetric, non-negative operator on H. The form

$$(62) \qquad \mathscr{E}(u, v) = \langle Au, v \rangle$$

is the hyperfinite representation of E we shall be working with.

Let us check that E really is the standard part of $\mathscr{E}$. If G_α is the resolvent of $\mathscr{E}$ and $f \in K$, the two functions from $\mathbb{R}_-$ to $\mathbb{R}$ given by

$$(63) \qquad \alpha \mapsto \langle R_\alpha f, f \rangle \quad \text{and} \quad \alpha \mapsto {}^\circ\langle G_\alpha {}^*f, {}^*f \rangle$$

are analytic. From the series expansions

$$(64) \qquad G_\alpha = \sum_{n=0}^{\infty} (\alpha - \alpha_0)^n G_{\alpha_0}^{n+1} = \sum_{n=0}^{\infty} (\alpha - \alpha_0)^n (P^* R_{\alpha_0} P)^{n+1}$$

$$(65) \qquad R_\alpha = \sum_{n=0}^{\infty} (\alpha - \alpha_0)^n R_\alpha^{n+1},$$

it is easy to see that they agree in a neighborhood of α_0, and hence they are equal for all negative α. If F is the standard part of $\mathscr{E}$, we know from 5.1.19 that

$$(66) \qquad F(f, f) = -\lim_{{}^\circ\alpha \to -\infty} {}^\circ(\alpha^2 \langle G_\alpha {}^*f, {}^*f \rangle + \alpha \langle P^*f, P^*f \rangle),$$

and by the corresponding standard result [see, e.g., Lemma 1.3.4 in Fukushima (1980)], we get

$$(67) \qquad E(f, f) = \lim_{\alpha \to -\infty} -(\alpha^2 \langle R_\alpha f, f \rangle + \alpha \langle f, f \rangle).$$

Since the two expressions are equal, E is the standard part of $\mathscr{E}$.

Returning to the setting of this section, from now on we assume that $K = L^2(X, m)$, where X is a Hausdorff space and m is a Radon measure. The following conditions are the standard counterparts of 6.2.3(i)–(ii):

6.2.8. CONDITIONS. Assume that the resolvent R_α has a symmetric kernel $R_\alpha(\cdot, \cdot)$ satisfying:

(i) $R_\alpha(\cdot, \cdot)$ is continuous off the diagonal; i.e., at all points (x, y) where $x \neq y$.

(ii)   For all $x \in X$ and $\varepsilon \in \mathbb{R}_+$, there is a compact K_ε such that $|R_\alpha(x, y)| < \varepsilon$ if $y \notin K_\varepsilon$.

REMARK. Note that these conditions imply that if x has no compact neighborhood, then $R_\alpha(x, y) = 0$ for all $y \neq x$, and that they therefore may seem rather worthless in nonlocally compact spaces. However, in the infinite-dimensional case, the situation can often be saved by choosing a weaker topology than the one originally given.

Recall that C is a closed subset of X of m measure zero, and that $\tilde{\rho}$ is a probability measure supported on C. We shall say that R_α is $\tilde{\rho}$-*dense* if for each $g \in L^1(C, \tilde{\rho})$ and each $\varepsilon \in \mathbb{R}_+$, there is a $u \in \mathscr{D}[E]$ such that

$$\int R_\alpha(\cdot, y) u(y) \, dm(y)$$

is bounded on C and differs from g with less than ε in $L^1(\tilde{\rho})$ norm. Note that by the resolvent equation

$$R_\alpha = R_{\alpha_0}(I - (\alpha - \alpha_0)R_\alpha),$$

all the R_α's are $\tilde{\rho}$-dense if and only if one of them is.

The following definition makes precise what we are looking for.

6.2.9. DEFINITION. A symmetric, lower bounded form $\tilde{E}$ is a *perturbation of E supported on C* if:

   (i) $\mathscr{D}[E] \subseteq \mathscr{D}[\tilde{E}]$;
   (ii) $E \neq \tilde{E}$;
   (iii) whenever $f, g \in \mathscr{D}[E]$ and there is a neighborhood of C where f is continuous and g vanishes, then $E(f, g) = \tilde{E}(f, g)$.

We can now translate 6.2.6 into the following theorem.

6.2.10. THEOREM. Let E be a densely defined, non-negative, symmetric form on $L^2(X, m)$, where X is a Hausdorff space and m is a Radon measure, and let $\tilde{\rho}$ be a Radon probability measure supported on a closed set $C \subset X$ of m measure zero. Assume that the resolvent kernel $R_\alpha(x, y)$ is non-negative, $\tilde{\rho}$-dense, and satisfies 6.2.8. If in addition

(68) $$\|\alpha R_\alpha f + f\|_{L^2(\tilde{\rho})} \to 0 \text{ as } \alpha \to -\infty$$

for all $f \in \mathscr{D}[E]$ which are continuous in a neighborhood of C, and there is an $\alpha_0 \in \mathbb{R}_-$ such that

(69) $$R_{\alpha_0}(x, y)R_{\alpha_0}(x, z) \in L^1(m \times \rho^2),$$

then E has a perturbation supported on C.

PROOF. We begin by making a careful choice of hyperfinite representations Y, μ, B, and ρ of X, m, C, and $\tilde{\rho}$, respectively. Recall from the proof of Corollary 3.4.10 that there is a hyperfinite equivalence relation $\equiv$ on $^*X$ such that each equivalence class is a *Borel set, and the equivalence class of each nearstandard point is contained in its monad. Refine $\equiv$ to an equivalence relation $\sim$ by letting $x \sim y$ if and only if $x \equiv y$ and either none or both of x and y belong to $^*C$. For each equivalence class p of $\sim$, choose an element $x_p \in p$, and define

$$Y = \{x_p \,|\, p \text{ is an equivalence class}\},$$

$$B = \{x_p \,|\, p \text{ is an equivalence class}, p \subset {}^*C\}.$$

Let μ be the hyperfinite measure on Y given by

$$\mu\{x_p\} = {}^*m(p)$$

and ρ the measure on B defined by

$$\rho\{x_p\} = {}^*\tilde{\rho}(p).$$

Note that $m = L(\mu) \circ \mathrm{st}^{-1}$ and $\tilde{\rho} = L(\rho) \circ \mathrm{st}^{-1}$, and that μ and ρ have disjoint supports.

If P denotes the natural "projection" of ${}^*L^2(X, m)$ on $L^2(Y, \mu)$, let

$$(70) \qquad\qquad G_{\alpha_0} = P^* R_{\alpha_0} P$$

as in (60). As already observed, E is the standard part of the hyperfinite form $\mathscr{E}$ defined in (62).

The key observation is that if $\tilde{\nu} = m + \tilde{\rho}$, then

$$(71) \quad G_{\alpha_0}(x_p, x_q) = {}^*\tilde{\nu}(p)^{-1}{}^*\tilde{\nu}(q)^{-1} \int_p \int_q {}^*R_{\alpha_0}(x, y)\, d^*\tilde{\nu}(x)\, d^*\tilde{\nu}(y)$$

is a kernel for G_{α_0} as an operator from $L^2(Y, \mu)$ to itself. Using this formula, it is easy to check that the $\tilde{\rho}$-denseness of R_α implies the ρ-denseness of G_α, and that the assumption of R_α satisfying 6.2.8 implies that G_α satisfies 6.2.3.

Let $\nu = \mu + \rho$. Then

$$\int G_{\alpha_0}(x, y) G_{\alpha_0}(x, z)\, d(\mu \times \rho^2)(x, y, z)$$

$$= \int_{Y-B} \left(\int_B G_{\alpha_0}(x, y)\, d\nu(y) \right)^2 d\nu(x)$$

$$\leq \int_{{}^*X - {}^*C} \left(\int_{{}^*C} {}^*R_{\alpha_0}(x, y)\, d^*\nu(y) \right)^2 d^*\tilde{\nu}(x)$$

$$= \int R_{\alpha_0}(x, y) R_{\alpha_0}(x, z)\, d(m \times \tilde{\rho}^2)(x, y, z),$$

and thus (69) implies the S-integrability of $G_{\alpha_0}(x, y) G_{\alpha_0}(x, z)$. But then all the assumptions of 6.2.6 are satisfied, and the standard form $\tilde{E} = {}^\circ\tilde{\mathscr{E}}$ is a nontrivial perturbation of E.

It only remains to check that the perturbation is supported on C in the sense of 6.2.9(iii). Assume that there is a neighborhood of C where f is continuous and g vanishes, and let $\bar{f}, \bar{g} \in L^2(Y, \nu)$ be liftings of f and g with respect to ν. From (68) we get that

$$(72) \qquad\qquad {}^\circ\|\alpha \hat{G}_\alpha \bar{f} + \bar{f}\|_{L^2(\rho)} \to 0$$

and

$$(73) \qquad\qquad {}^\circ\|\alpha \hat{G}_\alpha \bar{g}\|_{L^2(\rho)} \to 0$$

as $°\alpha \to -\infty$. By (26)

$$(74) \quad \tilde{E}(f, g) - E(f, g) = \lim_{°\alpha \to -\infty} \langle((1/\lambda) - G'_\alpha)^{-1}(-\alpha\hat{G}_\alpha)\bar{f}, (-\alpha\hat{G}_\alpha)\bar{g}\rangle_{L^2(\rho)},$$

where the right-hand side is zero by (72), (73), and the uniform boundedness of the operators $(1/\lambda - G'_\alpha)^{-1}$. The theorem is proved.

REMARK. Note that E will in general have many different perturbations supported on C since we are free to choose the difference

$$(1/\lambda(x)) - \int G_{\alpha_0}(x, y)\, d\rho(y)$$

in 6.2.6, and also, to a certain extent, the measure $\tilde{\rho}$ on C.

Using the methods above, it is easy to prove the following standard version of 6.2.7:

6.2.11. THEOREM. Let E be a Dirichlet form on $L^2(X, m)$, where X is a Hausdorff space and m is a Radon measure, and assume that the resolvent kernel $R_\alpha(x, y)$ satisfies 6.2.8. Given a closed set $C \subset X$ of m measure zero, a Radon probability measure $\tilde{\rho}$ supported on C, and a bounded Borel function $\tilde{\lambda}: C \to \mathbb{R}$, let E_0 be the form defined on the bounded and continuous elements of $\mathscr{D}[E]$ by

$$(75) \qquad E_0(f, g) = E(f, g) - \int \tilde{\lambda}fg\, d\tilde{\rho}.$$

Assume that

$$(76) \qquad \lim_{\alpha \to -\infty} \int\int R_\alpha(x, y)\, d\tilde{\rho}(x)\, d\tilde{\rho}(y) = 0;$$

that for all $f \in D[E_0]$

$$(77) \qquad \lim_{\alpha \to -\infty} \|f + \alpha R_\alpha f\|_{L^2(\tilde{\rho})} = 0;$$

and that there exist $\alpha_0 \in \mathbb{R}_-$, $\varepsilon \in \mathbb{R}_+$ such that for all x

$$(78) \qquad \tilde{\lambda}(x) \le \left(\int R_{\alpha_0}(x, y)\, d\tilde{\rho}(y) + \varepsilon\right)^{-1};$$

then E_0 is closable.

Although there are no continuity conditions on the resolvent kernel in Theorem 6.2.7, we are now assuming that R_α satisfies Conditions 6.2.8. This is just a convenient way of guaranteeing sufficient regularity for (76)–(78) to make sense.

In our applications we shall be studying perturbations of the free Hamiltonian form

$$E(f, g) = \int (-\Delta f)g \, dx$$

in $\mathbb{R}^d$. Some of the assumptions of Theorems 6.2.10 and 6.2.11 are automatically satisfied for this form, and to make clear exactly what has to be checked, we state the following corollary.

6.2.12. COROLLARY. Let Δ be the Laplacian and m the Lebesgue measure in $\mathbb{R}^d$. Let E_0 be the form defined on $C_0^2(\mathbb{R}^d)$ by

$$E_0(f, g) = \int (-\Delta f)g \, dm,$$

and let E be its closure. Assume that $\tilde{\rho}$ is a Radon probability measure supported on a closed set C of Lebesgue measure zero

(a) If for some $\alpha_0 \in \mathbb{R}_-$

$$(79) \qquad \int \int R_{\alpha_0} R_{\alpha_0}(x, y) \, d\tilde{\rho}(x) \, d\tilde{\rho}(y) < \infty,$$

then E has a perturbation supported on C.

(b) If $\tilde{\lambda}: C \to \mathbb{R}$ is a bounded Borel function and there exist $\alpha_0 \in \mathbb{R}_-$, $\varepsilon \in \mathbb{R}_+$ such that $\tilde{\lambda}(x) \le (\int R_{\alpha_0}(x, y) \, d\tilde{\rho}(y) + \varepsilon)^{-1}$ and

$$(80) \qquad \int \int R_{\alpha_0}(x, y) \, d\tilde{\rho}(x) \, d\tilde{\rho}(y) < \infty,$$

then the form $E_0(f, g) - \int \tilde{\lambda}fg \, d\tilde{\rho}$ is closable.

PROOF. That R_α satisfies 6.2.8 is well known; see, e.g., the explicit formulas given in Lemma 6.3.1 below. If $f \in \mathscr{D}[E_0]$ is continuous in a neighborhood of C, then $-\alpha R_\alpha f$ converges uniformly to f on C, and hence $\|f + \alpha R_\alpha f\|_{L^2(\rho)} \to 0$. Since $\{R_\alpha u \mid u \in \mathscr{D}[E_0]\} \supset C_0^4(\mathbb{R}^d)$ is dense in $L^1(\tilde{\rho})$, the resolvent R_α is $\tilde{\rho}$-dense. Finally, since $R_\alpha(x, y)$ decreases to zero at all nonsingular points, (80) implies (76). Hence all the conditions of Theorems 6.2.10 and 6.2.11 are satisfied, and the corollary is proved.

We shall take our leave of the general theory of singular perturbations here; as already announced, the next two sections will be dedicated to the study of two important special cases—point interactions and potentials supported on Brownian paths. The theory we have presented above is new, and, partly for that reason, this section is among the most open-ended in the book; although we have a fairly good understanding of when perturbations supported on null sets exist, we know almost nothing about their properties. Perhaps the work which has been done on point interactions (and which we review briefly in the next section) can serve as a guide for the general theory; among the topics people have studied are spectral properties, resonances, and convergence of approximating operators. It also seems evident that there must be a connection between singular perturbations and Hausdorff measure and dimension, which has not yet been exploited. Let us finally mention that in collaboration with Karwowski we have developed an alternative approach to Corollary 6.2.12 using Fourier transforms and ultraviolet cutoffs. We hope to give an account of this method in Albeverio *et al* (1986a) [see also the announcement Albeverio *et al.* (1984b)].

6.3. POINT INTERACTIONS

Point interactions (i.e., perturbations of the free Hamiltonian supported on a discrete set) have a long and venerable history. Kronig and Penney (1931) used operators of the form

$$(1) \qquad\qquad -\frac{d^2}{dx^2} - \sum_{i \in \mathbb{Z}} \lambda \delta_i$$

to model one-dimensional crystals, and a few years later, when Wigner (1933) showed that the diplon (the system consisting of a proton and a neutron) is held together by forces of extremely short range, a three-dimensional model based on point interactions was developed by Bethe and Peierls (1935). Fermi (1936) used a similar approach to study the motion of neutrons through hydrogenous substances, but although the work was continued and extended to N-body problems by, e.g., Huang, Lee, Luttinger, Yang, and Wu [see Huang and Yang (1957), Huang *et al.* (1957), Lee *et al.* (1957), and Wu (1959)] in the 1950s, no mathematical foundation for the theory existed until 1961. In that year, Berezin and Faddeev (1961) not only showed how to interpret one-point interactions as self-adjoint operators, but also gave a complete classification of the resulting Hamiltonians [see the excellent survey article by Flamand (1967) and the book by Demkov and Ostrovskii (1975) for accounts of this and related work]. Friedman (1971, 1972) showed that operators of this kind can be obtained as limits of operators defined by ordinary potentials, and in his AMS address on internal set theory, Nelson (1977) gave a nonstandard treatment of these results. Friedman and Nelson did not obtain the full classification of Berezin and Faddeev; the gap was filled by Albeverio *et al.* (1979b), who also extended the theory from perturbations in one point to perturbations in an arbitrary, finite number of points. The latest nonstandard contribution to the field is the thesis of Alonso y Coria (1978).

During the last few years, our understanding of point interactions and their properties has grown rapidly, and the rich and complex theory which has been created is described in detail in the monograph by Albeverio *et al.* (1986b). In this section our aim is much more modest; we just want to show you two nonstandard ways of constructing point interactions (recall that we have already taken a brief look at a third method in Section 5.6).

A. Application of the General Theory

The first method is based on the theory of the previous section; we simply want to apply Corollary 6.2.12. To obtain the necessary information about the kernels $R_\alpha(\cdot, \cdot)$ and $R_\alpha R_\alpha(\cdot, \cdot)$, note that

$$R_\alpha = \int_0^\infty T^t \cdot e^{\alpha t} \, dt$$

and

$$R_\alpha R_\alpha = \int_0^\infty t T^t \cdot e^{\alpha t} \, dt,$$

where T^t is the semigroup of $-\Delta$. Since T^t is given by the Gaussian kernel

$$T^t(x, y) = (4\pi t)^{-d/2} \exp\left(-\frac{\|x - y\|^2}{4t}\right),$$

we get

$$(2) \qquad R_\alpha(x, y) = \int_0^\infty (4\pi t)^{-d/2} \exp\left(-\frac{\|x - y\|^2}{4t}\right) \exp(\alpha t) \, dt$$

and

$$(3) \ (R_\alpha R_\alpha)(x, y) = \int_0^\infty \frac{1}{4\pi} (4\pi t)^{-d/2+1} \exp\left(-\frac{\|x - y\|^2}{4t}\right) \exp(\alpha t) \, dt.$$

Since $R_\alpha(x, y)$ and $(R_\alpha R_\alpha)(x, y)$ depend only on the distance $r = \|x - y\|$ between the points x and y, we can treat them as one-dimensional functions $R_\alpha^{(d)}(r)$ and $(R_\alpha R_\alpha^{(d)})(r)$, where we have introduced the superscript d to emphasize the dependence on the dimension. From (2) and (3) it follows immediately that

$$(4) \qquad\qquad (R_\alpha R_\alpha)^{(d)}(r) = \frac{1}{4\pi} R_\alpha^{(d-2)}(r).$$

Our first lemma summarizes well-known asymptotic properties of $R_\alpha^{(d)}(r)$ as r goes to zero and to infinity:

6.3.1. LEMMA. Let $R_\alpha^{(d)}(r)$ be as above, $\alpha \in \mathbb{R}_-$:

(i) If $d < 2$, then $R_\alpha^{(d)}(r)$ is continuous and bounded and converges uniformly to zero as $\alpha \to -\infty$.

(ii) When $d \geq 2$, $R_\alpha^{(d)}(r)$ is continuous off the diagonal, and the asymptotic expression as $r \to \infty$ is

$$R_\alpha^{(d)}(r) \sim \tfrac{1}{2}(2\pi)^{-(d-1)/2}(-\alpha)^{(d-3)/4}r^{-(d-1)/2}\,e^{-\sqrt{-\alpha}\,r}$$

(iii) When $d = 2$, the asymptotic behavior as $r \to 0$ is

$$R_\alpha^{(2)}(r) \sim -\frac{1}{2\pi}\ln(\sqrt{-\alpha}\,r)$$

(iv) When $d > 2$, the asymptotic behavior as $r \to 0$ is

$$R_\alpha^{(d)}(r) \sim \frac{\Gamma[(d-2)/d]}{4\pi^{d/2}}\,r^{-d+2}.$$

Most of the properties above should be known to the reader from the basic theory of partial differential equations (after all, R_α is nothing but the Green function), and we shall not prove the lemma here. To obtain the finer estimates, one may either integrate the right-hand side of (2) in terms of Bessel functions [see Gradshteyn and Ryshik (1965), Eq. 3.471.9], or use Fourier transforms as in Glimm and Jaffe (1981).

The important thing to notice about 6.3.1 is that the singularity at the origin gets worse as the dimension increases. By (4), the singularity of $(R_\alpha R_\alpha)$ grows two dimensions slower than the singularity of R_α, and this is why 6.2.12(a) is a more powerful tool than 6.2.12(b). Let us take a look at what happens for point interactions.

6.3.2. PROPOSITION. Let E be the closed form on $L^2(\mathbb{R}^d, m)$ (where m is the Lebesgue measure) generated by $-\Delta$. If $d \leq 3$ and $a_1, \ldots, a_n \in \mathbb{R}^d$, there is a nontrivial perturbation of E supported on $\{a_1, \ldots, a_n\}$. If $d = 1$ and $\beta_1, \ldots, \beta_n \in \mathbb{R}$, the form defined on the continuous and bounded elements of $\mathcal{D}[E]$ by

$$\tilde{E}(f, g) = E(f, g) - \sum_{i=1}^{n} \beta_i f(a_i)g(a_i) \tag{5}$$

is closable.

PROOF. By 6.2.12 combined with 6.3.1.

It is known that if $d \geq 4$, then $-\Delta$ is essentially self-adjoint even when restricted to functions vanishing in the vicinity of $\{a_1, \ldots, a_n\}$, and thus E has no perturbations supported on finite sets. In this sense 6.3.2 is the best possible result. On the other hand, we can use Theorem 6.2.6 and formula 6.2.26 to obtain more detailed information about the possible perturbations of E; if we, for instance, choose the λ and the ρ in 6.2.6 uniform, the perturbation in each point a_i is the same. But we shall leave the closer study of the properties of the perturbed operators to the reader; instead we shall take a look at a different nonstandard approach to point interactions.

B. An Alternative Approach

So far our strategy in this chapter has been to reformulate a given problem as a discrete problem in a hyperfinite setting. The technique is quite general and allows us to treat successfully a number of problems concerning singular coefficients and perturbations. It is often possible, however, to give a continuous nonstandard treatment; i.e., we may use all of $*X$ as the underlying space rather than a hyperfinite, S-dense subset Y. In the discrete approach one is first led to a problem in linear algebra, whose solution is later pushed down to the standard space by the theory of Chapter 5 or some similar device. In the continuous approach a singular standard problem in X is replaced by a sufficiently regular problem in $*X$, and again a meaningful standard answer may be forthcoming if one is only able to prove the necessary smoothness of the nonstandard solution. As an example, we shall see how point interactions in $\mathbb{R}^3$ can be treated by this method.

If χ_ε is the indicator function of the interval $[0, \varepsilon]$, we first consider self-adjoint operators on the *complex* space $L^2(\mathbb{R}_+, dr)$ of the form

$$(6) \qquad A = -(d^2/dr^2) + \lambda\chi_\varepsilon$$

with Dirichlet boundary conditions at $r = 0$, where λ is a constant. What this means is the following. Let E be the form defined on the functions in $C^\infty(\mathbb{R}_+)$ that vanish outside a compact subset of $(0, \infty)$, by

$$(7) \qquad E(f, g) = \int_0^\infty \left(-\frac{d^2}{dr^2}f + \lambda\chi_\varepsilon f \right) g \, dr.$$

If $\bar{E}$ is the closure of E, then A is the self-adjoint operator defined by $\bar{E}$. Similarly, we let A_0 be the self-adjoint operator on $L^2(\mathbb{R}_+, dr)$ given by

$$(8) \qquad A_0 = -\frac{d^2}{dr^2}$$

and Dirichlet boundary conditions at $r = 0$.

We shall choose $\varepsilon \approx 0$, and show that for certain infinitely large choices of λ, the nonstandard operator A induces a standard perturbation of A_0.

In a sense that we shall explain later, this solves the "radial part" of the problem of a point interaction at the origin in $\mathbb{R}^3$.

As in Section 6.2, our main tool will be the resolvent kernel of A. We fix an infinitesimal $\varepsilon > 0$, and let v_1, v_2 be solutions of the equation

$$(9) \qquad\qquad -u'' + \lambda\chi_\varepsilon u - \alpha u = 0$$

on $^*\mathbb{R}^+$, where $\alpha \in \mathbb{C} - \mathbb{R}_+$. If we choose v_1 and v_2 such that $v_1(0) = 0$ and $v_2(r) \to 0$ as $r \to \infty$, elementary Sturm-Liouville theory tells us that the quantity

$$(10) \qquad\qquad K = v_1'v_2 - v_1v_2'$$

is independent of r, and that the inverse of $(A - \alpha)$ is given by the integral kernel

$$(11) \qquad G_\alpha(x, y) = \frac{1}{K}\begin{cases} v_1(x)v_2(y) & \text{if } 0 \le x \le y \\ v_1(y)v_2(x) & \text{if } 0 \le y \le x \end{cases}$$

whenever $K \neq 0$.

Since $\lambda\chi_\varepsilon$ is constant on each of the intervals $[0, \varepsilon]$ and (ε, ∞), equation (9) can be solved explicitly. Assuming that $\text{Re}(\sqrt{\alpha}) > 0$, we see that v_1 and v_2 can be chosen as follows:

$$(12) \qquad v_1(r) = \begin{cases} \sin(\sqrt{\alpha - \lambda}\, r) & \text{if } 0 \le r \le \varepsilon \\ ae^{\sqrt{\alpha}\, r} + be^{-\sqrt{\alpha}\, r} & \text{if } \varepsilon \le r \end{cases}$$

and

$$(13) \qquad v_2(r) = \begin{cases} ce^{\sqrt{\lambda - \alpha}\, r} + de^{-\sqrt{\lambda - \alpha}\, r} & \text{if } 0 \le r \le \varepsilon \\ e^{-\sqrt{\alpha}\, r} & \text{if } \varepsilon \le r. \end{cases}$$

Since $v_1(r)$ and $v_1'(r)$ have to be continuous at $r = \varepsilon$, we get

$$(14) \qquad a = \tfrac{1}{2}e^{-\sqrt{\alpha}\,\varepsilon}\left(\sin\sqrt{\alpha - \lambda}\,\varepsilon + \sqrt{\frac{\alpha - \lambda}{\alpha}}\cos\sqrt{\alpha - \lambda}\,\varepsilon\right),$$

$$(15) \qquad b = \tfrac{1}{2}e^{\sqrt{\alpha}\,\varepsilon}\left(\sin\sqrt{\alpha - \lambda}\,\varepsilon - \sqrt{\frac{\alpha - \lambda}{\alpha}}\cos\sqrt{\alpha - \lambda}\,\varepsilon\right).$$

The corresponding expressions for c and d are

$$(16) \qquad c = \tfrac{1}{2}e^{-(\sqrt{\lambda-\alpha}+\sqrt{\alpha})\varepsilon}\left(1 - \sqrt{\frac{\alpha}{\lambda - \alpha}}\right),$$

$$(17) \qquad d = \tfrac{1}{2}e^{(\sqrt{\lambda-\alpha}-\sqrt{\alpha})\varepsilon}\left(1 + \sqrt{\frac{\alpha}{\lambda - \alpha}}\right).$$

Evaluating K as the limit of $v_1'v_2 - v_1v_2'$ as r goes to infinity, we find that

$$(18) \qquad\qquad K = 2a\sqrt{\alpha}.$$

Combining (12), (13), and (18), we can turn (11) into an explicit formula for $G_\alpha(x, y)$. Since A is a self-adjoint operator, $G_\alpha(x, y)$ is symmetric, and we need only consider the case $x \leq y$:

$$(19) \quad G_\alpha(x, y) = \begin{cases} \dfrac{\sin(\sqrt{\alpha - \lambda}\, x)}{2a\sqrt{\alpha}} (ce^{\sqrt{\lambda - \alpha}\, y} + d\bar{e}^{\sqrt{\lambda - \alpha}\, y}), & x \leq y \leq \varepsilon \\[2ex] \dfrac{1}{2a\sqrt{\alpha}} \sin(\sqrt{\alpha - \lambda}\, x)\, e^{-\sqrt{\alpha}\, y}, & x \leq \varepsilon \leq y \\[2ex] \dfrac{1}{2\sqrt{\alpha}} (e^{\sqrt{\alpha} x} + \dfrac{b}{a} e^{-\sqrt{\alpha} x}) e^{-\sqrt{\alpha} y}, & \varepsilon \leq x \leq y \end{cases}$$

On the other hand, the resolvent kernel of the unperturbed operator A_0 is

$$(20) \qquad R_\alpha(x, y) = \frac{1}{2\sqrt{\alpha}} (e^{\sqrt{\alpha} x} - e^{-\sqrt{\alpha} x}) e^{-\sqrt{\alpha} y}.$$

Our task is to find the λ's which make G_α different from R_α in standard part.

From (16) and (17), we get that if $°a$, $\alpha \neq 0$, then $G_\alpha(x, y)$ is finite when $x \approx y$. Hence $G_\alpha(x, y)$ has no singularity at the origin, and comparing (20) and the third clause in (19), we see that the only way G_α may differ from R_α in standard part is by

$$(21) \qquad \mathrm{st}(b/a) \neq -1.$$

We first observe that if λ is finite, then

$$(22) \qquad a \approx \frac{1}{2} \sqrt{\frac{\alpha - \lambda}{\alpha}}$$

and

$$(23) \qquad b \approx -\frac{1}{2} \sqrt{\frac{\alpha - \lambda}{\alpha}},$$

and thus $b/a \approx -1$. As for infinite λ, we see that if $\sqrt{(\alpha - \lambda)/\alpha} \cos \sqrt{\alpha - \lambda}\, \varepsilon$ is infinite, then again $b/a \approx -1$. This means that we have to choose λ infinite. but such that

$$(24) \qquad \sqrt{\frac{\alpha - \lambda}{\alpha}} \cos \sqrt{\alpha - \lambda}\, \varepsilon$$

is finite. For infinite λ

$$(25) \qquad \sqrt{\alpha - \lambda} \cos \sqrt{\alpha - \lambda}\, \varepsilon \approx \sqrt{-\lambda} \cos \sqrt{-\lambda}\, \varepsilon,$$

and thus it suffices to choose λ such that the quantity

$$(26) \qquad \tilde{\beta} = \sqrt{-\lambda} \cos(\sqrt{-\lambda}\, \varepsilon)$$

is finite. Hence we must let

(27) $$\sqrt{-\lambda}\,\varepsilon = (k + \tfrac{1}{2})\pi + \eta$$

for some $k \in {}^*\mathbb{Z}$ and some infinitesimal η. Solving for λ, we get

(28) $$\lambda = -\left(k + \frac{1}{2}\right)^2 \frac{\pi^2}{\varepsilon^2} - 2\frac{\pi}{\varepsilon}\left(k + \frac{1}{2}\right)\frac{\eta}{\varepsilon} - \left(\frac{\eta}{\varepsilon}\right)^2,$$

and combining (27) and (26), we see that

(29) $$\tilde{\beta} = \frac{\eta}{\varepsilon}\cos\left(\left(k + \frac{1}{2}\right)\pi + \eta\right) + \left(k + \frac{1}{2}\right)\pi\frac{\eta}{\varepsilon}\frac{\cos((k + \frac{1}{2})\pi + \eta)}{\eta}.$$

If we choose η/ε nearstandard and $k \in \mathbb{N}$, then $\tilde{\beta}$ has a standard part which is equal to the standard part of the second term in (29):

(30) $$\beta \equiv \mathrm{st}((-1)^k\tilde{\beta}) = -(k + \tfrac{1}{2})\pi\,\mathrm{st}(\eta/\varepsilon).$$

It turns out that without any loss of generality we may take η/ε to be a real number γ. Formulas (28) and (30) may then be rewritten as

(31) $$\lambda = -(k + \tfrac{1}{2})^2(\pi^2/\varepsilon^2) + (2/\varepsilon)\beta - \gamma^2$$

and

(32) $$\beta = -(k + \tfrac{1}{2})\pi\gamma.$$

The functions v_1 and v_2 are S-continuous and have S-continuous derivatives on ${}^*\mathbb{R}_+$. Hence their standard parts u_1 and u_2 exist and satisfy the equation

(33) $$-u'' - \alpha u = 0.$$

Moreover,

(34) $$\frac{u_1'(0)}{u_1(0)} = \mathrm{st}\frac{v_1'(\varepsilon)}{v_1(\varepsilon)},$$

and since $v_1(\varepsilon) = \sin(\sqrt{\alpha - \lambda}\,\varepsilon)$ and $v_1'(\varepsilon) = \sqrt{\alpha - \lambda}\cos(\sqrt{\alpha - \lambda}\,\varepsilon)$, we get

(35) $$\frac{u_1'(0)}{u_1(0)} = \mathrm{st}\left(\frac{\sqrt{\alpha - \lambda}\cos(\sqrt{\alpha - \lambda}\,\varepsilon)}{\sin(\sqrt{\alpha - \lambda}\,\varepsilon)}\right).$$

From (25), (26), and the fact that $\sin(\sqrt{\alpha - \lambda}\,\varepsilon) \approx (-1)^k$, we see that

(36) $$u_1'(0) = \beta u_1(0).$$

It follows that

(37) $$\tilde{R}_\alpha(x, y) = \begin{cases} u_1(x)u_2(y) & \text{if } x \le y \\ u_1(y)u_2(x) & \text{if } y \le x \end{cases}$$

is the resolvent of the operator $-d^2/dr^2$ on $\mathbb{R}_+$ with the boundary condition

(38) $$u'(0) = \beta u(0).$$

If $\tilde{A}_0$ is the restriction of $-d^2/dr^2$ to the C^∞ functions which vanish outside a compact subset of $(0, \infty)$, it is known that the only self-adjoint extensions of $\tilde{A}_0$ are the ones given by condition (38) when β runs through $\mathbb{R}$ [see, e.g., Reed and Simon (1975)]. Thus we have proved:

6.3.3. PROPOSITION. Consider the self-adjoint operator

(39) $$A_\beta = -\frac{d^2}{dr^2} + \lambda_\varepsilon(\beta)\chi_\varepsilon$$

with Dirichlet boundary conditions on the nonstandard Hilbert space $^*L^2(\mathbb{R}_+, dr)$, where $\lambda_\varepsilon(\beta)$ is a number of the form

(40) $$\lambda_\varepsilon(\beta) = -\left(k + \frac{1}{2}\right)^2 \frac{\pi^2}{\varepsilon^2} + \frac{2}{\varepsilon}\beta + \delta$$

with $\beta, \delta \in \mathbb{R}$, ε a positive infinitesimal, $k \in \mathbb{N}$, and χ_ε the indicator function of $[0, \varepsilon]$. Then A_β is nearstandard in the sense that its resolvent $(A_\beta - \alpha)^{-1}$ is nearstandard, and the standard part of $(A_\beta - \alpha)^{-1}$ is the resolvent of the self-adjoint operator $-d^2/dr^2$ with the boundary condition $u'(0) = \beta u(0)$. Thus when β runs through $\mathbb{R}$, the standard part of A_β runs through all self-adjoint extensions of $-d^2/dr^2$ independently of k and δ.

Comparing (31) and (40), we see that the term $(-\gamma^2)$—which depends on β through (32)—has been replaced by the independent term δ, but it is easy to check that this makes no difference.

Let us now explain the relationship between 6.3.3 and point interactions in $\mathbb{R}^3$. If $\tilde{\Delta}$ is the restriction of $\Delta = \sum_{i=1}^3 \partial^2/\partial x_i^2$ to the C^∞ functions that vanish outside a compact subset of $\mathbb{R}^3 - \{0\}$, we are interested in the self-adjoint extensions of $\tilde{\Delta}$. In polar coordinates Δ takes the form

(41) $$\Delta = \frac{\partial^2}{\partial r^2} + \frac{2}{r}\frac{\partial}{\partial r} + \frac{1}{r^2}B,$$

where B is the Laplace–Beltrami operator on $L^2(S^2)$. But B has discrete spectrum with eigenvalues $-l(l + 1)$, $l \in \mathbb{Z}^+$, of finite multiplicity, and thus it suffices to study the self-adjoint extensions of the restriction $\tilde{\Delta}_l$ of

(42) $$\Delta_l = \frac{d^2}{dr^2} + \frac{2}{r}\frac{d}{dr} - \frac{l(l + 1)}{r^2}$$

to those functions in $L^2(\mathbb{R}_+, r^2\, dr)$ which are C^∞ and vanish outside compact subsets of $\mathbb{R}_+ - \{0\}$.

The map $f(r) \to rf(r)$ is a unitary equivalence between $L^2(\mathbb{R}_+, r^2\, dr)$ and $L^2(\mathbb{R}_+, dr)$, which carries Δ_l into

$$(43) \qquad d^2/dr^2 - l(l+1)/r^2,$$

and since it also maps the class of C^∞ functions vanishing outside compact subsets of $\mathbb{R}_+ - \{0\}$ to itself, it suffices to study the self-adjoint extensions of (43). It is well known [see, e.g., Reed and Simon (1975)] that (43) is essentially self-adjoint for $l \geq 1$. We have therefore only to consider the self-adjoint extensions of $-d^2/dr^2$ on $L^2(\mathbb{R}_+, dr)$, and this is exactly what we did in 6.3.3.

We now fix a positive infinitesimal ε, and write Δ also for the self-adjoint Laplacian in the nonstandard space $^*H = {}^*L^2(\mathbb{R}^3)$. We want to know when the self-adjoint perturbation

$$(44) \qquad A_\beta = -\Delta + \lambda\chi_\varepsilon(\|x\|)$$

is nearstandard to a self-adjoint, nontrivial perturbation of the standard operator $-\Delta$. If we split $^*H$ into its rotationally symmetric part $^*H_S$ and the orthogonal complement $^*H_S^\perp$,

$$^*H = {}^*H_S \oplus {}^*H_S^\perp,$$

we get immediately that the restriction of A_β to $^*H_S^\perp$ is nearstandard and that its standard part is $-\Delta$ (this is because we have already remarked that the restriction of Δ to the C^∞ functions in $H_S^\perp$ that vanish outside a compact subset of $\mathbb{R}^3 - \{0\}$ is essentially self-adjoint). Hence we need only consider the restriction of A_β to $^*H_S$, and we have seen that this operator is unitary equivalent to (43). Combining our results, we have the following theorem.

6.3.4. THEOREM. Consider the self-adjoint operator

$$(45) \qquad A_\beta = -\Delta + \lambda_\varepsilon(B)\chi_\varepsilon$$

on $^*L^2(\mathbb{R}^3)$, where $\lambda_\varepsilon(\beta)$ is a number of the form

$$(46) \qquad \lambda_\varepsilon(\beta) = -(k + \tfrac{1}{2})^2(\pi^2/\varepsilon^2) + (2/\varepsilon)\beta + \delta$$

with $\beta, \delta \in \mathbb{R}$, $k \in \mathbb{N}$, ε a positive infinitesimal, and χ_ε the indicator function of the ball $\{\|x\| \leq \varepsilon\}$. Then the resolvent $(A_\beta - \alpha)^{-1}$ is nearstandard, and its standard part is independent of k and δ, but different for different choices of β. When β runs through $\mathbb{R}$, the standard part of A_β runs through all self-adjoint extensions of the restriction of $-\Delta$ to the C^∞ functions supported on compact subsets of $\mathbb{R}^3 - \{0\}$.

Theorem 6.3.4 gives an alternative description of the one-point perturbations of $-\Delta$ to the one we obtained in 6.3.2 by using the methods of Section 6.2. In Albeverio *et al.* (1979b), a third approach based on approximations by finite rank operators was also exploited, but we shall not discuss this here. Instead we shall give a short survey of recent contributions—standard as well as nonstandard—to the study of point perturbations.

We have already mentioned the nonstandard contributions by Nelson (1977), Albeverio *et al.* (1979b), and Alonso y Coria (1978). Using standard methods, Albeverio *et al.* (1977, 1980) gave the alternative description in terms of Dirichlet forms, which we described in Section 5.6, and Grossmann *et al.* (1980a,b) used approximations by finite rank operators to initiate the study of the finer properties of the perturbations such as eigenvalues and bound states, resonances, and infrared convergence. The reader should also consult contributions by Thomas (1979, 1980) and Zorbas (1980). The extension from perturbations in a finite number of points to discrete, countable sets was made in Grossmann *et al.* (1980a) [see also Svendsen (1981) for related models]. It is interesting to observe that this theory is not covered by Section 6.2 (since $\tilde{\rho}$ is a probability measure and $\tilde{\lambda}$ is a bounded function), but we should add that the infinite theory is obtained from the finite by a rather direct limit argument. In Albeverio and Høegh-Krohn (1981) point interactions were characterized as limits of smooth, local potentials, and this approach has been refined and extended by Albeverio *et al.* (1982c), Holden (1981), and Holden *et al.* (1983, 1984). Connections to the fashionable field of random potentials have been established in, e.g., Albeverio *et al.* (1982a). These are just a few of the many recent contributions to the field; the reader is referred to Albeverio *et al.* (1986b) and Demkov and Ostrovskii (1975) for additional information and references.

6.4. PERTURBATIONS BY LOCAL TIME FUNCTIONALS

We now turn to the study of perturbations of the Laplacian induced by potentials supported on Brownian paths. Given a Brownian motion

$$(1) \qquad\qquad b: \Omega \times [0, 1] \to \mathbb{R}^d,$$

we let C_ω denote the path

$$(2) \qquad\qquad C_\omega = \{b(\omega, t) \mid t \in [0, 1]\}$$

for each $\omega \in \Omega$. There is a natural measure on C_ω induced by b:

$$(3) \qquad\qquad \tilde{\rho}_\omega(A) = m_1\{t \in [0, 1] \mid b(\omega, t) \in A\};$$

where m_1 is the one-dimensional Lebesgue measure. We shall use the theory of Section 6.2 to study operators on $L^2(\mathbb{R}^d, m)$ given by forms

$$(4) \qquad \tilde{E}_\omega(f, g) = \int (-\Delta f) g \, dm - \int \lambda_\omega(x) f(x) g(x) \, d\tilde{\rho}_\omega(x)$$

(as usual, m is the Lebesgue measure on $\mathbb{R}^d$). Using the definition of $\tilde{\rho}_\omega$, we may rewrite (4) as

$$(5) \qquad \tilde{E}_\omega(f, g) = \int (-\Delta f) g \, dm - \int_0^1 \lambda_\omega(b(t)) f(b(t)) g(b(t)) \, dt,$$

and if δ is the delta function on $\mathbb{R}^d$, a formal calculation "shows" that the associated operator is

$$(6) \qquad H_\omega = -\Delta - V_\omega(\,\cdot\,),$$

where

$$(7) \qquad V_\omega(x) = \lambda_\omega(x) \int_0^1 \delta(x - b(t)) \, dt.$$

Comparing (7) and 3.3.8, we see that H_ω is a perturbation of $-\Delta$ by a local time functional. Although local time in the sense of 3.3.8 exists only in dimension one, we shall see that nontrivial perturbations of $-\Delta$ of the form (6) exist for $d \leq 5$. For $d = 4, 5$ it is necessary to choose λ_ω infinitesimal.

Before we turn to the technical part of the theory, we would like to give a brief sketch of the contents of this section. After having applied Corollary 6.2.12 to the form (4) to obtain the results just mentioned, we shall try to explain why local time perturbations are of interest to physicists. In this section we shall concentrate on models of polymer molecules, but in Section 7.5 we shall take a look at some applications to four-dimensional quantum field theory. As models of polymers, Brownian paths have some unrealistic features; e.g., a Brownian path in $\mathbb{R}^3$ will intersect itself infinitely often, while a polymer molecule does not intersect itself at all. To remedy this deficiency attempts have been made to replace the Brownian motion by a self-avoiding modification, but the probabilistic behavior of the process then becomes extremely complicated and little has yet been achieved. We shall end the section with brief accounts of two of the most promising of these attempts—Westwater's (1980, 1981, 1982, 1985) work on Edwards's polymer model and Lawler's (1980) on self-avoiding random walks.

A. Application of the General Theory

Our first lemma will show that the condition of Corollary 6.2.12 (a) is satisfied for almost all Brownian paths when the dimension d is less than or equal to 5. The proof consists in reducing the problem to a question about convergence of certain multiple integrals, and the basic observation is always the same—that for all $r > 0$

$$(8) \qquad \int_{\|x\| \leq r} \|x\|^{-p} \, dx \qquad \text{converges iff} \quad p < d,$$

(9) $\displaystyle\int_{\|x\|\geq r} \|x\|^{-p}\, dx$ converges iff $p > d,$

as is easily seen by changing to polar coordinates.

6.4.1. LEMMA. Let R_α be the resolvent of $-\Delta$ in $\mathbb{R}^d$. If $d \leq 5$ and $\alpha \in \mathbb{R}_-$, then

(10) $(R_\alpha R_\alpha)(\cdot, \cdot) \in L^1(\tilde{\rho}_\omega^2)$

for almost all ω.

PROOF. The integral kernel $(R_\alpha R_\alpha)(\cdot, \cdot)$ is described in detail by Lemma 6.3.1 and equation 6.3.4. When $d \leq 3$, the kernel is bounded and there is nothing to prove. We shall leave the case $d = 4$ to the reader and concentrate on dimension 5. To see what goes wrong in higher dimensions, let us try to carry out the proof for a general $d > 4$.

From Lemma 6.3.1 and formula 6.3.4 we know that there are positive constants C and β such that

(11) $(R_\alpha R_\alpha)^{(d)}(r) \leq C \dfrac{e^{-\beta r}}{r^{d-4}}.$

To prove the lemma it suffices to show that

$$I = E\left(\int\int \frac{\exp(-\beta\|x - y\|)}{\|x - y\|^{d-4}}\, d\tilde{\rho}_\omega(x)\, d\tilde{\rho}_\omega(y) \right)$$

$$= E\left(\int_0^1 \int_0^1 \frac{\exp(-\beta\|b(t) - b(s)\|)}{\|b(t) - b(s)\|^{d-4}}\, ds\, dt \right)$$

is finite. Introducing Gaussian kernels, we have

$$I = \int_{\mathbb{R}^d} \int_0^1 \int_0^1 \frac{\exp(-\beta\|u\|)}{\|u\|^{d-4}} \frac{\exp(-\|u\|^2/2|t - s|)}{(2\pi|t - s|)^{d/2}}\, ds\, dt\, du,$$

and substituting the new variable $S = (t - s)/\|u\|^2$ for s, this turns into

$$I = \int_{\mathbb{R}^d} \int_0^1 \int_{(t-1)/\|u\|^2}^{t/\|u\|^2} \frac{\exp(-\beta\|u\|)}{\|u\|^{2d-6}} \frac{\exp(-1/2|S|)}{(2\pi|S|)^{d/2}}\, dS\, dt\, du.$$

If we let

$$K = \int_{-\infty}^{\infty} (2\pi|S|)^{-d/2} \exp(-1/2|S|)\, dS$$

(the integral converges since $d > 2$), then

(12) $I \leq K \displaystyle\int_{\mathbb{R}^d} \frac{\exp(-\beta\|u\|)}{\|u\|^{2d-6}}\, du,$

which according to (8) converges if and only if $2d - 6 < d$, i.e., $d < 6$. The proof is complete.

Combining the lemma and Corollary 6.2.12 (a), we get:

6.4.2. THEOREM. Let $b : \Omega \times [0, 1] \to \mathbb{R}^d$ be a Brownian motion and m the Lebesgue measure in $\mathbb{R}^d$. If $d \le 5$, then for almost all $\omega \in \Omega$, the form

$$(13) \qquad E(f, g) = \int (-\Delta f) g \, dm$$

has a nontrivial, closed perturbation supported on the path

$$C_\omega = \{ b(\omega, t) \mid t \in [0, 1] \}.$$

The next problem we shall consider is when we can take the perturbation in the theorem to be of the form

$$(14) \qquad \tilde{E}_\omega(f, g) = \int (-\Delta f) g \, dm - \int_{C_\omega} \lambda_\omega f g \, d\tilde{\rho}_\omega$$

for some standard function $\lambda_\omega : C_\omega \to \mathbb{R}$. Combining the following lemma with 6.2.12 (b), we see that this is the case when $d \le 3$.

6.4.3. LEMMA. Let $R_\alpha(\cdot, \cdot)$ be the resolvent kernel of $-\Delta$ in $\mathbb{R}^d$. If $d \le 3$ and $\alpha < 0$, then

$$(15) \qquad R_\alpha(\cdot, \cdot) \in L^1(\tilde{\rho}_\omega^2)$$

for almost all ω.

PROOF. This is almost identical to the proof of the preceding lemma. By Lemma 6.3.1 there is nothing to prove for $d = 1$, and we shall leave the two-dimensional case to the reader. Assume that $d \ge 3$; then there are positive constants C and β such that for all $r > 0$

$$(16) \qquad R_\alpha^{(d)}(r) \le C \frac{e^{-\beta(r)}}{r^{d-2}},$$

and it therefore suffices to find those d for which

$$I = E\left(\int_0^1 \int_0^1 \frac{\exp(-\beta \|b(t) - b(s)\|)}{\|b(t) - b(s)\|^{d-2}} \, ds \, dt \right)$$

is finite. Since b is a Brownian motion

$$I = \int_{\mathbb{R}^d} \int_0^1 \int_0^1 \frac{\exp(-\beta \|u\|)}{\|u\|^{d-2}} \cdot \frac{\exp(-\|u\|^2/2|t - s|)}{(2\pi|t - s|)^{d/2}} \, ds \, dt \, du$$

$$= \int_{\mathbb{R}^d} \int_0^1 \int_{(t-1)/\|u\|^2}^{t/\|u\|^2} \frac{\exp(-\beta \|u\|)}{\|u\|^{2d-4}} \frac{\exp(-1/2|S|)}{(2\pi|S|)^{d/2}} \, dS \, dt \, du,$$

where we have substituted $S = (t - s)/\|u\|^2$ for s. Since $d > 2$, the integral

$$K = \int_{-\infty}^{\infty} \frac{\exp(-1/2|S|)}{(2\pi|S|)^{d/2}}\, dS$$

converges, and hence

(17) $$I \le K \int_{\mathbb{R}^d} \frac{\exp(-\beta\|u\|)}{\|u\|^{2d-4}}\, du,$$

which converges if and only if $2d - 4 < d$, i.e., for $d < 4$. This proves the lemma.

By Corollary 6.2.12b the form $\tilde{E}_\omega$ in (14) is closable if for some $\alpha \in \mathbb{R}_-$, $\varepsilon \in \mathbb{R}_+$, the inequality

(18) $$\lambda_\omega(x) \le 1 \bigg/ \left(\int R_\alpha(x, y)\, d\tilde{\rho}_\omega(y) + \varepsilon \right)$$

holds for $\tilde{\rho}_\omega$-a.a. x. From the lemma we see that for almost all ω, the function

$$\int R_\alpha(\cdot, y)\, d\tilde{\rho}_\omega(y)$$

is finite $\tilde{\rho}_\omega$-almost everywhere, and this implies the closability of $\tilde{E}_\omega$ for a large class of functions λ_ω. But the lemma is too weak to give a closer description of this class; e.g., it cannot tell us whether we are allowed to choose λ_ω positive and constant, which is the most interesting choice from the physical point of view. To settle this problem we shall analyze the situation with more care.

In dimension three the resolvent kernel of the Laplacian has a particularly simple form

(19) $$R_\alpha(x, y) = \frac{\exp(-\sqrt{-\alpha}\,\|x - y\|)}{4\pi\|x - y\|}.$$

For each $\alpha \in \mathbb{R}_-$, we define a stochastic process $X^\alpha : \Omega \times [0, 1] \to \mathbb{R}$ by

(20) $$X^\alpha(\omega, t) = 4\pi \int_{C_\omega} R_\alpha(b(t), y)\, d\tilde{\rho}_\omega(y)$$

$$= \int_0^1 \frac{\exp(-\sqrt{-\alpha}\,\|b(t) - b(s)\|)}{\|b(t) - b(s)\|}\, ds,$$

where the factor 4π is just for notational convenience. Observe that if we knew that for almost all ω

(21) $$X^\alpha(\omega, t) \to 0 \text{ uniformly in } t \text{ when } \alpha \to -\infty,$$

then we could conclude from Corollary 6.2.12b that the form (16) is closable for all bounded functions λ_ω.

To prove (21) we first observe that by Fatou's lemma

$$(22) \qquad X^\alpha(\omega, t) \leq \liminf_{s \to t} X^\alpha(\omega, s),$$

and thus the paths of X^α are lower semicontinuous. In a short while we shall establish the following estimate:

6.4.4. LEMMA. For each $\varepsilon \in \mathbb{R}_+$, there is a constant C_ε such that for all $\alpha \in \mathbb{R}_-$, $s, t \in [0, 1]$

$$(23) \qquad E(|X^\alpha(t) - X^\alpha(s)|^3) \leq C_\varepsilon |t - s|^{3/2 - \varepsilon}.$$

The conclusion of this lemma is the condition of Kolmogorov's continuity theorem [see, e.g., Simon (1979)], and hence the restriction of X^α to the dyadic rationals

$$D = \{(k/2^n) \mid k, n \in \mathbb{N}, 0 \leq k \leq 2^n\}$$

is continuous. We can define a continuous modification Y^α of X^α by

$$(24) \qquad Y^\alpha(\omega, t) = \lim_{\substack{s \to t \\ s \in D}} X^\alpha(\omega, s).$$

Since the paths of X^α are lower semicontinuous,

$$(25) \qquad X^\alpha(\omega, t) \leq Y^\alpha(\omega, t)$$

for all t, and the paths of X^α must be bounded.

Assuming Lemma 6.4.4, we may now prove:

6.4.5. THEOREM. Let Δ be the Laplace operator in $\mathbb{R}^3$. If $b : \Omega \times [0, 1] \to \mathbb{R}^3$ is a Brownian motion and m_1 is the Lebesgue measure on $[0, 1]$, then for each $\omega \in \Omega$ let the measure $\tilde{\rho}_\omega$ be defined by

$$\tilde{\rho}_\omega(A) = m_1\{t \in [0, 1] \mid b(\omega, t) \in A\}.$$

The form

$$\tilde{E}_\omega(f, g) = \int (-\Delta f) g \, dm - \int \lambda f g \, d\tilde{\rho}_\omega$$

is closable for all bounded functions λ.

PROOF. By Corollary 6.2.12 it is enough to prove (21). In fact, by (24) and (25) it suffices to show that for almost all ω

$$(26) \qquad X^\alpha(\omega, t) \to 0 \text{ uniformly for all } t \in D \text{ as } \alpha \to -\infty.$$

Note that since the paths of X^α are bounded, the limit in (26) holds pointwise by Lebesgue's convergence theorem.

Assume that (26) does not hold; then there is an internal set $\Omega_1 \subset \Omega$ of noninfinitesimal measure $P(\Omega_1) = \delta$ such that

(27) $$\forall \omega \in \Omega_1 \, \forall \alpha \in \mathbb{R}_- \, \exists t \in D(X^\alpha(\omega, t) \geq \varepsilon)$$

for some $\varepsilon \in \mathbb{R}_+$.

For each $n \in \mathbb{N}$ let

$$D_n = \{(k/2^n) \,|\, k \in \mathbb{N}, 0 \leq k \leq 2^n\}.$$

It follows from (27) that for each $\alpha \in \mathbb{R}_-$ there must be an integer $n(\alpha)$ such that

(28) $$P\{\omega \,|\, \exists t \in D_{n(\alpha)}(X^\alpha(\omega, t) \geq \varepsilon)\} \geq \delta/2.$$

Applying transfer, we see that for an infinitely large negative α, (28) holds for some $n(\alpha) \in {}^*\mathbb{N} - \mathbb{N}$. Since (26) is true when uniform convergence is replaced by pointwise convergence, ${}^*X^\alpha(\omega, t)$ must be infinitesimal at all standard points t. By the *-version of Lemma 6.4.4, the restriction of ${}^*X^\alpha$ to $D_{n(\alpha)}$ satisfies the condition of the hyperfinite Kolmogorov theorem, 4.8.5. Hence ${}^*X^\alpha \upharpoonright D_{n(\alpha)}$ is S-continuous, and since it is also infinitesimal at all standard points, it cannot possibly satisfy (28). Thus (26) holds and we are done.

B. The Basic Estimate

It still remains to prove Lemma 6.4.4, and we warn the reader at the very beginning that this is going to be quite messy. Our main tool will be the following estimate.

6.4.6. LEMMA. There is a constant C such that if $e \in \mathbb{R}^3$ has length $\frac{1}{2}$,

$$\int_{\mathbb{R}^3} \exp(-\beta \|k - a\|^2) \left| \frac{1}{\|k - e\|} - \frac{1}{\|k + e\|} \right| dk \leq \frac{C}{\sqrt{\beta}(\sqrt{\beta} + 1)}$$

for all $a \in \mathbb{R}^3$ and all $\beta \in \mathbb{R}_+$.

PROOF. Fix a, and introduce a new variable $y = \sqrt{\beta}\, k$:

$$I(\beta, a) = \int_{\mathbb{R}^3} \exp(-\beta \|k - a\|^2) \left| \frac{1}{\|k - e\|} - \frac{1}{\|k + e\|} \right| dk$$

$$= \frac{1}{\beta} \int_{\mathbb{R}^3} \exp(-\|y - \sqrt{\beta}\, a\|^2) \left| \frac{1}{\|y - \sqrt{\beta}\, e\|} - \frac{1}{\|y + \sqrt{\beta}\, e\|} \right| dy.$$

When β goes to infinity, the last integral goes to zero if $a \neq \pm e$, and it converges to

$$\int_{\mathbb{R}^3} \frac{e^{-\|z\|^2}}{\|z\|} \, dz < \infty$$

if $a = \pm e$. Hence $I(\beta, a)$ goes to zero at least as fast as $1/\beta$ when $\beta \to \infty$. To see what happens when $\beta \to 0$, we first observe that

$$I(\beta, a) = \frac{1}{\beta} \int_{\mathbb{R}^3} \exp(-\|y - \sqrt{\beta} \, a\|^2) \left| \frac{1}{\|y - \sqrt{\beta} \, e\|} - \frac{1}{\|y + \sqrt{\beta} \, e\|} \right| dy$$

$$\leq \frac{2\|e\|}{\sqrt{\beta}} \int_{\mathbb{R}^3} \frac{\exp(-\|y - \sqrt{\beta} \, a\|^2)}{\|y - \sqrt{\beta} \, e\| \, \|y + \sqrt{\beta} \, e\|} \, dy$$

by the triangle inequality, and that the last integral converges to

$$\int_{\mathbb{R}^3} \frac{e^{-\|z\|^2}}{\|z\|^2} \, dz < \infty$$

when $\beta \to 0$. Thus $I(\beta, a)$ goes to infinity as $1/\sqrt{\beta}$ when $\beta \to 0$.

It follows that for each a there is a constant $C(a)$ such that

$$I(\beta, a) \leq \frac{C(a)}{\sqrt{\beta} (\sqrt{\beta} + 1)},$$

and a little thought will make it clear that we can, in fact, choose $C(a)$ independent of a.

In our applications the β in Lemma 6.4.6 will be of the form $\beta = \|x\|^2/2t$, and we then have

$$(29) \qquad \frac{1}{\sqrt{\beta}(\sqrt{\beta} + 1)} = \frac{2t}{\|x\|(\|x\| + \sqrt{2t})} \leq \frac{2t}{\|x\|^2}.$$

We shall also use the algebraic identity

$$(30) \quad \alpha\|k - a\|^2 + \beta\|k - b\|^2 = (\alpha + \beta) \left\| k - \frac{\alpha a + \beta b}{\alpha + \beta} \right\|^2 + \frac{\alpha \beta}{\alpha + \beta} \|a - b\|^2$$

with $\alpha = \|x\|^2/2t_1$, $\beta = \|x\|^2/2t_2$:

$$(31) \qquad \frac{\|x\|^2}{2t_1} \|k - a\|^2 + \frac{\|x\|^2}{2t_2} \|k - b\|^2$$

$$= \frac{\|x\|^2(t_1 + t_2)}{2t_1 t_2} \left\| k - \frac{t_2 a + t_1 b}{t_1 + t_2} \right\|^2 + \frac{\|x\|^2}{2(t_1 + t_2)} \|a - b\|^2.$$

PROOF OF LEMMA 6.4.4. This argument is related to the proof of Lemma 6.4.1, but it is much more complicated. We shall estimate the integral

$$(32) \qquad I(s, t) = E(|X^0(t) - X^0(s)|^3)$$

$$= E\left(\int_0^1 \int_0^1 \int_0^1 \left|\frac{1}{\|b(t) - b(u_1)\|} - \frac{1}{\|b(s) - b(u_1)\|}\right|\right.$$

$$\times \left|\frac{1}{\|b(t) - b(u_2)\|} - \frac{1}{\|b(s) - b(u_2)\|}\right|$$

$$\times \left.\left|\frac{1}{\|b(t) - b(u_3)\|} - \frac{1}{\|b(s) - b(u_3)\|}\right| du_1 \, du_2 \, du_3\right).$$

Since

$$\left|\frac{e^{-\sqrt{-\alpha}\|x\|}}{\|x\|} - \frac{e^{-\sqrt{-\alpha}\|y\|}}{\|y\|}\right| \leq \left|\frac{1}{\|x\|} - \frac{1}{\|y\|}\right|,$$

it suffices to prove 6.4.4 for $\alpha = 0$ as the result is then obviously true for all negative α; i.e., it suffices to show that

$$(33) \qquad\qquad I(s, t) \leq C_\varepsilon |t - s|^{3/2 - \varepsilon}.$$

We shall prove (33) by introducing Gaussian kernels in (32). To treat the possible dependencies between the increments $b(t) - b(u_i)$, $b(s) - b(u_j)$, $1 \leq i, j \leq 3$, we must first split the domain of integration into several parts according to the ordering of u_1, u_2, u_3, s, and t. There are 120 different ways to order five elements, but many of these give rise to exactly the same calculations, and it is enough to consider the following six basic cases.

CASE 1. $u_1 < u_2 < u_3 < s < t.$

CASE 2. $u_1 < u_2 < s < u_3 < t.$

CASE 3. $u_1 < u_2 < s < t < u_3.$

CASE 4. $u_1 < s < u_2 < t < u_3.$

CASE 5. $u_1 < s < u_2 < u_3 < t.$

CASE 6. $s < u_1 < u_2 < u_3 < t.$

We cannot treat all the different cases here, but shall concentrate on the first and the last one. These are the extreme cases in the sense that in case 1 all the u's lie outside the interval (s, t), while in case 6 they all belong to the interval. For the intermediate cases 2–5 one needs only combine the methods used for the extreme cases in a rather straightforward way, and we leave this to the reader.

PROOF OF CASE 1. We must estimate

$$I_1(s, t) = E\left(\int_0^s \int_0^{u_3} \int_0^{u_2} \left| \frac{1}{\|b(t) - b(u_1)\|} - \frac{1}{\|b(s) - b(u_1)\|} \right| \right.$$

$$\times \left| \frac{1}{\|b(t) - b(u_2)\|} - \frac{1}{\|b(s) - b(u_2)\|} \right|$$

$$\left. \times \left| \frac{1}{\|b(t) - b(u_3)\|} - \frac{1}{\|b(s) - b(u_3)\|} \right| du_1 \, du_2 \, du_3 \right).$$

Note that $b(t) - b(s)$, $b(s) - b(u_3)$, $b(u_3) - b(u_2)$, $b(u_2) - b(u_1)$ are independent random variables. When we now introduce Gaussian kernels, x, x_3, x_2, and x_1 will be variables corresponding to these increments:

$$I_1(s, t) = \int_{\mathbb{R}^3} \int_{\mathbb{R}^3} \int_{\mathbb{R}^3} \int_{\mathbb{R}^3} \int_0^s \int_0^{u_3} \int_0^{u_2} \left(\frac{1}{\|x + x_1 + x_2 + x_3\|} - \frac{1}{\|x_1 + x_2 + x_3\|} \right)$$

$$\times \left(\frac{1}{\|x + x_2 + x_3\|} - \frac{1}{\|x_2 + x_3\|} \right)$$

$$\times \left(\frac{1}{\|x + x_3\|} - \frac{1}{\|x_3\|} \right) \frac{\exp(-\|x\|^2/2(t - s))}{[2\pi(t - s)]^{3/2}}$$

$$\times \frac{\exp(-\|x_3\|^2/2(s - u_3))}{[2\pi(s - u_3)]^{3/2}} \frac{\exp(-\|x_2\|^2/2(u_3 - u_2))}{[2\pi(u_3 - u_2)]^{3/2}}$$

$$\times \frac{\exp(-\|x_1\|^2/2(u_2 - u_1))}{[2\pi(u_2 - u_1)]^{3/2}} du_1 \, du_2 \, du_3 \, dx_1 \, dx_2 \, dx_3 \, dx.$$

We introduce new variables as follows: new time variables are

$$v_3 = s - u_3, \qquad v_2 = u_3 - u_2, \qquad v_1 = u_2 - u_1,$$

and new space variables are k_1, k_2, k_3 given by

$$x_1 + x_2 + x_3 = \|x\|k_1 - x/2,$$

$$x_2 + x_3 = \|x\|k_2 - x/2,$$

$$x_3 = \|x\|k_3 - x/2.$$

Note that

$$x_2 = \|x\|(k_2 - k_3), \qquad x_1 = \|x\|(k_1 - k_2).$$

If we let

$$e = \frac{x}{2\|x\|},$$

and extend the domain of the time integrals to be the whole unit interval, we get

$$
I_1(s, t) \le \frac{1}{(2\pi)^6} \int_{\mathbb{R}^3} \int_{\mathbb{R}^3} \int_{\mathbb{R}^3} \int_{\mathbb{R}^3} \int_0^1 \int_0^1 \int_0^1 \|x\|^6 \left| \frac{1}{\|k_1 + e\|} - \frac{1}{\|k_1 - e\|} \right|
$$

$$
\times \left| \frac{1}{\|k_2 + e\|} - \frac{1}{\|k_2 - e\|} \right| \left| \frac{1}{\|k_3 + e\|} - \frac{1}{\|k_3 - e\|} \right|
$$

$$
\times \frac{\exp(-\|x\|^2 \|k_1 - k_2\|^2 / 2v_1)}{v_1^{3/2}}
$$

$$
\times \frac{\exp(-\|x\|^2 \|k_2 - k_3\|^2 / 2v_2)}{v_2^{3/2}} \frac{\exp(-\|x\|^2 \|k_3 - e\|^2 / 2v_3)}{v_3^{3/2}}
$$

$$
\times \frac{\exp(-\|x\|^2 / 2(t - s))}{(t - s)^{3/2}} \, dv_1 \, dv_2 \, dv_3 \, dk_1 \, dk_2 \, dk_3 \, dx.
$$

We now carry out the integration with respect to k_1, k_2, and k_3 in that order, using Lemma 6.4.6 in each case. The result is

$$
I_1(s, t) \le \frac{(2C)^3}{(2\pi)^6} \int_{\mathbb{R}^3} \int_0^1 \int_0^1 \int_0^1 \frac{dv_1}{v_1^{1/2}(\|x\| + \sqrt{2v_1})} \frac{dv_2}{v_2^{1/2}(\|x\| + \sqrt{2v_2})}
$$

$$
\times \frac{dv_3}{v_3^{1/2}(\|x\| + \sqrt{2v_3})} \|x\|^3 \frac{\exp(-\|x\|^2 / 2(t - s))}{(t - s)^{3/2}} \, dx.
$$

By elementary calculus

$$
\int_0^1 \frac{dv}{v^{1/2}(\|x\| + \sqrt{2v})} = \sqrt{2} \ln\left(1 + \frac{\sqrt{2}}{\|x\|}\right)
$$

and thus

$$
I_1(s, t) \le \left(\frac{C}{\sqrt{2}\,\pi^2}\right)^3 \int_{\mathbb{R}^3} \|x\|^3 \ln\left(1 + \frac{\sqrt{2}}{\|x\|}\right)^3 \frac{\exp(-\|x\|^2 / 2(t - s))}{(t - s)^{3/2}} \, dx.
$$

Changing variables for the last time, we let

$$
y = x/\sqrt{t - s}
$$

and get

$$
I_1(s, t) \le \left(\frac{C}{\sqrt{2}\,\pi^2}\right)^3 |t - s|^{3/2} \int_{\mathbb{R}^3} \|y\|^3 \ln\left(1 + \frac{\sqrt{2}}{\sqrt{t - s}\,\|y\|}\right)^3 e^{-\|y\|^2 / 2} \, dy.
$$

It is easy to check that when $t - s$ goes to zero, the last integral goes to infinity more slowly than $|t - s|^{-\varepsilon}$ for any $\varepsilon > 0$. This completes the proof of case 1.

PROOF OF CASE 6. We shall estimate

$$I_6(s, t) = E\left(\int_s^t \int_0^{u_3} \int_0^{u_2} \left| \frac{1}{\|b(t) - b(u_1)\|} - \frac{1}{\|b(s) - b(u_1)\|} \right| \right.$$

$$\times \left| \frac{1}{\|b(t) - b(u_2)\|} - \frac{1}{\|b(s) - b(u_2)\|} \right|$$

$$\left. \times \left| \frac{1}{\|b(t) - b(u_3)\|} - \frac{1}{\|b(s) - b(u_3)\|} \right| du_1 \, du_2 \, du_3 \right).$$

In this case the independent increments are $b(u_1) - b(s)$, $b(u_2) - b(u_1)$, $b(u_3) - b(u_2)$, $b(t) - b(u_3)$, and when we introduce Gaussian densities, they will be represented by the variables x_1, x_2, x_3, x_4:

$$I_6(s, t) = \int_{\mathbb{R}^3} \int_{\mathbb{R}^3} \int_{\mathbb{R}^3} \int_{\mathbb{R}^3} \int_0^t \int_0^{u_3} \int_0^{u_2} \left| \frac{1}{\|x_2 + x_3 + x_4\|} - \frac{1}{\|x_1\|} \right|$$

$$\times \left| \frac{1}{\|x_3 + x_4\|} - \frac{1}{\|x_1 + x_2\|} \right| \left| \frac{1}{\|x_4\|} - \frac{1}{\|x_1 + x_2 + x_3\|} \right|$$

$$\times \frac{\exp(-\|x_1\|^2 / 2(u_1 - s))}{[2\pi(u_1 - s)]^{3/2}}$$

$$\times \frac{\exp(-\|x_2\|^2 / 2(u_2 - u_1))}{[2\pi(u_2 - u_1)]^{3/2}} \frac{\exp(-\|x_3\|^2 / 2(u_3 - u_2))}{[2\pi(u_3 - u_2)]^{3/2}}$$

$$\times \frac{\exp(-\|x_4\|^2 / 2(t - u_3))}{[2\pi(t - u_3)]^{3/2}} du_1 \, du_2 \, du_3 \, dx_1 \, dx_2 \, dx_3 \, dx_4.$$

We introduce new variables as follows: new time variables are

$$v_1 = u_1 - s, \qquad v_2 = u_2 - u_1, \qquad v_3 = u_3 - u_2,$$

and for notational convenience we shall write v_4 for the dependent variable

$$v_4 = t - u_3 = t - (s + v_1 + v_2 + v_3).$$

New space variables are x, k_1, k_2, k_3, given by

$$x = x_1 + x_2 + x_3 + x_4,$$
$$x_1 = \|x\| k_1 + x/2,$$
$$x_1 + x_2 = \|x\| k_2 + x/2,$$
$$x_1 + x_2 + x_3 = \|x\| k_3 + x/2.$$

Note that

$$x_2 = \|x\|(k_2 - k_1), \qquad x_3 = \|x\|(k_3 - k_2), \qquad x_4 = x/2 - \|x\| k_3.$$

Letting

$$e = x/2\|x\|,$$

we get

$$I_6(s,t) \leq \frac{1}{(2\pi)^6} \int_{\mathbb{R}^3} \int_{\mathbb{R}^3} \int_{\mathbb{R}^3} \int_{\mathbb{R}^3} \int_0^{t-s} \int_0^{t-s-v_3} \int_0^{t-s-v_3-v_2} \|x\|^6$$

$$\times \left| \frac{1}{\|k_1-e\|} - \frac{1}{\|k_1+e\|} \right| \left| \frac{1}{\|k_2-e\|} - \frac{1}{\|k_2+e\|} \right|$$

$$\times \left| \frac{1}{\|k_3-e\|} - \frac{1}{\|k_3+e\|} \right|$$

$$\times \frac{\exp(-\|x\|^2\|k_1+e\|^2/2v_1)}{v_1^{3/2}} \frac{\exp(-\|x\|^2\|k_2-k_1\|^2/2v_2)}{v_2^{3/2}}$$

$$\times \frac{\exp(-\|x\|^2\|k_3-k_2\|^2/2v_3)}{v_3^{3/2}} \cdot \frac{\exp(-\|x\|^2\|k_3-e\|^2/2v_4)}{v_4^{3/2}}$$

$$\times dv_1\, dv_2\, dv_3\, dk_1\, dk_2\, dk_3\, dx.$$

To this point we have followed the proof of case 1 closely, but we cannot carry out the integration now as we did then. The problem is that each k_i appears in two of the exponents. In case 1, the variable k_1 appeared in only one exponent, and this was what made it possible to apply Lemma 6.4.6. It is here that we need the algebraic identity (31). Applying it three times in succession we get:

$$\frac{\|x\|^2\|k_1+e\|^2}{2v_1} + \frac{\|x\|^2\|k_1-k_2\|^2}{2v_2} + \frac{\|x\|^2\|k_2-k_3\|^2}{2v_3} + \frac{\|x\|^2\|k_3-e\|^2}{2v_4}$$

$$= \frac{\|x\|^2(v_1+v_2)}{2v_1v_2} \left\| k_1 - \frac{v_1k_2-v_2e}{v_1+v_2} \right\|^2 + \frac{\|x\|^2}{2(v_1+v_2)} \|k_2+e\|^2$$

$$+ \frac{\|x\|^2\|k_2-k_3\|^2}{2v_3} + \frac{\|x\|^2\|k_3-e\|^2}{2v_4}$$

$$= \frac{\|x\|^2(v_1+v_2)}{2v_1v_2} \left\| k_1 - \frac{v_1k_2-v_2e}{v_1+v_2} \right\|^2$$

$$+ \frac{\|x\|^2(v_1+v_2+v_3)}{2(v_1+v_2)v_3} \left\| k_2 - \frac{(v_1+v_2)k_3-v_3e}{v_1+v_2+v_3} \right\|^2$$

$$+ \frac{\|x\|^2}{2(v_1+v_2+v_3)} \|k_3+e\|^2 + \frac{\|x\|^2\|k_3-e\|^2}{2v_4}$$

$$= \frac{\|x\|^2(v_1+v_2)}{2v_1v_2} \left\| k_1 - \frac{v_1k_2-v_2e}{v_1+v_2} \right\|^2$$

$$+ \frac{\|x\|^2(v_1+v_2+v_3)}{2(v_1+v_2)v_3} \left\| k_2 - \frac{(v_1+v_2)k_3-v_3e}{v_1+v_2+v_3} \right\|^2$$

$$+\frac{\|x\|^2(t-s)}{2(v_1+v_2+v_3)v_4}\left\|k_3-\frac{(v_1+v_2+v_3)e-v_4e}{t-s}\right\|^2+\frac{\|x\|^2}{2(t-s)}.$$

If we use the abbreviations

$$c_1=\frac{v_1k_2-v_2e}{v_1+v_2},$$

$$c_2=\frac{(v_1+v_2)k_3-v_3e}{v_1+v_2+v_3},$$

$$c_3=\frac{(v_1+v_2+v_3)e-v_4e}{t-s},$$

the expression for $I_6(s,t)$ can be rewritten as

$$I_6(s,t)\le\frac{1}{(2\pi)^6}\int_{\mathbb{R}^3}\int_{\mathbb{R}^3}\int_{\mathbb{R}^3}\int_{\mathbb{R}^3}\int_0^{t-s}\int_0^{t-s-v_3}\int_0^{t-s-v_3-v_2}\|x\|^6$$

$$\times\left|\frac{1}{\|k_1-e\|}-\frac{1}{\|k_1+e\|}\right|\left|\frac{1}{\|k_2-e\|}-\frac{1}{\|k_2+e\|}\right|$$

$$\times\left|\frac{1}{\|k_3-e\|}-\frac{1}{\|k_3+e\|}\right|$$

$$\times\frac{1}{(v_1v_2v_3v_4)^{3/2}}\exp\left(-\frac{\|x\|^2(v_1+v_2)}{2v_1v_2}\|k_1-c_1\|^2\right)$$

$$\times\exp\left(-\frac{\|x\|^2(v_1+v_2+v_3)}{2(v_1+v_2)v_3}\|k_2-c_2\|^2\right)$$

$$\times\exp\left(-\frac{\|x\|^2(t-s)}{2(v_1+v_2+v_3)v_4}\|k_3-c_3\|^2\right)$$

$$\times\exp\left(\frac{\|x\|^2}{2(t-s)}\right)dv_1\,dv_2\,dv_3\,dk_1\,dk_2\,dk_3\,dx.$$

We now evaluate the integrals with respect to k_1, k_2, and k_3 (in this order), using Lemma 6.4.6 and inequality (29):

$$I_6(s,t)\le\frac{(2C)^3}{(2\pi)^6}\int_{\mathbb{R}^3}\int_0^{t-s}\int_0^{t-s-v_3}\int_0^{t-s-v_3-v_2}\frac{\|x\|^6}{(v_1v_2v_3v_4)^{3/2}}$$

$$\times\frac{v_1v_2}{\|x\|^2(v_1+v_2)}\frac{(v_1+v_2)v_3}{\|x\|^2(v_1+v_2+v_3)}\frac{(v_1+v_2+v_3)v_4}{\|x\|^2(t-s)}$$

$$\times\exp\left(-\frac{\|x\|^2}{2(t-s)}\right)dv_1\,dv_2\,dv_3\,dx$$

$$=\frac{(2C)^3}{(2\pi)^6}\int_{\mathbb{R}^3}\int_0^{t-s}\int_0^{t-s-v_3}\int_0^{t-s-v_3-v_2}$$

$$\times \frac{\exp(-\|x\|^2/2(t-s))}{(v_1 v_2 v_3 v_4)^{1/2}(t-s)} \, dv_1 \, dv_2 \, dv_3 \, dx.$$

We change variables again, letting

$$y = \frac{x}{\sqrt{t-s}}, \qquad w_1 = \frac{v_1}{t-s}, \qquad w_2 = \frac{v_2}{t-s}, \qquad w_3 = \frac{v_3}{t-s},$$

and note that

$$v_4 = t - s - (v_1 + v_2 + v_3)$$

$$= t - s - (t-s)(w_1 + w_2 + w_3)$$

$$= (t-s)(1 - (w_1 + w_2 + w_3)).$$

The expression for $I_6(s, t)$ now becomes

$$I_6(s, t) \leq (t-s)^{3/2} \frac{(2C)^3}{(2\pi)^6}$$

$$\times \int_{\mathbb{R}^3} \int_0^1 \int_0^{1-w_3} \int_0^{1-w_3-w_2} \frac{dw_1 \, dw_2 \, dw_3}{[w_1 w_2 w_3 (1 - w_1 - w_2 - w_3)]^{1/2}}$$

$$\times \exp\left(-\frac{\|y\|^2}{2}\right) dy,$$

and since the integral converges, case 6 is proved (even for $\varepsilon = 0$).

As we have already said, we shall not prove the remaining cases 2–5 here, but just repeat that they can be obtained by combining the methods of the two proofs we have given; the main point is to realize that one has to treat the u_i's which lie in the interval (s, t) differently from those which lie outside.

C. Models of Polymers

After all these tedious technical calculations, we shall end this section by a brief and informal introduction to the mathematical physics of polymers and its relationship to the operators we have been studying. Polymers are long and flexible chain molecules composed of many repeat units (called monomers) occurring, e.g., in polyethylene (chains of $-CH_2-$ units), rubbers, plastics, and certain biological substances (biopolymers). The number of monomers in one molecule may be extremely large, up to 10^5–10^6 units. We shall only be concerned with linear chains where the centers of the repeat units are connected by a polygonal path, and ignore the branching network structure common in more complex polymers. If one ignores details below a certain characteristic magnitude called the persistence length, real-life polymers are almost perfectly flexible. This suggests that they can be modeled by Markov chains where the angles between two consecutive links in the polygonal path joining the monomer centers have a certain distribution. The idea is that the smaller the typical angle, the larger is the persistence length. For a detailed analysis of this model see Flory (1969) and Westwater (1980, 1981, 1982, 1985). If we ignore the persistence length altogether, the simple random walk can be used as a model, and if we also let the distance between consecutive centers go to zero, we end up with a Brownian motion. Hence Brownian paths may serve as a first approximation to polymer molecules.

If we accept this model, the $\tilde{E}_\omega$'s we have constructed above are just Hamiltonian forms describing the behavior of a quantum mechanical particle moving in the vicinity of a polymer molecule and only interacting with it through forces of extremely short range. Recalling the heuristic expression (6)

$$(34) \qquad H_\omega = -\Delta - \lambda_\omega(\cdot) \int_0^1 \delta(\cdot - b(t)) \, dt$$

for the operator generated by $\tilde{E}_\omega$, a formal application of the Feynman–Kac formula gives the expression

$$(35) \qquad \langle T_\omega^t f, g \rangle = \int_{\tilde{\Omega}} f(\tilde{b}(t)) g(\tilde{b}(0))$$

$$\times \exp\left(\int_0^t \int_0^1 \lambda_\omega(\tilde{b}(s)) \, \delta(b(r) - \tilde{b}(s)) \, dr \, ds \right) d\tilde{P},$$

where $\tilde{b}$ is a new Brownian motion independent of b, for the associated semigroup T_ω^t. As it stands, this formula makes no sense, but using the hyperfinite version of the Feynman–Kac formula in Theorem 5.3.11, we can interpret and prove it.

Let $\mathscr{E}$ be a hyperfinite representation of E on $L^2(Y, \mu)$, and let $\tilde{\delta} : Y^2 \to \mathbb{R}$ be the "delta function"

$$\tilde{\delta}(x, y) = \begin{cases} \mu(x)^{-1} & \text{if } x = y, \\ 0 & \text{otherwise.} \end{cases}$$

Choose Δt so small that $\mu(x) \log \Delta t$ is infinite for all $x \in Y$, and let T be the time line $\{0, \Delta t, 2\Delta t, \ldots\}$. If $X : \Omega \times T \to Y$ and $\tilde{X} : \tilde{\Omega} \times T \to Y$ are independent copies of the Markov process generated by $\mathscr{E}$, their standard parts b and $\tilde{b}$ are Brownian motions.

Consider the forms

$$(36) \qquad \tilde{\mathscr{E}}_\omega(f, g) = \mathscr{E}(f, g) - \int_0^1 \lambda_\omega(X(r)) f(X(r)) g(X(r)) \, dr.$$

If λ_ω is chosen according to Theorem 6.2.6, the hyperfinite Feynman–Kac formula, 5.3.11, tells us that

$$(37) \qquad Q_\omega^t f(x) \approx \tilde{E}_x\left(f(\tilde{X}(t)) \right.$$
$$\left. \times \exp\left(\int_0^t \int_0^1 \lambda_\omega(\tilde{X}(r)) \, \tilde{\delta}(X(r), \tilde{X}(s)) \, dr \, ds \right) \right)$$

(note that condition 5.3.11 (b) is satisfied since we have chosen $\mu(x) \log \Delta t$ infinite), where Q_ω^t is the semigroup generated by $\tilde{\mathscr{E}}_\omega$. This immediately gives us our interpretation of (35).

Recall that according to theorem 6.4.5, we can choose λ to be any bounded function as long as the dimension is less than or equal to three. With this in mind, it should come as no surprise to learn that Westwater (1980) has given a natural interpretation of

$$(38) \qquad T(t, s) = \int_0^t \int_0^s \delta(b_1(u) - b_2(v)) \, du \, dv$$

as a real-valued stochastic process when $d \le 3$; the idea of his construction is simply to replace δ by approximating δ-functions δ_ε and then look at the limit as $\varepsilon \to 0^+$. When $d > 3$ this limit does not exist, and if we want to give a standard interpretation of (35), we must also let the approximating λ_ε's go to zero as ε goes to zero. Theorem 6.4.2 tells us that when $d = 4$ or 5, it is possible to find such a family of λ_ε's, at least in the sense that T_ω^t is different from the free semigroup. What happens to the integral in the exponent of (35) in this case is an interesting question we shall leave to the reader to answer.

So far we have been using Brownian paths as models for polymers, and as far as the general influence of polymer molecules on their environment is concerned, this may not be totally unrealistic and uninteresting. But as an explanation of the polymers themselves, their structure and properties, the model is far too simple and idealized. Not only have we disregarded the persistence length of the molecules, but we have also not taken into account the fact that since two monomers cannot occupy the same part of space, a polymer molecule cannot intersect itself. This is usually referred to as the "excluded volume effect," and mathematically it has a grave impact on the class of reasonable models. If we think of the growth of a polymer molecule as a stochastic process, to require that the paths do not intersect themselves means that the possible positions at any time are dependent on the entire past of the process; the model can no longer be Markov. As a consequence, the analysis of the resulting process becomes extremely complicated, and very few mathematical results have been obtained in the physically interesting case $d = 3$. However, on the more heuristic level quite a lot of information has been obtained through computer simulation [see e.g., Barber and Ninham (1970), Domb (1969), Flory (1969), Freed (1981), and Hammersley and Morton (1954)]. In higher dimensions analytic methods have had greater success, and later in this section we shall give a brief introduction to Lawler's work on self-avoiding random walks. But first we want to discuss a slightly different approach, which was introduced by Edwards (1965, 1975) and given a solid mathematical basis by Westwater (1980, 1981, 1982, 1985).

Also in Edwards's model the polymer molecule is described by a stochastic process, but instead of the dynamic picture suggested by the term "self-avoiding random walk," the description in this case is quite static; the process is given in terms of a measure on the space $\mathscr{C}([0, 1], \mathbb{R}^d)$. If W is the Wiener measure on $\mathbb{R}^d$, this measure is formally described by

$$(39) \qquad d\rho_\lambda = Z^{-1}\left(\exp\left(\lambda \int_0^1 \int_0^1 \delta(x(u) - x(v))\, du\, dv\right) dW(x)\right),$$

where λ is a negative constant called the coupling constant, and Z is a normalization factor

$$(40) \qquad Z = \int \exp\left(\lambda \int_0^1 \int_0^1 \delta(x(u) - x(v))\, du\, dv\right) dW(x).$$

We shall not discuss here the physical assumptions underlying (39) [the survey article by Freed (1981) is strongly recommended for anybody who wants a better understanding of polymer models], but intuitively it seems quite reasonable; since we are interested in nonintersecting paths, we penalize them whenever they intersect themselves (remember that λ is negative).

In dimension one it is easy to give sense to (39) by making use of Brownian local time, and in two dimensions a solution was given by Varadhan in an appendix to Symanzik's (1969) paper. The case $d = 3$ required a new technique and was solved by Westwater (1980, 1982). His proof is too long and technical to be presented here, but we shall try to explain the main idea, which exploits the close relationship between (39) and (36), or more precisely between the exponents

$$(41) \qquad \int_0^1 \int_0^1 \delta(b_1(u) - b_2(v))\, du\, dv$$

and

$$(42) \qquad \int_0^1 \int_0^1 \delta(x(u) - x(v))\, du\, dv.$$

Recall that according to (38) we know how to make sense of (41), and the idea is to interpret (42) by somehow expressing it in terms of (41). The problem is to split the Brownian motion x in (42) into two independent Brownian motions b_1 and b_2 as in (41).

Here is the trick. Instead of integrating at once over the entire square $[0, 1] \times [0, 1]$, we break the integral in (42) down into smaller pieces. First we divide $[0, 1] \times [0, 1]$ into four smaller squares $[0, \frac{1}{2}] \times [0, \frac{1}{2}], [0, \frac{1}{2}] \times [\frac{1}{2}, 1]$, $[\frac{1}{2}, 1] \times [0, \frac{1}{2}]$, and $[\frac{1}{2}, 1] \times [\frac{1}{2}, 1]$, and integrate over the square in the upper left-hand corner $[0, \frac{1}{2}] \times [\frac{1}{2}, 1]$ and the one in the lower right-hand corner $[\frac{1}{2}, 1] \times [0, \frac{1}{2}]$. We repeat the procedure with each of the two remaining squares $[0, \frac{1}{2}] \times [0, \frac{1}{2}]$ and $[\frac{1}{2}, 1] \times [\frac{1}{2}, 1]$, split them into four pieces, and integrate over the upper left-hand and lower right-hand corners. Again we repeat the procedure with respect to the four unused subsquares, and so on to infinity. Figure 6.1 shows the domains of integration in each of the three first steps. The nth time we repeat the procedure we will be integrating over 2^n squares each of area 4^{-n}. If we sum all these contributions we get the integral over the unit square $[0, 1] \times [0, 1]$.

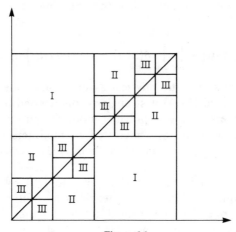

Figure 6.1

To understand the idea behind this procedure, let us see what happens to the integral over the first subsquare $[0, \frac{1}{2}] \times [\frac{1}{2}, 1]$. The two processes x_1, $x_2 : \Omega \times [0, 1] \to \mathbb{R}^3$ defined by

$$(43) \qquad x_1(s) = x(\tfrac{1}{2} - s) - x(\tfrac{1}{2})$$

and

$$(44) \qquad x_2(t) = x(\tfrac{1}{2} + t) - x(\tfrac{1}{2})$$

are independent Brownian motions. By a formal change of variables

$$\int_{1/2}^{1} \int_{0}^{1/2} \delta(x(s) - x(t)) \, ds \, dt = \int_{0}^{1/2} \int_{0}^{1/2} \delta(x_1(s) - x_2(t)) \, ds \, dt.$$

Hence $\int_{1/2}^{1} \int_{0}^{1/2} \delta(x(s) - x(t))\, ds\, dt$ has the same distribution as the quantity $T(\frac{1}{2}, \frac{1}{2})$ in (38). Similarly, the integral over any of the squares of area 4^{-n} occurring in the nth step of the procedure above will have the same distribution as $T(1/2^n, 1/2^n)$. This reduces the problem of interpreting (39) to a problem about the existence of the limit of a sum of random variables consisting of copies of $T(1/2^n, 1/2^n)$. Since the copies of $T(1/2^n, 1/2^n)$ are dependent, this is quite a difficult task, and we can only refer the reader to Westwater's papers for the solution.

Knowing that the measures ρ_λ exist, we may ask for their properties. Despite the formal expression (39), ρ_λ is singular with respect to Wiener measure when $\lambda < 0$; in fact, Westwater (1982) has shown that if $\lambda_1 \neq \lambda_2$, then ρ_{λ_1} and ρ_{λ_2} have disjoint support. The results of Kusuoka (1983) seem to imply that the paths still intersect for finite λ's, but nothing is known about the limit as λ goes to infinity. For other work on self-intersection, see Dynkin (1985), Lawler (1985), LeGall (1985), Yor (1985), and Rosen (1983).

By the work of Varadhan and Westwater we know that the measures ρ_λ in (39) exist when $d \leq 3$. Westwater's argument highlights the close relationship between (39) and (35), and since we have proved that (35) makes sense for $d \leq 5$, it is natural to ask what happens to (39) in dimensions four and five. This question is not only the result of an idle wish to generalize, but is also of the utmost importance in mathematical physics; through Symanzik's (1969) program it is intimately connected to the existence of nontrivial quantum fields in dimension four (the physically important dimension). We shall explain this connection in Section 7.5. An attempt to generalize Westwater's argument to four dimensions will probably meet a number of obstacles; not only is the existence of (39) much more precarious in this case, but also the fact that we have to choose λ positive may create new problems. Westwater's limit theorem is so complicated that it may be a good idea first to give a nonstandard treatment of it in dimension three [for $d = 2$, this has already been done by Stoll (1985)]. It is the kind of limit construction that seems to lend itself naturally to nonstandard methods. Once this has been achieved, an attempt should be made in dimension four.

As promised above we shall end this section with a few remarks on self-avoiding random walks. A *simple random walk* of length N is a sequence $\{x_1, \ldots, x_N\}$ of elements in $\mathbb{Z}^d$ such that $|x_i - x_{i-1}| = 1$ for all $i \leq N$, and a *self-avoiding random walk* of length N is a simple random walk $\{x_1, \ldots, x_N\}$ where $x_i \neq x_j$ when $i \neq j$. On the set Ω_N of all simple random walks of length N, the natural measure is the normalized counting measure

$$P_N\{x_1, \ldots, x_N\} = 2^{-Nd}.$$

It is not so obvious what is the "right" measure to put on the set S_N of all self-avoiding random walks of length N, but also here it has been customary to use the uniform measure.

$$Q_N\{x_1, \ldots, x_N\} = 1/|S_N|.$$

Precise results about the measures Q_N have been scarce [see Barber and Ninham (1970), Brydges *et al.* (1982), Brydges and Spencer (1985), Domb (1969), Flory (1969), Freed (1981), Hammersley and Morton (1954), and Kesten (1963)] and a few years ago Lawler (1980) introduced a new and more tractable measure on S_N. His idea was to construct self-avoiding random walks from simple ones by erasing all the loops. The construction can be described more precisely as follows.

Given a simple random walk $\{x_1, \ldots, x_N\}$, we define

$$\varphi = \min\{j | \exists i < j (x_i = x_j)\},$$

$$\tau = \text{the } i < \varphi \text{ for which } x_i = x_\varphi,$$

and let $\{y_1, \ldots, y_M\}$ be the path given by

$$y_j = \begin{cases} x_j & \text{if } j \leq \tau, \\ x_{j+(\varphi-\tau)} & \text{if } \tau \leq j \leq M. \end{cases}$$

Observe that $M = N - (\varphi - \tau)$ and that $\{y_1, \ldots, y_M\}$ is the path obtained from $\{x_1, \ldots, x_N\}$ by erasing the loop $\{x_\tau, x_{\tau+1}, \ldots, x_{\varphi-1}\}$. If $\{y_1, \ldots, y_M\}$ is self-avoiding, we stop the process; if not, we repeat the loop-erasing operation till we get a self-avoiding path. Figure 6.2 illustrates the process.

Let Ω be the set of simple random walks of infinite length, and let P be the natural measure on Ω. If $d \geq 3$, the simple random walk is transient, i.e., $\|x_n\| \to \infty$ with probability one as $n \to \infty$. Hence we can apply the procedure above to an infinite random walk $\{x_i\}$, and the result is an infinite, self-avoiding path $\{z_i\}$. We shall use the notation

$$D_N(\{x_i\}) = \{z_1, z_2, \ldots, z_N\}.$$

Lawler's measure on S_N is defined by

$$q_N(\{z_1, \ldots, z_N\}) = P(\{x_i\} | : \{z_1, \ldots, z_N\} = D_N\{x_i\}).$$

Let $X_N : \Omega_N \times \{1, 2, \ldots, N\} \to \mathbb{Z}^d$ be simple random walk considered as a stochastic process on (Ω_N, P_N). If we choose N infinite, we know that the standard part of the process $Y_N : \Omega_N \times \{0, 1/N, \ldots, 1\} \to {}^*\mathbb{R}^d$ defined by

$$Y_N(\omega, t) = \frac{d}{\sqrt{N}} X_N(\omega, Nt)$$

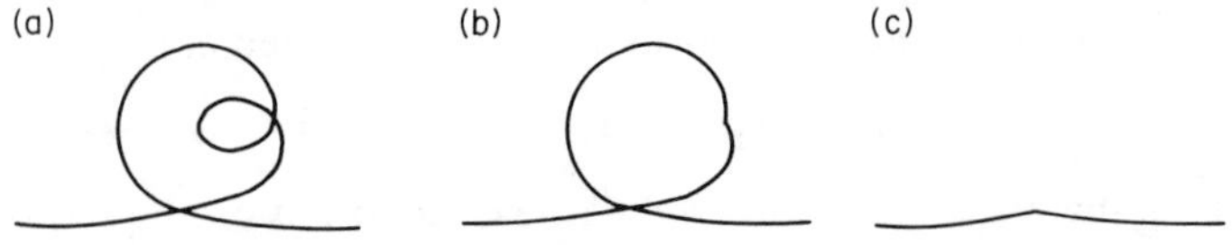

(a)   (b)   (c)

Figure 6.2

is a Brownian motion. We can also consider the loop-erased, self-avoiding random walk as a stochastic process $X_N: S_N \times \{1, 2, \ldots, N\} \to \mathbb{Z}^d$ on (S_N, q_N). Lawler's main result states that if $d \geq 5$, there is a positive, real number K_d such that the standard part of

$$Y_N(\omega, t) = \frac{d}{\sqrt{N}} X_N(\omega, K_d N t)$$

is a Brownian motion. On the basis of heuristic arguments and numerical evidence it has been conjectured that the limit distribution of a self-avoiding random walk is Gaussian when $d \geq 4$ but not when $d \leq 3$, and the result above is the first successful step toward a proof of this claim. The original conjecture is concerned with the uniform measure Q_N and not with Lawler's measure q_N, and in a later paper Lawler (1983) has studied the relationship between the two approaches. We would like to point out, however, that there is no reason to consider Q_N any more natural than q_N; indeed, Lawler (1980) has given the following characterization of the loop-erased random walk, which is perhaps more intuitive than the definition. Given a subset $A \subset \mathbb{Z}^d$, we may define a *random walk with taboo set A* by restricting our attention to those simple random walks which do not hit A. Lawler proved that given the n first values $x(1), x(2), \ldots, x(n)$ of a loop-erased random walk, then $x(n + 1)$ has the same distribution as $y(n + 1)$, where y is a random walk with taboo set $\{x(1), x(2), \ldots, x(n)\}$ and $y(n) = x(n)$.

For more information about the importance of self-avoiding random walks in polymer physics and quantum field theory, see Freed (1981) and Nelson (1983).

6.5. APPLICATIONS OF NONSTANDARD ANALYSIS TO THE BOLTZMANN EQUATION

A. The Equation

The (nonlinear) Boltzmann equation (Boltzmann, 1867) is the basic equation of gas kinetics. The aim is to deduce the macroscopic behavior of gases from the microscopic model of kinetic gas theory, in which a gas is described as consisting of rigid spheres (molecules) interacting by collisions, according to the laws of elastic collisions of classical mechanics. It is a basic model for nonequilibrium phenomena and for the study of the approach to equilibrium (asymptotic Maxwell–Boltzmann distribution) in gases.

To discuss the equation we need some notation. Consider a gas of identical point molecules supposed first to interact by a potential of finite range d. The molecules move in a region Λ of $\mathbb{R}^3$, with some suitable boundary conditions on the boundary of Λ, e.g., periodic ones. The velocities of the molecules can take all values in $\mathbb{R}^3$, so that the phase space of each molecule is $M = \Lambda \times \mathbb{R}^3$.

Let F be the density of the molecules in M, a quantity supposed *a priori* to exist and depend differentiably on time and space, so that an evolution equation can be set up. The number of molecules at time t in a region A of M is then

$$\int_A F(x, v, t) \, dx \, dv.$$

The Boltzmann equation expresses $\partial F / \partial t$ through a balance between the numbers of molecules entering a region of collision and leaving it.

Let us consider two molecules of initial velocities v_1 and v_2, initially separated in space at a distance larger than the range d of the potential. Let v_1' and v_2' be the respective velocities of the molecules after collision has taken place and the molecules again are separated by a distance larger than d. Conservation of momentum and conservation of energy give

$$v_1 + v_2 = v_1' + v_2' \qquad \text{and} \qquad v_1^2 + v_2^2 = (v_1')^2 + (v_2')^2.$$

These equations are not sufficient, in our three-dimensional case, to completely specify the collision. For such a complete specification one can, e.g., introduce a plane P orthogonal to $w \equiv v_2 - v_1$ and at rest with respect to the first molecule. In this plane, u is the vector from the first molecule to the point of intersection with the plane P of the straight line from the second molecule at time $-\infty$ in the direction of w. Note that in the case of finite range d one has collisions only if $|u| \leq d$. We denote in general by B the values of u for which one has collisions.

Boltzmann, using the conservation laws and a statistical hypothesis (the famous *Stosszahlansatz*), computed the number of collisions occurring in a short interval of time in small regions Δx of position space between molecules with incoming momenta in small regions Δv_1, Δv_2, respectively $\Delta v_1'$, $\Delta v_2'$, of velocity space. From this computation he then derived the equation

(1) $\quad (\partial F / \partial t)(x, v_1, t) + v_1 \nabla_x F(x, v_1, t) = (QF)(x, v_1, t), \qquad t > 0,$

where ∇_x is the gradient with respect to the position $x \in \Lambda$ and Q is the so-called collision operator

(2) $\qquad (QF)(x, v_1, t)$

$$= \int_B \left(\int_{\mathbb{R}^3} [F(x, v_1', t) F(x, v_2', t) \right.$$

$$\left. - F(x, v_1, t) F(x, v_2, t)] |v_2 - v_1| \, dv_2 \right) du.$$

This is the Boltzmann equation, which should be solved under an initial condition $F(x, v, 0)$. In the case where the range of the potential is infinite one has to take $B = \mathbb{R}^2$.

This equation has been derived from classical mechanics without any statistical hypothesis, e.g., in the case of a purely hard-core potential, a model equivalent to having hard spheres of radius $d/2$. This was done by Lanford (1975), starting with a gas of n such spheres in a fixed container Λ and letting $n \to \infty$, $d \to 0$ in such a way that $nd^2 = 1$. He showed that if the initial condition $F(x, v, 0)$ is such that $F(x, v, 0) = \lim_{d \to 0} \lim_{n \to \infty} n^{-1} F^{(n,d)}(x, v, 0)$, the $F^{(n,d)}$ being initial values for the density function for the n molecules of diameter d, then $\lim_{d \to 0} \lim_{n \to \infty} n^{-1} F^{(n,d)}(x, v, t)$ exists almost everywhere and the limit function $F(x, v, t)$ is a generalized solution of the Boltzmann equation, at least for sufficiently small times $t > 0$.

There is an extensive literature, both physical and mathematical, on the Boltzmann equation and its consequences; see, e.g., Truesdell and Muncaster (1980) and Cercignani (1975). Main mathematical problems which have been studied are the following:

(1) *Existence and Uniqueness of Solutions.* In the so-called spatially homogeneous case one looks at solutions which are independent of the space variable x. In this case existence and uniqueness of classical solutions were proved under suitable initial conditions by several workers, starting with Carleman (1933) and including Morgenstern, Wild, Truesdell, and Povzner. The strongest results in this case seem to be those obtained by Arkeryd (1972–1984) [for additional references see Truesdell and Muncaster (1980), Chapter 21]. We shall discuss this case further below, using methods of nonstandard analysis. For solutions depending on space (i.e., in the spatially inhomogeneous case), existence and uniqueness of weak solutions for a finite interval of time have been proved in the case of potentials of finite range starting with Grad (1958), and including work by Ukai, Shizuta, Nishida and Imai, Glickson, and Kaniel and Shinbrot [see Truesdell and Muncaster (1980), Chapter 20]. Whether these solutions are classical is an open problem.

Ukai, Shizuta, and Nishida and Imai have shown global in time existence near the equilibrium (of the Maxwell distribution). The only presently known global in time existence results far from equilibrium have been obtained recently by Arkeryd, using nonstandard analysis. We shall describe this below.

(2) *Study of the Asymptotic Behaviour in Time of the Solutions of the Boltzmann Equation, Approach to Equilibrium.* Proofs of the asymptotic approach to the Maxwell distribution under suitable initial conditions and

assumptions on the potential are available in the spatially homogeneous case with finite-range potentials (Carleman, Arkeryd, and others). The first proof of convergence from initial conditions far from equilibrium and for potentials of infinite range has been obtained by Arkeryd (1982), using methods of nonstandard analysis. This will also be discussed further below. For finite-range potentials and particular boundary conditions convergence has been proven for x-dependent solutions near equilibrium, by Ukai (1974), and then extended by Shizuta, Imai, and Nishida.

Let us introduce some notation that we shall need in our nonstandard study below. Together with the Boltzmann equation (1) it is convenient to consider the following modification of it:

$$(1') \qquad (\partial F / \partial t)(x, v_1, t) + v_1 \nabla_x F(x, v_1, t) = (Q_w F)(x, v_1, t),$$

where Q_w is defined as the collision operator Q with the quantity $|v_2 - v_1|$ replaced by a function $w(v_1, v_2, u)$ of v_1, v_2, and u. Below we shall consider together with w its upper and lower truncated version w_n^m, defined to be equal to w for $|u| \leq m$ and $v_1^2 + v_2^2 \leq n^2$, and equal to zero if either $|u| \leq m$, $v_1^2 + v_2^2 > n^2$, or $|u| > m$.

We remark at this point that the results we shall prove below for $(1')$ with w replaced by w_n^m actually also hold in the case $w(v_1, v_2, u) = |v_2 - v_1|^\alpha$ with $0 \leq \alpha < 2$.

We shall consider for simplicity the case where the molecules have positions x varying in Λ, with Λ a vessel with periodic boundary conditions; i.e., we take Λ to be the torus $\mathbb{R}^3 / \mathbb{Z}^3$. However, completely similar techniques yield corresponding results in many cases where Λ is a bounded region with other boundary conditions, and in the case where there is, in addition to the two-body forces, some external force acting upon the molecules.

B. Physically Natural Initial Conditions

Natural initial conditions for the Boltzmann equation are such that $F(x, v, 0) = F_0(x, v) \geq 0$ with no more severe restrictions than F_0, $v^2 F_0$, $F_0 \log F_0$ in $L^1(dx\, dv)$. In fact, $F_0 \geq 0$, $F_0 \in L^1(dx\, dv)$ expresses the fact that F_0 is a density, and $v^2 F_0 \in L^1$ comes from the fact that $\int v^2 F_0\, dx\, dv$ has the physical interpretation of the total energy of the molecules. Computing formally, one sees that $\int F_0\, dx\, dv = \int F\, dx\, dv$ and $\int v^2 F_0\, dx\, dv = \int v^2 F\, dx\, dv$ for all t, so that $\int F\, dx\, dv$, $\int v^2 F\, dx\, dv$ are conserved quantities. In a similar way one sees that the famous H quantity

$$HF(t) \equiv \int F \log F\, dx\, dv$$

decreases as t increases, i.e., $HF(t) \leq HF(0)$.

As is well known [see, e.g., Truesdell and Muncaster (1980) and Cercignani (1975)], HF can be interpreted as an entropy function. So the natural setting for studying the Boltzmann equation (1) is in a space of functions F which are such that $F \geq 0$, F, $v^2 F$, and $F \log F \in L^1$.

But there are serious difficulties due to the lack of linearity, boundedness, and continuity of Boltzmann's equation (1). This comes about in three different ways. First, QF is not linear in F. Second, at least in the case $B = \mathbb{R}^2$, the two terms in Q are not well defined, when taken separately. Third, Q as a function of x contains the square of a L^1 function. In the space-homogeneous case, where one looks at solutions F independent of x, the third problem, of course, disappears. This is the main reason why existence and uniqueness have been easier to obtain when x is absent than otherwise. We shall present below the nonstandard way of doing this. In the space-inhomogeneous case (x dependence) the only global existence results for the above physical conditions are for solutions in a Loeb L^1 setting by nonstandard analytical methods.

C. The Equation with a Truncated Collision Term

Before considering the full Eq. (1) we shall look first at the simpler case of the "truncated Boltzmann equation"

$$(3) \qquad (d/dt)F(x + v_1 t, v_1, t) = (Q_n^m F)(x + v_1 t, v_1, t)$$

with

$$(Q_n^m F)(x, v_1, t) = \int_B \int_{\mathbb{R}^3} [\langle F \otimes F \rangle(x, v_1', v_2', t)$$
$$- \langle F \otimes F \rangle(x, v_1, v_2, t)] w_n^m(v_1, v_2, u) \, dv_2 \, du,$$

where m, n are integers, and

$$\langle F \otimes F \rangle \equiv \begin{cases} F \otimes F, & \text{if } |F \otimes F| \leq n, \\ n \, \text{sign} \, F \otimes F, & \text{otherwise.} \end{cases}$$

w_n^m was defined above in Section 6.5.A.

In this case we have the following

6.5.1. THEOREM. There exists a unique non-negative solution of (3) in L^∞ when the initial value F_0 is in L_+^∞.

SKETCH OF PROOF. It is easy to see from the definition of $\langle F \otimes F \rangle$, that for any functions F, G:

$$|\langle F \otimes F \rangle(x, v_1, v_2, t) - \langle G \otimes G \rangle(x, v_1, v_2, t)|$$
$$\leq |F(x, v_1, t)| \cdot |F(x, v_2, t) - G(x, v_2, t)|$$
$$+ |G(x, v_2, t)| \cdot |F(x, v_1, t) - G(x, v_1, t)|.$$

Since the integration in Q_n^m is over bounded sets, this implies

$$\|Q_n^m F - Q_n^m G\|_\infty \le K \cdot (\|F\|_\infty + \|G\|_\infty) \cdot \|F - G\|_\infty,$$

for some constant K. Thus Q_n^m is locally Lipschitz in L^∞ and there exists a unique local solution F of the integrated version of (3)

$$F(x + v_1 t, v_1, t) = F_0(x, v_1) + \int_0^t Q_n^m F(x + v_1 s, v_1, s)\, ds.$$

But

$$\|Q_n^m F\|_\infty \le \pi m^2 \cdot 2n \cdot n \cdot \pi n^3 4/3 \equiv K'$$

and so

$$\|F(t)\|_\infty \le \|F_0\|_\infty + t K'.$$

Thus F exists for all $t > 0$. To complete the proof it remains to verify $F \ge 0$. It is easy to show that F is the only solution of the equation

(4)
$$G(x + v_1 t, v_1, t) = \exp(-H(x, v_1, t)) F_0(x, v_1)$$

$$+ \int_0^t \exp(-H(x, v_1, t) + H(x, v_1, s))$$

$$\times Q_n^{m'} G(x + v_1 s, v_1, s)\, ds$$

$$\equiv \tilde{Q} G(x, v_1, t),$$

where

$$H(x, v_1, t) \equiv \int_0^t \int_B \int_{\mathbb{R}^3} F(x + v_1 \tau, v_2, \tau) w_n^m(v_1, v_2, u)\, dv_2\, du\, d\tau$$

and

$$(Q_n^{m'} F)(x, v_1, s) = \int\int [\langle F \otimes F \rangle(x, v_1', v_2', s) + F \otimes F(x, v_1, v_2, s)$$

$$- \langle F \otimes F \rangle(x, v_1, v_2, s)] w_n^m(v_1, v_2, u)\, dv_2\, du.$$

We note that $Q_n^{m'}$ has a structure similar to Q_n^m but it has the nice property of being order preserving in the sense that $Q_n^{m'} G_1 \ge Q_n^{m'} G_2 \ge 0$ if $G_1 \ge G_2 \ge 0$. For small t, the solution G of (4) is the limit of the increasing sequence $G_1 = 0$, $G_j = \tilde{Q} G_{j-1}$, $j \ge 2$. It follows that the solution of (4), hence of (3), satisfies $F \ge 0$ for small t. By a continuation argument we then get $F \ge 0$ for all $t > 0$, which completes the proof of the theorem; for more details consult Arkeryd (1972).

6.5.2. REMARK. The solution of (3) given in Theorem 6.5.1 conserves mass, i.e., $\int F_0 \, dv = \int F \, dv$, and energy, i.e., $\int v^2 F_0 \, dv = \int v^2 F \, dv$, and the H function introduced in Section 6.5.A is decreasing. The reason for these properties is that the simple formal argument mentioned above is actually rigorously valid if F_0, $v^2 F_0 \in L^1$. From this the general case follows, since

$$F = F_0 \qquad \text{if} \quad |v_1| \geq n$$

and for $|v_1| \leq n$, F depends only on F_0 restricted to $|v_1| \leq n$, and the latter function evidently is in L^1.

D. The Nonstandard Tool

We shall now study the Boltzmann equation (1) in a bounded region Λ of x space with initial condition $F_0 \in L^1_+(A \times \mathbb{R}^3)$, under the assumption of a two-body potential of finite or infinite range, following Arkeryd (1981a,b, 1984).

Our starting point will be the truncated case of Theorem 6.5.1 but now within the enlarged universe $V(*\mathbb{R})$. Choose $n \in *\mathbb{N} - \mathbb{N}$ and let

$$f_0(x, v) \equiv \min(*F_0(x, v), n) + n^{-1} \exp(-v^2).$$

It is often easy to show that the $V(*\mathbb{R})$ solution f with initial value f_0 obtained by Theorem 6.5.1 defines a weak solution in $V(\mathbb{R})$

$$L_t : G \to \,^{\circ}\!\int f^* G,$$

for G belonging to some suitably nice space. In this way one can recover various results, previously proved directly in $V(\mathbb{R})$, i.e., by standard means.

This approach has two positive aspects. On the one hand our method for recovering these old results has advantages of a conceptual, expository, and pedagogical nature. On the other hand—and perhaps more importantly—new results are obtained in this way. The first proof of asymptotic convergence to the Maxwellian value of F when $t \to \infty$ of solutions of (1) in the space-homogeneous case with no cutoff was obtained by Arkeryd, using the fact that a method from the standard cutoff case can be directly applied in the nonstandard setting to a cutoff in $*\mathbb{N} - \mathbb{N}$. In Section 6.5.E we shall prove the existence of solutions and in Section 6.5.F asymptotic convergence.

As another interesting development, in the space-inhomogeneous case far from equilibrium, the nonstandard technique has suggested a *new solution concept*, which in fact has yielded the only existence results so far. This will be discussed in Section 6.5.G.

Both these cases show once more that nonstandard analysis can be an efficient additional—rather than alternative—tool for working mathematicians. In the discussion of the two cases that now follows, some arguments are only sketched. For further details the reader is referred to the original papers (Arkeryd, 1981a, b, 1982, 1984).

E. The Space-Homogeneous Case: Global Existence of Solutions

We shall discuss the space-homogeneous Boltzmann equation (1); i.e., we look for initial data F_0 and solutions F independent of the space variable x. The two-body potentials can be of finite or infinite range.

To (1) we can associate the weak form

$$\int F(v, t)g(v, t) \, dv = \int F_0(v)g(v, s) \, dv$$

$$+ \int_0^t \int_{\mathbb{R}^3} F(v, s) \, \partial_s g(v, s) \, ds \, dv$$

$$+ \int_0^t \int_{\mathbb{R}^3} (QF)(v, s)g(v, s) \, dv \, ds,$$

for $t > 0$ and all $g \in C^{1,\infty} \equiv \{G \in C^1([0, \infty) \times \mathbb{R}^3) \,|\, |G|_1 \equiv \sup|G(u, t)| + \sup|\partial_t G(v, t)| + \sup|\nabla_v G(v, t)| < \infty\}$.

Corresponding to the truncated version of Section 6.5.C we have a similar weak form. In the extended universe we can write this form as

$$(5) \qquad \int_{*\mathbb{R}^3} f(v, t)g(v, t) \, {}^*dv = \int_{*\mathbb{R}^3} f_0(v)g(v, 0) \, {}^*dv$$

$$+ \int_0^t \int_{*\mathbb{R}^3} f(v, s) \, \partial_s g(v, s) \, {}^*dv \, ds$$

$$+ \int_0^t \int_{\mathbb{R}^3} (Q'_n f)(v, s)g(v, s) \, {}^*dv \, ds,$$

where $Q'_n \equiv Q''_n$. In (5) and future formulas of the same kind our notation is a little sloppy as we should really have asterisks on all differentials, e.g., ${}^*dv \, {}^*ds$ and not just ${}^*dv \, ds$. We note that we have by change of variables

$$(Q'_n f(s), g(s)) \equiv \int (Q'_n f)(v_1, s)g(v_1, s) \, {}^*dv_1$$

$$= \int_{*\mathbb{R}^3 \times \mathbb{R}^3 \times B} [g(v'_1, s) - g(v_1, s)]$$

$$\times \langle f \otimes f \rangle(v_1, v_2, s)w_n^n(v_1, v_2, u) \, {}^*dv_1 \, dv_2 \, du.$$

We can now state the following theorem.

6.5.3. THEOREM. Suppose that $F_0 \geq 0$ and F_0, $v^2 F_0$, $F_0 \log F_0 \in L^1$. Then for $m = n \in {}^*\mathbb{N} - \mathbb{N}$ the solution f in $V({}^*\mathbb{R})$ of Theorem 6.5.1 with initial value

$$f_0(v) = \min({}^*F_0(v), n) + n^{-1} \exp(-v^2)$$

defines a mapping in $V(\mathbb{R})$

$$F : \mathbb{R}_t \to L^1_+$$

through

$$\int_{\mathbb{R}^3} F(v, t) G(v)\, dv = {}^\circ\!\int_{{}^*\mathbb{R}^3} f(v, t){}^*G(v)\, {}^*dv,$$

for all $G \in C^{1,\infty}$.

The function F is a weak solution of (1) in the sense that, with $F_t(v) \equiv F(v, t)$:

$$\int_{\mathbb{R}^3} F_t(v) G(v, t)\, dv = \int_{\mathbb{R}^3} F_0(v) G(v, 0)\, dv$$

$$+ \int_0^t \int_{\mathbb{R}^3} F_s(v)\, \partial_s G(v, s)\, dv\, ds + \int_0^t (QF_s, G(s))\, ds.$$

Here

$$(QF, G) \equiv \int_{\mathbb{R}^3 \times \mathbb{R}^3 \times B} [G(v_1') - G(v_1)] F(v_1) F(v_2) |v_1 - v_2|\, dv_1\, dv_2\, du$$

and potentials of infinite range are required to be of inverse kth power type, $k > 2$, or suitable generalizations.

REMARK. For a fuller discussion of such potentials giving convergence of (QF, G) see Arkeryd (1981a), p. 14. This paper also considers consequences of the present theorem not treated below, such as the behavior of higher moments of F.

PROOF. Since the real-valued mapping L_t from the space $C_0(\mathbb{R}^3)$ of continuous functions on $\mathbb{R}^3$ with compact support defined by

$$(L_t G)(t) \equiv {}^\circ\!\int_{{}^*\mathbb{R}^3} f(v, t){}^*G(v)\, {}^*dv$$

is linear with

$$|L_t(G)| \leq \left({}^\circ\!\int_{{}^*\mathbb{R}^3} f(v, t)\, {}^*dv \right) {}^\circ\!\sup_{{}^*\mathbb{R}^3} |{}^*G(v)| = \left(\int_{\mathbb{R}^3} F_0(v)\, dv \right) \sup_{\mathbb{R}^3} |G(v)|,$$

it follows that L_t defines a measure μ_t. Recall from Section 6.5.C that under the given assumptions on F_0, we have that $\int F_0\, dv = \int F\, dv$, $\int v^2 F_0\, dv = \int v^2 F\, dv$ (mass and energy conservation), and using this we get

$$\int_{|v|>j} |f(v, t)| \, ^*dv \le (1 + j^2)^{-1} \int_{^*\mathbb{R}^3} (1 + v^2)f(v, t) \, ^*dv$$

$$= (1 + j^2)^{-1} \int (1 + v^2)f_0(v) \, ^*dv \approx 0,$$

for all $j \in {}^*\mathbb{N} - \mathbb{N}$. From this it follows that for any sequence $k \in \mathbb{N}$, $k \to \infty$ we have

$$\lim_{k \to \infty} \mu_t\{|v| > k\} = 0.$$

From the remark at the end of Section 6.5.C we have by transfer

$$\int_{^*\mathbb{R}^3} f(v, t) \log f(v, t) \, ^*dv \le \int_{^*\mathbb{R}^3} f_0(v) \log f_0(v) \, ^*dv,$$

with the right member finite by assumption. Applying the elementary inequality $y \log y \ge -z + y \log z$ ($y \ge 0$, $z > 0$) to the case $y = f(v, t)$, $z = \exp(-v^2)$ and using the above inequality we have, with $\log^+ x \equiv \sup(0, \log x)$:

$${}^\circ\!\int_{^*\mathbb{R}^3} f \log^+ f \, ^*dv \le \int_{\mathbb{R}^3} F_0(v) \log F_0(v) \, dv + \int_{\mathbb{R}^3} \exp(-v^2) \, dv$$

$$+ \int_{\mathbb{R}^3} F_0(v)v^2 \, dv \equiv K.$$

From this we get for *-measurable subsets Ω in $^*\mathbb{R}^3$ and for any $N > 1$:

$$\int_\Omega f(v, t) \, ^*dv \le N \int_\Omega {}^*dv + (\log N)^{-1} \int_{|f|>N} f(v, t) \log^+ f(v, t) \, ^*dv$$

$$\le N \int_\Omega {}^*dv + (\log N)^{-1} \int_{^*\mathbb{R}^3} f(v, t) \log^+ f(v, t) \, ^*dv$$

$$\le N \int_\Omega {}^*dv + K(\log N)^{-1}.$$

It follows that

(a) $$\int_\Omega f(v, t) \, ^*dv \approx 0 \quad \text{if} \quad \int_\Omega {}^*dv \approx 0.$$

Now let Ω_k be a sequence of measurable sets in $\mathbb{R}^3$, which is such that

$$\lim_{k \to \infty} \int_{\Omega_k} dv = 0.$$

From (a) above we then get, using also the definition of μ_t, that

$$\lim_{k \to \infty} \mu_t(\Omega_k) = 0.$$

This implies that μ_t is absolutely continuous with respect to dv and its Radon–Nikodym derivative is F_t, i.e., $d\mu_t = F_t \, dv$, with $F_t \in L^1_+(\mathbb{R}^3)$. This proves the first part of the theorem. To prove that F is a weak solution of (1) in the sense stated in the theorem it suffices to remark that under the assumptions of the theorem Eq. (5) holds (rigorously, not only formally). Using this, one sees that F is a weak solution by arguments close to those just used to prove the first part of the theorem.

F. The Space-Homogeneous Case: Asymptotic Convergence to Equilibrium

In this section we only treat forces of inverse kth power type, $k \geq 5$, thereby avoiding a discussion of the extra technical restrictions needed to include a wider range of the interactions of the previous section. The discussion is based on Arkeryd (1982). We consider again the space-homogeneous case with initial conditions

$$F_0 \in L^1_+, \qquad v^2 F_0 \in L^1_+, \qquad F_0 \log F_0 \in L^1.$$

6.5.4. LEMMA.   Assume that for some $s > 2$

$$(1 + |v|)^s F_0 \in L^1(\mathbb{R}^3).$$

Assume moreover that the force K between the molecules is repulsive of the form $|K(x)| = \gamma/|x|^k$, for some $k \geq 5$. Then there exists a constant $C_{F_0,s}$, depending only on s and on

$$\int_{\mathbb{R}^3} (1 + |v|)^s F_0(v) \, dv,$$

such that

$$\int_{\mathbb{R}^3} (1 + |v|)^s F(v, t) \, dv \leq C_{F_0,s}, \qquad t > 0,$$

where F is the weak solution of the Boltzmann equation given by Theorem 6.5.3. Moreover, for $G = 1$, v, v^2 one has

$$\int_{\mathbb{R}^3} G(v) F(v, t) \, dv = \int_{\mathbb{R}^3} G(v) F_0(v) \, dv.$$

The proof of this result can be found in Elmroth (1983) to which we refer the reader, since the proof falls outside our present discussion of nonstandard analytical methods.

6.5.5. LEMMA. Let F be the solution of Theorem 6.5.3 and let t_n be a sequence such that for some $s \geq 2$ one has

$$\int_{\mathbb{R}^3} (1 + |v|)^s F(t_n, v) \, dv < \infty;$$

then for $G(v) = (1 + |v|)^{s'}$, $2 < s' < s$, the sequence $GF(t_n)$ contains a subsequence converging weakly toward GH for some $H \in L^1_+$.

SKETCH OF PROOF. The proof follows from Dunford and Petti's (1940) theorem. In fact, weak relative compactness is equivalent to

(a) uniformly bounded total masses, i.e., $\sup_n \int \tilde{F}(t_n, v) \, dv < \infty$, with $\tilde{F}(t, v) = (1 + |v|)^s F(t, v)$, which holds by assumption;

(b) uniformly small masses for large velocities, i.e., for all $\varepsilon > 0$ there is some r_0 such that $\sup \int_{|v|>r_0} \tilde{F}(t_n, v) \, dv < \varepsilon$, which holds by Lemma 6.5.4;

(c) uniformly small masses on small sets, i.e., for all $\varepsilon > 0$ there is some $\delta(\varepsilon) > 0$ such that $\sup \int_A \tilde{F}(t_n, v) \, dv < \varepsilon$ if $\int_A dv < \delta(\varepsilon)$, which holds as in the proof of Theorem 6.5.3, since

(6) $$\int_{*\mathbb{R}^3} f(v, t) \log f(v, t) \, {}^* dv \leq \int_{*\mathbb{R}^3} f_0(v) \log f_0(v) \, {}^* dv.$$

In the next lemma we give an estimate on the H function Hf from below, which goes back to Gibbs.

6.5.6. LEMMA. Consider the class $\mathscr{C}$ of functions $E \in L^1_+(\mathbb{R}^3)$ which satisfy

(7) $$\int_{\mathbb{R}^3} E(v) \, dv = A, \qquad \int_{\mathbb{R}^3} E(v) v \, dv = B, \qquad \int_{\mathbb{R}^3} E(v) |v|^2 \, dv \leq C$$

for some finite constants A, B, C. Let

$$E_0(v) = \exp(-a|v|^2 + b \cdot v + c),$$

with $a > 0$, $b \in \mathbb{R}^3$, $c \in \mathbb{R}$, be the unique exponential function satisfying the relations in (7) with equality also in the third relation. Then every $E \in \mathscr{C}$ satisfies

$$\int_{\mathbb{R}^3} E(v) \log E(v) \, dv \geq \int_{\mathbb{R}^3} E_0(v) \log E_0(v) \, dv,$$

and the equality holds only for $E = E_0$.

SKETCH OF PROOF. It is enough to consider the case $b = 0$, since we can reduce to this case by translation. Set $q(E) = E \log E + av^2E - (c + 1)E$. $q(E)$ has the minimum $E = E_0$ with $q(E_0) = -E_0$, which implies that $E \log E \geq -av^2E + (c + 1)E - E_0$, with equality iff $E = E_0$. From this the lemma follows by integration.

We can now study the asymptotic properties for $t \to \infty$ of the solution F of Theorem 6.5.3. We show that we have weak convergence of F to the equilibrium Maxwell distribution.

6.5.7. THEOREM. Assume that F_0 is x-independent and $F_0 \geq 0$, F_0, v^2F_0, $F_0 \log F_0 \in L^1$, as well as $(1 + |v|)^s F_0 \in L^1_+$ for some $s > 2$. Then for $G(v) = (1 + |v|)^{s'}$, $0 \leq s' < s$, the family $(GF(t))_{t>0}$ converges as $t \to \infty$ in the weak L^1 sense to GE_0, where E_0 is the Maxwell distribution

$$E_0(v) = \exp(-a|v|^2 + bv + c), \qquad a > 0,$$

defined as in Lemma 6.5.6 by choosing the constants

$$A = \int F_0(v) \, dv, \qquad B = \int F_0(v)v \, dv, \qquad \text{and} \qquad C = \int F_0(v)|v|^2 \, dv.$$

PROOF. In this proof we use $\log f$. This is well defined since $f_0(v) > n^{-1} \exp(-v^2) > 0$ implies, as we see from the solution formula (4) for f, that $f(t, v) > 0$ in the whole of $*\mathbb{R}_+ \times \mathbb{R}^3$; i.e., $\log f$ is defined everywhere. But information about the domain of $\log F$ requires further arguments. We also need an estimate on $(QF, \log F)$. For this we actually would like $\log F \in C^{1,\infty}$, which does not hold in general. On the other hand, since f is a solution in the sense of Theorem 6.5.1 [and not only in the sense of (5)], i.e., f is a "strong nonstandard solution," the corresponding quantity is well defined for f (under the relevant nonstandard cutoff). The estimates for f mentioned in Section 6.5.B can this time be carried through rigorously for f, not only formally. This, together with Lemma 6.5.6 with a Gaussian E_0' corresponding to f_0 (in the sense of Lemma 6.5.6), implies that

$$HE_0' \leq Hf(t) = Hf_0 + \int_0^t Nf(s) \,{}^*ds$$

with

$$(Nf)(t) = 4^{-1} \int_{*\mathbb{R}^3 \times \mathbb{R}^3 \times B} [\langle f \otimes f \rangle(v_1', v_2', t)$$
$$- \langle f \otimes f \rangle(v_1, v_2, t)]w_n^n(v_1, v_2, u)$$
$$\times \log[f \otimes f(v_1, v_2, t)/f \otimes f(v_1', v_2', t)] \,{}^*dv_1 \, dv_2 \, du.$$

We recall that (v_1', v_2') are determined by v_1, v_2, u by the laws of collisions, as mentioned in Section 6.5.A.

We notice that the integrand in Nf is nonpositive. Evidently $-Nf(s) > k^{-1} > 0$ at most on a set of measure $k(Hf_0 - HE_0')$. So there is an increasing sequence $(t_j)_{j \in \mathbb{N}}$ of nearstandard reals, such that

$$(8) \qquad \lim_{j \to \infty} {}^{\circ}Nf(t_j) = 0.$$

By Lemmas 6.5.4 and 6.5.5 we can choose the sequence so that

$$(9) \qquad (1 + |v|)^{s'} F(t_j) \to (1 + |v|)^{s'} E$$

weakly, with $E \in L_+^1$ satisfying

$$\int_{\mathbb{R}^3} G(v) E(v)\, dv = \int_{\mathbb{R}^3} G(v) F_0(v)\, dv,$$

for $G(v) = 1, v, v^2$. In particular,

$$\int_{\mathbb{R}^3} G(v) E(v)\, dv = \lim_{j \to \infty} \int_{\mathbb{R}^3} G(v) F(t_j, v)\, dv$$

$$= {}^{\circ}\!\int_{{}^*\mathbb{R}^3} G(v) f_0(v)\, {}^*dv.$$

Define

$$\Delta E(v_1, v_2, u) \equiv E \otimes E(v_1', v_2') - E \otimes E(v_1, v_2).$$

We will see in Lemma 6.5.8, that $\Delta E = 0$ for a.e. $(v_1, v_2, u) \in \mathbb{R}^3 \times \mathbb{R}^3 \times B$. Using this and a classical computation [see Arkeryd (1972)], one concludes that $E = E_0$.

We shall now prove that $(1 + |v|)^{s'} E_0$, for $s' < s$, is the weak L^1 limit of $(1 + |v|)^{s'} F(t)$ as $t \to \infty$. In fact, make the *ad absurdum* assumption that $(1 + |v|)^{s'} F(t) \not\to (1 + |v|)^{s'} E_0$ for some $s' < s$. We know by Lemma 6.5.5 that there is a sequence $(t_j')_{j \in \mathbb{N}}$ such that for some $E' \neq E_0$, $(1 + |v|)^{s'} F(t_j') \to (1 + |v|)^{s'} E'$. Take $G \in C^{1,\infty}$ such that

$$(10) \qquad \int_{\mathbb{R}^3} E'(v) G(v)\, dv \neq \int_{\mathbb{R}^3} E_0(v) G(v)\, dv.$$

(This is possible since $E \neq E_0$.) By Theorem 6.5.3

$$\left| \int [F(v, t_2) - F(v, t_1)] G(v)\, dv \right|$$

$$\leq |G|_1 \cdot |t_2 - t_1| \cdot \left[\int_{\mathbb{R}^3} F_0(v)\, dv + C \left(\int_{\mathbb{R}^3} (1 + |v|^2) F_0(v)\, dv \right)^2 \right].$$

In particular, for some $\tau_0 > 0$ we have

(11)
$$\left| \int_{\mathbb{R}^3} [F(v, t + \tau) - F(v, t)] G(v) \, dv \right|$$
$$\leq \frac{1}{2} \left| \int_{\mathbb{R}^3} [E'(v) - E_0(v)] G(v) \, dv \right|, \qquad 0 < \tau \leq \tau_0.$$

By the above inequality $HE_0' \leq Hf(t) = Hf_0 + \int_0^t Nf(s) \, {}^*ds$ we get for some finite C and some $j_0 \in {}^*\mathbb{N} - \mathbb{N}$:

$$0 \leq \int_0^{\tau_0} \sum_{j \leq j_0} (-Nf)(t_j' + \tau) \, {}^*d\tau \leq C.$$

Since $Nf \leq 0$, there are some $\tau' \leq \tau_0$ and some finite C' such that

$$\sum_{j \leq j_0} (-Nf)(t_j' + \tau) < C',$$

thus $\lim_{j \to \infty} {}^\circ Nf(t_j + \tau') = 0$.

As in the case of (8) and (9) above we can choose $(t_j')_{j \in \mathbb{N}}$ such that, moreover, $F(t_j' + \tau') \to E_0$ in the weak L^1 sense. Then

(12)
$$\lim_{j \to \infty} \int_{\mathbb{R}^3} [F(v, t_j' + \tau') - F(v, t_j')] G(v) \, dv$$
$$= \int_{\mathbb{R}^3} [E_0(v) - E'(v)] G(v) \, dv.$$

This is $\neq 0$ according to (10), which, however, contradicts (11). Thus the *ad absurdum* assumption must be rejected and we have $(1 + |v|)^{s'} F(t) \to (1 + |v|)^{s'} E_0$ for all $s' < s$. To complete the proof it remains to establish the following lemma.

6.5.8. LEMMA. With the notations in the proof of Theorem 6.5.7 we have

$$\Delta E \equiv E \otimes E(v_1', v_2') - E \otimes E(v_1, v_2) = 0 \qquad \text{for a.e.} \quad v_1, v_2, u.$$

PROOF. Assume *ad absurdum* that $\Delta E \neq 0$ on a set of positive $dv_1 \, dv_2 \, du$ measure. Then there is a bounded measurable set $\Omega \subset \{|v_1| + |v_2| + |u| \leq r_0\} \subset \mathbb{R}^3 \times \mathbb{R}^3 \times B$ of measure m for some $r_0, m > 0$ such that

$$\int_\Omega \Delta E(v_1, v_2, u) |v_2 - v_1| \, dv_1 \, dv_2 \, du = C_0$$

for some real C_0 with $0 < C_0 < 1$.

We shall now derive a contradiction of this with Eq. (8). Let $K > 1$ be given together with an arbitrary subset $\Omega' \subset \Omega$ of Lebesgue measure

$\int dv_1 \, dv_2 \, du < K^{-2}$. Then for the solution F of Theorem 6.5.3 we have, defining ΔF similarly to ΔE,

$$\int_{\Omega'} \Delta F(v_1, v_2, u, t) |v_2 - v_1| \, dv_1 \, dv_2 \, du$$

$$\leq 2r_0 \overset{\circ}{\int} f \otimes f(v_1, v_2, u, t) \, {}^*du \, dv_1 \, dv_2$$

$$\leq 2r_0 (K^{-1} + (1/\log K) 2\pi r_0^2)$$

$$\times \left[Hf_0 + \int_{{}^*\mathbb{R}^3} \exp(-|v|^2) \, dv + \int_{{}^*\mathbb{R}^3} f_0(v)|v|^2 \, dv \right] \int_{{}^*\mathbb{R}^3} f_0(v) \, dv,$$

the latter estimate being similar to the one used in the proof of (c) in Lemma 6.5.5. By this estimate there is a standard K_0 such that

$$\int_\Omega \Delta F(v_1, v_2, u, t) |v_2 - v_1| \, dv_1 \, dv_2 \, du < C_0/4$$

if $\Omega' \subset \Omega$, $t > 0$, and $\int_{\Omega'} dv_1 \, dv_2 \, du < \pi r_0^2 / K_0^2$. The subset of Ω with $F \otimes F(v_1, v_2, t) > K_0^2 (\int_{\mathbb{R}^3} f_0(v) \, dv)^2 \equiv C_1$ has measure smaller than $\pi r_0^2 / K_0^2$, since

$$(13) \qquad \int_{\mathbb{R}^3} F(t, v) \, dv = \int_{\mathbb{R}^3} f_0(v) \, dv.$$

Together with the argument in connection with (9) above, this implies that there is an index n_0, and given $n > n_0$ a set $\Omega_j \subset \Omega$ of measure larger than $m - \pi r_0^2 / K_0^2$, such that for the sequence t_j as in the proof of Theorem 6.5.7

$$F \otimes F(v_1, v_2, t_j) \leq K_0^2 \left(\int_{\mathbb{R}^3} f_0(v) \, dv \right)^2$$

on Ω_j and

$$(14) \qquad \overset{\circ}{\int}_{{}^*\Omega_j} \Delta f(v_1, v_2, u, t_j) |v_2 - v_1| \, {}^*dv_1 \, dv_2 \, du$$

$$= \int_{\Omega_j} \Delta F(v_1, v_2, u, t_j) |v_2 - v_1| \, dv_1 \, dv_2 \, du > C_0/2.$$

From this it follows, with $\Delta f^+ = \max(0, \Delta f)$, by a proof of Arkeryd (1982) carried over to our present nonstandard setting, that there is a standard $\varepsilon > 0$ and, for each $n > n_0$, a partition ${}^*\Omega_j = \Omega_{j1} \cup \Omega_{j2}$ such that

$$\overset{\circ}{N} f(t_j) \leq -(4mC_1 r_0)^{-1} \left(\int_{\Omega_{j1}} \Delta^+ f(t_j) |v_2 - v_1| \, {}^*dv_1 \, dv_2 \, du \right)^2$$

$$-\log(1 + \varepsilon) \overset{\circ}{\int}_{\Omega_{j2}} \Delta^+ f(t_j) |v_2 - v_1| \, {}^*dv_1 \, dv_2 \, du$$

$$\leq -\min[(4mC_1r_0)^{-1}, \log(1 + \varepsilon)]C_0^2/16,$$

which contradicts (8) above. This shows that the *ad absurdum* assumption has to be rejected, which proves the lemma.

6.5.9. REMARK. For other results in the space-homogeneous case see Arkeryd (1981a).

G. The Space-Inhomogeneous Case: A Loeb-Measure Approach

In this section we look for x-dependent solutions of the Boltzmann equation (1) in the case of periodic boundary conditions, i.e., $x \in \Lambda \equiv \mathbb{R}^3/\mathbb{Z}^3$. We present essentially the results in the papers where Arkeryd (1981b, 1984) managed by nonstandard techniques to extend to the space-inhomogeneous case the strong results first obtained for the space-homogeneous case. We first consider Eq. (3) extended by transfer for $m \in \mathbb{N}$, $n \in {}^*\mathbb{N} - \mathbb{N}$. We remark that this corresponds to the sole assumption of having a cutoff in the impact parameter (which is justified in the case of finite-range potentials), but no cutoff in velocity space. Take as initial value

$$f_0(v) = \min({}^*F_0(v), n) + n^{-1}\exp(-v^2)$$

with F_0 satisfying $F_0 \geq 0$, F_0, v^2F_0, $F_0 \log F_0 \in L^1$.

We recall from Chapter 3 that an internal function g from an internal measure space $(X, \mathcal{A}, \nu)$ is S-integrable if

(i) g is $\mathcal{A}$-measurable;
(ii) $\int_X |g|\, d\nu$ is finite, thus nearstandard;
(iii) if $\Omega \in A$ and $g(\Omega) \approx 0$ [in the sense that $g(a) \approx 0$, for all $a \in \Omega$], then $\int_\Omega |g|\, d\nu \approx 0$ (this is needed since ν is not finite);
(iv) if $\Omega \in A$ and $\nu(\Omega) \approx 0$, then $\int_\Omega |g|\, d\nu \approx 0$.

Let $(X, L(\mathcal{A}), L(\nu))$ be the Loeb space associated with $(X, \mathcal{A}, \nu)$. If g is an S-integrable function, then one has that ${}^\circ g : X \to \mathbb{R}$ is Loeb integrable, and that

$${}^\circ\!\int_\Omega g\, d\nu = \int_\Omega {}^\circ g\, dL(\nu)$$

for all $\Omega \in A$.

If, moreover, $h : X \to {}^*\mathbb{R}$ is an $\mathcal{A}$-measurable function and for some finite hyperreal K we have

$$|h(x)| \leq K|g(x)| \qquad \text{for all} \quad x \in X,$$

then h is S-integrable.

6.5.10. LEMMA. Let f be the solution of (3) described above, with initial condition f_0. Then $^\circ f$ is Loeb integrable with respect to the measure $L(dx\,dv)$ generated from the internal Lebesgue measure $^*dx\,dv$.

PROOF. We have to verify (i)–(iv) above. But (i) and (ii) are immediate; (iv) follows as in the proof of Theorem 6.5.3. In fact, we need only replace in the reasoning starting from

$$\int_{^*\mathbb{R}^3} f \log f \, ^*dv \le \int_{^*\mathbb{R}^3} f_0 \log f_0 \, ^*dv$$

all $^*dv$ integrations by $^*dx\,dv$ integrations and we arrive at $\int_\Omega f \, ^*dv\,dx \simeq 0$ for any internal measurable set Ω of infinitesimal measure in $^*\Lambda \times {}^*\mathbb{R}^3$, which proves (iv).

To prove (iii) it suffices to realize that, as in the first part of the proof of Theorem 6.5.3,

$$\int_{^*\Lambda \times \{|v|>j\}} |f(x, v, t)| \, ^*dv\,dx \le (1+j^2)^{-1} \int_{^*\Lambda \times {}^*\mathbb{R}^3} (1+v^2) f(x, v, t) \, ^*dv\,dx$$

$$= (1+j^2)^{-1} \int (1+v^2) f_0(x, v) \, ^*dv\,dx$$

$$= (1+j^2)^{-1} C,$$

with $C \in {}^*\mathbb{R}_+$ and nearstandard. Now let A be *Lebesgue measurable and suppose $f(A) \approx 0$. Then it follows from the above inequality that $\int_A f(x, v, t) \, ^*dv\,dx \approx 0$, which proves (iii).

Let ns $^*\mathbb{R}^3$ denote the nearstandard points of $^*\mathbb{R}^3$.

6.5.11. LEMMA. For Loeb a.e. $(x, v_1) \in {}^*\Lambda \times \text{ns}\, {}^*\mathbb{R}^3$ the function $f(x + v_1\tau, v_2, \tau) w_n^m(v_1, v_2)$ is S-integrable in (v_2, τ), and for $|u| \le m$

$$^\circ \int_0^t \int_{^*\mathbb{R}^3} f(x + v_1\tau, v_2, \tau) w_n^m(v_1, v_2) \, ^*dv_2\,d\tau$$

$$= \int_0^t \int_{\text{ns}\, {}^*\mathbb{R}^3} {}^\circ f(x + v_1\tau, v_2, \tau)^\circ |v_2 - v_1| L(dv_2\,d\tau).$$

PROOF. We find for finite v_1, using energy conservation, that

$$\left| ^\circ \int_0^t \int_{|v_2|>j} f(x + v_1\tau, v_2, \tau) w_n^m(v_1, v_2) \, ^*dx\,dv_2\,d\tau \right|$$

$$\le (1+j^2)^{-1/2} K_{v_1} \int_{^*\Lambda \times \text{ns}\, {}^*\mathbb{R}^3 \times [0,t]} f_0(x, v_2)(1+v_2^2) \, ^*dx\,dv_2\,d\tau,$$

for some constant K_{v_1}.

This implies that for Loeb a.e. $(x, v_1) \in {}^*\Lambda \times \text{ns} \, {}^*\mathbb{R}^3$

(α)
$$\lim_{j \to \infty} {}^{\circ} \int_0^t \int_{|v_2| \le j} f(x + v_1\tau, v_2, \tau) w_n^m(v_1, v_2) \, {}^* dv_2 \, d\tau$$

$$= {}^{\circ} \int_0^t \int_{{}^*\mathbb{R}^3} f(x + v_1\tau, v_2, \tau) w_n^m(v_1, v_2) \, {}^* dv_2 \, d\tau.$$

Similarly, by the boundedness of the H function Hf we obtain for Loeb a.e. $(x, v_1) \in {}^*\Lambda \times \text{ns} \, {}^*\mathbb{R}^3$:

(β)
$$\int_{B_{j,c'}} f(x + v_1\tau, v_2, \tau)|v_2 - v_1| \, {}^* dv_2 \, d\tau \approx 0,$$

where

$$B_{j,c'} = \{(v_2, \tau) \in \{v_2 | |v_2| \le j\} \times {}^*[0, t] \,|\, f(x + v_1\tau, v_2, \tau) \ge c'\},$$

c' is a fixed number in ${}^*\mathbb{N} - \mathbb{N}$.

From (α) and (β) it follows that for Loeb a.e. $(x, v_1) \in {}^*\Lambda \times \text{ns} \, {}^*\mathbb{R}^3$ the function $f(x + v_1\tau, v_2, \tau) w_n^m(v_1, v_2)$ is S-integrable in (v_2, τ). For these values of (x, v_1) we have that ${}^{\circ}f(x + v_1\tau, v_2, \tau)^{\circ}|v_2 - v_1|$ is Loeb integrable in (v_2, τ) with

$$ {}^{\circ} \int_0^t \int_{{}^*\mathbb{R}^3} f(x + v_1\tau, v_2, \tau) w_n^m(v_1, v_2) \, {}^* dv_2 \, d\tau$$

$$= \int_0^t \int_{\text{ns} \, {}^*\mathbb{R}^3} {}^{\circ}f(x + v_1\tau, v_2, \tau)^{\circ}|v_2 - v_1| L(dv_2 \, d\tau).$$

This proves the lemma.

6.5.12. LEMMA. For Loeb a.e. $(x, v_1) \in {}^*\Lambda \times \text{ns} \, {}^*\mathbb{R}^3$ and $|u| \le m$

$$ {}^{\circ} \int_0^t \int_{{}^*\mathbb{R}^3} \langle f \otimes f \rangle (x + v_1 s, v_1, v_2, s) w_n^m(v_1, v_2) \, {}^* dv_2 \, ds$$

$$= \int_0^t \int_{\text{ns} \, {}^*\mathbb{R}^3} {}^{\circ}f \otimes {}^{\circ}f(x + v_1 s, v_1, v_2, s)^{\circ}|v_1 - v_2| L(dv_2 \, ds).$$

PROOF. The product of a bounded *Lebesgue-measurable function and an S-integrable one is S-integrable. From (4) it is easy to see that $f(x + v_1 s, v_1, s)$ is such a bounded *Lebesgue-measurable function and from Lemma 6.5.11 we see that $f(x + v_1 s, v_2, s) w_n^m(v_1, v_2)$ is S-integrable in (v_2, s) for Loeb a.e. $(x, v_1) \in {}^*\Lambda \times \text{ns} \, {}^*\mathbb{R}^3$. We conclude that $f \otimes f(x + v_1 s, v_1, v_2, s) w_n^m(v_1, v_2)$ is S-integrable in (v_2, s) for Loeb a.e. $(x, v_1) \in {}^*\Lambda \times \text{ns} \, {}^*\mathbb{R}^3$. This also holds for $\langle f \otimes f \rangle w_n^m$, since $0 \le \langle f \otimes f \rangle \le f \otimes f$. It follows that $°[\langle f \otimes f \rangle(x + v_1 s, v_1, v_2, s)w_n^m(v_1, v_2)]$ is Loeb integrable on $\{v_2; |v_2| \leq j\} \times [0, t]$ for Loeb a.e. $(x, v_1) \in {}^*\Lambda \times \text{ns}\, {}^*\mathbb{R}^3$. Moreover,

$$(\gamma) \qquad °\int_0^t \int_{{}^*\mathbb{R}^3} \langle f \otimes f \rangle(x + v_1 s, v_1, v_2, s)w_n^m(v_1, v_2)\, {}^*dv_2\, ds$$

$$= \int_0^t \int_{\text{ns}\, {}^*\mathbb{R}^3} °[\langle f \otimes f \rangle(x + v_1 s, v_1, v_2, s)|v_1 - v_2|]L(dv\, ds).$$

But for a.e. $(x, v_1) \in {}^*\Lambda \times \text{ns}\, {}^*\mathbb{R}^3$ we have

$$°(\langle f \otimes f \rangle w_n^m) = °f \otimes °f °|v_2 - v_1|$$

for a.e. $(v_2, s) \in \text{ns}\, {}^*\mathbb{R}^3 \times [0, t)$, which is easy to see, since for a.e. (x, v_1) each factor of $f \otimes f$ is finite a.e. in $\text{ns}\, {}^*\mathbb{R}^3 \otimes [0, t]$. This in turn is a consequence of the fact that f is S-integrable. Thus for Loeb a.e. $(x, v_1) \in {}^*\Lambda \times \text{ns}\, {}^*\mathbb{R}^3$ we have from (γ) the equality in the lemma.

6.5.13. LEMMA. For Loeb a.e. $(x, v_1) \in {}^*\Lambda \times \text{ns}\, {}^*\mathbb{R}^3$

$$(15) \qquad °\int_0^t \int_{{}^*B} \int_{{}^*\mathbb{R}^3} \langle f \otimes f \rangle(x + sv_1, v_1', v_2', s)w_n^m(v_1, v_2, u)\, {}^*dv_2\, du\, ds$$

$$= \int_0^t \int_{{}^*B} \int_{\text{ns}\, {}^*\mathbb{R}^3} °f \otimes °f(x + sv_1, v_1', v_2', s)°|v_1 - v_2|L(dv_2\, du\, ds).$$

PROOF. The following discussion holds for Loeb a.e. (x, v_1). Let χ be the characteristic function of any set $v_1^2 + v_2^2 \leq k^2$. Since the integrals

$$\int \chi f(x + sv_1, v_j', s)\, {}^*dv_1\, dv_2\, du\, ds, \qquad j = 1, 2,$$

both are finite, it follows that

$$\chi f(x + sv_1, v_1', s) \quad \text{and} \quad \chi f(x + sv_1, v_2', s)$$

are finite for Loeb a.e. (v_2, u, s). And so for Loeb a.e. (v_2, u, s)

$$°\langle f \otimes f \rangle(x + sv_1, v_1', v_2', s)w_n^m = °f \otimes °f(x + sv_1, v_1', v_2', s)°|v_1 - v_2|.$$

To prove the S-integrability of $\langle f \otimes f' \rangle w_n^m(v_2, u, s) \equiv \langle f \otimes f \rangle$ $(x + sv_1, v_1', v_2', s) w_n^m$ it suffices to show that for some finite $j > 1$ that $\langle f \otimes f' \rangle w_n^m$ is S-integrable on the set

$$\Omega_j = \{(v_2, u, s)|\langle f \otimes f' \rangle \geq j\langle f \otimes f \rangle\}.$$

But on Ω_2

$$0 \leq \langle f \otimes f' \rangle w_n^m \leq 2(\langle f \otimes f' \rangle - \langle f \otimes f \rangle)w_n^m,$$

and it suffices to consider the S-integrability of the right member on Ω_2. We only have to check (ii), (iii), and (iv). Now

$$f \otimes f' \geq jf \otimes f$$

on Ω_j for $j > 1$, and

$$(16) \quad 0 \leq \int_{\Omega_j} (\langle f \otimes f' \rangle - \langle f \otimes f \rangle) w_n^m {}^* dv_2 \, du \, ds$$

$$\leq (\log j)^{-1} \int_{\Omega_j} (\langle f \otimes f' \rangle - \langle f \otimes f \rangle) \log(f \otimes f'/f \otimes f) w_n^m {}^* dv_2 \, du \, ds$$

$$\leq C/\log j,$$

with C finite. In particular this proves (ii), if we take $j = 2$.

The possibility of majorizing a.e. by a finite C in the last member of (16) follows from the usual proof of the H theorem, more precisely from the positivity of the integrand, and from

$$\int_A (\langle f \otimes f' \rangle - \langle f \otimes f \rangle) w_n^m \log(f \otimes f'/f \otimes f) {}^* dx \, dv_1 \, dv_2 \, du \, ds$$

$$\leq \left\{ \int_{*M} (f_0 \log f_0 + \exp(-v^2) + v^2 f_0) {}^* dx \, dv \right\} \in \text{ns} \, {}^*\mathbb{R}_+,$$

when

$$A = {}^*(M \times R^3 \times B \times R_+).$$

To prove (iii) we let $k \in {}^*N - N$. It follows from $f_0 > 0$ and the S-integrability of $\langle f \otimes f \rangle w_n^m$ that

$$0 < j_0^{-2} \equiv \int_{\Omega_2} \langle f \otimes f \rangle w_n^m \wedge k^{-1} {}^* dv_2 \, du \, ds \approx 0.$$

Hence

$$\int_{\Omega_2 - \Omega_{j_0}} \langle f \otimes f' \rangle w_n^m \wedge k^{-1} {}^* dv_2 \, du \, ds \approx 0.$$

As for the remaining part of Ω_2, by (16)

$$0 \leq \int_{\Omega_{j_0}} \langle f \otimes f' \rangle w_n^m \wedge k^{-1} {}^* dv_2 \, du \, ds$$

$$\leq \int_{\Omega_{j_0}} \langle f \otimes f' \rangle w_n^m {}^* dv_2 \, du \, ds < \frac{2C}{\log j_0} \approx 0.$$

This proves (iii).

To prove (iv) we let $k \in {}^*N - N$ and set

$$\Omega = \{(v_2, u, s) \in \Omega_2 | \langle f \otimes f' \rangle w_n^m > k\}.$$

It follows from (ii) that

$$\int_\Omega {}^* dv_2 \, du \, ds \approx 0,$$

and from $f_0 > 0$ together with (4), that

$$\int_\Omega \langle f \otimes f \rangle w_n^m \, {}^* dv_2 \, du \, ds > 0 \qquad \text{if} \quad \int_\Omega {}^* dv_2 \, du \, ds > 0.$$

Then

$$0 < j_0^{-2} \equiv \int_\Omega \langle f \otimes f \rangle w_n^m \, {}^* dv_2 \, du \, ds \approx 0,$$

since $\langle f \otimes f \rangle$ is S-integrable. Hence

$$\int_{\Omega - \Omega_{j_0}} \langle f \otimes f' \rangle w_n^m \, {}^* dv_2 \, du \, ds \approx 0.$$

As for the remaining part of Ω

$$\int_{\Omega_{j_0}} (\langle f \otimes f' \rangle - \langle f \otimes f \rangle) w_n^m \, {}^* dv_2 \, du \, ds \approx 0$$

by (16). This proves (iv) and so the S-integrability.

Finally, to obtain (15) we shall also check that

$$\lim_{\nu \to \infty} {}^\circ \int_{A_\nu} \langle f \otimes f' \rangle w_n^m \, {}^* dv_2 \, du \, ds = 0,$$

where

$$A_\nu = \{(v_2, u, s) \in \Omega ; |v_2| \geq \nu\}.$$

Suppose the limit equals $\varepsilon > 0$. By (16) we can choose $j \in N$ such that

$${}^\circ \int_{\Omega_j} (\langle f \otimes f' \rangle - \langle f \otimes f \rangle) w_n^m \, {}^* dv_2 \, du \, ds < \varepsilon/2.$$

This, however, leads to a contradiction, since by Lemma 6.5.12

$$\lim_{\nu \to \infty} {}^\circ \int_{A_\nu} \langle f \otimes f \rangle w_n^m \, {}^* dv_2 \, du \, ds = 0.$$

From the above Lemmas 6.5.10–13 we then get the following existence theorem for the space-inhomogeneous Boltzmann equation (1) on a torus

$\Lambda \equiv \mathbb{R}^3/\mathbb{Z}^3$ with a two-body potential of finite range, so that the impact parameter is in $B = \{u \in \mathbb{R}^2 | |u| \le m\}$, for some $m < \infty$. The initial conditions are the natural ones from the physical point of view, as in all theorems above.

6.5.14. THEOREM. Consider the Boltzmann equation (1) with initial condition $F_0 \ge 0$ Lebesgue a.e., and such that F_0, $v^2 F_0$, $F_0 \log F_0 \in L^1(\Lambda \times \mathbb{R}^3)$. Then the solution f of the equation

$$(d/dt)f(x + v_1 t, v_1, t) = (Q_n^m f)(x + v_1 t, v_1, t)$$

with Q_n^m given in Section 6.5.C for $m \in \mathbb{N}$, $n \in {}^*\mathbb{N} - \mathbb{N}$, with initial condition

$$f_0(x, v) = \min({}^*F_0(x, v), n) + n^{-1} \exp(-v^2),$$

is nearstandard for Loeb a.e. $(x, v_1) \in {}^*\Lambda \times \text{ns }{}^*\mathbb{R}^3$, and its standard part $^\circ f$ is a solution of the Boltzmann equation (1) with initial value F_0 in the following sense: for a.e. $(x, v_1) \in {}^*\Lambda \times \text{ns }{}^*\mathbb{R}^3$

$$^\circ f(x + v_1 t, v_1, t) = {}^{\circ *}F_0(x, v_1)$$

$$+ \int_0^t \int_{\text{ns }{}^*\mathbb{R}^3 \times {}^*B} {}^\circ f \otimes {}^\circ f(x + v_1 s, v_1', v_2', s)^\circ |v_2 - v_1| L(dv_2 \, du \, ds)$$

$$- \int_0^t \int_{\text{ns }{}^*\mathbb{R}^3 \times {}^*B} {}^\circ f \otimes {}^\circ f(x + v_1 s, v_1, v_2, s)^\circ |v_2 - v_1| L(dv_2 \, du \, ds).$$

PROOF. The integral form of (3) is

$$f(x + v_1 t, v_1, t) = f_0(x, v_1)$$

$$+ \int_0^t \int_{{}^*B \times {}^*\mathbb{R}^3} \langle f \otimes f \rangle(x + v_1 s, v_1', v_2', s) w_n^m(v_1, v_2) {}^*du \, dv_2 \, ds$$

$$- \int_0^t \int_{{}^*B \times {}^*\mathbb{R}^3} \langle f \otimes f \rangle(x + v_1 s, v_1, v_2, s) w_n^m(v_1, v_2) {}^*du \, dv_2 \, ds.$$

Taking the standard part and using Lemmas 6.5.12 and 6.5.13 for Loeb a.e. $(x, v_1) \in {}^*\Lambda \times \text{ns }{}^*\mathbb{R}^3$, we get

$$^\circ f(x + v_1 t, v_1, t) = {}^{\circ *}F_0(x, v_1)$$

$$+ \int_0^t \int_{\text{ns }{}^*\mathbb{R}^3 \times {}^*B} {}^\circ f \otimes {}^\circ f(x + v_1 s, v_1', v_2', s)^\circ |v_2 - v_1| L(du \, dv_2 \, ds)$$

$$- \int_0^t \int_{\text{ns }{}^*\mathbb{R}^3 \times {}^*B} {}^\circ f \otimes {}^\circ f(x + v_1 s, v_1, v_2, s)^\circ |v_2 - v_1| L(dv_2 \, du \, ds).$$

We shall now deduce some consequences of Theorem 6.5.14.

6.5.15. COROLLARY. The solution of Theorem 6.5.14 is t-continuous in the sense that for Loeb a.e. (x, v_1), given $t \geq 0$ and $\varepsilon \in \mathbb{R}_+$, there is a $\delta \in \mathbb{R}_+$ such that

$$|{}^{\circ}f(x + v_1 t, v_1, t) - {}^{\circ}f(x + v_1 t', v_1, t')| < \varepsilon$$

if $|t - t'| < \delta$.

For a discussion of the meaning of t continuity see Truesdell and Muncaster (1980, p. 343).

PROOF. If a function g on ns $*\mathbb{R}_+$ is Loeb integrable, then by the definition of the Loeb integral we have

$$\int \chi_\Delta(s) g(s) L(ds) = 0$$

for every characteristic function χ_Δ of a set $\Delta \subset$ ns $*\mathbb{R}_+$ with infinitesimal *Lebesgue measure. From this and the equality in Theorem 6.5.14 the corollary follows easily.

We shall now compare standard solutions, in case they exist, with Loeb solutions. We have the following

6.5.16. THEOREM. If $F: \mathbb{R}_+ \to L^1_+(\Lambda \times \mathbb{R}^3)$ is a solution of the Boltzmann equation in the sense that

$$F(x + v_1 t, v_1, t) = F_0(x, v_1) + \int_0^t \int_{\mathbb{B} \times \mathbb{R}^3} [F \otimes F(x + v_1 s, v_1', v_2', s)$$

$$- F \otimes F(x + v_1 s, v_1, v_2, s)]|v_2 - v_1| \, du \, dv_2 \, ds$$

for a.e. $(x, v_1) \in \Lambda \times \mathbb{R}^3$, then $\tilde{f} \equiv F \circ$ st is a Loeb solution satisfying for Loeb a.e. $(x, v_1) \in *\Lambda \times *\mathbb{R}^3$:

$$\tilde{f}(x + v_1 t, v_1, t) = {}^{\circ *}F_0(x, v_1)$$

$$+ \int_0^t \int_{\text{ns } *\mathbb{R}^3 \times *B} \tilde{f} \otimes \tilde{f}(x + v_1 s, v_1', v_2', s)^{\circ}(|v_2 - v_1|) L(dv_2 \, du \, ds)$$

$$- \int_0^t \int_{\text{ns } *\mathbb{R}^3 \times *B} \tilde{f} \otimes \tilde{f}(x + v_1 s, v_1, v_2, s)^{\circ}(|v_2 - v_1|) L(dv_2 \, du \, ds).$$

PROOF. This depends essentially on the fact that if f is Lebesgue integrable, then $\tilde{f} = f \circ$ st is Loeb integrable with $\int \tilde{f} L(dq) = \int f \, dq$.

6.5.17. REMARK. If a standard solution F of the Boltzmann equation (1) exists and *if* uniqueness holds in our corresponding Loeb problem of finding an ${}^{\circ}f$ which satisfies Theorem 6.5.14, then by using Theorem 6.5.16 we conclude that the Loeb solution $^\circ f$ is equal to $\tilde{f} = F \circ \mathrm{st}$ and is thus an extension to a "denser space" of the standard solution. For example, this situation arises in the spatially homogeneous case with finite fourth moments.

6.5.18. REMARK. The results obtained by nonstandard methods by Arkeryd have several consequences, for which we refer to Arkeryd's publications. In particular, as shown by Arkeryd (1981c), one can prove that the Loeb solutions of the Boltzmann equation discussed above are limits (in the weak sense) of such solutions for a Boltzmann equation with discretized time.

Summarizing, we have seen how nonstandard tools are quite powerful in handling a typical nonlinear Cauchy problem. In the space-homogeneous case the first proof, for potentials of infinite range, of convergence to the Maxwell distribution was by nonstandard methods. Moreover, the only existence proof so far for solutions in the space-inhomogeneous case far from equilibrium is also by nonstandard methods. Above all we look at this approach as very encouraging for handling other nonlinear problems. Nonstandard methods permit us to use—by transfer—conservation laws and monotone quantities such as the entropy, which would be otherwise difficult to exploit due to the nonlinearities and nonsmoothness involved. We would expect that this and similar ideas will turn out to be useful also in other problems of the theory of integral and differential equations.

6.6. A FINAL REMARK ON THE FEYNMAN PATH INTEGRAL AND OTHER MATTERS

In lecturing on nonstandard methods the question of the Feynman path integral is inevitably brought up. It is possible that the "new numbers" of nonstandard theory can also make this notion precise?

The Feynman path integral was introduced by Feynman to give an alternative formulation of the theory of quantum dynamics [see Feynman (1948) and Feynman and Hibbs (1965)]. The heuristic, but powerful, idea of Feynman has been given several mathematical expressions; see, e.g., the surveys in Albeverio *et al.* (1979a). Here we shall use nonstandard analysis to give an interpretation of Feynman's ideas, in a spirit close to the original motivation.

REMARK. The reader should not confuse the Feynman path integral with the "Euclidean" version, the Feynman–Kac integral; for the latter see Section 5.3.

Let us first briefly recall Feynman's heuristic argument. Consider a quantum mechanical particle of mass m moving in $\mathbb{R}^s$ under the action of a potential $V(x)$. The dynamics is given by the operator

$$e^{-(i/\hbar)tH}$$

in the Hilbert space $L^2(\mathbb{R}^s)$, where $\hbar$ is Planck's constant divided by 2π and $H = H_0 + V$, where $H_0 \equiv -(\hbar^2/2m)\,\Delta$.

Using the Lie–Trotter formula we can write

$$(1) \qquad \exp\left(-\frac{i}{\hbar}tH\right) = \lim_{n\to\infty}\left(\exp\left(-\frac{i}{\hbar}\frac{t}{n}V\right)\exp\left(-\frac{i}{\hbar}\cdot\frac{t}{n}H_0\right)\right)^n$$

[where the limit is taken in the sense of strong convergence of operators] if V is real-valued and $H_0 + V$ is essentially self-adjoint on $D(H_0) \cap D(V)$, i.e. the domains of H_0 and V as self-adjoint operators.

The kernel of $\exp(-(i/\hbar)\cdot(t/n)H_0)$ is

$$\exp\left(-\frac{i}{\hbar}\cdot\frac{t}{n}H_0\right)(x,y) = \left(2\pi i\frac{\hbar}{m}\cdot\frac{t}{n}\right)^{-s/2}\exp\left[\frac{im}{(2\hbar(t/n))}(x-y)^2\right]$$

with proper determination of the $z^{-s/2}$ part.

The if φ is a smooth function of compact support

$$(2) \qquad \left(\exp\left(-\frac{i}{\hbar}\cdot\frac{t}{n}V\right)\exp\left(-\frac{i}{\hbar}\cdot\frac{t}{n}H_0\right)\right)^n \varphi(x)$$

$$= \int_{\mathbb{R}^s}\cdots\int_{\mathbb{R}^s}\exp\left(\frac{i}{\hbar}S_t(x,x_{n-1},\ldots,x_0)\right)\varphi(x_0)\prod_{j=0}^{n-1}\tilde{d}x_j,$$

where $\tilde{d}x_j = (2\pi i\hbar t/m\cdot n)^{-s/2}\,dx_j$, dx_j being the Lebesgue measure on $\mathbb{R}^s$, and

$$(3) \qquad S_t(x_n, x_{n-1}, \ldots, x_0) \equiv \sum_{j=1}^{n}\left[\frac{m}{2}\frac{(x_j - x_{j-1})^2}{(t/n)^2} - V(x_j)\right]\frac{t}{n}.$$

Here the integral on the right-hand side of (2) exists as an improper Lebesgue integral.

Following Feynman, it is natural to look at (2) as an integral along piecewise linear continuous paths in the following way. Let $t_j = j\cdot(t/n)$, $j = 0,\ldots,n$, and let $\gamma_n(\tau)$ be the piecewise linear path $[0, t] \to \mathbb{R}^s$ which is equal to x_j for $\tau = t_j$. Define

$$S_t(\gamma_n) \equiv \frac{m}{2}\int_0^t \dot{\gamma}_n^2(\tau)\,d\tau - \int_0^t \tilde{V}(\gamma_n(\tau))\,d\tau,$$

where $(m/2)\int_0^t \dot{\gamma}_n^2(\tau)\,d\tau$ is the kinetic energy

$$\frac{m}{2}\sum_{j=0}^{n-1}\int_{t_j}^{t_{j+1}}\dot{\gamma}_n^2(\tau)\,d\tau = \frac{m}{2}\sum_{j=1}^{n}\frac{(x_j - x_{j-1})^2}{(t/n)^2}\frac{t}{n}$$

associated with γ_n, and

$$\int_0^t \tilde{V}(\gamma_n(\tau))\,d\tau,$$

with $\tilde{V}(\gamma_n(\tau)) \equiv V(\gamma_n(j(t/n)))$ for $\tau \in [j(t/n), (j+1)(t/n)]$, is the approximate potential energy associated with γ_n. Then $S_t(\gamma_n) = S_t(x, x_{n-1}, \ldots, x_0)$ and the right-hand side of (2) appears as the path integral

$$(4)\qquad \int_{\Gamma_x}\exp\left(\frac{i}{\hbar}S_t(\gamma_n)\right)\varphi(\gamma_n(0))\,d\gamma_n$$

with Γ_x the space of all piecewise linear continuous paths γ_n such that $\gamma_n(t) = x$ and γ_n has discontinuities in the derivatives only at the points $j(t/n), j = 1, \ldots, n-1$. Note that $\Gamma_x = \Gamma_0 + x$, where $\Gamma_0 \cong \mathbb{R}^{ns}$. The measure $d\gamma_n$ is given by

$$d\gamma_n \equiv \prod_{j=0}^{n-1}\left(\frac{2\pi i\hbar t}{mn}\right)^{-s/2}d\gamma_n\left(j\frac{t}{n}\right) = \prod_{j=0}^{n-1}\tilde{d}x_j.$$

Feynman's idea is now to take n "big" and to look at (4) as an integral over "the space Γ of all continuous paths ending at time t in x." *Formally* (1), (2), and (4) give

$$(5)\qquad \exp\left(-\frac{i}{\hbar}tH\right)\varphi(x) = \int_{\gamma(t)=x}\exp\left(\frac{i}{\hbar}S_t(\gamma)\right)\varphi(\gamma(0))\,d\gamma.$$

We have stressed "formally," although the right-hand side can be interpreted in certain cases as the limit as $n \to \infty$ of (4). But there is no standard measure $d\gamma$ on the space of paths which gives the right-hand side of (5) as a genuine integral. This difficulty can, at least up to a point, be overcome by using nonstandard analysis.

REMARK. In the literature on Feynman path integrals the definition by the above limiting procedure or related ones is sometimes called "sequential limit definition"; see, e.g., Albeverio *et al.* (1979a), Kallianpur and Bromley (1984) and references therein.

Since $\Gamma_0 \cong \mathbb{R}^{sn}$ we can rewrite (4) as

$$(6)\qquad \int_{\mathbb{R}^{sn}}\exp\left(\frac{i}{\hbar}S_t(\gamma_n + x)\right)\varphi(\gamma_n(0) + x)\,d\gamma_n.$$

Extending from $n \in \mathbb{N}$ to an $n \in {}^*\mathbb{N}$ we can now introduce the *internal Feynman path integral* by exactly the same definition as for (6). In a very intuitive sense we now integrate over *hyperfinite* piecewise linear paths; i.e., we let $\gamma_n(\tau)$ be a piecewise linear path $^*[0, t] \to {}^*\mathbb{R}^s$, where we now have divided $^*[0, t]$ into a hyperfinite number of points $t_j = j \cdot (t/n), j = 0, \ldots, n$, and $n \in {}^*\mathbb{N}$. Thus, on the ordinary scale γ_n "looks continuous," but on the infinitesimal scale it is a hyperfinite polygonal path.

Thus our construction captures some of the formal properties of Feynman's original proposal. But it is an *internal* quantity; $\exp((i/\hbar)S_t(\gamma_n + x))\,{}^*d\gamma_n$ is an unbounded, wildly oscillating $^*\mathbb{C}$-valued measure which cannot be turned into a Loeb measure by the techniques described in Chapter 3. And even if there were a "Loeb" construction we would have difficulties in passing from the hyperfinite and highly irregular path space to the continuous version.

On the other hand, it is not difficult to exhibit classes of potentials V for which the internal Feynman path integral is nearstandard and its standard part solves the Schrödinger equation; i.e., the solution of the Schrödinger equation

$$i\hbar \frac{\partial}{\partial t}\psi = H\psi$$

with a suitable initial condition $\varphi \in L^2(\mathbb{R}^s)$ is given by

$$(7) \quad \psi = \exp\left(-i\frac{t}{\hbar}H\right)\varphi \approx \int_{{}^*\mathbb{R}^{sn}} \exp\left(\frac{i}{\hbar}{}^*S_t(\gamma_n + x)\right){}^*\varphi(\gamma_n(0) + x)\,{}^*d\gamma_n,$$

for some $n \in {}^*\mathbb{N} - \mathbb{N}$.

REMARK. This includes H which are strong resolvent limits as $k \to \infty$ of Hamiltonians H_k for which the Lie–Trotter formula holds.

For related work, in particular, on a "nonstandard Lie–Trotter" formula, see Sloan (1977, 1978).

In solving the Schrödinger equation we have assumed that $\hbar$ is standard positive. What happens in (7) when $\hbar \neq 0$ is infinitesimal? It was the heuristic insight of Dirac and Feynman that in order to recover classical mechanics from quantum mechanics one should exploit the fact that Planck's constant is small, hence find the asymptotics of the solution of the Schrödinger equation for $\hbar \to 0$. Writing this solution as an oscillating integral as above suggests an application of a method of stationary phase, and formally this works since the stationary points of the phase are seen to be the points where the first variation of $S_t(\gamma + x)$ vanishes, and by Hamilton's principle these are the classical orbits ending at time t.

Whereas in finite dimensions the method of stationary phase is a classical tool, which has its origin in methods of Stokes and Kelvin and has enjoyed a steady refinement up to recent years [see, e.g., Guillemin and Sternberg (1977)], it took a long time before a corresponding method of stationary phase could be developed in infinite dimensions, sufficiently powerful to handle certain classes of Feynman path integrals. This was done in Albeverio and Høegh-Krohn (1977) and Rezende (1985) by using the definition of Feynman path integrals given in Albeverio and Høegh-Krohn (1976).

But it is also possible, indeed even "natural," to take (7) as a point of departure and use the explicit definition of the internal Feynman path integral in the calculations; i.e., we could start from

$$(8) \qquad \left(\exp\left(-\frac{i}{\hbar} \cdot \frac{t}{n}(H_0) \right) \exp\left(-\frac{i}{\hbar} \cdot \frac{t}{n} {}^*V \right) \right)^n {}^*\varphi(x)$$

$$\approx \int_{{}^*\mathbb{R}^{sn}} \exp\left(\frac{i}{\hbar} {}^*S_t(\gamma_n + x) \right) {}^*\varphi(\gamma_n(0) + x) \, {}^*d\gamma_n,$$

for $n \in {}^*\mathbb{N} - \mathbb{N}$ and $\hbar \approx 0$.

We shall not pursue this approach further here, since it would essentially consist in reproducing the "hard core" of the standard approach of Hepp (1974) and Hagedorn (1980). We have only wanted to emphasize that the internal Feynman path integral is a precise, well-defined mathematical notion, which captures a significant part of the original heuristics and which can be effectively used in "hyperfinite" calculations. But one has to be careful; the integral is internal, and remains internal!

REMARK. Harthong has developed another interesting nonstandard extension of the classical method of stationary phase for oscillatory integrals; see Harthong (1981, 1984). In particular, he has applied his results to a revealing discussion of certain optical phenomena ("moiré patterns") and to the study of wave propagation.

REFERENCES

S. Albeverio and R. Høegh-Krohn (1976). *Mathematical theory of Feynman path integrals.* Lect. Notes Math. **523**. Springer-Verlag, Berlin.

S. Albeverio and R. Høegh-Krohn (1977). Oscillatory integrals and the method of stationary phase in infinitely many dimensions, with applications to the classical limit of quantum mechanics, I. *Invent. Math.* **40**.

S. Albeverio and R. Høegh-Krohn (1981). Point interactions as limits of short range interactions. *J. Operator Theory* **6**.

S. Albeverio, R. Høegh-Krohn, and L. Streit (1977). Energy forms, Hamiltonians, and distorted Brownian paths. *J. Math. Phys.* **18**.

S. Albeverio, Ph. Combe, R. Høegh-Krohn, G. Rideau, M. Sirugue-Collin, M. Sirugue, and R. Stora (eds.) (1979a). *Feynman path integrals. Lect. Notes Phys.* **106**. Springer-Verlag, Berlin.

S. Albeverio, J. E. Fenstad, and R. Høegh-Krohn (1979b). Singular perturbations and nonstandard analysis. *Trans. Amer. Math. Soc.* **252**.

S. Albeverio, R. Høegh-Krohn, and L. Streit (1980). Regularization of Hamiltonians and processes. *J. Math. Phys.* **21**.

S. Albeverio, R. Høegh-Krohn, W. Kirsch, and F. Martinelli (1982a). The spectrum of the three-dimensional Kronig-Penney model with random point defects. *Adv. Appl. Math.* **3**.

S. Albeverio, Ph. Blanchard, and R. Høegh-Krohn (1982b). Some applications of functional integration. In *Mathematical Problems in Theoretical Physics. Lecture Notes in Phys.* (R. Schrader, R. Seiler, D. A. Uhlenbrock, eds.), **153**.

S. Albeverio, F. Gesztesy, and R. Høegh-Krohn (1982c). The low energy expansion in nonrelativistic scattering theory. *Ann. Inst. H. Poincaré Sect. A* **37**.

S. Albeverio, F. Gesztesy, R. Høegh-Krohn, and H. Holden (1984a). Some exactly solvable models in quantum mechanics and the low energy expansions. In *Proc. Leipzig Conf. Operator Algebras, 1983, Teubner, Stuttgart.*

S. Albeverio, J. E. Fenstad, R. Høegh-Krohn, W. Karwowski, and T. Lindstrøm (1984b). Perturbations of the Laplacian supported by null sets, with applications to polymer measures and quantum fields. *Phys. Lett.* **104**.

S. Albeverio, J. E. Fenstad, R. Høegh-Krohn, W. Karwowski, and T. Lindstrøm (1986a). (In preparation.)

S. Albeverio, F. Gesztesy, R. Høegh-Krohn, and H. Holden (1986b). *Solvable Models in Quantum Mechanics* (in preparation).

A. Alonso y Coria (1978). *Shrinking potentials in the Schrödinger equation.* Ph.D. thesis. Princeton Univ., Princeton, New Jersey.

L. Arkeryd (1972a). On the Boltzmann equation I, II. *Arch. Rational Mech. Anal.* **45**, 1–34.

L. Arkeryd (1972b). An existence theorem for a modified space-inhomogeneous, non-linear Boltzmann equation. *Bull. Amer. Math. Soc.* **78**, 610–614.

L. Arkeryd (1981a). Intermolecular forces of infinite range and the Boltzmann equation. *Arch. Rational Mech. Anal.* **77**, 11–23.

L. Arkeryd (1981b). A non-standard approach to the Boltzmann equation. *Arch. Rational Mech. Anal.* **77**, 1–10.

L. Arkeryd (1981c). A time-wise approximated Boltzmann equation. *IMA J. Appl. Math.* **27**, 373–383.

L. Arkeryd (1982). Asymptotic behavior of the Boltzmann equation with infinite range forces. *Comm. Math. Phys.* **86**, 475–484.

L. Arkeryd (1984). Loeb solutions of the Boltzmann equation. *Arch. Rational Mech. Anal.* **86**.

M. N. Barber and B. W. Ninham (1970). *Random and Restricted Walks.* Gordon and Breach, New York.

F. A. Berezin and L. D. Faddeev (1961). A remark on Schrödinger's equation with a singular potential. *Soviet Math. Dokl.* **2**.

H. Bethe and R. Peierls (1935). Quantum theory of the diplon. *Proc. R. Soc. London A* **148**.

B. Birkeland (1980). A singular Sturm-Liouville problem treated by nonstandard analysis. *Math. Scand.* **47**.

D. Brydges and T. Spencer (1985). Self-avoiding walk in 5 or more dimensions, *Comm. Math. Phys.* **97**.

D. Brydges, J. Fröhlich, and T. Spencer (1982). The random walk representation of classical spin systems and correlation inequalities. *Comm. Math. Phys.* **83**.

C. Cercignani (1975). *Theory and Application of the Boltzmann Equation.* Scottish Academic Press.

Y. N. Demkov and V. N. Ostrovskii (1975). *The Use of Zero-Range Potentials in Atomic Physics.* Nauka, Moscow (in Russian).

C. Domb (1969). Self-avoiding walks on lattices. *In* K. E. Shuler (ed.), *Stochastic Processes in Chemical Physics.* Wiley, New York.

N. Dunford and B. J. Pettis (1940). Linear operations on summable functions. *Trans. Amer. Math. Soc.* **47**, 323-392.

E. B. Dynkin (1985). Random fields associated with multiple points of the Brownian motion. *J. Funct. Anal.* **62**.

T. Elmroth (1983). Global boundedness of moments of solutions of the Boltzmann equation for infinite range forces. *Arch. Rational Mech. Anal.* **82**, 1-2.

S. F. Edwards (1965). The statistical mechanics of polymers with excluded volume. *Proc. Phys. Soc. London* **85**.

S. F. Edwards (1975). A note on the convergence of perturbation theory in polymer problems. *J. Phys. A.* **8**.

E. Fermi (1936). Sul moto dei neutroni nelle sostance idrogenate. *Ric. Sci.* **7**.

R. P. Feynman (1948). Space-time approach to non-relativistic quantum mechanics. *Rev. Mod. Phys.* **20**.

R. P. Feynman and A. R. Hibbs (1965). *Quantum Mechanics and Path Integrals,* McGraw-Hill, New York.

G. Flamand (1967). Mathematical theory of non-relativistic two- and three-particle systems with point interactions. *In* F. Lurcat (ed.), *Cargese Lectures in Theoretical Physics 1967.* Gordon and Breach, New York.

P. J. Flory (1969). *Statistical Mechanics of Chain Molecules.* Wiley (Interscience), New York.

K. F. Freed (1981). Polymers as self-avoiding random walks. *Ann. Probab.* **9**.

C. N. Friedman (1971). Perturbations of the Schrödinger equation by potentials with small support, semigroup product formulas, and applications to quantum mechanics. Ph.D thesis. Princeton Univ., Princeton, New Jersey.

C. N. Friedman (1972). Perturbations of the Schrödinger equation by potentials with small support. *J. Funct. Anal.* **10**.

M. Fukushima (1980). *Dirichlet Forms and Markov Processes.* North-Holland Publ., Amsterdam.

J. Glimm and A. Jaffe (1981). *Quantum Physics—A Functional Integral Point of View.* Springer-Verlag, New York and Berlin.

I. S. Gradshteyn and I. M. Ryshik (1965). *Table of Integrals, Series and Products.* Academic Press, New York.

A. Grossmann, R. Høegh-Krohn, and M. Mebkhout (1980a). A class of explicitly soluble, local, many-center Hamiltonians for one-particle quantum mechanics in two and three dimensions. *J. Math. Phys.* **21**.

A. Grossmann, R. Høegh-Krohn, and M. Mebkhout (1980b). The one-particle theory of periodic point interaction. *Comm. Math. Phys.* **77**.

V. Guillemin and S. Sternberg (1977). *Geometric Asymptotics.* Amer. Math. Soc., Providence, Rhode Island.

G. A. Hagedorn (1980). Semiclassical quantum mechanics. *Comm. Math. Phys.* **71**.

J. Hammersley and K. W. Morton (1954). Poor Man's Monte Carlo. *J.R. Statist. Soc.* **16**.

J. Harthong (1981). Le moiré. *Adv. Appl. Math.* **2**.

J. Harthong (1984). Etudes sur la mécanique quantique. *Astérisque* 111.

K. Hepp (1974). The classical limit for quantum mechanical correlation functions. *Comm. Math. Phys.* **35**.

H. Holden (1981). *Konvergens mot punkt-interaksjoner,* cand. real. thesis, Math. Inst., Univ. of Oslo.

H. Holden, R. Høegh-Krohn, and S. Johannesen (1983). The short range expansion. *Adv. Appl. Math.* **4.**

H. Holden, R. Høegh-Krohn, and S. Johannesen (1984). The short-range expansion in solid state physics. *Ann. Inst. H. Poincaré, Sect. A* **41.**

R. Høegh-Krohn, H. Holden, S. Johannesen, and T. Wentzel-Larsen (1986). The Fermi surface for point interactions *J. Math. Phys.* **27.**

K. Huang and C. N. Yang (1957). Quantum-mechanical many-body problem with hard-sphere interaction. *Phys. Rev.* **105.**

K. Huang, C. N. Yang, and J. M. Luttinger (1957). Imperfect Bose gas with hard-sphere interaction. *Phys. Rev.* **105.**

G. Kallianpur and G. Bromley (1984). Generalized Feynman integrals using analytic continuation in several complex variables. In *Stochastic Analysis and Applications* (M. Pinsky, ed.). Dekker, New York.

T. Kato (1976). *Perturbation Theory for Linear Operators,* 2nd ed., Springer-Verlag, Berlin and New York.

H. Kesten (1963). On the number of self-avoiding walks I. *J. Math. Phys.* **4.**

R. D. L. Kronig and W. G. Penney (1931). Quantum mechanics of electrons in crystal lattices. *Proc. R. Soc. London, Ser. A* **130.**

S. Kusuoka (1985a). On the path property of Edward's model for long polymer chains in three dimensions. In *Proceedings of the USP Meeting on Stochastic Processes and Infinite Dimensional Analysis.* (S. Albeverio, ed.). Res. Notes. Math., Pitman, London.

S. Kusuoka (1985b). Asymptotics of polymer measures in one dimension. In *Proceedings of the Bielefeld Conference on Infinite Dimensional Analysis and Stochastic Processes.* (S. Albeverio, ed.). Res. Notes Math., Pitman, London.

O. E. Lanford, III (1975). Time evolution of large classical systems. In *Dynamical Systems, Theory and Applications* (J. Moser, ed.) pp. 1–111, *Lecture Notes in Physics* **38,** Springer-Verlag, Berlin and New York.

G. A. Lawler (1980). A self-avoiding random walk. *Duke Math. J.* **47.**

G. F. Lawler (1983). A connective constant for loop-erased self-avoiding random walk. *J. Appl. Probab.* **20.**

G. F. Lawler (1985). Intersections of random walks in four dimensions II. *Comm. Math. Phys.* **97.**

T. D. Lee, K. Huang, and C. N. Yang (1957). Eigenvalues and eigenfunctions of a Bose system of hard spheres and its low-temperature properties. *Phys. Rev.* **106.**

J. F. LeGall (1985). Sur le temps local d'intersection du movement Brownien plan, et la méthode de renormalization de Varadhan. *In* (J. Azema and M. Yor, eds.) *Sém. Prob.* XIX, 83/84. Lect. Notes Math. 1123, Springer-Verlag, Berlin and New York.

A. L. MacDonald (1976). Sturm-Liouville theory via nonstandard analysis. *Indiana Univ. Math. J.* **25.**

E. Nelson (1977). Internal set theory: A new approach to nonstandard analysis. *Bull. Amer. Math. Soc.* **83.**

E. Nelson (1983). A remark on the polymer problem in four dimensions. Studies in applied mathematics, 1–5. *Adv. Math. Suppl. Stud.* **8,** Academic Press, New York and London.

J. Persson (1981). Second order linear ordinary differential equations with measures as coefficients. *Matematiche* (Catania) **36.**

J. Persson (1984). Linear distribution differential equations. *Comment. Math. Univ. St. Pauli* **33.**

M. Reed and B. Simon (1975). *Methods of Modern Mathematical Physics II: Fourier Analysis, Self-Adjointness.* Academic Press, New York and London.

J. Rezende (1985). The method of stationary phase for oscillatory integrals on Hilbert space. *Comm. Math. Phys.* **101**.

J. Rosen (1983). A local time approach to the self-intersections of Brownian paths in space. *Comm. Math. Phy.* **88**.

B. Simon (1979). *Functional Integration and Quantum Physics.* Academic Pres , New York and London.

B. Simon (1982). Schrödinger semigroups. *Bull. Amer. Math. Soc.* (*N.S.*) **7**.

A. Sloan (1977). An application of the nonstandard Trotter product formula. *J. Math. Phys.* **18**.

A. Sloan (1978). A note on the exponential of distributions. *Pacific J. Math.* **79**.

A. Stoll (1985). Doctoral dissertation, Bochum.

A. Stoll (1986). Self-repellent random walks and polymer measures in two dimensions. *In* (S. Albeverio, Ph. Blanchard, and L. Streit, eds.). Proc. II Bibos Symp., Lect. Notes Math., Springer-Verlag, New York and Berlin.

E. C. Svendsen (1981). The effect of submanifolds upon essential self-adjointness and deficiency indices. *J. Math. Anal. Appl.* **80**.

K. Symanzik (1969). Euclidean quantum field theory. *In* (R. Jost, ed.), *Local Quantum Theory.* Academic Press, New York and London.

L. E. Thomas (1979). Birman-Schwinger bounds for the Laplacian with point interactions. *J. Math. Phys.* **20**.

L. E. Thomas (1980). Scattering from point interactions. *In* (De Santo, Saenz, and Zachary, eds.) *Mathematical Methods and Applications of Scattering Theory. Proc. 1979.* Springer-Verlag, Berlin and New York.

C. Truesdell and R. G. Muncaster (1980). *Fundamentals of Maxwell's Kinetic Theory of a Simple Monatomic Gas.* Academic Press, New York.

S. Ukai (1974). On the existence of global solutions of mixed problem for non-linear Boltzmann equation. *Proc. Jpn. Acad.* **50**, 179–184.

F. Wattenberg (1977). Nonstandard measure theory—Hausdorff measure. *Proc. Amer. Math. Soc.* **65**.

J. Westwater (1980). On Edwards' model for long polymer chains. *Comm. Math. Phys.* **72**.

J. Westwater (1981). On Edwards' model for polymer chains: II. The self-consistent potential. *Comm. Math. Phys.* **79**.

J. Westwater (1982). On Edwards' model for polymer chains: III. Borel summability. *Comm. Math. Phys.* **84**.

J. Westwater (1985). On Edwards' model for polymer chains. *In* (S. Albeverio and Ph. Blanchard, eds.) *Trends and Developments in the Eighties.* Proc. Bielefeld. Enc. Math. Phys. IV. World Scientific, Singapore.

E. Wigner (1933). On the mass defect of Helium. *Phys. Rev.* **43**.

T. T. Wu (1959). Ground state of a Bose system of hard spheres. *Phys. Rev.* **115**.

M. Yor (1985). Renormalisation et convergence en loi pour les temps locaux d'intersections du mouvement brownien dans $\mathbb{R}^3$. *In* (J. Azema and M. Yor, eds.) *Sém Prob.* XIX 83184. Lect. Notes in Math. 1123. Springer-Verlag, Berlin and New York.

J. Zorbas (1980). Perturbation of self-adjoint operators by Dirac distribution. *J. Math. Phys.* **21**.

CHAPTER 7

HYPERFINITE LATTICE
MODELS

The common thread of this chapter is the study of hyperfinite lattice models. In Sections 7.1–7.3 our lattices are hyperfinite, but the lattice spacing or distance between neighboring points in the lattice is kept fixed and finite. In Section 7.1 we study the stochastic evolution of lattice systems; in Section 7.2 we use the hyperfinite model to study the classical equilibrium theory; and in Section 7.3 we discuss the global Markov property under a variety of assumptions. In the second part of the chapter we introduce hyperfinite lattices with infinitesimal spacing as models for field theories. In Section 7.4 we discuss a number of models for quantum field theories, including some brief remarks on gauge fields; in Section 7.5 we investigate the connection between fields and polymers, extending the discussion of Section 6.4.

7.1. STOCHASTIC EVOLUTION OF LATTICE SYSTEMS

The study of the stochastic evolution of lattice systems was initiated in the case of the Ising model by Glauber. The idea is to look at the ferromagnetic system obtained by immersing the usual Ising model [see Faris (1979)] in a heat bath, keeping the temperature fixed, but varying the energy by making the spins (elementary ferromagnets) flip up and down at a given rate. The observed motion is then described by a continuous Markov process, and the model is so constructed as to have the Gibbs states of the Ising system as equilibrium states; cf. Section 7.2.

In this section we shall study the stochastic evolution of these systems, using a hyperfinite lattice system with noninfinitesimal spacing. As usual we start with a brief résumé of the standard theory.

A d-dimensional system is specified by a set of *sites* $\Lambda \subseteq \mathbb{Z}^d$. At each site $i \in \Lambda$ there is a *spin* which can be either "up" $(+1)$ or "down" (-1); i.e., associated with Λ we have a *configuration space* $\Omega_\Lambda = \{-1, +1\}^\Lambda$. An element $q \in \Omega_\Lambda$ is called a configuration, and $q(i)$, which is equal to either $+1$ or -1, is called the spin at site i with respect to the configuration q. Often we shall write q_i for $q(i)$.

For $i, j \in \Lambda$ we let $|i - j|$ denote the usual Euclidean distance and we let $B_{i,r} = \{j \in \Lambda \,|\, |i - j| \le r\}$ be the intersection of Λ with the ball of radius r around i. We note that Ω_Λ is compact in the product topology and as usual we let $C(\Omega_\Lambda)$ stand for the set of real-valued continuous functions on Ω_Λ.

When $\Lambda = \mathbb{Z}^d$ we drop subscripts; i.e., we write Ω for $\{-1, +1\}^{\mathbb{Z}^d}$ and $C(\Omega)$ for the corresponding set of continuous functions. A function $f : \Omega \to \mathbb{R}$ is called *tame* if there is a *finite* $\Lambda \subseteq \mathbb{Z}^d$ such that $f(q) = f(q')$ for all q, $q' \in \Omega$ which agree on Λ, i.e., such that $q \restriction_\Lambda = q' \restriction_\Lambda$. We call Λ a *base* for the tame function f. There is a simple identification between functions in $C(\Omega_\Lambda)$, Λ finite, and tame functions in $C(\Omega)$ with base Λ; namely, to each $f \in C(\Omega_\Lambda)$ we define $f' \in C(\Omega)$ by the equation $f'(q) = f(q \restriction_\Lambda)$. Finally, observe that the set of tame functions is dense in $C(\Omega)$ as a consequence of the Stone–Weierstrass theorem.

We shall study certain *diffusion* or *drift systems* on the infinite lattice $\mathbb{Z}^d$. We assume that the stochastic evolution of such systems is controlled by a speed function $c = \mathbb{Z}^d \times \Omega \to \mathbb{R}$ which satisfies the following two conditions:

(1)$\quad$ There is an $M \in \mathbb{R}$ such that $0 < c(i, q) \le M$ for all $i \in \mathbb{Z}^d$, $q \in \Omega$.

(2)$\quad$ There is an $L \in \mathbb{R}$ such that $c(i, q) = c(i, q')$ for all $i \in \mathbb{Z}^d$ and all q, q' satisfying $q \restriction_{B_{i,L}} = q' \restriction_{B_{i,L}}$.

In standard terminology this means that we study systems with *finite-range interaction*. The prime example is the classical Ising model with speed function

(3)$\qquad c(i, q) = \exp\left[-\dfrac{\beta}{2} \sum_{|i-j|=1} q_i q_j - h q_i \right], \qquad \beta > 0, \quad h \in \mathbb{R};$

i.e., we have a system with nearest-neighbor interaction and an external field h.

We want the speed function to govern the stochastic evolution in the following way: given that the system is in configuration q at time t_0, then the probability that the spin at a single site $i \in \mathbb{Z}^d$ will be reversed at time $t_0 + \Delta t$ shall be $c(i, q) \Delta t + o(\Delta t)$, while the probability that the spins will be reversed at two or more sites in the time interval Δt shall be $o(\Delta t)$.

Translated into the language of semigroups, this means that we want to construct a Markov semigroup of operators T_t, $t \geq 0$, on $C(\Omega)$ having an infinitesimal generator which is the extension of the following operator A defined on the set of tame functions in $C(\Omega)$:

$$(4) \qquad (Af)(q) = \sum_{i \in \mathbb{Z}^d} c(i, q)(f(q^{(i)}) - f(q)),$$

where $q^{(i)}$ is the configuration obtained from q by reversing the spin at site i.

7.1.1. REMARK. This approach to interacting particle systems is due to Spitzer (1970); however, a particular case had been discussed earlier by Glauber (1963). There is a large literature on this topic; for an introduction see Kindermann and Snell (1980), which has an extensive bibliography; see also Martin (1977). In this section we give an exposition of the hyperfinite approach of Helms and Loeb (1979); see also Helms and Loeb (1982).

With the standard approach there is no direct way to extend definition (4) from a tame to an arbitrary continuous function. Going from finite to hyperfinite we shall see how to preserve the explicit form of (4). This gives an internal *bounded* operator and, hence, a semigroup. Add the Loeb construction and some standard-part arguments and the semigroup T_t on $C(\Omega)$ stands in front of you. We proceed to the details.

Choose an $N \in {}^*\mathbb{N} - \mathbb{N}$ and let $\Gamma = B_{0,N}$ be a hyperfinite subset of ${}^*\mathbb{Z}^d$. We let Ω_Γ be the set of internal mappings from Γ to $\{-1, +1\}$ and $C(\Omega_\Gamma)$ be the set of internal hyperreal-valued functions on Ω_Γ in the maximum norm. Note that for finite Λ, $C(\Omega_\Lambda)$ consists of all functions from Ω_Λ to $\mathbb{R}$. If f is any function on ${}^*\Omega$ and q is any configuration in ${}^*\Omega - \Omega_\Gamma$, we let f_q be the function on Ω_Γ defined by $f_q(q') = f(q' \times q)$, $q' \in \Omega_\Gamma$. Following the usual terminology we call q an *external configuration*. This terminology may be confusing; "external" here means external relative to the hyperfinite set Ω_Γ, *not* external as opposed to internal in the language of nonstandard theory.

For later purposes we shall make use of the standard part map st_Γ defined by $\mathrm{st}_\Gamma q = q \upharpoonright_{\mathbb{Z}^d}$, for $q \in \Omega_\Gamma$. But first the hyperfinite construction.

Fix an external configuration q_0 and set

$$(5) \qquad (A_{\Gamma,q_0}f)(q) = \sum_{i \in \Gamma} {}^*c(i, q \times q_0)[f(q^{(i)}) - f(q)],$$

where $f \in C(\Omega_\Gamma)$, $q \in \Omega_\Gamma$, and ${}^*c$ is the standard extension of the given speed function c. Our first simple but basic observation is that A_{Γ,q_0} is an internal bounded operator on $C(\Omega_\Gamma)$; in fact

$$\|A_{\Gamma,q_0}\| \leq 2 \cdot M \cdot |\Gamma|,$$

where M is the constant specified in (1) and $|\Gamma|$ is the internal cardinality (or "volume") of Γ. We also note that $A_{\Gamma,q_0}1 = 0$. By transfer of standard theory we may introduce an internal semigroup of operators $S_{\Gamma,q_0}(t)$, $t \geq 0$, on $C(\Omega_\Gamma)$ by the explicit formula

$$(6) \qquad\qquad S_{\Gamma,q_0}(t) = \exp(tA_{\Gamma,q_0}).$$

The sought-for semigroup T_t, $t \geq 0$, will in a suitable sense be the *standard part* of the internal semigroup $S_{\Gamma,q_0}(t)$.

7.1.2. REMARK. We shall use the power-series definition of the exponential function

$$\exp(tA_{\Gamma,q_0}) = \sum_{n \in {}^*\mathbb{N}} \frac{(tA_{\Gamma,q_0})^n}{n!}.$$

Internally, the time evolution is governed by the semigroup

$$Q(t) = (1 + A_{\Gamma,q_0}\,\Delta t)^{t/\Delta t}$$

(see Section 5.3), and it might have been more natural to use this representation for $\exp(tA_{\Gamma,q_0})$. However, $Q(t)$ and S_{Γ,q_0} have the same standard parts; we are free to choose and shall in this section follow the approach of Helms and Loeb (1979).

The basic estimate is contained in the following lemma.

7.1.3. LEMMA. Let $f \in C(\Omega)$ be a tame function with base Λ and let $n \in {}^*\mathbb{N}$; then

$$\|A_{\Gamma,q_0}^n {}^*f_{q_0}\| \leq \|f\|2^n M^n n! \exp[|\Lambda| + n(2L + 1)^d],$$

where $\|\ \|$ is the supremum norm.

This looks more impressive than it really is. Estimates of this form are well known from standard theory [see, e.g., Holley (1970, 1972)]; a detailed proof in the hyperfinite setting is given by Helms and Loeb (1979). Since there is nothing new *in principle* we omit the verification.

We remind the reader of the notion of S-continuity on Ω_Γ that goes with the standard part map st_Γ: f is *S-continuous* if it is internal and $f(q) = f(q')$ whenever $\mathrm{st}_\Gamma q = \mathrm{st}_\Gamma q'$. We now come to the basic result on finiteness and S-continuity.

7.1.4. PROPOSITION. If t is a finite non-negative hyperreal and $f \in C(\Omega_\Gamma)$ is a finite-valued S-continuous function on Ω_Γ, then $S_{\Gamma,q_0}(t)f$ is finite-valued and S-continuous on S_Γ.

The proposition is proved in several steps. First we show that it is true for functions of the form $^*f_{q_0}$, where f is tame, provided t is chosen such that $2Mt \exp(2L + 1)^d < 1$. Since tame is dense, we immediately extend to functions $f \in C(\Omega)$. Next we observe that if we can prove 7.1.4 with the above restriction on t, then 7.1.4 holds for all finite non-negative t by the semigroup property. To complete the proof we observe that given a finite-valued S-continuous internal function f on Ω_Γ, the standard part $^\circ f$, defined by setting $^\circ f(q) = \,^\circ(f(^*q \upharpoonright_\Gamma))$, $q \in \Omega$, is continuous on Ω and $\|^*(^\circ f)_{q_0} - f\| \approx 0$. From this observation it is not difficult to infer the S-continuity of $S_{\Gamma, q_0}(t)f$ from the S-continuity of $S_{\Gamma, q_0}(t)^*(^\circ f)_{q_0}$. And the finiteness part follows since $\|S_{\Gamma, q_0}(t)f\| \leq \|f\|$ and $\|f\|$ is finite since by assumption the set $\{n \in {}^*\mathbb{N} \mid |f(q)| \leq n, \text{ all } q \in \Omega_\Gamma\}$ contains all hyperfinite integers, hence some finite $n_0 \in \mathbb{N}$.

It remains to prove 7.1.4 for f tame with base Λ and t satisfying the condition that $2Mt \exp(2L + 1)^d < 1$. In fact, we shall prove something stronger. Let q_0, q_0' be external configurations, let q, $q' \in \Omega_\Gamma$, and assume that $\mathrm{st}_\Gamma q = \mathrm{st}_\Gamma q'$; then

$$(7) \qquad S_{\Gamma, q_0}(t)^* f_{q_0}(q) \approx S_{\Gamma, q_0'}(t)^* f_{q_0'}(q').$$

Since f is tame we see from the explicit definition in (5) that $A^n_{\Gamma, q_0}{}^* f_{q_0}(q) = A^n_{\Gamma, q_0'}{}^* f_{q_0'}(q')$ for all $n \in \mathbb{N}$. By the internal definition principle there must be some $\omega \in {}^*\mathbb{N} - \mathbb{N}$ such that the equality is true for all $n \leq \omega$. Since

$$S_{\Gamma, q_0}(t)^* f_{q_0}(q) = \sum_{n=0}^{\omega} \frac{t^n}{n!} A^n_{\Gamma, q_0}{}^* f_{q_0}(q) + \sum_{n \geq \omega + 1} \frac{t^n}{n!} A^n_{\Gamma, q_0}{}^* f_{q_0}(q),$$

we get

$$S_{\Gamma, q_0}(t)^* f_{q_0}(q) - S_{\Gamma, q_0'}(t)^* f_{q_0'}(q')$$

$$\approx \sum_{n \geq \omega + 1} \frac{t^n}{n!} [A^n_{\Gamma, q_0}{}^* f_{q_0}(q) - A^n_{\Gamma, q_0'}{}^* f_{q_0'}(q')].$$

Using the estimate from Lemma 7.1.3, we see that the right-hand side is dominated by

$$2\|f\| e^{|\Lambda|} \sum_{n \geq \omega + 1} t^n 2^n M^n \exp n(2L + 1)^d.$$

Using the restriction on t, we immediately conclude that the difference is infinitesimal, proving 7.1.4 for tame f. And, as we have explained, this suffices to prove all of 7.1.4.

7.1.5. REMARK. The proof of 7.1.4 tells us that the external configuration q_0 and the hyperfinite lattice Γ have only an infinitesimal effect on $S_{\Gamma, q_0}(t)^* f_{q_0}$. This corresponds to the standard fact that the limit thermodynamic quantities are independent of how we go to the limit and how we choose the external configurations as long as we go to the limit in the sense of van Hove; see Section 7.2.

It remains to take the standard part of the semigroup $S_{\Gamma,q_0}(t)$. To this end we note that for each $t \in {}^*[0, \infty\rangle$ and $q \in \Omega_\Gamma$, $S_{\Gamma,q_0}(t)$ defines a transition function $U_t^{q_0}(q, \cdot)$ which in turn gives us an internal probability measure $P(E) = U_t^{q_0}(q, E)$ on the algebra $\mathscr{A}$ of internal subsets of Ω_Γ. Let $\langle \Omega_\Gamma, L(\mathscr{A}), L(P) \rangle$ be the associated Loeb space; see Section 3.2. Note that whenever $E \in \mathscr{B}(\Omega)$ then $\mathrm{st}_\Gamma^{-1}(E) \in L(\mathscr{A})$. Thus we can define a probability measure P_S on $\mathscr{B}(\Omega)$ by the equation $P_S(E) = L(P)(\mathrm{st}_\Gamma^{-1} E)$; this is by now a familiar technique. We also know from Section 3.2 that if $f \in C(\Omega)$, then

$$(8) \qquad \int_\Omega f \, dP_S = \int_{\Omega_\Gamma} {}^\circ({}^*f_{q_0}) \, dL(P) \approx \int_{\Omega_\Gamma} {}^*f_{q_0} \, dP.$$

This general construction can be used to define a transition function $T_t^{q_0}$ in the following way. Let $q \in \Omega$ and $t \in [0, \infty\rangle$; this gives a particular transition function $U_t^{q_0}({}^*q \restriction_\Gamma, \cdot)$. The general construction then delivers a measure on $\mathscr{B}(\Omega)$; we let $T_t^{q_0}(q, \cdot)$ denote this measure. And the measure introduces an operator $T_t^{q_0} f$ on the bounded Borel measurable functions on Ω by the equation $T_t^{q_0} f = \int_\Omega f(q) T_t^{q_0}(\cdot, dq)$.

7.1.6. LEMMA. If $f \in C(\Omega)$, then $T_t^{q_0} f \in C(\Omega)$. Moreover,

$$T_{t+s}^{q_0}(q, E) = \int T_t^{q_0}(q', E) T_s^{q_0}(q, dq')$$

for all $q \in \Omega$, $E \in \mathscr{B}(\Omega)$, and $s, t \in [0, \infty\rangle$.

The proof is rather simple using general facts of the Loeb construction such as (8) and the corresponding semigroup properties of the hyperfinite entities. We show how to prove the continuity assertion. Let $f \in C(\Omega)$; now Ω is compact so f is uniformly continuous on Ω, hence ${}^*f_{q_0}$ is finite-valued and S-continuous on Ω_Γ, which by 7.1.4 implies that $S_{\Gamma,q_0}(t){}^*f_{q_0}$ is also finite-valued and S-continuous. Using (8), we have the following calculation:

$$S_{\Gamma,q_0}(t){}^*f_{q_0}({}^*q \restriction_\Gamma) = \int_{\Omega_\Gamma} {}^*f_{q_0}(q') U_t^{q_0}({}^*q \restriction_\Gamma, dq')$$

$$\approx \int_{\Omega_\Gamma} {}^\circ[{}^*f_{q_0}(q')] L(P_{t,{}^*q \restriction_\Gamma})(dq')$$

$$= \int_\Omega f(q') T_t^{q_0}(q, dq')$$

$$= T_t^{q_0} f(q).$$

We conclude that $T_t^{q_0} f \in C(\Omega)$. To prove the second part we use the corresponding property of the transition function $U_t^{q_0}(q, \cdot)$.

Observe that we have dropped the Γ as an index to the transition function $U_t^{q_0}$; using Remark 7.1.5, we may as well drop the q_0 too. In fact the transition function $T_t^{q_0}(q, \cdot)$ and the semigroup $T_t^{q_0}$, $t \geq 0$, are both independent of the set Γ and the configuration q_0. Thus $T_t^{q_0}$ will from now on simply be denoted by T_t.

7.1.7. THEOREM. The family of operators T_t, $t \geq 0$, is the unique Feller semigroup whose infinitesimal generator is an extension of the operator A defined in (4).

To prove strong continuity we need only prove strong continuity at $t = 0$, and this is easy using the estimate in 7.1.3. We show that for f tame

$$\lim_{t \to 0^+} \frac{T_t f - f}{t} = Af.$$

Let f have the finite set $\Lambda \subseteq \mathbb{Z}^d$ as base. Then for any $q \in \Omega_\Gamma$ and small t

$$\left| \frac{S_{\Gamma, q_0}(t) {}^* f_{q_0}(q) - {}^* f_{q_0}(q)}{t} - A_{\Gamma, q_0} {}^* f_{q_0}(q) \right|$$

$$\leq \|f\| e^{|\Lambda|} t^{-1} \sum_{\substack{n \in {}^* \mathbb{N} \\ n \geq 2}} [2Mt \exp(2L+1)^d]^n.$$

The right-hand side can be made less than any positive number by choosing t sufficiently small. Uniqueness follows as in the standard case; see, e.g., Holley (1972).

We have thus completed our restricted task of constructing the semigroup T_t via a hyperfinite extension of the infinite lattice system. The theory does not stop here; the next topic of interest is the existence of invariant measures for the T_t semigroup and their connection to the equilibrium states to be considered in the next section; some references are Holley and Stroock (1976, 1977), Liggett (1985), Stroock (1978), and Sullivan (1975).

We have discussed the case of discrete fiber $\{-1, +1\}$. Extension to the case of compact fiber, i.e., to configuration spaces of the form $I^{\mathbb{Z}^d}$, where I is some compact subset of $\mathbb{R}$, is rather immediate. The extension to the case of continuous fiber, i.e., to configuration spaces of the form $\mathbb{R}^\Lambda$, $\Lambda \subset \mathbb{Z}^d$, has also been studied; a reference is Faris (1979).

In Section 7.3 we shall discuss both lattice systems with compact fibers and continuous fibers in connection with the global Markov property. But first we turn to classical equilibrium theory.

7.2. EQUILIBRIUM THEORY

In this section we shall give a brief exposition of a hyperfinite approach to classical equilibrium theory. But first a résumé of the classical theory for a finite system.

REMARK. For the standard theory consult Ruelle (1983), Israel (1979), Preston (1976), Gross (1982), and Sinai (1982).

An *interaction* Φ is a map from finite nonempty subsets $X \subseteq \mathbb{Z}^d$ to real-valued continuous functions on Ω_X, i.e., $\Phi(X) \in C(\Omega_X)$. Let Λ be a *finite* subset of $\mathbb{Z}^d$; the *Hamiltonian* or energy function for Λ defined by the interaction Φ is the function

$$(1) \qquad H_\Lambda^\Phi = \sum_{X \subset \Lambda} \Phi(X).$$

The classical Ising model is given by the following interaction: let $\Phi(X)(q) = kq_iq_j$ if X consists of two sites i, j with distance 1; let $\Phi(X)(q) = hq_i$ if X consists of exactly one site i; otherwise let $\Phi(X)(q) = 0$. We then get

$$(2) \qquad H_\Lambda^\Phi(q) = k \sum_{\substack{i,j \in \Lambda \\ |i-j|=1}} q_iq_j + h \sum_{i \in \Lambda} q_i.$$

In Section 7.1 we introduced the Ising model via a speed function $c(i, q)$; see formula (3) of 7.1. These are obviously equivalent procedures; in the present section we stick to interactions and their associated Hamiltonians.

The Hamiltonian introduced in (1) allows no interaction between spins inside Λ and spins outside Λ; i.e., we have a situation with free boundary conditions. For many purposes other boundary conditions are important. We discuss one type: an *external configuration* for Λ is an element in Ω_{Λ^c}, where Λ^c is the complement of Λ in $\mathbb{Z}^d$. (As pointed out in Section 7.1, we are here in a terminological quandary. In nonstandard analysis "external" means not internal. In statistical mechanics "external" is defined with respect to the complement of a given set Λ; in this sense an external configuration may well be internal as an entity of nonstandard analysis. We trust that the reader can live with this dilemma.)

An external configuration $q \in \Omega_{\Lambda^c}$ defines a Hamiltonian

$$(3) \qquad {}_qH_\Lambda^\Phi(q') = \sum_{X \cap \Lambda \neq \varnothing} \Phi(X)(q' \times q),$$

where q' is any configuration in Ω_Λ. Here $(q' \times q)(i) = q'(i)$ if $i \in \Lambda$ and $(q' \times q)(i) = q(i)$ if $i \in \Lambda^c$.

A Hamiltonian of the form H_Λ^Φ always makes sense; the sum in (3) may diverge. But if Φ is a *finite-range interaction*, i.e., there exists some number

l such that $\Phi(X) = 0$ if the diameter of X is larger than l, then (3) is well defined. We shall in this brief exposition mostly restrict ourselves to finite-range interactions, but the theory has wider scope; see Ostebee *et al.* (1976) for results involving infinite-range interactions (Coulomb systems).

Let $\mathcal{B}_0$ be the linear space of finite-range interactions. Completing $\mathcal{B}_0$ with respect to the norm

$$\|\|\Phi\|\| = \sum_{X \ni 0} \frac{\|\Phi(X)\|_\infty}{|X|}$$

we get a Banach space $\mathcal{B}$, and completing $\mathcal{B}_0$ with respect to

$$\|\Phi\|_{\sim} = \sum_{X \ni 0} \|\Phi(X)\|_\infty$$

we get a Banach space $\tilde{\mathcal{B}}$. We see that $\mathcal{B}_0 \subset \tilde{\mathcal{B}} \subset \mathcal{B}$ and that $\mathcal{B}_0$ is dense in both $\tilde{\mathcal{B}}$ and $\mathcal{B}$. Much of the theory extends from interactions in $\mathcal{B}_0$ to $\tilde{\mathcal{B}}$ or sometimes to all of $\mathcal{B}$. We see that $_qH_\Lambda^\Phi$ is finite if $\Phi \in \tilde{\mathcal{B}}$.

We shall also assume that we are dealing with *translation-invariant interactions*: given $i \in \mathbb{Z}^d$ there is a natural map $t_i : \Omega_{X+i} \to \Omega_X$ obtained by setting $(t_i q)(j) = q(j + i)$. This map can be lifted to a map $t_i : C(\Omega_X) \to C(\Omega_{X+i})$. We say that Φ is translation-invariant if $\Phi(X + i) = t_i \Phi(X)$ for all finite X and all $i \in \mathbb{Z}^d$.

From interactions and Hamiltonians we pass to the important thermodynamic entities such as pressure, entropy, and mean energy. These are obtained by a suitable averaging process. Since our basic space is $\Omega_\Lambda = \Omega_0^\Lambda$, where $\Omega_0 = \{-1, +1\}$, a natural *a priori* measure on it is the normalized counting measure; i.e., each configuration $q \in \Omega_\Lambda$ is given the weight $2^{-|\Lambda|}$. Given any function $F \in C(\Omega_\Lambda)$, we let $\langle F \rangle_0$ denote the expected value of F with respect to the normalized counting measure.

The *partition function* associated with the interaction Φ and the finite set Λ is the function

$$(4) \qquad Z_\Lambda = \langle e^{-H_\Lambda^\Phi} \rangle_0$$

A *state* of the finite system is a probability measure on the space of configurations Ω_Λ. The *Gibbs state* or *equilibrium state* is defined by giving each $q \in \Omega_\Lambda$ the weight

$$(5) \qquad \rho_1(q) = \frac{e^{-H_\Lambda^\Phi(q)}}{\sum_{q' \in \Omega_\Lambda} e^{-H_\Lambda^\Phi(q')}} = Z_\Lambda^{-1} \cdot e^{-H_\Lambda^\Phi(q)} \cdot 2^{-|\Lambda|}.$$

The *thermal average* is the expected value with respect to the equilibrium state ρ_1. Thus given $F \in C(\Omega_\Lambda)$, the thermal average $\langle F \rangle_1$ is given by

$$(6) \qquad \langle F \rangle_1 = Z_\Lambda^{-1} \langle e^{-H_\Lambda^\Phi} F \rangle_0.$$

REMARK. In the classical case with fiber $\Omega_0 = \{-1, +1\}$ it is more usual to use the counting measure rather than the normalized counting measure; i.e., we would have replaced $Z_\Lambda = \langle e^{-H_\Lambda^\Phi} \rangle_0$ by $Z_\Lambda = \sum_{q \in \Omega_\Lambda} e^{-H_\Lambda^\Phi(q)}$. This would not change the Gibbs measure. We have chosen the normalized counting measure in order to facilitate the comparison with recent standard expositions; see, e.g., Israel (1979).

We can now introduce the *pressure* and *mean energy* by

$$(7) \qquad P_\Lambda(\Phi) = |\Lambda|^{-1} \ln Z_\Lambda, \qquad \langle H_\Lambda^\Phi \rangle_1 = Z_\Lambda^{-1} \langle e^{-H_\Lambda^\Phi} H_\Lambda^\Phi \rangle_0.$$

Here we have used a Hamiltonian with free boundary conditions; the same definitions extend to Hamiltonians $_q H_\Lambda^\Phi$ with external boundary conditions $q \in \Omega_{\Lambda^c}$.

Let ρ be a state for the finite system Λ; i.e., ρ is a probability measure on the space Ω_Λ that gives a weight $\rho(q)$ to the point $q \in \Omega_\Lambda$. Let $\rho^{(\Lambda)}(q) = \rho(q) \cdot 2^{|\Lambda|}$. In fancy language $\rho^{(\Lambda)}$ is the Radon–Nikodym derivative of ρ with respect to the counting measure on Ω_Λ. And this is the appropriate language to use when the fiber Ω_0 is an arbitrary compact metric space provided with some *a priori* probability measure μ_0. The *entropy* of ρ in Λ is defined as

$$(8) \qquad\qquad S_\Lambda(\rho) = -\langle \rho^{(\Lambda)} \ln \rho^{(\Lambda)} \rangle_0.$$

We see that $-|\Lambda| \ln 2 \leq S_\Lambda(\rho) \leq 0$, and for the Gibbs state in Λ we obtain

$$(9) \qquad S_\Lambda(\rho) = -\langle Z_\Lambda^{-1} e^{-H_\Lambda^\Phi} \ln Z_\Lambda^{-1} e^{-H_\Lambda^\Phi} \rangle_0 = |\Lambda| P_\Lambda(\Phi) + \langle H_\Lambda^\Phi \rangle_1.$$

We have the following variational principle.

7.2.1. PROPOSITION. For any state ρ of the finite system Λ with Hamiltonian H_Λ^Φ,

$$S_\Lambda(\rho) - \rho(H_\Lambda^\Phi) \leq |\Lambda| P_\Lambda(\Phi).$$

Equality holds if and only if ρ is the Gibbs state ρ_1.

Here $\rho(F)$ is the expected value with respect to the measure ρ. The proof is a simple application of the well-known Jensen inequality.

We have now completed our sketch of the finite classical theory. But finite systems have limited physical interest; e.g., no finite system can exhibit the phenomenon of phase transition. In order to obtain a more meaningful physical model we must pass to the *thermodynamic* or *bulk* limit.

The configuration space of the infinite system will be $\Omega = \{-1, +1\}^{\mathbb{Z}^d}$. An interaction for the infinite system will—as for the finite systems—be a map Φ from finite subsets X of $\mathbb{Z}^d$ to real-valued continuous functions on Ω_X. For the rest we cannot give direct explicit formulas but have to resort to limiting procedures; e.g., the pressure $P(\Phi)$ for the infinite system should be obtained as

$$P(\Phi) = \lim_{\Lambda \to \infty} P_\Lambda(\Phi) = \lim_{\Lambda \to \infty} |\Lambda|^{-1} \ln Z_\Lambda$$

as Λ converges to $\mathbb{Z}^d$. The same should be true for entropy and mean energy.

The crucial notion is here the idea of *limit in the sense of van Hove*. For each positive integer a we partition the space $\mathbb{Z}^d$ into a family $\mathscr{C}_a$ of cubes of the form

$$\{i \in \mathbb{Z}^d \mid n_j a \le i_j < (n_j + 1)a, \qquad j = 1, \ldots, d\}$$

for integers $n_1, \ldots, n_d$. Let Λ be a finite subset of $\mathbb{Z}^d$; we define:

$N_a^+(\Lambda) = $ number of cubes in $\mathscr{C}_a$ which intersect Λ,

$N_a^-(\Lambda) = $ number of cubes in $\mathscr{C}_a$ contained in Λ.

We then say that a sequence $\langle \Lambda_n \rangle_{n \in \mathbb{N}}$ of finite subsets of $\mathbb{Z}^d$ *converges to infinity in the sense of van Hove* if for all a

   (i) $N_a^-(\Lambda_n) \to \infty$.

   (ii) $N_a^-(\Lambda_n)/N_a^+(\Lambda_n) \to 1$.

In order to formulate the basic convergence results we need the notion of a *state of the infinite system*: ρ is a state of the infinite system if ρ is a probability measure on the configuration space Ω. Note that ρ induces a state on every finite system Λ since there is a natural embedding t_Λ of $C(\Omega_\Lambda)$ into $C(\Omega)$, namely $(t_\Lambda f)(q) = f(q \upharpoonright_\Lambda)$, $q \in \Omega$. Thus the probability that the restriction of ρ assigns to an element q of the finite set Ω_Λ is simply $\rho(t_\Lambda \chi_q) = \rho(\{q \times q' \mid q' \in \Omega_{\Lambda^c}\})$, where χ_A is the characteristic function of the subset $A \subseteq \mathbb{Z}^d$.

Given a state ρ of the infinite system we have for each finite Λ a well-defined entity $S_\Lambda(\rho)$, the entropy of the state ρ in Λ.

Since ρ is a probability measure on Ω we have an immediate notion of *translation-invariant state*; we let E^I denote the set of translation-invariant states for the infinite system.

The basic convergence results of the classical theory can now be stated.

7.2.2. PROPOSITION. (i)   Let $\Phi \in \mathscr{B}$; then $\lim_{\Lambda_n \to \infty} P_{\Lambda_n}(\Phi)$ converges for all sequences Λ_n that go to infinity in the sense of van Hove, and the limit is independent of the particular van Hove sequence.

(ii)   Let $\rho \in E^I$; then $\lim_{\Lambda_n \to \infty} |\Lambda_n|^{-1} S_{\Lambda_n}(\rho)$ converges for all sequences Λ_n that go to infinity in the sense of van Hove, and the limit is independent of the particular van Hove sequence.

We use $P(\Phi)$ to denote the infinite-volume pressure and $S(\rho)$ to denote the mean entropy of the infinite system in state ρ.

The next topic would be to discuss the appropriate notion of *equilibrium state for the infinite system* and to provide the necessary existence results. But here we shall take leave of the classical theory and turn to a hyperfinite version of the limit theory.

In the *hyperfinite approach* we choose an infinite integer $N \in {}^*\mathbb{N} - \mathbb{N}$ and choose a hyperfinite rectangle Γ of diameter $2N$ as the set of sites for the hyperfinite limit system. We recall from Section 7.1 that Ω_Γ is the set of internal mappings from Γ to $\{-1, +1\}$ and that $C(\Omega_\Gamma)$ is the set of all hyperreal-valued functions on Ω_Γ in the maximum norm. We shall for simplicity restrict our attention to interactions ${}^*\Phi$, where Φ *is a standard finite-range interaction.* Let q be an external configuration for the hyperfinite system (in the sense explained in Section 7.1). We then have a Hamiltonian

$$(10) \qquad {}_q H_\Gamma^{*\Phi}(q') = \sum_{\Gamma \cap X = \varnothing} {}^*\Phi(X)(q' \times q),$$

where X is hyperfinite and $q' \in \Omega_\Gamma$. We also have a well-defined partition function

$$(11) \qquad Z_{\Gamma, q} = \langle e^{-{}_q H_\Gamma^{*\Phi}} \rangle_0,$$

where we take the average with respect to the internal normalized counting measure on Ω_Γ.

An internal *state of the hyperfinite system* Γ is an internal probability measure on the space of configurations Ω_Γ. By analogy with the finite theory we can introduce:

7.2.3. DEFINITION. The *internal equilibrium state with boundary condition* q_0 is the internal probability measure

$$(12) \qquad \rho_{q_0}(q) = Z_{\Gamma, q_0}^{-1} \exp(-{}_{q_0} H_\Gamma^{*\Phi}(q)) 2^{-|\Gamma|}.$$

This is in complete analogy with definition (5) and introduces ρ_{q_0} as an internal finitely additive probability measure on the algebra of internal subsets of Ω_Γ. As usual we can use the Loeb construction to obtain a σ-additive probability structure $\langle \Omega_\Gamma, L(\mathscr{A}), L(\rho_{q_0}) \rangle$, where $\mathscr{A}$ denotes the algebra of internal subsets of Ω_Γ.

7.2.4. DEFINITION. An *equilibrium state* with boundary condition q_0 for the infinite limit system Ω_Γ is any probability measure of the form $L(\rho_{q_0})$, where ρ_{q_0} is the internal measure given by Definition 7.2.3.

Thus in the hyperfinite case the existence of equilibrium states for the limit systems is trivial. It remains to verify that they are well behaved with respect to the finite-dimensional approximations. We shall also discuss how the equilibrium states in 7.2.4 are related to the equilibrium states of the standard theory.

In the standard theory equilibrium states may be characterized as those states which satisfy the *Dobrushin-Lanford-Ruelle equations.* In our hyperfinite setting the definition is particularly simple. Let ρ be an internal state of the Ω_Γ. Let Λ be finite and $q \in \Omega_\Lambda$. The conditional probability for the configuration q in Λ given the external configuration $q_0 \in \Omega_{\Gamma-\Lambda}$ is

$$(13) \qquad \rho(q\,|\,q_0) = \frac{\rho(q \times q_0)}{\sum_{q' \in \Omega_\Lambda} \rho(q' \times q_0)}.$$

Since Ω_Λ is finite $\rho(q\,|\,q_0)$ is well defined as a *hyperreal* number.

7.2.5. DEFINITION. The internal state ρ satisfies the DLR equations for the interaction Φ if for all finite $\Lambda \subset \mathbb{Z}^d$, all $q_1, q_2 \in \Omega_\Lambda$, and all $q_0 \in \Omega_{\Gamma-\Lambda}$ we have

$$(14) \qquad \rho(q_1\,|\,q_0) = \exp({}_{q_0}H_\Lambda^\Phi(q_2) - {}_{q_0}H_\Lambda^\Phi(q_1))\rho(q_2\,|\,q_0).$$

Since Φ is of finite range, ${}_{q_0}H_\Lambda^\Phi(q_i)$, $i = 1, 2$, are the classical entities. (To be pedantic we should have written $q_0 \restriction_{\mathbb{Z}^d - \Lambda}$.)

7.2.6. PROPOSITION. Let ρ_{q_0} be an internal equilibrium state for the infinite system. Then ρ_{q_0} satisfy the DRL equations.

The proof is very simple; given a finite Λ, $q \in \Omega_\Lambda$, and $q_0' \in \Omega_{\Gamma-\Lambda}$, we observe that

$$
\begin{aligned}
{}_{q_0}H_\Gamma^{*\Phi}(q \times q_0') &= \sum_{\Gamma \cap X \neq \varnothing} {}^*\Phi(q \times q_0' \times q_0) \\
&= \sum_{\substack{\Lambda \cap X \neq 0 \\ \mathrm{diam}(X) < l}} \Phi(q \times q_0') + K = {}_{q_0'}H_\Lambda^\Phi(q) + K,
\end{aligned}
$$

where K is independent of q and l is such that if $\mathrm{diam}(X) \geq l$ then ${}^*\Phi(X) = 0$.

The definition of *pressure* and *entropy* for the hyperfinite system Γ is straightforward. But whereas Z_{Γ,q_0} is an infinite hyperreal number, we want the pressure and entropy to be finite. To this end we quote a few immediate inequalities of the finite theory. A simple calculation shows that for finite Λ

$$(15) \qquad \|H_\Lambda^\Phi\|_\infty \leq |\Lambda| \cdot \|\!\|\Phi\|\!\|.$$

Hence

$$(16) \qquad |P_\Lambda(\Phi)| \leq |\Lambda|^{-1}\|H_\Lambda^\Phi\|_\infty \leq \|\!\|\Phi\|\!\|, \qquad |P_\Lambda(\Phi) - P_\Lambda(\Psi)| \leq \|\!\|\Phi - \Psi\|\!\|.$$

If we use a Hamiltonian with external configuration q_0 we have

$$(17) \qquad \|{}_{q_0}H_\Lambda^\Phi\|_\infty \leq |\Lambda| \cdot \|\Phi\|_\sim,$$

from which we get the analog of (15) with $\|\!\|\cdot\|\!\|$ replaced by $\|\cdot\|_\sim$.

Let Φ be a finite-range interaction, $^*\Phi$ its extension, and q_0 an external configuration for Γ. We note that $\|^*\Phi\|_\sim = \|\Phi\|$. By transfer on (17) we get $\|_{q_0}H_\Gamma^{*\Phi}\| \leq |\Gamma| \cdot \|^*\Phi\|_\sim$, which implies that $|P_\Gamma(^*\Phi)| \leq \|\Phi\|_\sim$; i.e., the nonstandard pressure is finite and thus has a standard part.

7.2.7. PROPOSITION. Let $_{q_0}P_\Gamma(^*\Phi) = |\Gamma|^{-1} \ln Z_{\Gamma,q_0}$ be the nonstandard pressure in the hyperfinite system Γ associated with the finite-range interaction Φ and the external configuration q_0. Then $_{q_0}P_\Gamma(^*\Phi)$ is finite and

$$\mathrm{st}(_{q_0}P_\Gamma(^*\Phi)) = \lim_{\Lambda_n \to \infty} P_{\Lambda_n}(\Phi)$$

for any sequence Λ_n that converges to infinity in the sense of van Hove.

The proof is a simple adaptation of the classical convergence arguments as set out, e.g., in Israel (1979), Theorems I.2.3–5, and need not be repeated here. We emphasize that from our point of view we have a direct definition of the pressure for the hyperfinite system as $\mathrm{st}(_{q_0}P_\Gamma(^*\Phi))$; this is a finite real number which by Proposition 7.2.7 corresponds to the standard one.

The notion of entropy for the hyperfinite system also presents no difficulties; we simply use formula (8) above in order to define $S_\Gamma(\rho)$. And in a way similar to Proposition 7.2.7 we shall identify the nonstandard entropy with the usual van Hove limiting sequence. But recall that Proposition 7.2.2 requires translation-invariant states; we have the following result:

7.2.8. PROPOSITION. Let ρ be a standard translation-invariant state on Ω; then $|\Gamma|^{-1}S_\Gamma(^*\rho)$ is a finite hyperreal number and

$$S(^*\rho) = \mathrm{st}(|\Gamma|^{-1}S_\Gamma(^*\rho)) = \lim_{\Lambda_n \to \infty} |\Lambda|_n^{-1}S_{\Lambda_n}(\rho)$$

for any sequence Λ_n which converges to infinity in the sense of van Hove.

Finiteness is immediate from the fact that $-\ln 2 \leq |\Gamma|^{-1}S_\Gamma(^*\rho) \leq 0$; see (8) above. The convergence result, which identifies the hyperfinite entity with the standard one, is an adaptation of the standard convergence result; see Theorem II.2.2 in Israel (1979). Note that $S(\rho)$ is well defined for any internal state on Ω_Γ; it is for the convergence result that we require a state of the type $^*\rho$, where $\rho \in E^I$.

7.2.9. REMARK. Let us summarize what we have obtained so far. We have chosen a hyperfinite rectangle Γ as the set of sites for the limit system. We have introduced the notion of an internal equilibrium state for the system Ω_Γ, and we have verified that these states satisfy the DRL equations for finite sets $\Lambda \subseteq \mathbb{Z}^d$. We have further introduced pressure and entropy for the hyperfinite system and showed that they are equal to the limit entities of the standard theory. Thus we have a clean and satisfactory definition of the thermodynamic limit.

But the usual limit space would be $\Omega = \{-1, +1\}^{Z^d}$. There is no problem, however, in using the standard part map st_Γ to obtain an equilibrium state for the infinite system Ω from an internal state ρ_{q_0} on Ω_Γ. This is a simple application of the Loeb construction. From ρ_q we obtain a σ-additive measure $L(\rho_{q_0})$ on Ω_Γ with respect to the σ-algebra $L(\mathcal{A})$. From $L(\rho_{q_0})$ we introduce a state $\tilde{\rho}_{q_0}$ on Ω by setting

$$(18) \qquad\qquad \tilde{\rho}(B) = L(\rho_{q_0})(\mathrm{st}_\Gamma^{-1} B)$$

for Borel sets B in Ω. It is not difficult to show that $\tilde{\rho}_{q_0}$ will satisfy the DLR equations for finite Λ; thus $\tilde{\rho}_{q_0}$ is an equilibrium state for the infinite system Ω.

Do we get all standard equilibrium states from our hyperfinite model? The answer is essentially yes, but we have to complicate Definitions 7.2.3 and 7.2.4 a bit. And for simplicity we stick to a nearest-neighbor interaction Φ for the remainder of this section.

7.2.10. DEFINITION. Let μ be a hyperfinite probability measure on $\Omega_{\partial\Gamma}$, where $\partial\Gamma$ is the boundary of Γ. The *internal equilibrium state with boundary distribution* μ is the internal probability measure

$$(19) \qquad\qquad \rho_\mu(q) = \sum_{q' \in \Omega_{\partial\Gamma}} \rho_{q'}(q)\mu(q'),$$

where $\rho_{q'}$ is given by (12) in Definition 7.2.3.

From this we get the corresponding extension of Definition 7.2.4.

7.2.11. DEFINITION. An *equilibrium state* for the infinite system Ω_Γ is any probability measure of the form $L(\rho_\mu)$, where ρ_μ is the internal measure given by Definition 7.2.10.

Using the format of (18) in Remark 7.2.9, we may associate with ρ_μ an equilibrium state $\tilde{\rho}_\mu$ on the classical limit system Ω. And now we can argue that every standard equilibrium state ρ on Ω can be written in the form $\tilde{\rho}_\mu$ for some suitable measure μ on $\Omega_{\partial\Gamma}$. This follows from the well-known fact that the restriction of the equilibrium state ρ of Ω to a finite-volume lattice Λ can be written in the form

$$(20) \qquad\qquad \rho_\Lambda = \sum_{q \in \Omega_{\partial\Gamma}} \rho_q(\cdot)\mu(q),$$

where ρ_q is the Gibbs measure on Ω_Λ with external configuration q on $\Omega_{\partial\Gamma}$ and μ is a probability measure on $\Omega_{\partial\Gamma}$. Via transfer it follows that the given ρ is indeed of the form $\tilde{\rho}_\mu$ for some suitable internal measure μ on $\partial\Gamma$. The result is not restricted to nearest-neighbor interactions; see Hurd (1981), who gives a nonstandard version of the approach using *projective families of specifications* [consult Preston (1976) for the standard theory].

7.2.12. REMARK. Combining our remarks above with formulas (18) and (19) we have the following representation of equilibrium states in terms of "external" states

$$(21) \qquad \tilde{\rho}_\mu = \int_{\Omega_{\partial\Gamma}} \tilde{\rho}_q L(\mu)(dq);$$

i.e., given any standard equilibrium state ρ we can find a measure μ on $\Omega_{\partial\Gamma}$ such that $\rho = \tilde{\rho}_\mu$ can be written as in (21).

This would have been a "correct" representation in terms of (standard) extremal points if we knew that "extremal" or "pure" in the standard sense coincided with the notion of being of the form $\tilde{\rho}_q$ for some $q \in \Omega_{\partial\Gamma}$.

Half of this statement is true; if ρ is standard pure then the representation in (21) implies that ρ must be of the form $\rho = \tilde{\rho}_q$ for some $q \in \Omega_{\partial\Gamma}$.

There are examples, however, which show that the converse may fail. To understand the following argument the reader should refer ahead to our discussion of phase transitions at the end of this section. It is known that in the two-dimensional Ising model with no external field there are exactly two pure states provided the "inverse temperature" β is small enough, namely $\tilde{\rho}_+ = \tilde{\rho}_{+,\beta,0}$ and $\tilde{\rho}_- = \tilde{\rho}_{-,\beta,0}$.

Let Γ be a two-dimensional rectangle with an even number of sites on each edge (see Fig. 7.1). Let $q \in \Omega_{\partial\Gamma}$ be the configuration which assigns $+1$ to every second site on $\partial\Gamma$ and -1 to the rest as indicated by pluses and minuses in Fig. 7.1. Consider the state $\tilde{\rho}_q$. If it were standard pure it had to be either $\tilde{\rho}_+$ or $\tilde{\rho}_-$. Assume that $\tilde{\rho}_q = \tilde{\rho}_+$. If we reflect about the line L the effect is to interchange pluses and minuses on the boundary $\partial\Gamma$. This

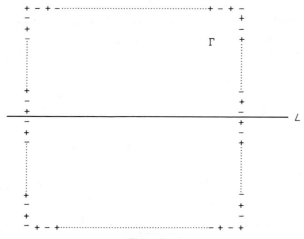

Figure 7.1

gives a new state $\tilde{\rho}_{q'}$, where q' is obtained from q by reversing the spin at each site in $\partial\Gamma$. Reversing spins changes $\tilde{\rho}_+$ into $\tilde{\rho}_-$, hence $\tilde{\rho}_{q'}$ would have to be equal to $\tilde{\rho}_-$. But the new state can also be obtained from the old one by a rotation, and $\tilde{\rho}_+$ is obviously unchanged by a rotation. Thus $\tilde{\rho}_{q'}$ would also have to be equal to $\tilde{\rho}_+$, a contradiction which shows that $\tilde{\rho}_q$ cannot be standard pure. In fact, it is not difficult to see that

$$\tilde{\rho}_q = \tfrac{1}{2}\tilde{\rho}_+ + \tfrac{1}{2}\tilde{\rho}_-.$$

The reader could also discuss this example in terms of finite approximations with appropriate boundary values.

We now state a *variational principle* for the hyperfinite theory. Let $\rho_q(^*\Phi) = \mathrm{st}(|\Gamma|^{-1}\rho(_qH_\Gamma^{^*\Phi}))$ for ρ_q an internal state on Ω_Γ.

7.2.13. PROPOSITION. Let Φ be a finite-range interaction and ρ_q an internal equilibrium state for Ω_Γ. Then

$$S(\rho_q) = P(^*\Phi) + \rho_q(^*\Phi).$$

The proof is simple. It follows by transfer from Proposition 7.2.1 that

$$S_\Gamma(\rho_q) = |\Gamma|_q P_\Gamma(^*\Phi) + \rho_q(_qH_\Gamma^{^*\Phi}).$$

And that is all we need to know. Note that both $S(\rho_q)$ and $\rho_q(^*\Phi)$ are independent of q.

7.2.14. REMARK. Macroscopic thermodynamics is to a large extent based on the relation

$$T\,dS = dE + P\,dV.$$

This is essentially the variational principle, 7.2.13, for the hyperfinite system. To make the proper identification we note that for any function f such that $\lim_{x\to\infty} df(x)/dx$ exists, $\lim_{x\to\infty} f(x)/x$ exists, and the two limits are equal. When Γ is hyperfinite we can thus identify

$$P = \frac{1}{\beta}\frac{\partial \ln Z}{\partial V} \quad \text{and} \quad P(^*\Phi) \approx |\Gamma|^{-1}\ln Z_\Gamma,$$

$$\frac{dS}{dV} \quad \text{and} \quad S(\rho) \approx |\Gamma|^{-1}S_\Gamma(\rho),$$

$$\frac{dE}{dV} \quad \text{and} \quad \rho(^*\Phi) \approx |\Gamma|^{-1}\rho(H_\Gamma^{^*\Phi}),$$

where in the hyperfinite system the temperature has been absorbed into the energy function.

We conclude our exposition of the equilibrium theory by a brief discussion of *phase transitions*. Phase transitions do not occur in finite lattice systems. We shall briefly explain how they may occur in hyperfinite systems as a kind of "change of scale" phenomenon, namely as a consequence of the distinction between *internal* or *-differentiability versus *external* or S-differentiability. We shall not develop a general theory but restrict attention to the classical Ising model, and in this we shall follow closely the exposition in Helms and Loeb (1979).

Let Γ be a hyperfinite lattice, q a configuration in Ω_Γ, and q_0 an external boundary condition, i.e., $q_0 \in {}^*\Omega - \Omega_\Gamma$. We recall the notation $q \times q_0$ for the combined configuration, i.e.,

$$q \times q_0(i) = \begin{cases} q(i) & \text{if } i \in \Gamma, \\ q_0(i) & \text{if } i \notin \Gamma. \end{cases}$$

In the *Ising model* the unique internal equilibrium state with boundary condition q_0 (see Definition 7.2.3) is given by

(22) $\rho_{q_0}(q) = Z_{\Gamma, q_0}^{-1}$

$$\times \exp\left(-\frac{\beta}{2} \sum_{\substack{i \in \Gamma \\ |i-j|=1}} (q \times q_0)(i)(q \times q_0)(j) - h \sum_{i \in \Gamma} q(i)\right) 2^{-|\Gamma|}.$$

Here β is a positive real parameter, the *inverse temperature*, and h is a real parameter representing *the external field*. We thus see that the Ising model on the hyperfinite lattice Γ is specified by two parameters, β and h.

We write $\rho_{q_0, \beta, h}$ to indicate the dependence of ρ_{q_0} on the parameters β and h. We use $\tilde{\rho}_{q_0, \beta, h}$ to denote the corresponding state of the classical system. We may now prove the following.

(i) If $h \neq 0$, i.e., if there is an external field, then all the states $\tilde{\rho}_{q_0, \beta, h}$ coincide; that is, we have but one equilibrium state of the classical system. This means that there is only one pure phase of the system, and there is no possibility for a phase transition.

(ii) If there is no external field, i.e., if $h = 0$, then for sufficiently small β all states $\tilde{\rho}_{q_0, \beta, 0}$ coincide, but for sufficiently large β there is more than one extremal state; i.e., phase transition does occur.

We shall indicate part of the proof for (i). Let $E_{\beta, h}^q$ denote the expected value with respect to the measure $\rho_{q, \beta, h}$ on Ω_Γ. There are two boundary conditions that we single out for special attention, namely $q \equiv +1$ and $q \equiv -1$; we write $\rho_{\beta, h}^\pm$ for the associated measures and use a corresponding notation for the expectations. In the same way we write the partition function as $Z(q, \beta, h)$.

It is customary to regard the set Ω_Γ as a lattice introducing the pointwise ordering relation $q \le q'$ iff $q(i) \le q'(i)$ for all $i \in \Gamma$. One may show [see Helms and Loeb (1979)] that if f is increasing on Ω_Γ and $-1 \le q_1 \le q_2 \le +1$, then

$$(23) \qquad E^-_{\beta,h}[f] \le E^{q_1}_{\beta,h}[f] \le E^{q_2}_{\beta,h}[f] \le E^+_{\beta,h}[f].$$

We also note for further use the following lemma [see Helms and Loeb (1979), Lemma 7]:

7.2.15. LEMMA. If β and h are standard real numbers, then

$$E^\pm_{\beta,h}[q(0)] \approx E^\pm_{\beta,h}\left[|\Gamma|^{-1} \sum_{i \in \Gamma} q(i)\right],$$

for all $q \in \Omega_\Gamma$.

It is the behavior of the internal pressure $_qP(\beta, h)$ which determines whether or not we have a phase transition. Let us slightly adapt our notation and write

$$(24) \qquad P_\pm(\beta, h) = |\Gamma|^{-1} \ln Z(\pm, \beta, h).$$

A simple estimate shows that $P_+(\beta, h) \approx P_-(\beta, h)$ for each $h \in {}^*\mathbb{R}$. It is also straightforward to show that

$$(25) \qquad P'_\pm(\beta, h) = E^\pm_{\beta,h}\left[|\Gamma|^{-1} \sum_{i \in \Gamma} q(i)\right] \approx E^\pm_{\beta,h}[q(0)],$$

where we use Lemma 7.2.15 to obtain the last equality.

It is now a basic fact that if $h \ne 0$ and β is kept fixed, then $P_\pm$ are both S-differentiable; see Helms and Loeb (1979), who closely follows the standard expositions of Lebowitz and Martin–Löf (1972) and Ruelle (1983). This means that the standard function

$$(26) \qquad G(h) = \mathrm{st}(P_\pm(h))$$

is well defined; it will be convex and differentiable at any standard $h \ne 0$, and furthermore

$$G'(h) \approx P'_\pm(h)$$

for any standard $h \ne 0$. We may then call upon (25) to conclude that $E^+_{h,\beta}[q(0)] \approx E^-_{h,\beta}[q(0)]$ for all $h \ne 0$. By translation invariance we may conclude that $E^+_{h,\beta}[q(i)] \approx E^-_{h,\beta}[q(i)]$ for all standard $i \in \Gamma$. Using (23), we see that

$$E^{q_0}_{\beta,h}[q(i)] \approx E^+_{\beta,h}[q(i)] \approx E^-_{\beta,h}[q(i)]$$

for all external configurations q_0. From this we may conclude that $L(\rho_{q_0,\beta,h}) = L(\rho_{+,\beta,h}) = L(\rho_{-,\beta,h})$ for all external boundary conditions q_0 on $\partial\Gamma$ and all $h \neq 0$. Hence all states $\tilde{\rho}_{q_0,\beta,h}$ coincide and we have no phase transition.

If there is no external field, i.e., if $h = 0$, then the above argument collapses at one point. The functions P_+ will, since the lattice Γ is hyperfinite, be *-differentiable, but now they turn out *not* to be S-differentiable for large enough β. We may as in (26) introduce a well-defined function $G = \text{st}(P_\pm)$, but the non-S-differentiability of $P_\pm$ means that G is not differentiable for $h = 0$ and sufficiently large β. And this suffices, using the standard arguments, to show that for large enough β, $\tilde{\rho}_{+,\beta,0} \neq \tilde{\rho}_{-,\beta,0}$. Since one may show for sufficiently small standard β that all states $\tilde{\rho}_{q,\beta,0}$ coincide, we do have a phase transition as β varies.

Our aim in this section has been to review a small part of classical equilibrium theory from a hyperfinite point of view, in particular to point out how a hyperfinite lattice gives a very natural model for the thermodynamic limit, a model which preserves faithfully all the explicit algebra and combinatorics of the finite models.

7.3. THE GLOBAL MARKOV PROPERTY

In this section we shall use the hyperfinite model Ω_Γ to discuss the global Markov property for the classical limit system Ω. We recall that the importance of the global Markov property lies in the fact that it allows us to introduce a certain semigroup which yields a rather complete probabilistic description of the system; see, e.g., Albeverio and Høegh-Krohn (1984a,b) and Albeverio *et al.* (1981). We also remind the reader of the importance of the global Markov property in the case of Euclidean field theory, where the infinitesimal generator obtained from the semigroup gives us the Hamiltonian of the associated Gårding–Wightman theory; see, e.g., Nelson (1973), Simon (1974), and Albeverio and Høegh-Krohn (1984b).

7.3.1. REMARK. The reader should note that the semigroup discussed in connection with the global Markov property is *not* the same as the "external" semigroup introduced in Section 7.1 to describe the evolution of a stochastic lattice system.

In rough terms we can describe the Markov property in the following way. Let Φ be a finite-range interaction and let $\mathbb{Z}^d$ be written as a disjoint union

$$\mathbb{Z}^d = \Lambda_1 \cup C \cup \Lambda_2,$$

where Λ_1 and Λ_2 are subsets of $\mathbb{Z}^d$ with distance larger than the range of the interaction Φ; i.e., C "insulates" Λ_1 from Λ_2 with respect to the interaction Φ. Let ρ be an equilibrium state for the infinite system and let $E_\rho(\cdot)$ denote the expectation with respect to ρ. We let $E_\rho(\cdot\,|\,C)$ denote the conditional expectation with respect to the σ-algebra $\mathscr{B}_C$ generated by the projection maps $f_i(q) = q(i)$, $i \in C$.

If for any *finite* Λ_1 one has that

(1) $$E_\rho(f_1 f_2 | C) = E_\rho(f_1 | C) E_\rho(f_2 | C),$$

where f_i, $i = 1, 2$, are bounded $\mathscr{B}_{\Lambda_i}$-measurable functions, then the Gibbs state ρ is said to have the *local Markov property*. We say that ρ has the *global Markov property* if (1) holds for *all* Λ_1, Λ_2.

In the final part of this section we shall extend some of the above discussion to the case of unbounded fiber with $\mathbb{R}$ replacing $\{-1, +1\}$ in the definition of the configuration space.

A. Hyperfinite Markov Property

We shall start out by discussing the Markov property on a hyperfinite lattice Γ. The global Markov property for the classical system is intimately tied up with the "behavior at infinity" of the system. We believe that the hyperfinite approach, which allows us to *see* what happens at infinity, offers an interesting alternative to the standard limit approach.

For simplicity we shall restrict attention to a *nearest-neighbor* interaction Φ. There are no difficulties in extending the results to an arbitrary finite-range interaction. Let Γ be a hyperfinite lattice and let ρ_{q_0} be a pure internal equilibrium state on Ω_Γ with respect to the nearest-neighbor interaction $^*\Phi$, see Definition 7.2.3; here q_0 is a boundary or external condition on $\Gamma^c = {}^*\mathbb{Z}^d - \Gamma$ fixing the state. Let Γ be written as a disjoint union $\Gamma = \Lambda_1 \cup C \cup \Lambda_2$, where C is a "curve" dividing Λ_1 and Λ_2; i.e., C is the common boundary of Λ_1 and Λ_2 inside Γ.

Let f_1 be an internal function on Ω_Γ which is independent of Λ_2; i.e., if q and q' differ only on sites in Λ_2, then $f_1(q) = f_1(q')$. Let f_2 in a similar way be independent of Λ_1. We let $E(\cdot)$ denote the (hyperfinite) expectation with respect to the internal measure ρ_{q_0} and let $E(\cdot\,|\,C)$ denote the corresponding conditional expectation. We want to prove the following *hyperfinite Markov property*:

7.3.2. PROPOSITION. $E(f_1 f_2 | C) = E(f_1 | C) E(f_2 | C)$.

We verify the proposition by a simple hyperfinite calculation. We start out by writing down an explicit formula for the conditional expectation. Let q_C be a configuration on the dividing curve C, i.e., $q_C \in \Omega_C$. Let

$q_C \times \Omega_{\Gamma-C}$ denote the set of all configurations $q \in \Omega_\Gamma$ which agree with q_C on C. We can write

$$\Omega_\Gamma = \bigcup_{q_C \in \Omega_C} q_C \times \Omega_{\Gamma-C}, \tag{2}$$

where the union is disjoint. The conditional expectation can now be written

$$E(f \mid C) = \sum_{q_C \in \Omega_C} \frac{\sum_{q \in q_C \times \Omega_{\Gamma-C}} f(q)\rho_{q_0}(q)}{\rho_{q_0}(q_C \times \Omega_{\Gamma-C})} \chi_{q_C \times \Omega_{\Gamma-C}}, \tag{3}$$

where $\chi_{q_C \times \Omega_{\Gamma-C}}$ is the characteristic function of the set $q_C \times \Omega_{\Gamma-C}$. Note that $E(f \mid C)$ as a function of $q \in \Omega_\Gamma$ only depends upon the behavior of q on C; thus for any $q_C \in \Omega_C$ the expression $E(f \mid C)(q_C)$ is well defined, and we see that

$$E(f \mid C)(q_C) = \sum_{q \in q_C \times \Omega_{\Gamma-C}} f(q)\rho_{q_0}(q \mid q_C) = E_{q_C}(f), \tag{4}$$

where $\rho_{q_0}(q \mid q_C)$ is the conditional probability measure derived from ρ_{q_0}, and E_{q_C} denotes expectation with respect to this measure.

Any element $q \in \Omega_\Gamma$ can be written as $q = q_1 \times q_2 \times q_C$, where $q_i \in \Omega_{\Lambda_i}$ and $q_C \in \Omega_C$. The measure $\rho_{q_0}(\cdot \mid q_C)$ induces measures on Ω_{Λ_1} and Ω_{Λ_2} via the formulas

$$\begin{aligned}
\rho_{\Lambda_1}(q_1) &= \rho_{q_0}(q_1 \times \Lambda_2 \times q_C \mid q_C), \\
\rho_{\Lambda_2}(q_2) &= \rho_{q_0}(\Lambda_1 \times q_2 \times q_C \mid q_C),
\end{aligned} \tag{5}$$

where, e.g., $q_1 \times \Lambda_2 \times q_C$ is the set of all configurations q in Ω_Γ such that $q \restriction_{\Lambda_1 \cup C} = q_1 \times q_C$. An explicit calculation gives the following splitting property:

$$\rho_{q_0}(q_1 \times q_2 \times q_C \mid q_C) = \rho_{\Lambda_1}(q_1)\rho_{\Lambda_2}(q_2). \tag{6}$$

In order to prove the proposition it suffices by (4) to show that $E_{q_C}(f_1 f_2) = E_{q_C}(f_1)E_{q_C}(f_2)$ for all $q_C \in \Omega_C$. But this is a straightforward calculation using (6):

$$\begin{aligned}
E_{q_C}(f_1 f_2) &= \sum_{q \in q_C \times \Omega_{\Gamma-C}} f_1(q)f_2(q)\rho_{q_0}(q \mid q_C) \\
&= \sum_{\substack{q_1 \in \Omega_{\Lambda_1} \\ q_2 \in \Omega_{\Lambda_2}}} f_1(q_1)f_2(q_2)\rho_{\Lambda_1}(q_1)\rho_{\Lambda_2}(q_2) \\
&= E_{q_C}(f_1)E_{q_C}(f_2).
\end{aligned}$$

This completes the proof of Proposition 7.3.2.

REMARK. The *local Markov property* for the limit space Ω is an immediate corollary of the hyperfinite Markov property 7.3.2. In the next section we shall discuss what the hyperfinite version has to say about the global Markov property.

B. Lifting and the Global Markov Property

We shall now explain how this simple hyperfinite calculation can be used to discuss the global Markov property on Ω.

We are now given a nearest-neighbor interaction Φ on Ω. By transfer we have a nearest-neighbor interaction $^*\Phi$ on Ω_Γ, where $\Gamma = \Gamma_\varkappa = \{x \in {}^*\mathbb{Z}^d \,||x_i| < \varkappa, i = 1, \ldots, d\}$ for some $\varkappa \in {}^*\mathbb{N} - \mathbb{N}$. Let ρ_{q_0} be an internal state for Ω_Γ, where $q_0 \in \Omega_{\Gamma^c}$. From ρ_{q_0} we derive via the Loeb construction a state $L(\rho_{q_0})$. As explained in Section 7.2, $L(\rho_{q_0})$ induces a measure $\tilde{\rho}_{q_0}$ on Ω by setting

$$(7) \qquad \tilde{\rho}_{q_0}(B) = L(\rho_{q_0})(\mathrm{st}_\Gamma^{-1} B),$$

for Borel sets B in Ω. $\tilde{\rho}_{q_0}$ is an equilibrium state of the classical system Ω.

Write $\mathbb{Z}^d$ as a disjoint union $\mathbb{Z}^d = \Lambda_1 \cup C \cup \Lambda_2$, where C is the common boundary of Λ_1 and Λ_2. C has an extension $^*C_\Gamma = {}^*C \cap \Gamma$ that splits Γ in the form $\Gamma = \Lambda_1' \cup {}^*C_\Gamma \cup \Lambda_2'$ for suitable Λ_1', Λ_2', where there is no interaction between Λ_1' and Λ_2' with respect to $^*\Phi$. Let $f \in C(\Omega)$ and let $^*f_\Gamma$ be the extension to Γ given by $^*f_\Gamma(q) = {}^*f(q \times q_0)$, $q \in \Omega_\Gamma$.

As above let $E(\cdot | {}^*C_\Gamma)$ denote the internal hyperfinite conditional expectation with respect to the measure ρ_{q_0} and the internal algebra $\mathscr{A}_{^*C_\Gamma}$ generated by the disjoint family of sets $q_{^*C_\Gamma} \times \Omega_{\Gamma - {}^*C_\Gamma}$, $q_{^*C_\Gamma} \in \Omega_{^*C_\Gamma}$. We shall let $E^{(s)}(\cdot | {}^*C_\Gamma)$ denote the conditional expectation with respect to the measure $L(\rho_{q_0})$ and the Loeb algebra $L(\mathscr{A}_{^*C_\Gamma})$. From Proposition 3.2.12 we know that $E(^*f_\Gamma | {}^*C_\Gamma)$ is a *lifting* of $E^{(s)}(^\circ(^*f_\Gamma) | {}^*C_\Gamma)$, i.e.,

$$(8) \qquad {}^\circ E(^*f_\Gamma | {}^*C_\Gamma) = E^{(s)}(^\circ(^*f_\Gamma) | {}^*C_\Gamma)$$

for a set of $L(\rho_{q_0})$ measure one. We have thus the picture shown in Fig. 7.2. Let f and g be tame functions with supports in D_1 and D_2, respectively. From Section 7.3.A we know that we have the following splitting:

$$E(^*f_\Gamma \cdot {}^*g_\Gamma | {}^*C_\Gamma) = E(^*f_\Gamma | {}^*C_\Gamma) \cdot E(^*g_\Gamma | {}^*C_\Gamma),$$

which by (8) implies a "global" Markov property on Γ with respect to the curve $^*C_\Gamma$. But we are interested in the global Markov property on $\mathbb{Z}^d$ and it is not at all clear how the results on Γ can be "pushed down" to $\mathbb{Z}^d$. If the interaction Φ is strong enough there could be an influence between D_1 and D_2 through the infinite part. In technical terms, it is not at all clear that the conditional expectation $E^{(s)}(\cdot | {}^*C_\Gamma)$ on Ω_Γ is "essentially the same" as the conditional expectation $E(\cdot | C)$ on Ω with respect to the measure $\tilde{\rho}_{q_0}$ and the algebra $\mathscr{B}_C$.

There are various ways of controlling the behavior "at infinity." We start by discussing the notion of lifting:

7.3.3. DEFINITION. The internal conditional expectation $E(\cdot | {}^*C_\Gamma)$ is called a *C-lifting* if there exists a set $N \subset \Omega_\Gamma$ which is $L(\mathscr{A}_{^*C_\Gamma})$ measurable

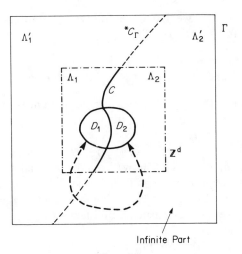

Figure 7.2

and with $L(\rho_{q_0})(N) = 0$ such that if q_Γ, $q'_\Gamma \notin N$ and $\mathrm{st}_\Gamma q_\Gamma = \mathrm{st}_\Gamma q'_\Gamma$, then

(9)
$$E(*f_\Gamma | *C_\Gamma)(q_\Gamma) \approx E(*f_\Gamma | *C_\Gamma)(q'_\Gamma),$$

for all tame functions f on Ω.

We have to show that if $E(\cdot | *C_\Gamma)$ is a C-lifting in the sense of Definition 7.3.3, then it is a lifting of the conditional expectation $E(\cdot | C)$ on Ω.

For $q \in \Omega$, let $\mu(q) = \{q'_\Gamma \in \Omega_\Gamma | \mathrm{st}_\Gamma q'_\Gamma = q\}$ be the *monad* of q. Let $h(q_\Gamma)$ be the function

(10)
$$h(q_\Gamma) = E(*f_\Gamma | *C_\Gamma)(q_\Gamma),$$

where $f \in C(\Omega)$ is tame. As remarked above, it follows from 3.2.12 that $h(q_\Gamma) \approx E^{(s)}(°(*f_\Gamma) | *C_\Gamma)(q_\Gamma)$ for $L(\rho_{q_0})$ almost all $q_\Gamma \in \Omega_\Gamma$.

We now assume that $E(\cdot | *C_\Gamma)$ is a C-lifting in the sense of 7.3.3. We may then introduce a function g on Ω:

(11)
$$g(q) = \begin{cases} \mathrm{st}(h(q'_\Gamma)) & \text{if } \mu(q) \not\subset N \text{ and } q'_\Gamma \in \mu(q) - N, \\ 0 & \text{if } \mu(q) \subset N. \end{cases}$$

Because of (9) the value of $g(q)$ is independent of the choice of representative $q'_\Gamma \in \mu(q) - N$.

7.3.4. PROPOSITION. The function g is a version of the conditional expectation on Ω with respect to the measure $\tilde{\rho}_{q_0}$ and the algebra $\mathscr{B}_C$.

We know that $E^{(s)}(\cdot | *C_\Gamma)$ is measurable with respect to the Loeb algebra $L(\mathscr{A}_{*C_\Gamma})$. As the reader may either directly verify or conclude via the general theory of Section 3.4, we have the following relationship between the σ-algebras $\mathcal{B}_C$ and $L(\mathcal{A}_{*C_\Gamma})$:

(i) If $B \in \mathcal{B}_C$, then $\mathrm{st}_\Gamma^{-1}(B) \in L(\mathcal{A}_{*C_\Gamma})$.

(ii) If $B \subseteq \Omega$ and $\mathrm{st}_\Gamma^{-1}(B) \in L(\mathcal{A}_{*C_\Gamma})$, then $B \in \mathcal{B}_C$.

In order to show that g is $\mathcal{B}_C$-measurable, it suffices to show that if q_1, $q_2 \in \Omega$ have the same restriction to the curve C, i.e., $q_1 \restriction C = q_2 \restriction C$, then $\mu(q_1) \subset N$ iff $\mu(q_2) \subset N$.

The proof is simple. Let $\hat{q}_1 \in \mu(q_1) - N$; we must find some $\hat{q}_2 \in \mu(q_2) - N$. Since $\hat{q}_1 \in \mu(q_1)$ it follows from the internal definition principle that there is some hyperfinite $\Gamma_1 \subset \Gamma$ such that $\hat{q}_1 \restriction C_{\Gamma_1} = {}^*q_1 \restriction C_{\Gamma_1}$. But ${}^*q_1$ and ${}^*q_2$ are equal on C_{Γ_1}; hence by setting $\hat{q}_2$ equal to ${}^*q_2$ on C_{Γ_1} and equal to $\hat{q}_1$ otherwise, we have constructed an element $\hat{q}_2 \in \mu(q_2) - N$. The rest goes by symmetry.

For the final part of the proof of 7.3.4, let B be any tame set in $\mathcal{B}_C$. Then

$$\int_B g\tilde{\rho}_{q_0}(dq) = \int_{\mathrm{st}_\Gamma^{-1}B} E^{(s)}({}^\circ({}^*f_\Gamma) | {}^*C_\Gamma) L(\rho_{q_0})(dq)$$

$$= \int_{\mathrm{st}_\Gamma^{-1}B} {}^\circ({}^*f_\Gamma) L(\rho_{q_0})(dq)$$

$$= \int_B f\tilde{\rho}_{q_0}(dq).$$

Thus we may take g as a version of the conditional expectation $E(\cdot | C)$.

Combining Propositions 7.3.2 and 7.3.4, we obtain a proof of the following theorem.

7.3.5. THEOREM. $\tilde{\rho}_{q_0}$ has the global Markov property with respect to C if $E(\cdot | {}^*C_\Gamma)$ is a C-lifting.

This theorem was proved by Kessler (1984) generalizing a previous result, Theorem 7.3.7, where the lifting condition is replaced by the stronger condition of S-continuity. We will discuss this in the next subsection. We shall say that $\tilde{\rho}_{q_0}$ satisfies the *lifting condition* if $E(\cdot | {}^*C_\Gamma)$ is a C-lifting for all C.

But before we continue let us remark that *lifting is not a necessary condition for the global Markov property*. Goldstein (1980) introduced a "condition C" which represents another way of controlling the behavior at infinity and which yields the global Markov property; see also Föllmer (1980). Kessler (1984) has shown that the lifting condition 7.3.3 and condition C are incomparable, i.e., neither implies the other.

Kessler (1984) gave a hyperfinite version of condition C which we shall briefly discuss. Choose $\varkappa, \lambda \in {}^*\mathbb{N} - \mathbb{N}$ such that $\varkappa - \lambda \in {}^*\mathbb{N} - \mathbb{N}$ and consider the hyperfinite lattices $\Gamma = \Gamma_\varkappa$ and $\Gamma' = \Gamma_\lambda$. We are in the "standard" situation; i.e., $\mathbb{Z}^d$ is written as a disjoint union $\mathbb{Z}^d = \Lambda_1 \cup C \cup \Lambda_2$, where C is a "curve" separating Λ_1, Λ_2 with respect to the nearest-neighbor interaction Φ. Ω_Γ is our basic configuration space and we let ρ_{Γ, q_0} denote the internal equilibrium state on Ω_Γ determined by the boundary condition q_0.

Let (q_0, q') denote a configuration which is equal to q' on ${}^*C_{\Gamma'} = {}^*C_\Gamma \cap \Gamma'$ and is equal to q_0 on $(\Gamma - \Gamma') \cap \Lambda_2'$, where, as above, Λ_2' is the extension of Λ_2 to Γ. $\rho_{\Gamma' \cap \Lambda_2', (q_0, q')}$ is then the internal equilibrium measure on $\Omega_{\Gamma' \cap \Lambda_2'}$ with external condition (q_0, q'). We define:

$$(12) \qquad \rho_{\Gamma, \Gamma', \Lambda_2', q_0}(B_1 \cap B_2) = \sum_{q' \in B_1} \rho_{\Gamma, q_0}(q') \rho_{\Gamma' \cap \Lambda_2', (q_0, q')}(B_2),$$

where B_2 is $\mathscr{A}_{\Gamma' \cap \Lambda_2'}$-measurable and B_1 is $\mathscr{A}_{\Gamma - (\Gamma' \cap \Lambda_2')}$-measurable.

REMARK. Notice that if $B_1 = q'_{\Gamma - (\Gamma' \cap \Lambda_2')} \times \Omega_{\Gamma' \cap \Lambda_2'}$ and $B_2 = \Omega_{\Gamma - (\Gamma' \cap \Lambda_2')} \times q_{\Gamma' \cap \Lambda_2'}$, then $B_1 \cap B_2$ is a singleton $q' = q'_{\Gamma - (\Gamma' \cap \Lambda_2')} \times q_{\Gamma' \cap \Lambda_2'}$ and

$$(13) \qquad \rho_{\Gamma, \Gamma', \Lambda_2', q_0}(q') = \rho_{\Gamma, q_0}(q'_{\Gamma - (\Gamma' \cap \Lambda_2')} \times \Omega_{\Gamma' \cap \Lambda_2'}) \rho_{\Gamma' \cap \Lambda_2', (q_0, q')}(q'_{\Gamma' \cap \Lambda_2'}).$$

Thus we see that $\rho_{\Gamma, \Gamma', \Lambda_2', q_0}$ is "close" to ρ_{Γ, q_0}, but represents a different way of controlling what happens at infinity. And if these measures do not differ "too much," they could have the same standard parts.

7.3.6. DEFINITION. Let $\Gamma = \Gamma_\varkappa$ for some infinite $\varkappa$ and let ρ_{Γ, q_0} be an internal equilibrium measure on Ω_Γ. $\tilde{\rho}_{\Gamma, q_0}$ satisfies *condition* C if there exists $\lambda \in {}^*\mathbb{N} - \mathbb{N}$ such that $\varkappa - \lambda \in {}^*\mathbb{N} - \mathbb{N}$ and such that

$$\tilde{\rho}_{\Gamma, q_0} = \tilde{\rho}_{\Gamma, \Gamma', \Lambda_2', q_0},$$

where $\Gamma' = \Gamma_\lambda$.

Goldstein showed that any measure $\tilde{\rho}_{\Gamma, q_0}$ satisfying condition C has the global Markov property. Kessler (1984, 1985) has constructed examples of measures of the form $\tilde{\rho}_{\Gamma, q_0}$ that do not satisfy the global Markov property and, hence, do not satisfy condition C. For related work see von Weizsäcker (1980) and Higuchi (1984).

We shall give a brief indication why condition C implies the global Markov property. Even if we do not have a lifting in the sense of 7.3.3, we can show that for any tame g there exist infinite η's in ${}^*\mathbb{N}$, $\eta \leq \varkappa$, such that $E_{\rho_{\Gamma, q_0}}(g \mid {}^*C_{\Gamma_\eta})$ is a lifting (notice that the conditional expectation as a function depends only on sites in ${}^*C_{\Gamma_\eta} = {}^*C_\Gamma \cap \Gamma_\eta$), and that the set of such η's is an initial segment.

Condition C can now be used (Kessler, 1984) to show that for any $\lambda_0 \in {}^*\mathbb{N} - \mathbb{N}$ there is some $\lambda \in {}^*\mathbb{N} - \mathbb{N}$, $\lambda \leq \lambda_0$, such that we have the right kind of splitting

$$E_{\rho_{\Gamma,q_0}}(f_1 \cdot f_2 | {}^*C_{\Gamma_\lambda}) \approx E_{\rho_{\Gamma,q_0}}(f_1 | {}^*C_{\Gamma_\lambda}) \cdot E_{\rho_{\Gamma,q_0}}(f_2 | {}^*C_{\Gamma_\lambda}).$$

Putting these facts together and adding some measure theory yields the global Markov property; see Kessler (1984) for full details.

And to obtain the right splitting from condition C we use the hyperfinite Markov property, 7.3.2, with an appropriate choice of the measures $\rho_{\Gamma,\Gamma_\xi \wedge_2',q_0}$, for some ξ.

C. S-Continuity, Dobrushin's Condition and the Global Markov Property

We recall that the S-continuity of $E(\cdot | {}^*C_\Gamma)$ means that

(14) $$E({}^*f_\Gamma | {}^*C_\Gamma)(q_\Gamma) \approx E({}^*f_\Gamma | {}^*C_\Gamma)(q_\Gamma'),$$

whenever $\mathrm{st}_\Gamma q_\Gamma = \mathrm{st}_\Gamma q_\Gamma'$. Thus S-continuity implies the lifting property and we have the following theorem.

7.3.7. THEOREM. $\tilde{\rho}_{q_0}$ has the global Markov property with respect to C if $E(\cdot | {}^*C_\Gamma)$ is S-continuous.

The converse is not true; the lifting property does not imply the S-continuity; a counterexample can be found in Kessler (1984). The measures he constructs are not translation-invariant and it remains an interesting task to decide the relationship between lifting, condition C, and S-continuity in the translation-invariant case.

Lifting and S-continuity are rather "abstract" properties of the measure ρ_{q_0}. We shall in this subsection give an example due to Dobrushin (1968a,b), where we infer the S-continuity from a condition imposed on the interaction Φ. We continue the notation from the last subsection.

Let $\Gamma' = \Gamma - {}^*C_\Gamma$ and fix a configuration q_{*C_Γ} on ${}^*C_\Gamma$. We shall consider internal equilibrium states ρ_q' on $\Omega_{\Gamma'}$ with boundary conditions equal to q_0 on Γ^c and q_{*C_Γ} on ${}^*C_\Gamma$. Any tame function $f \in C(\Omega)$ has an extension to an internal function on both Ω_Γ and $\Omega_{\Gamma'}$; denote these extensions by ${}^*f_\Gamma$ and ${}^*f_{\Gamma'}$, respectively. Note that for $q \in \Omega_{\Gamma'}$ we have ${}^*f_{\Gamma'}(q) = {}^*f_\Gamma(q \times q_{*C_\Gamma}) = {}^*f(q \times q_{*C_\Gamma} \times q_0)$. Observe further [see (4) above] that:

(15) $$E_{\rho_q'}({}^*f_{\Gamma'}) = E({}^*f_\Gamma | {}^*C_\Gamma)(q_{*C_\Gamma}).$$

We shall now describe a way to calculate the expected value $E_{\rho_q'}({}^*f_{\Gamma'})$ which, under suitable assumptions on the interaction potential Φ, does not depend upon the behavior at infinity; thus we have the S-continuity of $E(\cdot | {}^*C_\Gamma)$.

7.3.8. REMARK. Uniqueness of Gibbs or equilibrium measures was first proved by Dobrushin (1968a,b). The connection between global Markov property and uniqueness has been remarked upon on several occasions [see, e.g., Albeverio *et al.* (1981) and Föllmer (1980)] and has been used to prove the global Markov property for lattice systems. The present exposition gives a somewhat different arrangement of the facts.

For each $i \in \mathbb{Z}^d - C$ and $q^* \in q_{*C_\Gamma} \times \Omega_{\Gamma'}$, let $\rho_i(\cdot \mid q^*)$ be the equilibrium measure at site i with boundary condition q^*. Further, let $t_i(f)(q^*)$ denote the expected value of f with respect to the measure $\rho_i(\cdot \mid q^*)$.

Fix an enumeration $i_1, i_2, \ldots$ of the sites in $\mathbb{Z}^d - C$. It is easy to see that for each $f \in C(\Omega_{\mathbb{Z}^d - C})$

$$(16) \qquad Tf = \lim_{n \to \infty} t_{i_1} \cdots t_{i_n} f$$

exists (and, indeed, will be independent of how the sites are enumerated).

Consider the norm

$$(17) \qquad \|f\| = \sum_{i \in \mathbb{Z}^d - C} \sup\{|f(q) - f(q')|\},$$

where the sup is taken over all pairs $q, q' \in q_{*C_\Gamma} \times \Omega_{\Gamma'}$ such that $q = q'$ off i, i.e., $q(j) = q'(j)$ for all $j \neq i$.

7.3.9. PROPOSITION. Let

$$\rho_{ij} = \tfrac{1}{2} \sup\{\|\rho_i(\cdot \mid q) - \rho_j(\cdot \mid q')\|_{\mathrm{var}}\},$$

where the sup is taken over all pairs $q, q' \in q_{*C_\Gamma} \times \Omega_{\Gamma'}$ such that $q = q'$ off i, and define

$$\alpha = \sup_i \sum_j \rho_{ij},$$

where $i, j \in \mathbb{Z}^d - C$. If $\alpha < 1$, then

$$\|Tf\| < \alpha \cdot \|f\|.$$

The following simple observation is behind the inequality. Let μ_1 and μ_2 be probability measures on some common space and let E_{μ_1} and E_{μ_2} be the associated expectations; then

$$|E_{\mu_1}(f) - E_{\mu_2}(f)| \le \tfrac{1}{2} \|\mu_1 - \mu_2\|_{\mathrm{var}} \cdot |\sup f - \inf f|.$$

The route from this inequality to the proposition can be found in Gross (1979).

Iterating the T transform shows that $\|T^n f\| \le \alpha^n \|f\|$. It is not too difficult [see Gross (1979)] to see that $T^n f$ will converge uniformly to some number $E(f)$.

We can now put the various bits together. Associated with ρ'_q we have a measure $\tilde{\rho}'_q$ on $\Omega_{\mathbb{Z}^d - C}$ [see (7)] such that for any $f \in C(\Omega)$,

$$(18) \qquad E_{\rho'_q}(^*f_{\Gamma'}) \approx E_{\tilde{\rho}'_q}(f).$$

Since ρ'_q is an internal equilibrium measure it follows that for all $i \in \mathbb{Z}^d - C$

$$(19) \qquad E_{\tilde{\rho}_q}(t_i f) = E_{\tilde{\rho}_q}(f).$$

From this we may conclude that

$$(20) \qquad E_{\rho'_q}(^*f_{\Gamma'}) \approx E_{\tilde{\rho}_q}(T^n f),$$

and since $T^n f$ converges uniformly to the *constant* $E(f)$, it follows from (15) and (20) that

$$(21) \qquad E(^*f_\Gamma | ^*C_\Gamma)(q_{*C_\Gamma}) \approx E(f).$$

Since $E(f)$ does not depend upon the infinite part of q_{*C_Γ}, the S-continuity is proved under the assumption of Proposition 7.3.9.

7.3.10. REMARK. The condition $\alpha < 1$ is a condition on the interaction Φ. Let us make this explicit. First recall that for each finite $X \subseteq \mathbb{Z}^d$, $\Phi(X) \in C(\Omega_X)$. Let us denote by $\|\Phi\|_\infty$ the norm

$$(22) \qquad \|\Phi\|_\infty = \sup_i \sum_{X \ni i} (|X| - 1)\|\Phi(X)\|_\infty.$$

Next let us introduce a measure of the total strength of the interaction between sites $i, j, i \neq j$:

$$(23) \qquad \zeta(i,j) = \sup_q \left| \sum_{X \ni \{i,j\}} \Phi(X)(q) \right|.$$

Observe that

$$\sum_{\substack{j \\ j \neq i}} \zeta(i,j) \leq \sum_{X \ni i} (|X| - 1)\|\Phi(X)\|_\infty.$$

The following inequality is simple but basic. Let $i \neq j$ and $q = q'$ off j; then

$$\|\rho_i(\cdot|q) - \rho_j(\cdot|q')\|_{\text{var}} \leq e^{4\zeta(i,j)} - 1.$$

From this it is not too difficult to conclude that

$$\rho_{ij} \leq 2e^{4\|\Phi\|_\infty}\zeta(i,j).$$

Recalling the definition of α in Proposition 7.3.9, we see that

$$(24) \qquad \alpha \leq 2e^{4\|\Phi\|_\infty}\|\Phi\|_\infty.$$

Note that if $\|\Phi\|_\infty < (2e)^{-1}$, then $\alpha < 1$ and the S-continuity of $E(\cdot | ^*C_\Gamma)$ follows.

If the interaction Φ satisfies the inequality $\|\Phi\|_\infty < (2e)^{-1}$, then there is a unique equilibrium measure on the classical system Ω. In this case the boundary condition q_0 in the measure ρ_{q_0} on Ω_Γ has at most an infinitesimal action in the space Ω; i.e., the measures $\tilde{\rho}_{q_0}$, $q_0 \in \Omega_{\Gamma^c}$, all coincide.

7.3.11. THEOREM. Let Φ be a finite-range interaction satisfying the inequality $\|\Phi\|_\infty < (2e)^{-1}$. Then the unique equilibrium measure on Ω satisfies the global Markov property.

The simple hyperfinite calculation of Proposition 7.3.2, the S-continuity which we have verified under the stated condition on the interaction Φ, and Theorem 7.3.7 add up to a proof of the theorem.

7.3.12. REMARK. The assumption in Theorem 7.3.11 is a special case of the "strong uniqueness condition" for the interaction Φ. Let α be an arbitrary product measure on Ω and let α_i, $i \in \mathbb{Z}^d$, be the ith component of α; i.e., α_i is a measure on a fiber $\{-1, +1\}$. Φ is said to have the *strong uniqueness property* if for all product measures α there exists at most one equilibrium state for the interaction $\Phi + \Psi_\alpha$, where Ψ_α is the interaction associated with the measure α.

Strong uniqueness implies S-continuity; for details see Kessler (1984).

D. Maximal and Minimal Gibbs States

We shall describe another situation where we have the global Markov property. The configuration space $\Omega = \{-1, +1\}^{\mathbb{Z}^d}$ has a natural ordering

$$(25) \qquad q \leq q' \text{ iff } q(i) \leq q'(i), \quad \text{all } i \in \mathbb{Z}^d.$$

We let K_+ denote the set of bounded increasing measurable functions on Ω. K_+ is a convex cone and determines the order on Ω in the sense that

$$(26) \qquad q \leq q' \text{ iff } F(q) \leq F(q'), \quad \text{all } F \in K_+.$$

We shall need the following (standard) lemma.

7.3.13. LEMMA. Let μ_1 and μ_2 be probability measures on Ω and assume that $\mu_1(F) = \mu_2(F)$ for all $F \in K_+$. Then $\mu_1 = \mu_2$.

The proof is a simple density argument. A dual order is defined on the set of probability measures on Ω by

$$(27) \qquad \mu_1 \leq \mu_2 \text{ iff } \mu_1(F) \leq \mu_2(F), \quad \text{all } F \in K_+.$$

This order is in a suitable sense a "linear extension" of the order defined in (25). If $\mu_1 \leq \mu_2$ and $\mu_2 \leq \mu_1$, then Lemma 7.3.13 tells us that $\mu_1 = \mu_2$.

Let Γ be a hyperfinite rectangle. The above definitions immediately carry over to the space Ω_Γ and to internal probability measures on Ω_Γ.

Let μ_1 and μ_2 be internal probability measures on Ω_Γ, $L(\mu_1)$ and $L(\mu_2)$ the associated Loeb measures, and $\tilde{\mu}_1$ and $\tilde{\mu}_2$ the "pull-backs" to Ω defined by (7) above. The following lemma is an immediate consequence of the Loeb construction.

7.3.14. LEMMA. Let μ_1 and μ_2 be internal probability measures on Ω_Γ and assume that $\mu_1 \le \mu_2$. Then $\tilde{\mu}_1 \le \tilde{\mu}_2$ on Ω.

Let Φ be a finite-range interaction on Ω and $^*\Phi$ its extension by transfer. Let Γ be a hyperfinite rectangle and $\rho_{\Gamma,\mu}$ the corresponding internal equilibrium state with boundary condition μ; see Definitions 7.2.3 and 7.2.10.

7.3.15. DEFINITION. The interaction Φ is called *attractive* if the map $\mu \mapsto \rho_{\Gamma,\mu}$ is increasing for all Γ.

We have chosen a rather abstract form of the notion of an attractive interaction in order to highlight the general ideas and to avoid getting involved in too detailed analytical considerations at this point. The class of attractive interactions includes many important examples; e.g., the interaction defining the Ising model [see (22) of Section 7.2] is attractive for $\beta \ge 0$ and h arbitrary.

7.3.16. THEOREM. If Φ is an attractive interaction of finite range on Ω, then there is a unique maximal Gibbs state $\tilde{\rho}_+$ and a unique minimal Gibbs state $\tilde{\rho}_-$ on Ω.

With our choice of definitions the proof is simple. Choose Γ hyperfinite and let $\rho_{\Gamma,+}$ be the internal equilibrium state on Ω_Γ with respect to the interaction $^*\Phi$ and with boundary condition $q \equiv +1$. Since Φ is attractive, $\rho_{\Gamma,+}$ is maximal on Ω_Γ. We will show that $\tilde{\rho}_+ \equiv \tilde{\rho}_{\Gamma,+}$ is the unique maximal Gibbs state on Ω. Indeed, if ρ is another Gibbs state on Ω, then ρ would be of the form $\tilde{\rho}_{\Gamma,\mu}$ for an internal state $\rho_{\Gamma,\mu}$ on Ω_Γ. But on Ω_Γ we have that $\rho_{\Gamma,\mu} \le \rho_{\Gamma,+}$; thus by Lemma 7.3.14, $\rho \le \tilde{\rho}_+$ on Ω. This shows that $\tilde{\rho}_+$ is maximal; uniqueness follows from Lemma 7.3.13; see the remark following (27). Choosing the boundary condition $q \equiv -1$ we obtain the minimal Gibbs state $\tilde{\rho}_-$.

We now aim toward the following theorem; for a related discussion see Föllmer (1980).

7.3.17. THEOREM. Let Φ be an attractive interaction of finite range. Then the unique maximal and minimal Gibbs states $\tilde{\rho}_+$ and $\tilde{\rho}_-$ both have the global Markov property.

To prove the appropriate lifting property we need the following basic estimate due to Kessler (1984).

7.3.18. PROPOSITION. Let Γ and Γ' be hyperfinite rectangles, $\Gamma \subseteq \Gamma'$, and let f be a positive increasing tame function. Then

$$L(\rho_{\Gamma',+})\{E_{\rho_{\Gamma',+}}(f\,|{}^*C_{\Gamma'})(q_{\Gamma'}) \ll E_{\rho_{\Gamma',+}}(f\,|{}^*C_{\Gamma})(q_{\Gamma})\} = 0.$$

REMARK. $a \ll b$ means that $a < b$ but $a \not\approx b$.

The proof is based on the following two facts. First, we have for finite rectangles $\Gamma \subseteq \Gamma'$ positive increasing functions f with support in Γ and configurations $q \in \Omega_{\mathbf{Z}^d}$

$$(28) \qquad E_{\rho_{\Gamma,+}}(f\,|\,C_{\Gamma})(q_{\Gamma}) \geq E_{\rho_{\Gamma',+}}(f\,|\,C_{\Gamma'})(q_{\Gamma'}) \geq 0.$$

The idea behind this inequality is that when we "project" down from Γ' to Γ we get something smaller than what we obtain in the attractive case from the maximal state. By transfer (28) remains true for hyperfinite rectangles and standard tame functions—in fact, even for internal functions. Next we observe that for tame positive increasing functions the non-negative sequence of the $E_{\rho_{\Gamma,+}}(f)$ decreases when the size of Γ increases. Thus we have

$$(29) \qquad\qquad E_{\rho_{\Gamma,+}}(f) \approx E_{\rho_{\Gamma',+}}(f)$$

for hyperfinite $\Gamma \subseteq \Gamma'$.

Assume now that 7.3.18 is false. Then there are ε, $\delta \in \mathbb{R}_+$ such that $\rho_{\Gamma',+}(B_\varepsilon) \geq \delta$, where $B_\varepsilon = C_\varepsilon \times \Omega_{\Gamma'-{}^*C_{\Gamma'}}$ and where

$$C_\varepsilon = \{(q_1 \times q_2) \in \Omega_{{}^*C_{\Gamma'}} \,|\, E_{\rho_{\Gamma',+}}(f\,|{}^*C_{\Gamma'})(q_1 \times q_2) \leq E_{\rho_{\Gamma',+}}(f\,|{}^*C_{\Gamma})(q_1) - \varepsilon\}.$$

REMARK. We have written a configuration $q_{\Gamma'}$ in $\Omega_{\Gamma'}$ as $q_1 \times q_2 \times q_3$, where $q_1 \in \Omega_{{}^*C_{\Gamma}}$, $q_2 \in \Omega_{{}^*C_{\Gamma'}-{}^*C_{\Gamma}}$, and $q_3 \in \Omega_{\Gamma'-{}^*C_{\Gamma'}}$.

We have the following expression for the expected value $E_{\rho_{\Gamma',+}}(f)$:

$$E_{\rho_{\Gamma',+}}(f) = \sum_{q_1 \times q_2 \in \Omega_{{}^*C_{\Gamma'}}} E_{\rho_{\Gamma',+}}(f\,|{}^*C_{\Gamma'})\rho_{\Gamma',+}(q_1 \times q_2 \times \Omega_{\Gamma'-{}^*C_{\Gamma'}}).$$

The sum splits in two, $q_1 \times q_2 \in C_\varepsilon$ or $q_1 \times q_2 \notin C_\varepsilon$, and using (27) and the definition of C_ε we obtain the following chain of inequalities:

$$\begin{aligned}
E_{\rho_{\Gamma',+}}(f) &\leq \sum_{q_1 \times q_2 \in C_\varepsilon} (E_{\rho_{\Gamma',+}}(f\,|{}^*C_{\Gamma})(q_1) - \varepsilon)\rho_{\Gamma',+}(q_1 \times q_2 \times \Omega_{\Gamma'-{}^*C_{\Gamma'}}) \\
&\quad + \sum_{q_1 \times q_2 \notin C_\varepsilon} (E_{\rho_{\Gamma',+}}(f\,|{}^*C_{\Gamma})(q_1))\rho_{\Gamma',+}(q_1 \times q_2 \times \Omega_{\Gamma'-{}^*C_{\Gamma'}}) \\
&\leq \sum_{q_1 \times q_2 \in \Omega_{{}^*C_{\Gamma'}}} E_{\rho_{\Gamma',+}}(f\,|{}^*C_{\Gamma})(q_1)\rho_{\Gamma',+}(q_1 \times q_2 \times \Omega_{\Gamma'-{}^*C_{\Gamma'}}) - \varepsilon \cdot \delta \\
&= E_{\rho_{\Gamma',+}}(E_{\rho_{\Gamma,+}}(f\,|{}^*C_{\Gamma})) - \varepsilon \cdot \delta \\
&\leq E_{\rho_{\Gamma,+}}(E_{\rho_{\Gamma,+}}(f\,|{}^*C_{\Gamma})) - \varepsilon \cdot \delta \\
&= E_{\rho_{\Gamma,+}}(f) - \varepsilon \cdot \delta,
\end{aligned}$$

but this contradicts (29), and the proposition is proved.

From Proposition 7.3.18 we can now prove that the conditional expectation is a *C-lifting* in the sense of Definition 7.3.3. And from this we can, as in Section 7.3.B, complete the proof of Theorem 7.3.17.

It remains to verify the C-lifting property. Let $\varkappa \in {}^*\mathbb{N} - \mathbb{N}$ be the size of Γ, and for $\lambda \le \varkappa$ let Γ_λ be the rectangle of size λ. For $\lambda < \varkappa$ and $\varepsilon \in \mathbb{R}_+$ we define

$$N_{\lambda,\varepsilon} = \{q \,|\, E_{\rho_{\Gamma,+}}(f\,|\,{}^*C_\Gamma)(q_\Gamma) \le E_{\rho_{\Gamma_\lambda,+}}(f\,|\,{}^*C_{\Gamma_\lambda})(q_{\Gamma_\lambda}) - \varepsilon\} \in \mathcal{A}_{*C_\Gamma}.$$

Since the sets are internal Proposition 7.3.18 implies that the least $\lambda \in {}^*\mathbb{N}$ (call it λ_ε) such that $\rho_{\Gamma,+}(N_{\lambda,\varepsilon}) \le \varepsilon$ is actually in $\mathbb{N}$; we also see that as λ increases the sequence $\rho_{\Gamma,+}(N_{\lambda,\varepsilon})$ decreases; finally, we observe that as ε decreases, λ_ε increases. We conclude from the Borel–Cantelli lemma that the set

$$N = \bigcup_{n \in \mathbb{N}} \bigcap_{\lambda \ge \lambda_{1/2^n}} N_{\lambda,1/2^n} \in L(\mathcal{A}_{*C_\Gamma})$$

has $L(\rho_{\Gamma,+})$ measure 0.

Now let q, q' be two configurations in Ω_Γ and assume that $q, q' \notin N$ but $\mathrm{st}_\Gamma q = \mathrm{st}_\Gamma q'$. From the internality of q, q' and the fact that they have the same standard part it follows that there is some $\lambda \in {}^*\mathbb{N} - \mathbb{N}$ such that $q_{\Gamma_\lambda} = q'_{\Gamma_\lambda}$. We then have

$$E_{\rho_{\Gamma,+}}(f\,|\,{}^*C_\Gamma)(q_\Gamma) \approx E_{\rho_{\Gamma_\lambda,+}}(f\,|\,{}^*C_{\Gamma_\lambda})(q_{\Gamma_\lambda})$$
$$= E_{\rho_{\Gamma_\lambda,+}}(f\,|\,{}^*C_{\Gamma_\lambda})(q'_{\Gamma_\lambda})$$
$$\approx E_{\rho_{\Gamma,+}}(f\,|\,{}^*C_\Gamma)(q'_\Gamma).$$

Here the two $\approx$-assertions follow from the fact that $q, q' \notin N$. The proof of the lifting property is now complete.

This proves Theorem 7.3.17. Before turning to the case of continuous fiber we summarize the situation so far in the accompanying diagram. The arrows in the diagram have all been discussed above except the one leading from strong uniqueness to extremality. A pure state is extremal if it is not a convex combination of other pure states; strong uniqueness thus trivially implies extremality.

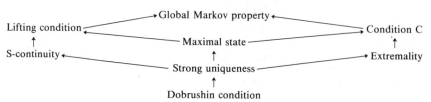

Kessler (1984, 1985) has shown that in the nontranslation-invariant case, the arrows cannot be reversed. Contrary to a conjecture of Goldstein (1980), he has also shown that extremality does not imply the global Markov property. It also follows from Kessler's work that lifting and condition C are incompatible; he also has an example showing that S-continuity does not imply condition C.

Some version of the hyperfinite Markov property and some adaptation of the lifting idea are behind the arrows above. We feel that the hyperfinite picture is an intuitive and effective way of organizing the material.

E. The Case of Unbounded Fiber

In the final part of this section we discuss the unbounded, continuous case. More precisely, this means that we shall replace the fiber $\{-1, +1\}$ by $\mathbb{R}$ and study configuration spaces of the form $\Omega_\Lambda = \mathbb{R}^\Lambda$, where $\Lambda \subseteq \mathbb{Z}^d$. This goes beyond the theory of Section 7.2 and we shall, therefore, first give a brief résumé of the basic standard theory.

An *interaction* Φ is a map from finite nonempty subsets $X \subseteq \mathbb{Z}^d$ to real-valued continuous functions on Ω_X. Let Λ be a finite subset of $\mathbb{Z}^d$; we introduce the *Gibbs* or *equilibrium measure* μ_Λ on Ω_Λ by the definition

$$(30) \qquad d\mu_\Lambda(q) = Z_\Lambda^{-1} e^{-H_\Lambda^\Phi(q)} \, dq,$$

where

$$(31) \qquad H_\Lambda^\Phi(q) = \sum_{X \subseteq \Lambda} \Phi(X)(q), \qquad q \in \Omega_\Lambda,$$

and

$$(32) \qquad Z_\Lambda = \int_{\Omega_\Lambda} e^{-H_\Lambda^\Phi(q)} \, dq.$$

In this section we shall restrict attention to interactions Φ of *finite range*, i.e., interactions Φ for which there exists a natural number l such that $\Phi(X) = 0$ if the diameter of $X \subseteq \mathbb{Z}^d$ is larger than l. With this restriction the following construction is well defined.

Let Λ be a finite subset of $\mathbb{Z}^d$. An *external configuration* for Λ is an element in Ω_{Λ^c}. Let q_0 be an external configuration for Λ; we introduce the *Gibbs* or *equilibrium measure with boundary condition* q_0 on Ω_Λ by

$$(33) \qquad d\mu_{\Lambda, q_0}(q) = Z_{\Lambda, q_0}^{-1} \exp(-_{q_0}H_\Lambda^\Phi(q)) \, dq,$$

where

$$(34) \qquad {}_{q_0}H_\Lambda^\Phi(q) = \sum_{X \cap \Lambda \neq \varnothing} \Phi(X)(q \times q_0), \qquad q \in \Omega_\Lambda,$$

and Z_{Λ, q_0} is defined in the same way as Z_Λ; see (30) above. We note that since Λ is finite and Φ has finite range the sum in (34) exists.

Let $E_{\mu_\Lambda}(\cdot\,|\,\Lambda_1^c)$ denote the conditional expectation with respect to the measure μ_Λ and the σ-algebra $\mathscr{B}_{\Lambda_1^c}$. For a large class of interactions the limit

$$(35) \qquad\qquad E_{\Lambda_1^c}^q(F) = \lim_{\Lambda\uparrow\mathbb{Z}^d} E_{\mu_\Lambda}(F\,|\,\Lambda_1^c)(q)$$

exists. When Φ is of finite range we have the following explicit formula:

$$(36) \qquad\qquad E_{\Lambda^c}^q(F) = Z_{\Lambda,q}^{-1} \int F(p)\exp(-_qH_\Lambda^\Phi(p))\,dp.$$

Let μ be a probability measure on $\Omega = \mathbb{R}^{\mathbb{Z}^d}$; μ is called a *Gibbs measure or equilibrium measure* on Ω if

$$(37) \qquad\qquad E_\mu(F\,|\,\Lambda^c)(q) = E_{\Lambda^c}^q(F)$$

for all finite $\Lambda \subseteq \mathbb{Z}^d$.

This concludes our brief sketch of the standard theory. For the *hyperfinite* version we start with a hyperfinite rectangle Γ and fix a boundary condition $q_0 \in \Omega_{\Gamma^c}$. By transfer we have an internal measure corresponding to (33). If ν is a probability measure on Ω_{Γ^c}, we can introduce a measure $\mu_{\Gamma,\nu}$ by

$$(38) \qquad\qquad d\mu_{\Gamma,\nu}(q) = \int_{\Omega_{\Gamma^c}} \mu_{\Gamma,q_0}(q)\,d\nu(q_0).$$

This is an internal measure, but by adjoining the Loeb construction and the standard part map we may introduce a measure μ on Ω by the equation

$$(39) \qquad\qquad \mu(B) = L(\mu_{\Gamma,\nu})(\mathrm{st}^{-1}B),$$

where B is a Borel set in Ω. By a straightforward adaptation of the theory in Section 7.2 (see also the extension theory of Sections 3.4 and 3.5) we see that *the measure μ of* (39) *is a Gibbs measure on Ω and that every Gibbs measure can be represented in this form for suitable Γ and ν.*

There is no difficulty in adapting previous definitions to the present case. So assume that Φ is of finite range and *attractive* (see Definition 7.3.15). In the unbounded case this is not enough to ensure the existence of a maximal and a minimal Gibbs measure; we need to impose a growth condition on the interaction Φ.

7.3.19. DEFINITION. A probability measure μ on Ω is called *tempered* if there are integers $k,\ N \in \mathbb{N}$ such that

$$(40) \qquad\qquad \mu(|q_i|) \le k(1+|i|)^N$$

for all $i \in \mathbb{Z}^d$.

An interaction Φ is called *tempered* if for all finite $\Lambda \subseteq \mathbb{Z}^d$ the measure $\mu_{\Lambda,\nu}$ is tempered whenever ν is tempered. Φ is called *uniformly tempered* if for any tempered ν the family $\{\mu_{\Lambda,\nu} || \Lambda| < \infty\}$ is uniformly tempered in the sense that there are k, $N \in \mathbb{N}$ such that

$$(41) \qquad \mu_{\Lambda,\nu}(|q_i|) \le k(1 + |i|)^N$$

for all finite $\Lambda \subseteq \mathbb{Z}^d$ and all $i \in \mathbb{Z}^d$.

We shall not in this exposition prove general results about tempered interactions; see Bellissard and Høegh-Krohn (1982) for detailed information. As an important example we mention

$$(42) \qquad H_\Lambda^\Phi(q) = \frac{\beta}{2} \sum_{\substack{i,j \in \Lambda \\ |i-j|=1}} |q_i - q_j|^2 + \beta \sum_{i \in \Lambda} V(q_i),$$

which is uniformly tempered under rather weak conditions on the potential V.

This interaction satisfies the following important inequality:

$$(43) \qquad E_{\{i\}^c}^q(|q_i|) \le a + \sum_{j \in \mathbb{Z}^d} r_{ij}|q_j|,$$

where $a > 0$ and $\sum_j r_{ij} \le c_1 < 1$. [We can always assume that $r_{ii} = 0$, and in the present simple case, (42), we need only sum over the nearest neighbors of i.]

To the proof we remark that (43) follows from (42) by standard techniques for the evaluation of the asymptotic behavior of Laplace transforms; see Bellissard and Høegh-Krohn (1982).

From (43) we can derive the following crucial inequality.

7.3.20. LEMMA. Let Φ be a uniformly tempered interaction of finite range satisfying the inequality (43). Then there is a constant c such that

$$(44) \qquad E_\mu(|q_i|) < c$$

for all $i \in \mathbb{Z}^d$ and all tempered Gibbs measures μ on Ω.

Using the fact that μ is a Gibbs measure and taking expectations on both sides of (43), we obtain

$$E_\mu(|q_i|) \le a + \sum_{j \in \mathbb{Z}^d} r_{ij} E_\mu(|q_j|),$$

which implies that

$$(45) \qquad \sum_j (\delta_{ij} - r_{ij}) E_\mu(|q_j|) \le a.$$

Let $(1 - r)$ be the matrix with entries $(1 - r)_{ij} = \delta_{ij} - r_{ij}$ and let X be the vector defined by $X_j = E_\mu(|q_j|)$; then (45) can be restated as the set of inequalities $((1 - r)X)_i \le a$. And since $(1 - r)_{ij}^{-1} = \delta_{ij} + r_{ij} + \sum_k r_{ik} r_{kj} + \cdots$

has only positive terms and $\sum_j (1 - r)_{ij}^{-1}$ by assumption converges and is bounded by some number <1 independent of i, it follows that there is some constant c such that $E_\mu(|q_i|) < c$ for all $i \in \mathbb{Z}^d$ and all tempered Gibbs measures μ.

We are now in a position to construct a *maximal tempered Gibbs measure* on Ω. Our starting point is a uniformly tempered interaction Φ satisfying Lemma 7.3.20. To control the growth we fix two integers $N_1, N_2 \in \mathbb{N}$ and choose a configuration $q^+ \in \Omega_{*\mathbb{Z}^d}$ such that

$$(46) \qquad\qquad |i|^{N_1} \leq |q_i^+| < |i|^{N_2}, \qquad \text{all } i \in {}^*\mathbb{Z}^d$$

and such that

$$(47) \qquad\qquad \sum_{i \in \mathbb{Z}^d} \frac{1}{q_i^+} < \infty.$$

The choice of N_1 will depend on the dimension d of $\mathbb{Z}^d$.

Let Γ be a hyperfinite rectangle and consider the measure μ_+ on Ω defined by

$$(48) \qquad\qquad \mu_+ = L(\mu_{\Gamma,q^+}) \circ \mathrm{st}^{-1}.$$

The first thing to note is that this is a tempered Gibbs measure on Ω; see (46) above. We must show that it is maximal, i.e.,

$$(49) \qquad\qquad E_\mu(F) \leq E_{\mu_+}(F)$$

for all $F \in K_+$ [see (27)] and all tempered Gibbs measures μ on Ω. Equation (49) also implies the uniqueness of μ_+.

We make the following preliminary calculation:

$$(50) \qquad E_\mu(F) = E_\mu(E_\mu(F \mid \Lambda^c)(q))$$

$$\leq \int_{B_{\Lambda,q^+}} E_{\Lambda^c}^q(F) \, d\mu(q) + \|F\|_\infty \mu(B_{\Lambda,q^+}^c),$$

where

$$(51) \qquad B_{\Lambda,q^+} = \{q \in \Omega \mid q_i \leq q_i^+, \text{ all } i \in \Lambda^c\} = \bigcap_{i \in \Lambda^c} \{q \mid q_i \leq q_i^+\}.$$

From the bound in Lemma 7.3.20 we obtain

$$(52) \quad \mu(B_{\Lambda,q^+}^c) \leq \sum_{i \in \Lambda^c} \mu(\{q \mid q_i > q_i^+\}) \leq \sum_{i \in \Lambda^c} \frac{1}{q_i^+} E_\mu(|q_i|) \leq \sum_{i \in \Lambda^c} \frac{c}{q_i^+}.$$

It follows from (47) that $\mu(B_{\Lambda,q^+}^c)$ can be made arbitrarily small by choosing Λ large.

Since Φ is attractive we have on B_{Λ,q^+} that $E_\Lambda^q{}^c(F) \le E_\Lambda^{q^+}{}^c(F)$, hence $\int_{B_{\Lambda,q^+}} E_\Lambda^q{}^c(F) \, d\mu(q) \le E_\Lambda^{q^+}{}^c(F)$. We conclude that given any $\varepsilon > 0$ there is a "large enough" finite $\Lambda \subseteq \mathbb{Z}^d$ such that

$$(53) \qquad E_\mu(F) \le E_\Lambda^{q^+}{}^c(F) + \varepsilon.$$

Passing to the hyperfinite picture we see that $E_\Gamma^{q^+}{}^c(F)$ is the internal expectation with respect to the measure μ_{Γ,q^+}. From (48) and standard Loeb theory we conclude that

$$(54) \qquad E_\mu(F) \le E_{\mu_+}(F)$$

for all tempered Gibbs measures μ.

7.3.21. THEOREM. Let Φ be an attractive and uniformly tempered interaction of finite range satisfying the inequality of (43); then there is a unique maximal Gibbs measure μ_+ and a unique minimal Gibbs measure μ_- with respect to Φ on Ω.

We conclude our story here. The global Markov property of $\mu_\pm$ has been proved by Zegarlinski (1984), using condition C; we invite the reader to look at this proof from the hyperfinite point of view; see Definition 7.3.6.

7.3.22. REMARK. The discussion of lattice models with "unbounded spins" has been largely motivated by the study of the corresponding problems for continuous quantum fields. In fact, existence, uniqueness, and the global Markov property were first proved for quantum field models (with trigonometric interactions) by Albeverio and Høegh-Krohn (1979). The case of lattice models with unbounded spins was analyzed under the Dobrushin uniqueness condition by Bellissard and Picco (1979). This was extended to more general settings by Bellissard and Høegh-Krohn (1982) and Zegarlinski (1984). For some recent work in the continuum quantum field case see Gielerak (1983), Zegarlinski (1985), and Röckner (1985). We invite the reader to look at this work in the context of the hyperfinite models with infinitesimal spacing that we construct in the next section.

7.4. HYPERFINITE MODELS FOR QUANTUM FIELD THEORY

In this section we will present the Euclidean quantum field theory as a continuous spin system on a hyperfinite lattice with infinitesimal spacing. We shall use this formulation in the next section to discuss the Φ_d^4 polymer representation.

A. The Program

We shall start by outlining the standard probabilistic construction of the free Euclidean field. This will be a special kind of Gaussian random field.

7.4.1. DEFINITION. Let $(Q, \mathcal{B}, \mu)$ be a probability space and H a real Hilbert space. The (unique) *Gaussian random field indexed by* H is the map $\Phi: H \to L^2(Q, d\mu)$ satisfying

   (i). Φ is linear;
   (ii) $\{\Phi(v) | v \in H\}$ is full, i.e., generates the measure algebra;
   (iii) each $\Phi(v)$ is a Gaussian random variable; and
   (iv) $\langle \Phi(v)\Phi(w) \rangle_{L^2} = \frac{1}{2}\langle v, w \rangle_H$.

A few comments may be in order: the measure algebra in (ii) is, of course, $\mathcal{B}/\mathcal{I}_0$, where $\mathcal{I}_0$ is the σ-ideal of sets of measure zero; fullness is clearly a necessary condition for uniqueness. A measurable function $f: Q \to \mathbb{R}$ determines a measure μ_f on $\mathbb{R}$ by the equation $\mu_f(B) = \mu(f^{-1}(B))$, and f is called a *Gaussian random variable* if μ_f has the form

$$(1) \qquad d\mu_f(x) = (2\pi a)^{-1/2} \exp(-\tfrac{1}{2}a^{-1}x^2)\, dx,$$

where $a = \int f^2\, d\mu$ is the variance of f. (Note that we only consider Gaussian random variables with mean 0.) Finally, note that $\langle\, , \rangle_H$ is the inner product in H and that $\langle \Phi(v)\Phi(w) \rangle_{L^2} = \int \Phi(v)\Phi(w)\, d\mu$.

We fix some terminology. Given the Hilbert space H we know that the Gaussian random field indexed by H is unique up to measure isomorphisms. We therefore write Q_H for the underlying measure space and $d\mu_{0,H}$ for the measure.

It is now possible to give a quick description of the free Euclidean field. First let N_m be the Hilbert space of all real distributions $f \in S'(\mathbb{R}^d)$ whose Fourier transforms are functions with finite norm, assuming $m > 0$ if $d = 1, 2$ and $m \geq 0$ if $d \geq 3$,

$$(2) \qquad \|f\|_{N_m}^2 = 2 \int_{\mathbb{R}^d} \frac{|\hat{f}(k)|^2\, dk}{k^2 + m^2} < \infty.$$

In the true physical case d, which is the number of space-time dimensions, is equal to 4.

7.4.2. DEFINITION. The *free Euclidean field of mass* m is the Gaussian random process Φ_0 indexed by N_m.

Let $d\mu_0$ denote the measure $d\mu_{0,N_m}$. The *Schwinger functions* S_n associated with the free Euclidean field Φ_0 are given by

$$(3) \qquad S_n(f_1, \ldots, f_n) = \int \Phi_0(f_1) \cdots \Phi_0(f_n)\, d\mu_0.$$

In the free case where only the mean (which we always take to be zero) and the covariance matter, we may restrict our interest to the free *two-point* functions S_2. We note that

$$(4) \qquad S_2(f, g) = \tfrac{1}{2}\langle f, g \rangle_{N_m} = \langle f, (-\Delta + m^2)^{-1}g \rangle_{L^2}.$$

From the Euclidean point of view the Schwinger functions are nothing but the moments of a certain random field. It may therefore not be unreasonable if we briefly explain how these entities arise from the physical theory of quantum fields. Classically a field Φ_F is a real-valued function. Thus in the quantum case the basic object should be an operator-valued function. But $\Phi_F(x)$, the field at the space-time point x, may be a too singular object; we are therefore led to the following point of view.

A *Hermitian (boson) scalar quantum field* is an operator-valued distribution Φ_F. The *state space* is some suitable separable Hilbert space H with a distinguished state Ω, the *vacuum*. The *observables* are the operators $\Phi_F(f), f \in S(\mathbb{R}^d)$. Various properties of relativistic invariance, microscopic causality, and regularity of the field must be assumed. It will suffice, however, for our coarse sketch just to introduce the *dramatis personae* but to leave their canonical text, i.e., the Wightman–Gårding axioms, unspoken.

We shall only focus for a moment on the following objects of the physical theory, the *Wightman distributions* or the *vacuum expectation values*:

$$(5) \qquad W_n(f_1, \ldots, f_n) = (\Omega, \Phi_F(f_1) \cdots \Phi_F(f_n)\Omega).$$

The point is that from the Wightman distribution we can pass to the Schwinger functions in the following way. A point $(z_1, \ldots, z_n) \in \mathbb{C}^{dn}$ is called *Euclidean* if each z_j is of the form

$$(6) \qquad z_j = (is_j, \mathbf{x}_j)$$

where s_j and the $\mathbf{x}_j$'s are real (points with pure imaginary time). With z_j we associate $y_j = (s_j, \mathbf{x}_j) \in \mathbb{R}^d$ and we parameterize the Euclidean points in $\mathbb{C}^{dn}$ by vectors $y = (y_1, \ldots, y_n) \in \mathbb{R}^{dn}$. We denote by $\mathscr{E}_n$ the set of *noncoincident Euclidean points*, i.e., Euclidean points $z = (z_1, \ldots, z_n)$ such that $y_i - y_j \neq 0$, $i \neq j$, for the associated y's.

A fundamental result of the Wightman–Gårding theory is that the Wightman functions can be extended to a domain in $\mathbb{C}^{dn}$ which includes the set $\mathscr{E}_n$; or, in somewhat more precise terms, the Wightman distribution W_n is the boundary value of an analytic function, which we also denote by W_n, defined on a domain in $\mathbb{C}^{dn}$ which includes the set $\mathscr{E}_n$.

7.4.3. DEFINITION. The restriction of W_n to $\mathscr{E}_n$ is called the *n-point Schwinger function*.

We shall write S_n for the Schwinger functions viewed as functions of the associated y's. The main point of this story is that one can write down a set of properties for the Schwinger functions which suffices to reconstruct the underlying physical field; i.e., given a set $\{S_n\}$ of functions satisfying these properties there exists an essentially unique Wightman–Gårding field Φ_F such that the Schwinger functions of this field are precisely the given functions S_n. And the story is completed by noting that the Schwinger functions as introduced in (3) satisfy the assumptions of *the reconstruction theorem*; thus the free Euclidean field via the Schwinger functions (3) gives us a model of a (noninteracting) scalar quantum field.

This is in outline one version of the general story; see Simon (1974) on how to fill in the details. But the Euclidean field can also be viewed in the following way; see, e.g., Glimm and Jaffe (1981). It follows from the Minlos theorem that it is possible to represent the free measure μ_0 as a probability measure on the distribution space $S'(\mathbb{R}^d)$; i.e., we can always take $Q_{N_m} = S'(\mathbb{R}^d)$. In this case the field as a map $\Phi_0: N_m \to L^2(Q_{N_m})$ can be represented as

$$(7) \qquad \Phi_0(f)(T) = T(f),$$

where $f \in S(\mathbb{R}^d)$ and $T \in S'(\mathbb{R}^d)$. And we may equally well read off the properties of the measure μ_0 in the generating functional

$$(8) \qquad S\{f\} = \int e^{iT(f)} \, d\mu_0(T)$$

as in the Schwinger or moment functions

$$(9) \quad S_n(f_1, \ldots, f_n) = \int \Phi_0(f_1) \cdots \Phi(f_n) \, d\mu_0 = \int T(f_1) \cdots T(f_n) \, d\mu_0(T).$$

We are led to the following general definition of a Euclidean random field.

7.4.4. DEFINITION. An *Euclidean field theory* is given by a probability measure μ on (the σ-algebra generated by the cylinder sets of) $S'(\mathbb{R}^d)$ whose generating functional

$$S\{f\} = \int e^{iT(f)} \, d\mu(T)$$

satisfies the following properties:

ANALYTICITY. The functional $S\{f\}$ is entire and analytic.

REGULARITY. For some p, $1 \le p \le 2$, some constant c, and all $f \in S(\mathbb{R}^d)$,

$$|S\{f\}| \le \exp c(\|f\|_1 + \|f\|_p^p).$$

INVARIANCE. $S\{f\}$ is invariant under Euclidean symmetries (translations, rotations, and reflections).

OSTERWALDER–SCHRADER REFLECTION POSITIVITY. Let $\mathbb{R}_+^d$ be the set of points $(t, \mathbf{x}) \in \mathbb{R}^d$ such that $t > 0$ and $\mathcal{A}_+$ the set of functions of the form

$$\sum_{j=1}^n c_j \exp(T(f_j)),$$

where $c_j \in \mathbb{C}$, $f_j \in C_0^\infty(\mathbb{R}_+^d)$, and $T \in S'(\mathbb{R}^d)$; then $(\theta A, A) \geq 0$, where $(\,,\,)$ is the $L^2(d\mu)$ inner product, $A \in \mathscr{A}_+$, and θ is time reflection, $\theta: (t, \mathbf{x}) \mapsto (-t, \mathbf{x})$.

ERGODICITY. For all $L^1(d\mu)$ functions F one has

$$\int F \, d\mu = \lim_{t \to \infty} \frac{1}{t} \int_0^t F(T_s) \, ds,$$

where $T_s(s', \mathbf{x}) = T(s' + s, \mathbf{x})$, for all $(s', \mathbf{x}) \in \mathbb{R}^d$.

The list of properties may look formidable, but each translates back to some physically meaningful property for the Wightman-Gårding field Φ_F. Euclidean invariance translates into relativistic invariance of Φ_F; ergodicity ensures the uniqueness of the vacuum state Ω; and the reflection positivity allows us to reconstruct a Hamiltonian of the physical field. Finally, some regularity must be imposed to control the physical theory, hence the first two properties of the definition.

The free Euclidean field as constructed in 7.4.2 is easily seen to satisfy the properties of 7.4.4. But it is of rather limited physical interest; there is no particle interaction. To rectify this situation we must search for some non-Gaussian random field which in some suitable way incorporates an interaction potential U.

We shall try to do this in the spirit of Definition 7.4.4 by keeping the process Φ_0 and the underlying space Q_{N_m} of the free Euclidean field but replacing the free measure $d\mu_0$ by an interaction measure

$$(10) \qquad\qquad dv = \frac{\exp(-U) \, d\mu_0}{\int \exp(-U) \, d\mu_0},$$

where U is the added *interaction potential*. Provided dv can be given a meaning, we can immediately write down the associated generating functional or the associated Schwinger functions and hope to reconstruct a physically nontrivial theory.

B. Free Scalar Fields

To construct a physically nontrivial theory is no small task; one way of approaching it would be to start with a *free lattice field*.

Let $\delta > 0$ be a fixed positive real number and define the lattice $\mathscr{L}_\delta$ with spacing or mesh δ to be the set

$$(11) \qquad\qquad \mathscr{L}_\delta = \{n\delta \mid n \in \mathbb{Z}^d\}.$$

On this space we consider the function space $l^2(\mathscr{L}_\delta)$ with the norm

$$(12) \qquad\qquad \|f\|_2^2 = \sum_{n \in \mathbb{Z}^d} \delta^d |f(n\delta)|^2 < \infty.$$

With our hyperfinite background we understand the role of the normalization constant δ^d; when δ is infinitesimal $\|f\|_2^2 \approx \int |f|^2 \, dx$.

It follows from (4) that the free Euclidean field has a "covariance matrix" $(-\Delta + m^2)^{-1}$; for the lattice approximation we shall have to study the discretization of this operator. As is well known from previous chapters $-\Delta_\delta$ on $l^2(\mathscr{L}_\delta)$ is given by the expression

$$(13) \qquad (-\Delta_\delta f)(n\delta) = \delta^{-2}\left[2df(n\delta) - \sum_{|n'-n|=1} f(n'\delta)\right].$$

We introduce the matrix C by the definition

$$(14) \qquad C_{n,n'} = \delta^{-d}(-\Delta_\delta + m^2)^{-1}_{n,n'}.$$

Let $\mathscr{T}_\delta = [-\pi/\delta, \pi/\delta]^d$ be the dual space of $\mathscr{L}_\delta$. We have the following explicit formula:

$$(15) \qquad C_{n,n'} = (2\pi)^{-d} \int_{\mathscr{T}_\delta} e^{ik(n-n')\delta} \mu_\delta(k)^{-2} \, dk,$$

where

$$(16) \qquad \mu_\delta(k)^2 = \delta^{-2}\left[2d - 2\sum_{i=1}^{d} \cos(k_i\delta)\right] + m^2.$$

Note that for $\delta \approx 0$ and k finite we have $\mu_\delta(k)^2 \approx \mu(k)^2 = k^2 + m^2$.

The free lattice field can be defined with respect to the full lattice $\mathscr{L}_\delta$. But from many points of view it is as natural to start with a "finite space cutoff." Let Λ be a bounded region in $\mathbb{R}^d$.

7.4.5. DEFINITION. With $\Lambda \subseteq \mathbb{R}^d$ we associate

  (i) $\Lambda_\delta = \Lambda \cap \mathscr{L}_\delta$;
  (ii) $\Lambda_\delta^{\text{int}} = \{n\delta \in \Lambda_\delta \mid \forall m(|n - m| = 1 \rightarrow m\delta \in \Lambda_\delta)\}$; and
  (iii) $\partial\Lambda_\delta = \Lambda_\delta - \Lambda_\delta^{\text{int}}$.

If M is any matrix indexed by $\mathscr{L}_\delta$ we denote by M^Λ its restriction to Λ_δ. We say that a matrix M is concentrated on $\partial\Lambda_\delta$ if $M_{m,n} \neq 0$ implies that $m\delta, n\delta \in \partial\Lambda_\delta$.

For the matrix C we have the following important result:

$$(17) \qquad [C^\Lambda]^{-1} = [C^{-1}]^\Lambda - B_{\partial\Lambda_\delta},$$

where $B_{\partial\Lambda_\delta}$ is a matrix concentrated on $\partial\Lambda_\delta$ with non-negative elements.

We shall now discuss a version of the free Euclidean field in the lattice Λ_δ. Note that Λ_δ is a finite set; let $l = |\Lambda_\delta|$ be the number of elements in

Λ_δ. Our truncated *measure space* will be the finite product $Q_{\Lambda_\delta} = \mathbb{R}^{\Lambda_\delta}$ with the *free measure*

(18) $d\mu_{0,\Lambda_\delta}(q) = (2\pi)^{-1/2}[\det(C^\Lambda)]^{-1/2}$

$$\times \exp\left(-\frac{1}{2} \sum_{\substack{n\delta, \\ n'\delta \in \Lambda_\delta}} (C^\Lambda)^{-1}_{n,n'} q_{n\delta} q_{n'\delta}\right) dq,$$

$q \in Q_{\Lambda_\delta}$, and $dq = \prod_{n\delta \in \Lambda_\delta} dq_{n\delta}$.

7.4.6. DEFINITION. The random field Φ_δ indexed by Λ_δ is the map

$$\Phi_\delta : \Lambda_\delta \times Q_{\Lambda_\delta} \to \mathbb{R}$$

given by

$$\Phi_\delta(n)(q) = q_{n\delta}, \qquad n\delta \in \Lambda_\delta.$$

It is called *the free lattice field of mass m* in Λ_δ.

Using formula (17) for $(C^\Lambda)^{-1}$, we get the expression

(19) $\displaystyle\sum_{n\delta, n'\delta \in \Lambda_\delta} (C^\Lambda)^{-1}_{n,n'} q_{n\delta} q_{n'\delta} = (2d\delta^{d-2} + m^2\delta^d) \sum_{n\delta \in \Lambda_\delta} q^2_{n\delta}$

$$- \sum_{\substack{|n-n'|=1 \\ n\delta, n'\delta \in \Lambda_\delta}} \delta^{d-2} q_{n\delta} q_{n'\delta} - \sum_{n\delta, n'\delta \in \partial\Lambda_\delta} B_{n,n'} q_{n\delta} q_{n'\delta}.$$

We see that the free measure $d\mu_{0,\Lambda_\delta}$ describes an Ising model with continuous Gaussian spins and nearest-neighbor interaction between spins inside Λ_δ but with extra couplings on the boundary.

We shall now explain how the free lattice field of Definition 7.4.6 is related to the free Euclidean field constructed in 7.4.2. The trick is to represent the random variables $\Phi_\delta(n)$, $n\delta \in \Lambda_\delta$, on the space Q_{N_m} by setting

(20) $\Phi_\delta(n) = \Phi_0(f_{n\delta})$,

where $f_{n\delta}$ is the function on $\mathbb{R}^d$ with Fourier transform satisfying

(21) $\hat{f}_{n\delta}(k) = \begin{cases} e^{-ikn\delta} \mu(k)/2\pi\mu_\delta(k), & \text{if } k \in \mathcal{T}_\delta, \\ 0, & \text{otherwise.} \end{cases}$

An explicit calculation shows that $\int \Phi_\delta(f_{n\delta}) \, d\mu_0 = 0$ and

(22) $\langle \Phi_\delta(f_{n\delta}) \Phi_\delta(f_{n'\delta}) \rangle_{L^2} = \frac{1}{2}\langle f_{n\delta}, f_{n'\delta}\rangle_{N_m} = C_{nn'}$,

with L^2 standing for $L^2(d\mu_0)$. Let $g \in C_0^\infty(\mathbb{R}^d)$ have support in the interior of $\Lambda \subseteq \mathbb{R}^d$. We set

(23) $\Phi_\delta(g) = \displaystyle\sum_{n\delta \in \Lambda_\delta} \delta^d g(n\delta) \Phi_\delta(n)$.

Using the representation $\Phi_\delta(n) = \Phi_\delta(f_{n\delta})$ one may show [see Simon (1974)] that $\Phi_\delta(g)$ converges to $\Phi_0(g)$ as $\delta \downarrow 0$ in each $L^p(Q_{N_m}, d\mu_0)$, $1 \le p < \infty$. Thus we can look upon the free lattice field as an approximation "from the inside" to the free Euclidean field.

We now pass to *the hyperfinite picture*, which will approximate the free Euclidean field "from the outside." Let $\delta > 0$ be infinitesimal and let Λ_δ be a hyperfinite lattice in $*\mathbb{R}^d$ with spacing δ. We use transfer on Definition 7.4.6, to construct the hyperfinite free field.

7.4.7. DEFINITION. Let $(*\mathbb{R}^{\Lambda_\delta}, L(\mathscr{B}), L(\mu_{0,\Lambda_\delta}))$ be the Loeb space defined from the internal Borel algebra $\mathscr{B}$ on $*\mathbb{R}^{\Lambda_\delta}$ and the internal measure μ_{0,Λ_δ} obtained by transfer on (18). *The hyperfinite lattice field* is the random field

$$\Phi_\delta : \Lambda_\delta \times *\mathbb{R}^{\Lambda_\delta} \to *\mathbb{R}$$

given by

$$\Phi_\delta(n)(q) = q_{n\delta},$$

for $n\delta \in \Lambda_\delta$ and $q \in *\mathbb{R}^{\Lambda_\delta}$.

We have the following regularity result.

7.4.8. LEMMA. Let $g \in C_0^\infty(\mathbb{R}^d)$; then

(24) $$\Phi_\delta(*g) = \sum \delta^d \, *g(n\delta)\Phi_\delta(n)$$

is nearstandard on a set of $L(\mu_{0,\Lambda_\delta})$ measure one in $*\mathbb{R}^{\Lambda_\delta}$.

For the proof we observe that we have the following formula for the covariance with respect to the internal measure μ_{0,Λ_δ}

(25) $$E(\Phi_\delta(*g)\Phi_\delta(*g)) = \sum \delta^{2d} \, *g(n\delta)\,*g(n'\delta)C_{nn'},$$

and this is finite since g has compact support.

Let D be a countable set of $C_0^\infty(\mathbb{R}^d)$ functions that is dense in N_m. It follows from the lemma that there is a set $\Omega \subseteq *\mathbb{R}^{\Lambda_\delta}$ of $L(\mu_{0,\Lambda_\delta})$-measure one such that $\Phi_\delta(*g)(q)$ is nearstandard for all $g \in D$ and all $q \in \Omega$.

7.4.9. THEOREM. The random field Φ_δ indexed by Λ_δ, where δ is a positive infinitesimal and Λ_δ is hyperfinite, is a "realization" of the free Euclidean field on the measure space $(*\mathbb{R}^{\Lambda_\delta}, L(\mathscr{B}), L(\mu_{0,\Lambda_\delta}))$.

What we mean, quite precisely, is that $\mathrm{st}(\Phi_\delta(*g))$, for $g \in D$, is a set of Gaussian random variables with mean zero and with the correct covariance,

$$\langle \mathrm{st}(\Phi_\delta(*g)), \mathrm{st}(\Phi_\delta(*g)) \rangle_{L^2} = \tfrac{1}{2}\langle g, g \rangle_{N_m}.$$

Since D is dense in N_m we can extend this equality to all of the latter space. Thus $\Phi_0(g)$ for $g \in N_m$ can be represented by $\mathrm{st}(\Phi_\delta(*g))$.

The proof consists in taking standard parts in formula (25); just as in Section 6.2 this gives us a standard integral with respect to the resolvent kernel of the Laplacian. Then use formula (4) above.

7.4.10. REMARK. We see that the representation in Theorem 7.4.9 is independent of the choice of infinitesimal δ and hyperfinite lattice Λ_δ. This corresponds to the convergence of $\Phi_\delta(g)$ to $\Phi_0(g)$ as $\delta \downarrow 0$ in the case of finite lattice approximations; see the discussion in connection with formula (23).

We have thus constructed the free field on a hyperfinite lattice. Formula (24) gives the field as an internal construction. The noteworthy aspect here is that the field is *pointwise defined*; $\Phi_\delta(n)$ makes computational sense for all points $n\delta \in \Lambda_\delta$. This is not possible in the standard approach of Definition 7.4.2. One cannot give a coherent sense to $\Phi_\delta(f_{n\delta})$ for δ infinitesimal, for if $n\delta$ is chosen so that $x = \text{st}(n\delta)$ exists as a point in $\mathbb{R}^d$, then $\Phi_\delta(f_{n\delta})$ would correspond to the field at the point x. But this is, in general, a nonexisting entity.

Thus in the standard approach one is led to the concept of a random field indexed by some space of distributions. In the hyperfinite version we use the sites of the lattice Λ_δ as our primary index set and extend by using formula (24). We will still have "infinities," but they can be controlled through a consistent algebra, hence lead to unambiguous and meaningful results.

REMARK. $(*\mathbb{R}^{\Lambda_\delta}, L(\mathcal{B}), L(\mu_{0,\Lambda_\delta}))$ is a hyperfinite realization of a *generalized random field*. Kessler (1984) has developed a general hyperfinite approach to distribution-valued random fields. It seems that the nonstandard approach offers certain technical advantages; in particular, when the underlying probability space is a Loeb space it turns out that the Gelfand and Urbanik characterizations of generalized fields of order n are equivalent; see Kessler's paper for full details.

While we are digressing, let us also take the opportunity to insert a brief note on the status of generalized functions and distribution theory versus nonstandard analysis. That nonstandard analysis can be a suitable framework for the study of distributions was realized quite early, e.g., with the Dirac δ-function $\delta(x)$ represented as $(2\pi\varepsilon)^{-d/2}e^{-|x|^2/2\varepsilon}$, ε infinitesimal (Robinson, 1966; Laugwitz, 1978). More generally, the definition of distributions as limits of sequences [Mikusinski and Sikorski (1973)] is easily taken over in nonstandard theory, replacing sequences and limits by infinitesimals and standard parts; see, e.g., Richter (1982). However, a systematic study, especially in connection with partial differential equations, is lacking [see, however, Li Bang-He (1978) for distributions defined in a nonstandard way through analytic functions]. In Section 6.5 we saw that generalized solutions in the nonstandard sense might indeed give new insights in problems of partial differential equations, where generalized functions in the standard sense failed. Kessler has developed a quite systematic approach to distributions by using a hyperfinite approach. Formula (24) above gives a hint; the distribution $\Phi_0(g)$, formally given by $\int \Phi_0(x)g(x)\,dx$, is realized as the standard part of

$$\sum \delta^{d\,*}g(n\delta)\Phi_\delta(n),$$

which makes perfect sense. More generally, Kessler replaces test functions, distributions, and differential operators by their hyperfinite analogs. In particular, elements of the distribution space $\mathscr{D}'$ are lifted to internal functions on a hyperfinite lattice Γ, S-dense in $\mathbb{R}^d$. The result mentioned above about n-order generalized random fields is but a special case of the nice control on generalized functions which can be achieved by using the hyperfinite picture.

REMARK.   There have been other attempts at using nonstandard methods to discuss quantum fields, e.g., Blanchard and Tarski (1978), Fittler (1984), Kelemen and Robinson (1972) and Nagamachi and Nishimura (1984).

C. Interacting Scalar Fields

Enough has now been said about the free field; the time has come to return to interactions. We shall follow the program hinted at in connection with formula (10).

One way of obtaining interactions is to construct suitable "local additive functionals" of the free field; ultimately we shall use our hyperfinite realization of Definition 7.4.7 to do this. But first we outline the standard procedure. Let $\delta > 0$ be a standard real and consider the lattice Λ_δ obtained from a bounded domain $\Lambda \subseteq \mathbb{R}^d$. Let g be a positive function with support in Λ and let u_δ be any continuous real function. We will study interactions of the form

$$(26) \qquad U_g^\delta = \lambda_\delta \sum \delta^d g(n\delta) u_\delta(\Phi_\delta(n)),$$

where λ_δ is a real constant, the "coupling constant." In accordance with (10) we introduce

$$(27) \qquad d\mu_{g,\Lambda_\delta} = \frac{\exp(-U_g^\delta)\,d\mu_{0,\Lambda_\delta}}{\int \exp(-U_g^\delta)\,d\mu_{0,\Lambda_\delta}}$$

We are in the finite case, and under a number of reasonable conditions μ_{g,Λ_δ} turns out to be a well-defined probability measure.

The function g in (26) represents a kind of "space cutoff"; e.g., it could be the characteristic function of some domain $\Lambda_\delta \subseteq \Lambda$. In order to obtain a nontrivial field we now let δ tend to zero while at the same time letting $\Lambda_\delta \uparrow \mathbb{R}^d$. To remove the space cutoff we let g converge to the constant function 1 on $\mathbb{R}^d$.

In the hyperfinite version this means that we want to choose $\delta > 0$ infinitesimal, Λ_δ a hyperfinite lattice, and g an internal function such that $g(n\delta) = 1$ for all nearstandard $n\delta \in \Lambda_\delta$. By transfer (26) and (27) still make sense. But we would like to extract from the internal construct μ_{g,Λ_δ} a non-Gaussian measure satisfying the requirements of Definition 7.4.4.

We shall discuss this problem in several steps. First we make some calculations in the truly finite case. Next we choose $\delta > 0$ infinitesimal and Λ_δ hyperfinite but keep a cutoff function g of compact support. We then retreat to the finite case to establish some inequalities, which in the final stage are used to remove the space cutoff in the hyperfinite model.

This is the general program which we now shall discuss in some detail in the case of exponential interaction where we choose $u(y) = \exp(\alpha y)$, independent of δ, for some real parameter α; see Albeverio and Høegh-Krohn (1974).

We start out by doing the following calculation, where $\delta \in \mathbb{R}_+$, Λ_δ is a finite lattice, and $g \geq 0$ has support in Λ:

$$(28) \qquad \int (U_g^\delta)^2 \, d\mu_{0,\Lambda_\delta} = \lambda_\delta^2 \sum_{n\delta, n'\delta \in \Lambda_\delta} \delta^{2d} g(n\delta) g(n'\delta)$$

$$\times \int_{\mathbb{R}^{\Lambda_\delta}} e^{\alpha(\Phi_\delta(n) + \Phi_\delta(n'))} \, d\mu_{0,\Lambda_\delta}.$$

From well-known properties of Gaussian integrals we see that

$$(29) \qquad \int_{\mathbb{R}^{\Lambda_\delta}} e^{\alpha(\Phi_\delta(n) + \Phi_\delta(n'))} \, d\mu_{0,\Lambda_\delta} = (A_\alpha^\delta)^2 e^{\alpha^2 C_{nn'}^\delta},$$

where we have now put a suffix δ on C_{nn} [see (14)] to indicate the dependence on δ and where

$$(30) \qquad A_\alpha^\delta = \exp\left(\frac{\alpha^2}{2} C_{nn}^\delta\right) = \exp\left(\frac{\alpha^2}{2} (2\pi)^{-d} \int_{T_\delta} \mu_\delta(k)^{-2} \, dk\right).$$

Thus

$$(31) \qquad \int (U_g^\delta)^2 \, d\mu_{0,\Lambda_\delta} = \lambda_\delta^2 (A_\alpha^\delta)^2 \sum_{n\delta, n'\delta \in \Lambda_\delta} \delta^{2d} g(n\delta) g(n'\delta) e^{\alpha^2 C_{nn'}^\delta}.$$

And since we are in the finite case everything is well defined.

Passing to the next step, we now choose $\delta \approx 0$ and Λ_δ as a hyperfinite lattice, but we keep a cutoff g of compact support. By transfer (31) still

makes sense, but it may be infinite. We must determine for which values
of d, α, and λ_δ the right-hand side of (31) is nearstandard.

For $d = 1$, the sum

$$(32) \qquad \sum_{n\delta, n'\delta \in \Lambda_\delta} \delta^{2d} {}^*g(n\delta)^*g(n'\delta)\, e^{\alpha^2 C_{nn'}^\delta}$$

is finite for all α since $C_{nn'}^\delta$ is finite for all n, n' and g has compact support.

For $d = 2$, $C_{nn'}^\delta$ is modulus finite terms, $-(1/2\pi)\ln|n\delta - n'\delta|$ when
$|n\delta - n'\delta| \approx 0$. Thus if $\alpha^2 < 4\pi$ and g is of compact support the sum (32)
is finite.

For $d \geq 3$ or $d = 2$ and $\alpha^2 \geq 4\pi$ the sum is not finite. [For more analytic
details on this point the reader may consult Albeverio and Høegh-Krohn
(1974).]

Thus in the case $d = 1$ or $d = 2$ and $\alpha^2 < 4\pi$ the right-hand side of (31)
would be finite if $\lambda_\delta^2(A_\alpha^\delta)^2$ were finite. We observe from (30) that A_α^δ is
infinite for $\delta \approx 0$ and $d = 2$. But we have freedom of choice; the coupling
constant λ_δ has so far been left unspecified. So let us choose

$$(33) \qquad \lambda_\delta = \lambda (A_\alpha^\delta)^{-1},$$

with $\lambda \in \mathbb{R}_+$, independent of δ, α. With this choice the right-hand side of
(31) is nearstandard. And it is not difficult to see that the standard part is

$$(34) \qquad \lambda^2 \int_{\mathbb{R}^d} g(x)g(y)\, e^{\alpha^2 G(x-y)}\, dx\, dy,$$

where $G(x - y)$ is the kernel of the operator $(-\Delta + m^2)^{-1}$; for similar
arguments see Section 6.2 and Theorem 7.4.9.

Thus we may conclude that U_g^δ is a positive function which has finite
$L^2(d\mu_{0,\Lambda_\delta})$ norm for the above choice of λ_δ; moreover, we easily see that
$\mathrm{st}(U_g^\delta) \neq 0$. This shows that for $d = 1$ and all $\alpha \in \mathbb{R}$ or $d = 2$ and $\alpha^2 < 4\pi$,
the function $\exp(-U_g^\delta)$ is nearstandard and not identically one. We thus
have the following result:

7.4.11. THEOREM. For $d = 1$ and all $\alpha \in \mathbb{R}$, or $d = 2$ and $\alpha^2 < 4\pi$, the
Loeb measure $L(\mu_{g,\Lambda_\delta})$ associated with the internal measure μ_{g,Λ_δ} as defined
in (27), and with coupling constant for $d = 2$ chosen as in (33), is absolutely
continuous with respect to the free field measure $L(\mu_{0,\Lambda_\delta})$ for all g of
compact support, the Radon–Nikodym derivative being the L^∞ function

$$\mathrm{st}\left(e^{-U_g^\delta} \Big/ \int e^{-U_g^\delta}\, d\mu_{0,\Lambda_\delta}\right).$$

We have thus completed the second stage of the program. Before we
proceed we note that the interaction U_g^δ is more commonly written

$$U_g^\delta = \lambda \sum \delta^d g(n\delta) : e^{\alpha \Phi_\delta(n)} :,$$

where $:e^{\alpha\Phi_\delta(n)}: = (A_\alpha^\delta)^{-1} e^{\alpha\Phi_\delta(n)}$ is the so-called Wick renormalization of $e^{\alpha\Phi_\delta(n)}$. In the Wick notation the identity in (29) is nothing but an instance of the identity

$$\langle :\exp \alpha f: :\exp \beta g: \rangle = \exp(\alpha\beta\langle fg\rangle).$$

We have chosen a more direct approach to highlight the choice of infinitesimal coupling constant in (33).

7.4.12. REMARK. One can obviously replace U_g^δ above by $\tilde{U}_g^\delta = \int U_g^\delta \, d\nu(\alpha)$ for any positive finite measure ν [with support in $(-\sqrt{4\pi}, \sqrt{4\pi})$ for $d = 2$]. This is the case that was studied by standard methods in Albeverio and Høegh-Krohn (1974); see also the references in Albeverio and Høegh-Krohn (1984b).

We now return to the discrete case in order to establish certain inequalities of crucial importance for the final space cutoff removal. Let $\delta \in \mathbb{R}_+$, let Λ_δ be a finite lattice, and let g have support in Λ. We shall use the interaction $\tilde{U}_g^\delta$, where in order to simplify matters we now choose a measure ν satisfying $\nu(\alpha) = \nu(-\alpha)$. The Schwinger functions associated with the measure μ_{g,Λ_δ} are given by

$$(35) \quad S_g^\delta(n_1\delta, \ldots, n_k\delta) = \int \Phi_\delta(n_1) \cdots \Phi_\delta(n_k) \, d\mu_{g,\Lambda_\delta}$$

$$= \left(\int \exp(-\tilde{U}_g^\delta) \, d\mu_{0,\Lambda_\delta} \right)^{-1}$$

$$\times \int \Phi_\delta(n_1) \cdots \Phi_\delta(n_k) \exp(-\tilde{U}_g^\delta) \, d\mu_{0,\Lambda_\delta}.$$

Let g, g' both have support in Λ; assume that supp $g \subseteq$ supp g' and that $g = g'$ on supp g. We want to show that

$$(36) \quad S_{g'}^\delta \le S_g^\delta.$$

In order to do this, introduce

$$(37) \quad S_{g',g,\beta}^\delta(n_1\delta, \ldots, n_k k)$$

$$= \frac{\int \Phi_\delta(n_1) \cdots \Phi_\delta(n_k) \exp(-\beta(\tilde{U}_{g'}^\delta - \tilde{U}_g^\delta)) \exp(-\tilde{U}_g^\delta) \, d\mu_{0,\Lambda_\delta}}{\int \exp(-\beta(\tilde{U}_{g'}^\delta - \tilde{U}_g^\delta)) \exp(-\tilde{U}_g^\delta) \, d\mu_{0,\Lambda_\delta}}.$$

Note that $S_{g'}^\delta = S_{g',g,1}^\delta$ and $S_g^\delta = S_{g',g,0}^\delta$. Hence in order to establish (36) we must compute $\partial S_{g',g,\beta}^\delta / \partial\beta$ and show that $\partial S_{g',g,\beta}^\delta / \partial\beta \le 0$ for $0 \le \beta \le 1$. This will follow by an application of the Griffiths–Kelly–Sherman inequalities; e.g., see Simon (1974). In brief outline the argument goes as follows.

Referring to the explicit form of the measure $d\mu_{g,\Lambda_\delta}$ [see (27) and (18)], we may immediately conclude from the GKS inequalities that

$$(38) \quad \int \Phi_\delta(n_1) \cdots \Phi_\delta(n_k) \Phi_\delta(n)^l \, d\mu_{g,\Lambda_\delta}$$

$$\geq \left(\int \Phi_\delta(n_1) \cdots \Phi_\delta(n_k) \, d\mu_{g,\Lambda_\delta} \right) \int \Phi_\delta(n)^l \, d\mu_{g,\Lambda_\delta},$$

for all l. Then, by series expansion and the fact that the measure $d\nu$ is even, one shows that

$$(39) \quad \int \Phi_\delta(n_1) \cdots \Phi_\delta(n_k)(\tilde{U}_{g'}^\delta - \tilde{U}_g^\delta) \, d\mu_{g,\Lambda_\delta}$$

$$\geq \left(\int \Phi_\delta(n_1) \cdots \Phi_\delta(n_k) \, d\mu_{g,\Lambda_\delta} \right) \int (\tilde{U}_{g'}^\delta - \tilde{U}_g^\delta) \, d\mu_{g,\Lambda_\delta},$$

which is precisely what is needed to prove that $\partial S_{g',g,\beta}^\delta / \partial \beta \leq 0$. In a similar way it also follows from the GKS inequalities that

$$(40) \quad\quad\quad\quad\quad\quad\quad\quad S_g^\delta \geq 0$$

for all g of support in Λ.

We are now ready for the final stage. Let us once more move to the hyperfinite model where $\delta > 0$ is infinitesimal and Λ_δ is hyperfinite. By transfer, the moments S_g^δ are well defined with respect to the appropriate internal measure.

Let us first consider the case $g = 0$ on Λ_δ. In this case we get the Schwinger functions of the free field. In more detail, let $n_1 \delta$ and $n_2 \delta$ be finite points and assume that $x = \mathrm{st}(n_1\delta) \neq \mathrm{st}(n_2\delta) = y$, $x, y \in \mathbb{R}^2$. Then for $d = 2$,

$$(41) \quad\quad {}^\circ(S_0^\delta(n_1\delta, n_2\delta)) = \int {}^\circ(\Phi_\delta(n_1)\Phi_\delta(n_2)) \, dL(\mu_{0,\Lambda_\delta})$$

$$= S_0(x, y)$$

$$= (2\pi)^{-2} \int \frac{e^{ik(x-y)} \, dk}{k^2 + m^2}.$$

And, as is well known in the free case, an arbitrary Schwinger function is a sum of products of the "two-point functions" given in (41). Thus if we have noncoincident points $x_1 = \mathrm{st}(n_1\delta) \neq x_2 = \mathrm{st}(n_2\delta) \neq \cdots \neq x_k = \mathrm{st}(n_k\delta)$, then ${}^\circ S_0^\delta(n_1\delta, \ldots, n_k\delta)$ is well defined and finite.

Now let g_n be the characteristic function of a cube centered at the origin and with sides of length $2n$. By transfer we have for all n that

$$(42) \quad\quad\quad\quad\quad\quad 0 \leq S_{g_n}^\delta \leq S_0^\delta.$$

Let g_ω be an internal function which is the characteristic function of a cube of side length 2ω and with supp g_ω bounded by Λ_δ; i.e., if $g_\omega(n\delta) = 1$, then $n\delta \in \Lambda_\delta$. We see that $S^\delta_{g_\omega}$, as an internal moment function with respect to the internal measure $\mu_{g_\omega, \Lambda_\delta}$, will also satisfy the inequalities (42). Furthermore, we notice that if $n_i \delta \approx n_i' \delta$, $i = 1, \ldots, k$, and the points $\mathrm{st}(n_i\delta) \in \mathbb{R}^2$ for $n = 1, \ldots, k$ are noncoincident, then $S^\delta_{g_\omega}(n_1\delta, \ldots, n_k\delta) \approx S^\delta_{g_\omega}(n_1'\delta, \ldots, n_k'\delta)$. This follows from the explicit definition of $S^\delta_{g_\omega}$ and the fact that $g_\omega(n_i\delta) = g_\omega(n_i'\delta)$, $i = 1, \ldots, k$. Thus $S^\delta_{g_\omega}$ has a well-defined standard part, and it follows from the inequalities

(43) $$0 \le S^\delta_{g_\omega} \le S^\delta_{g_n}, \qquad \text{all } n,$$

that

(44) $$°(S^\delta_{g_\omega}(n_1\delta, \ldots, n_k\delta)) = \int °(\Phi_\delta(n_1) \cdots \Phi_\delta(n_k)) \, dL(\mu_{g_\omega, \Lambda_\delta})$$

$$= \lim_{n \to \infty} \int °(\Phi_\delta(n_1) \cdots \Phi_\delta(n_k)) \, dL(\mu_{*g_n, \Lambda_\delta})$$

$$= \lim_{n \to \infty} °(S^\delta_{*g_n}(n_1\delta, \ldots, n_k\delta)).$$

We may formulate this as a theorem.

7.4.13. THEOREM. Let $\delta > 0$ be infinitesimal and Λ_δ hyperfinite, and let g_ω be an internal function such that $g_\omega(n\delta) = 1$ for all finite $n\delta$. Then $L(\mu_{g_\omega, \Lambda_\delta})$ is a non-Gaussian probability measure on $(*\mathbb{R}^{\Lambda_\delta}, L(\mathcal{B}))$, where $\mathcal{B}$ is the internal Borel algebra on $*\mathbb{R}^{\Lambda_\delta}$. The Schwinger functions are finite and S-continuous for finite arguments and satisfy the inequalities

$$0 \le S^\delta_{g_\omega} \le S_0,$$

where the S_0 are the (standard) Schwinger functions of the free Euclidean field.

REMARK. In the hyperfinite picture we seem to effect the double limit passage in a single step by choosing $\delta > 0$ infinitesimal (continuum limit) and Λ_δ hyperfinite (infinite-volume limit). Does this mean that in nonstandard analysis one need not worry about the order in which one takes iterated limits? Not at all, as our discussion in Section 1.3 showed; in the present model, however, we need not worry because of the strong monotonicity properties of the associated Schwinger functions.

In Albeverio and Høegh-Krohn (1974) a nontrivial lower bound for $\mathrm{st}\, S^\delta_{g_\omega}$ is given. It remains to verify that the measure $L(\mu_{g_\omega, \Lambda_\delta})$ is non-Gaussian. We recall that if μ is Gaussian, then

(45) $$\langle f_1 \cdots f_{2n} \rangle_\mu = \sum \langle f_{i_1} f_{j_1} \rangle_\mu \cdots \langle f_{i_n} f_{j_n} \rangle_\mu,$$

where $\langle \cdot \rangle_\mu$ denotes expectation with respect to the measure μ and the sum is taken over all ways of writing $\{1, \ldots, 2n\}$ as $i_1, \ldots, i_n, j_1, \ldots, j_n$ with $i_1 < i_2 < \cdots < i_n$, $i_1 < j_1, \ldots, i_n < j_n$. To verify the non-Gaussian character of $\mu = L(\mu_{g_{\omega'}, \Lambda_\delta})$ it suffices to choose $n = 2$, all the f's equal to $\Phi_\delta(^*h)$, for some suitable h, and by a direct computation see that the identity (45) is violated; for details in the standard case see Albeverio and Høegh-Krohn (1979).

It remains to discuss the axioms of Definition 7.4.4. There are two ways to proceed. In the standard approach, where one does not have the hyperfinite model but must work with the approximations given by the sequence g_n, one defines a family of generating functionals by the equations

$$(46) \qquad S_n\{f\} = \int \exp(i \operatorname{st} \Phi_\delta(f)) \, dL(\mu_{g_n, \Lambda_\delta}), \qquad n = 1, 2, \ldots,$$

where $f \in C_0^\infty(\mathbb{R}^2)$. One then shows that

$$(47) \qquad\qquad S_\infty\{f\} = \lim_{n \to \infty} S_n\{f\}$$

exists. An application of the Minlos theorem yields a probability measure $d\mu_\infty$ on $S'(\mathbb{R}^2)$, and one is back to the setting of Definition 7.4.4.

But having the hyperfinite model and a well-defined measure $L(\mu_{g_\omega, \Lambda_\delta})$ on $\{f\}$, we may introduce the required generating functional $S\{f\}$ directly; we set

$$(48) \qquad\qquad S\{f\} = \int \exp(i \operatorname{st} \Phi_\delta(f)) \, dL(\mu_{g_\omega, \Lambda_\delta}),$$

where $f \in C_0^\infty(\mathbb{R}^2)$. Notice that Definition 7.4.4 makes perfect sense in this setting and we may proceed to verify the axioms.

7.4.14. THEOREM. The probability measure $L(\mu_{g_\omega, \Lambda_\delta})$ in the space $(^*\mathbb{R}^{\Lambda_\delta}, L(\mathscr{B}))$ defines a non-Gaussian Euclidean field theory; i.e., the generating functional

$$S\{f\} = \int \exp(i \operatorname{st} \Phi_\delta(f)) \, dL(\mu_{g_\omega, \Lambda_\delta})$$

satisfies the properties of 7.4.4 (analyticity, regularity, invariance, reflection positivity, and ergodicity).

We shall not establish this in detail; we note that analyticity and regularity easily follow from the bounds established in Theorem 7.4.13. For invariance we have to show that if E is a Euclidean symmetry then $S\{f \circ E\} = S\{f\}$. But in the approximation g_n we have assumed that g_n is the characteristic function of some lattice $\Lambda_\delta^{(n)} = \Lambda^{(n)} \cap \mathscr{L}_\delta$, where $\Lambda^{(n)}$ is a suitable bounded domain in $\mathbb{R}^d$. Performing the symmetry E on the domains $\Lambda^{(n)}$, we get a new sequence g'_n and an infinite internal g'_ω which is easily seen to define the same measure as the given g_ω; this follows from the general form of the inequalities in (36). The Osterwalder–Schrader reflection positivity follows from the fact that the internal limit measure $\mu_{g_\omega, \Lambda_\delta}$ satisfies the hyperfinite global Markov property; see Section 7.3. The bounds of Theorem 7.4.13 are again essential for the verification of ergodicity; however, the details are somewhat laborious, and we ask the reader to study the original reference [Albeverio and Høegh-Krohn (1974)].

We have treated a particularly simple example to illustrate how one can exploit the hyperfinite picture to construct models for quantum fields.

Let us observe that the hyperfinite construction can be given in all space–time dimensions. From an interaction

$$(49) \qquad U_g^\delta = \lambda_\delta \sum \delta^d g(n\delta) u_\delta(\Phi_\delta(n)),$$

one can derive an internal measure μ_{g,Λ_δ}. The associated Loeb measure will always exist and some of the axioms, the translation and reflection property of Euclidean invariance and reflection positivity, will be seen immediately to be verified.

As for the remaining axioms, ways can be found to cope with them; the real difficulty lies in the fact that the measure $L(\mu_{g,\Lambda_\delta})$ may turn out to be trivial, i.e., equal to the free measure; compare the discussion in Albeverio and Høegh-Krohn (1974, 1980, 1984b) and Albeverio et al. (1979).

We add some remarks on "fields with boundary conditions" which will be used in Section 7.5. Let μ_{0,Λ_δ}^B be the Gaussian measure given by

$$(50) \quad d\mu_{0,\Lambda_\delta}^B(q) = (2\pi)^{-1/2} [\mathrm{Det}\, C^\Lambda]^{-1/2} \exp\left(-\frac{1}{2} \sum_{n,n'} (C^\Lambda)_{n,n'}^{-1} q_{n\delta} q_{n'\delta}\right) dq,$$

with $(C^\Lambda)^{-1} \equiv ((C^{-1})^\Lambda - B_{\partial\Lambda})$, $C^{-1} = \delta^d(-\Delta_\delta + m^2)$, $(C^{-1})^\Lambda$ the restriction of C^{-1} to Λ_δ, and $B_{\partial\Lambda}$ a measure concentrated on $\partial\Lambda_\delta$; see (17) above. We call μ_{0,Λ_δ}^B the free lattice field measure with boundary condition B.

Sometimes it is convenient to assume that Λ is a "hypercube" of the form $\prod_{i=1}^d (-l_i\delta, l_i\delta)$ and that $\Lambda_\delta = \Lambda \cap {}^*\mathbb{Z}_\delta^d$.

The free lattice field Φ_δ^B with boundary condition B indexed by Λ_δ is introduced as in 7.4.6 and 7.4.7. It is a Gaussian random field with mean zero and covariance

$$(51) \qquad E(\Phi_\delta^B(n)\Phi_\delta^B(n')) = (C^\Lambda)_{n,n'}^{-1},$$

where E now denotes expectation with respect to μ_{0,Λ_δ}^B.

In Section 7.5 we shall have a particular use of the case of "Neumann boundary conditions," which corresponds to setting

$(B_{\partial\Lambda})_{n,m} = \delta_{n,m}i,$

where $\begin{cases} i = 2 & \text{if} \quad n\delta \in \Lambda_\delta \text{ is a corner site of } \Lambda_\delta, \\ i = 1 & \text{if} \quad n\delta \in \Lambda_\delta \text{ is not a corner site of } \Lambda_\delta, \\ i = 0 & \text{otherwise.} \end{cases}$

Notice that this corresponds to dropping the terms $(q_{n\delta} - q_{n'\delta})^2$ across $\partial\Lambda_\delta$ in the expression for the density of μ_{0,Λ_δ}^B. Thus we have a discrete version of setting the normal derivative equal to zero.

Another frequent choice is the "Dirichlet boundary conditions" obtained by setting $B_{\partial\Lambda} = 0$.

We can use $(\Phi_\delta^B, \mu_{0,\Lambda_\delta}^B)$ in (26) and (27) to construct models of non-Gaussian Euclidean random fields. This has been carried out in a number of cases in constructive quantum field theory; see, e.g., Simon (1974) and Glimm and Jaffe (1981) and references therein.

We conclude this section with some remarks on the φ_d^4 model seen from our hyperfinite point of view. The φ_d^4 model has an interaction of the form (49) with u_δ a polynomial of fourth degree. We shall give an outline of an approach by Brydges *et al.* (1983) in the cases $d = 2, 3$ and indicate how their construction fits into the present framework. In fact, by using a hyperfinite lattice we shall overcome the "somewhat distasteful" construction using compactness and subsequences which the authors need to complete their program.

We start out with the finite-volume theory. Let Λ_δ be a finite lattice in $\mathbb{R}^d$ with spacing $\delta > 0$. On $\mathbb{R}^{\Lambda_\delta}$ we introduce the measure

$$(52) \qquad d\mu(\varphi) \equiv \prod_{x \in \Lambda_\delta} d\varphi_x \exp(-S_{\Lambda_\delta}(\varphi))/Z_{\Lambda_\delta},$$

where Z_{Λ_δ} is a suitable normalization factor and

$$(53) \qquad S_{\Lambda_\delta}(\varphi) \equiv \frac{1}{2} \sum_{\langle xy \rangle} \delta^{d-2}(\varphi_x - \varphi_y)^2 + \frac{1}{2}a \sum_{x \in \Lambda_\delta} \delta^d \varphi_x^2 + \frac{\lambda}{4} \sum_{x \in \Lambda_\delta} \delta^d \varphi_x^4,$$

where $\lambda \geq 0$ and $a \in \mathbb{R}$. Here $\langle xy \rangle$ means that we sum over all nearest neighbors in $\mathbb{Z}_\delta^d$, setting $\varphi_x \equiv 0$ for $x \notin \Lambda_\delta$ (which means that we impose Dirichlet boundary conditions).

REMARK. We follow the exposition of Brydges *et al.* (1983). Thus (52) contains both the "free" and the "interacting" part of the measure. The interacting part corresponding to (49) is

$$\frac{1}{2}a \sum_{x \in \Lambda_\delta} \delta^d \varphi_x^2 + \frac{\lambda}{4} \sum_{x \in \Lambda_\delta} \delta^d \varphi_x^4.$$

The authors reverse the usual procedure and first take the infinite-volume limit. Thus, let $\langle F \rangle_{\Lambda_\delta}^{(\delta)}$ denote expectation with respect to the measure (52).

Known inequalities and uniform bounds [see Brydges *et al.* (1983)] imply that the following limit is well defined:

$$(54) \qquad \langle F \rangle^{(\delta)} \equiv \lim_{\Lambda_\delta \nearrow \infty} \langle F \rangle^{(\delta)}_{\Lambda_\delta};$$

note that we keep δ fixed. In our language this means that we pass to a hyperfinite lattice Γ_δ with finite spacing δ and that we have

$$(55) \qquad \langle F \rangle^{(\delta)} \equiv \mathrm{st} \langle F \rangle^{(\delta)}_{\Gamma_\delta}.$$

It remains to take the continuum limit, i.e., to choose δ positive infinitesimal. To obtain something finite and nontrivial we need strong and uniform inequalities.

As above, where we needed to choose $\lambda_\delta = \lambda (A_\alpha^\delta)^{-1}$ [see (33)], in order to cancel certain infinities when passing to the continuum limit ($\delta \approx 0$), we also need here to introduce suitable mass counterterms. In fact, choose

$$(56) \qquad a(\delta) = \begin{cases} m_0^2 + \delta m_1^2(\delta), & d = 2, \\ m_0^2 + \delta m_1^2(\delta) + \delta m_2^2(\delta), & d = 3, \end{cases}$$

where $\delta m_1^2 \equiv -3\lambda C^{(\delta)}(0)$ and $\delta m_2^2 \equiv 6\lambda^2 \int C^{(\delta)}(0 - z) d^d z$ and $C^{(\delta)}(x, y)$ is the free two-point function in the infinite-volume limit, i.e., obtained by setting $\lambda = 0$ and $a = m_0^2$.

Let $S^{(\delta)}(x, y) \equiv \langle \varphi_x \varphi_y \rangle^{(\delta)}$ be the interacting two-point function in the infinite-volume limit. And let $\|| \cdot \||$ denote the norm

$$(57) \qquad \|| f \|| \equiv \| f \|_1 + \| f \|_\infty.$$

The key to the successful passage to the continuum limit is the following inequality, which crucially depends upon the "renormalization" introduced in (56): there exist universal constants $\lambda_0 > 0$ and c such that if $0 \le \lambda \le \lambda_0$, then

$$(58) \qquad \|| S^{(\delta)} - C^{(\delta)} \|| \le c\lambda^2;$$

universal here, of course, means independent of δ. As soon as this inequality is established the usual arguments allow us to complete the construction. In our setting we shall arrive at a nontrivial field theory based on the hyperfinite lattice Γ_δ with $\delta \approx 0$.

We conclude with a few remarks on (58). Three main ingredients go into the proof: (i) the field equation, which in the setting of Γ_δ is obtained by transfer from the finite-volume case; (ii) the so-called skeleton inequalities, which are proved by using the random-walk representation of a classical spin system [see Brydges *et al.* (1983)]; and (iii) the continuity of $\|| S^{(\delta)} - C^{(\delta)} \||$ in the parameter λ.

There is no success story to recount in the case $d = 4$. Glimm and Jaffe (1981) have obtained some positive results. There have also been some speculations that φ_4^4 with a positive λ is trivial, i.e., that the associated measure collapses to the free measure; the possibility of a nontrivial φ_4^4 with $\lambda < 0$ has also been discussed; see Albeverio *et al.* 1982, 1984a, and the references therein.

In Section 7.5 we shall return to the discussion of the φ_4^4 model in connection with its "polymer representation."

D. Some Concluding Remarks on Gauge Fields

There has been a great interest, both in mathematics and in physics, in *gauge fields*; e.g., see Fröhlich (1980) and Seiler (1982) and references therein. Many physicists have expressed the hope that quantized gauge fields might provide the appropriate physical description of interactions in elementary particle physics, and mathematicians have studied gauge fields as possible candidates for nontrivial models of relativistic quantum fields, as well as for the rich mathematical structure they exhibit at the classical level. In this concluding part we shall indicate how one can describe in nonstandard terms the continuum gauge field in two dimensions, starting from the model in Sections 7.4.B and 7.4.C of a hyperfinite lattice with infinitesimal lattice spacing. Moreover, we get a construction of "Markov cosurfaces" and "Markov covector fields" which are natural extensions of Abelian gauge fields to arbitrary space–time dimension.

Let us start by recalling some of the standard theory. Let Γ be a bounded domain in $\mathbb{R}^d$ and $\delta > 0$; as usual we set

$$\Gamma_\delta = \Gamma \cap \mathbb{Z}_\delta^d.$$

A *cell* A of Γ_δ is the hypercube obtained by translating $\{x \in \mathbb{R}^d \mid 0 \le x_i \le \delta\}$ by $n\delta$ for some $n \in \mathbb{Z}^d$. The cell as a point set is, of course, a subset of $\mathbb{R}^d$. It is determined by a set of vertices that are constrained to lie in Γ_δ.

We denote by $F_1, F_2, \ldots, F_{2d}$ the *faces* of A, considered as $(d - 1)$-dimensional hypersurfaces, oriented in such a way that the basis vector orthogonal to a face points outward. We then have, as sets, $\partial A = \bigcup_{i=1}^{2d} F_i$, with ∂A denoting the boundary of A. We shall call two cells *adjacent* if they have a common face, which, of course, will have opposite orientations as a face of the respective cells.

For $d = 2$ we define the *product* $F_1 F_2$ of two faces such that the endpoint of F_1 is the starting point of F_2 as the set $F_1 \cup F_2$ with the orientation being the one inherited from F_1 and F_2. In the same way, for $d > 2$ we define the product $F_1 F_2$ for faces for which $F_1 \cap F_2$ is $(d - 2)$-dimensional and with the orientation inherited from F_1 and from F_2 opposite to each other.

We define recursively

$$F_1 \cdot \cdots \cdot F_{n+1} = (F_1 \cdot \cdots \cdot F_n) F_{n+1}$$

and we let $\Sigma = \Sigma_{\Gamma_\delta}$ be the set of all such products.

To each element $S \in \Sigma$ we associate an element $C(S)$ in a fixed group G, supposed to be Abelian if $d > 2$, but not necessarily Abelian for $d = 2$. We assume that the map C satisfies

$$C(S_1 S_2) = C(S_1) C(S_2)$$

whenever $S_1 S_2$ is a well-defined element of Σ and the product on the right-hand side is the group operation in G. Further, we assume

$$C(S^{-1}) = C(S)^{-1},$$

where S^{-1} denotes the same set as S but with opposite orientation.

We call the map C a *cosurface* on Γ_δ with values in G; or, in different terminology, C is a G-valued $(d-1)$-cochain. We denote by $\Gamma_G = \Gamma_{G,\Gamma_\delta}$ the set of all cosurfaces on Γ_δ.

REMARK. In the case $d = 2$ the cells are squares having vertices in points of the lattice Γ_δ which are nearest or next-nearest neighbors. The faces are segments joining nearest-neighbor points. In the terminology of lattice gauge field theory the cells are "plaquettes" and the faces are "links" or "bonds."

Having established the basic geometric terminology, we now endow the group G with a measurable structure. We may then introduce a notion of a *stochastic cosurface* as a measurable map S from some probability space $(\Omega, \mathcal{B}, P)$ into Γ_G. For $d = 1$ there is no genuine product of faces and Σ reduces to the set of lattice points Γ_δ; thus in this case a stochastic cosurface is simply a G-valued discrete stochastic process indexed by Γ_δ.

We shall only marginally touch upon the notion of stochastic cosurface in these remarks; see Albeverio *et al.* (1984b, 1985) for a more extensive discussion.

We are now all set to introduce the basic notion of *interaction*, appropriate for lattice gauge field theories. Let U be a real-valued function on G such that $U(gh) = U(hg)$ for all $h, g \in G$. Let β be a real constant. An *interaction* on Γ_δ given by U and β is a family of functions $\{W_\Lambda\}$, where Λ runs over all finite unions of cells of Γ_δ and

$$W_\Lambda(C) \equiv -\beta \sum_{A \subset \Lambda} U(C(\partial A)),$$

the sum being over all cells A of Λ and where C is a cosurface in $\Gamma_{G,\Lambda}$;

actually, $W_\Lambda(C)$ depends only on cosurfaces of the form $C(\partial A)$, $A \subset \Lambda$. Also note that because of our assumption on U, $U(C(\partial A)) = U(C(F_1) \cdot \cdots \cdot C(F_{2d}))$ is independent of the order in which we are given the faces $F_1, \ldots, F_{2d}$ of A, provided that the order is compatible with the orientation.

The group G has a measure structure; more specifically, we assume that G has a G-invariant probability measure dg, and we further assume that β, U are such that the integral

$$c(\beta) \equiv \int_G \exp(\beta U(g)) \, dg$$

exists. We can then introduce a probability measure $d\mu_\Lambda$ on $\Gamma_{G,\Lambda}$ by the formula

$$d\mu_\Lambda(C) \equiv Z_\Lambda^{-1} \exp[-W_\Lambda(C)] \prod_{A \subset \Lambda} \prod_{F \in \partial A} dC(F),$$

where

$$Z_\Lambda \equiv \int \exp[-W_\Lambda(C)] \prod_{A \subset \Lambda} \prod_{F \in \partial A} dC(F),$$

the product being over all faces F of cells A in Λ. Since $U(C(\partial A))$ is invariant under cyclic permutations of the faces $F_1, \ldots, F_{2d}$ of ∂A and $W_\Lambda(C)$ is invariant under permutation of the cells, we see that the measure is independent of the order of the cells and of the order in which the faces in a cell are taken, provided this order is such that $\prod_{F \in \partial A} C(F) = C(\partial A)$.

REMARK. Some people may be confused by the notation $dC(F)$, but we may think of $C(F)$ as a *variable* taking value in the group G and that $C(\partial A)$ and, more generally, $C(S)$, $S \in \Sigma$, are composite terms built up according to the "defining relations" $C(S_1 S_2) = C(S_1)C(S_2)$ and $C(S^{-1}) = C(S)^{-1}$.

For further reference let us note that

$$Z_\Lambda = E_{\mu_\Lambda}(1) = c(\beta)^{N_\Lambda},$$

where N_Λ is the number of cells in Λ.

Let f be a function on $\Gamma_{G,\Lambda}$ of the form

$$f(C) = f_0(g_1, \ldots, g_l),$$

where f_0 is a real-valued bounded continuous or nonnegative measurable function on G^l and where each $g_j = C(F_j)$, with F_j a face of some cell A

in Λ such that $F_j \cap \partial\Lambda = \varnothing$. Thus f is "supported" in the interior of Λ, and we write for simplicity $\operatorname{supp} f = (F_1, \ldots, F_l)$. For such f we obtain the following formula for the expectation of f with respect to the measure μ_{Γ_δ} (recalling that Γ_δ itself is a finite union of cells):

$$(59) \quad E_{\mu_{\Gamma_\delta}}(f) = \int f \prod_{A \subset \operatorname{supp} F} \left[\frac{e^{\beta U(C(\partial A))}}{c(\beta)} \right] \prod_{A \subset \operatorname{supp} f} \prod_{F \in \partial A} dC(F).$$

By *transfer* we can also let Γ_δ be a *hyperfinite lattice* with *infinitesimal lattice spacing* extending beyond standard space $\mathbb{R}^d$. This gives us a well-defined internal quantity $E_{\mu_{\Gamma_\delta}}(f)$ and an associated Loeb measure $L(\mu_{\Gamma_\delta})$, which we may take as our starting point for constructing models of gauge field theories.

REMARK 1. μ_{Γ_δ} is a "Gibbs state" for the interaction W on the hyperfinite lattice Γ_δ, in the same spirit as the hyperfinite Gibbs states constructed in Section 7.2.

REMARK 2. For $d = 2$, $U(y) = \chi(y)$, where $y \in G$, G a compact Lie group, and χ is some character of an irreducible unitary representation of G, we can show that $L(\mu_{\Gamma_\delta})$ in the hyperfinite case realizes the probability measure describing a lattice gauge field theory (pure lattice Yang–Mills theory) on the lattice with spacing δ. The Gibbs state given by $L(\mu_{\Gamma_\delta})$ will be invariant under gauge transformations in the sense that if F is the face with vertices $x, y \in \Gamma_\delta$ and $\gamma(z) = \gamma_z$ is any measurable map from Γ_δ to G, then μ_{Γ_δ} is invariant under the map $C(F) \mapsto \gamma_x C(F) \gamma_y^{-1}$.

In common terminology $\exp(\beta U(C(\partial A)))$ is called a *Wilson loop* and thus μ_{Γ_δ} gives the distribution of products of Wilson loops.

Let $G = U(1)$ and let A_μ, $\mu = 1, 2$, be smooth mappings from $\mathbb{R}^2$ into the Lie algebra of G. This determines a 1-form $A = \sum A_\mu \, dx_\mu$; physically one looks upon A_μ as "gauge fields" with "field strengths" $F_{\mu\nu} = \partial_\mu A_\nu - \partial_\nu A_\mu$, $\mu, \nu = 1, 2$. The relation of the 1-form A with our cosurface is as follows.

Let $\hat{A}$ be the cell of Γ_δ with vertices $v_1 \equiv \delta z$, $v_2 \equiv \delta(z + e_1)$, $v_3 \equiv \delta(z + e_1 + e_2)$, $v_4 \equiv \delta(z + e_2)$, with $z \in \mathbb{Z}^2$ and e_1, e_2 an orthogonal basis of $\mathbb{Z}^2$. Then

$$\int_{\partial\hat{A}} A = \int_{\partial\hat{A}} \sum_\mu A_\mu(x) \, dx_\mu \approx \delta A_1(\tfrac{1}{2}(v_1 + v_2)) - \delta A_1(\tfrac{1}{2}(v_3 + v_4))$$

$$+ \delta A_2(\tfrac{1}{2}(v_2 + v_3)) - \delta A_2(\tfrac{1}{2}(v_1 + v_4)),$$

where $\partial\hat{A} = (v_1, v_2) \cup (v_2, v_3) \cup (v_3, v_4) \cup (v_4, v_1)$, and (v_i, v_j) denotes the link between v_i, v_j, oriented positively from v_i to v_j. Now define the cosurface

C by setting

$$C(v_1, v_2) = \exp[-ig\, \delta A_1(\tfrac{1}{2}(v_1 + v_2))],$$

$$C(v_2, v_3) = \exp[-ig\, \delta A_2(\tfrac{1}{2}(v_2 + v_3))],$$

$$C(v_3, v_4) = \exp[ig\, \delta A_1(\tfrac{1}{2}(v_3 + v_4))],$$

$$C(v_4, v_1) = \exp[ig\, \delta A_2(\tfrac{1}{2}(v_1 + v_4))],$$

where g is some real parameter. Then

$$C(\partial\hat{A}) = C(v_1, v_2)C(v_2, v_3)C(v_3, v_4)C(v_4, v_1)$$

$$\approx \exp\left[-ig \int_{\partial\hat{A}} A\right].$$

For δ infinitesimal we see that giving a probability distribution to the cosurfaces $C(\partial\hat{A})$ amounts to giving a distribution to the flow $\int_{\partial\hat{A}} A$ of the 1-form A across $\partial\hat{A}$. As we shall see below, our choice μ_{Γ_δ} of distribution for $C(\partial\hat{A})$ yields a distribution for $\int_{\partial\hat{A}} A$ that coincides with the usual physical one in the cases considered in the theory of gauge fields.

REMARK 3. For $d > 2$ the usual Yang–Mills lattice field theory studies random variables associated with two-dimensional plaquettes, whereas we study random variables associated with $(d - 1)$-dimensional surfaces. Thus only for $d = 2$ do our models include the usual Yang–Mills models.

It is easy to write down the quantity $E_{\mu_{\Gamma_\delta}}(f)$; it is far more difficult to understand what it means. We shall try to gain some insight by discussing some examples. A unit cell A of Γ_1 will, by a process of subdivision, split into a family of smaller cells. To be specific, let $\delta = 2^{-n}$ for some $n \in \mathbb{N}$. In Γ_δ the cell A will be divided into 2^{dn} smaller cells A_n^i, $i = 1, \ldots, 2^{dn}$. A face F_i of A will be divided into $2^{(d-1)n}$ faces $F_{i,1}, \ldots, F_{i,2^{(d-1)n}}$, each $F_{i,j}$ being the face of some cell $A_n^{i'}$; but, of course, not every face of a cell $A_n^{i'}$ will be of the form $F_{i,j}$. We shall use $F_{i',j}^n$, $j = 1, \ldots, 2d$, to enumerate the faces of $A_n^{i'}$.

Let us for simplicity consider functions f on Γ_{G,Γ_δ} of the special form

$$f(C) = f_0(C(F_1), \ldots, C(F_{2d})),$$

where f_0 is a bounded real-valued continuous function on G^{2d} and each F_i is a face of A in Γ_1. To be precise, if C and C' are cosurfaces in Γ_{G,Γ_δ} satisfying

$$C(F_{i,1}) \cdot \cdots \cdot C(F_{i,2^{(d-1)n}}) = C'(F_{i,1}) \cdot \cdots \cdot C'(F_{i,2^{(d-1)n}})$$

for $i = 1, \ldots, 2d$, then $f(C) = f(C')$. Using formula (1) to calculate $E_{\mu_{\Gamma_\delta}}(f)$, we obtain

(60)
$$E_{\mu_{\Gamma_\delta}}(f) = \int f_0(C(F_1), \ldots, C(F_{2d}))$$

$$\times \prod_{i=1}^{2^{dn}} \left\{ \frac{\exp[\beta U(C(\partial A_n^i))]}{c(\beta)} \right\} \prod dC(F_{i,j}^n)$$

$$= c(\beta)^{-2^{dn}} \int f_0(C(F_1), \ldots, C(F_{2d}))$$

$$\times \prod_i \exp\left[\beta U\left(\prod_j C(F_{i,j}^n) \right) \right] \prod_j dC(F_{i,j}^n)$$

$$= c(\beta)^{-2^{dn}} \int f_0(C(F_1), \ldots, C(F_{2d}))$$

$$\times q_\beta^{*(2^{dn})} \left(\prod_{i=1}^{2d} C(F_i) \right) \prod_{i=1}^{2d} dC(F_i),$$

where $q_\beta(g) \equiv \exp[\beta U(g)]$ for $g \in G$ and $q_\beta^{*(k)}$ means k-fold convolution of q_β with itself. Notice that this calculation reduces the given integral $E_{\mu_{\Gamma_\delta}}(f)$ to an integral of fixed dimensionality $2d$, and where the C occurring in the reduced integral can be interpreted as a cosurface in Γ_1.

A calculation always extends by transfer. Thus let $n \in {}^*\mathbb{N} - \mathbb{N}$ and take $\delta = 2^{-n}$. In a hyperfinite lattice Γ_δ the unit cell A is subdivided into a hyperfinite number of cells A_n^i. We assume that the group G is compact and carries a normalized Haar measure. Let f be an internal function of the form

(61)
$$f(C) = {}^*f_0(C(F_1), \ldots, C(F_{2d})),$$

where C is a cosurface on Γ_δ with values in ${}^*G$ and f_0 is a bounded continuous real-valued function on G^{2d}. Notice that in this case $C(F_i)$ is the hyperfinite product

$$C(F_i) = C(F_{i,1}) \cdot \cdots \cdot C(F_{i,2^{(d-1)n}}),$$

which as a group element belongs to ${}^*G$. Since G is compact, $C(F_i)$ is nearstandard in ${}^*G$. We may thus conclude that f as a function on $\Gamma_{{}^*G,\Gamma_\delta}$ is S-integrable with respect to the internal measure μ_{Γ_δ} and that we have

(62)
$$E_{L(\mu_{\Gamma_\delta})}({}^\circ f) = \mathrm{st}(E_{\mu_{\Gamma_\delta}}(f)),$$

where, to emphasize, f is of the form (61) and $L(\mu_{\Gamma_\delta})$ is the Loeb measure on $\Gamma_{{}^*G,\Gamma_\delta}$ associated with μ_{Γ_δ}.

But (62) is of limited use unless we can determine the quantity

$$c(\beta)^{-2^{dn}} q_\beta^{*(2^{dn})} \left(\prod_{i=1}^{2d} C(F_i) \right).$$

A general answer is difficult; let us take $G = U(1)$ with $U(g) = \operatorname{Re} \chi(g)$, where χ is the character $\chi(e^{i\varphi}) = e^{i\varphi}$, where $e^{i\varphi}$, $\varphi \in [0, 2\pi)$ gives the natural parameterization of G.

Choosing $n \in {}^*\mathbb{N} - \mathbb{N}$, a short calculation shows that with the choice $\beta = 2^{dn}/\sigma^2$,

(63)
$$c(\beta)^{-2^{dn}} q_\beta^{*(2^{dn})} \left(\prod_{i=1}^{2d} C(F_i) \right) \approx Q_1(C(\partial A)),$$

where the C on the right-hand side can be considered as a (standard) cosurface on Γ_1 and Q_1 is the standard entity

$$Q_1(e^{i\varphi}) = e^{-\varphi^2/2\sigma^2}.$$

If we define $Q_t(e^{i\varphi}) = \exp(t(-\varphi^2/2\sigma^2))$, we see that Q_t is a semigroup on $U(1)$; in fact, Q_t is the semigroup describing Brownian motion with diffusion coefficient σ on $U(1)$, which "explains" why we added σ^2 to the choice of β.

From (60), (62), and (63), we now conclude that

$$E_{L(\mu_{\Gamma_\delta})}({}^\circ f) = \int f_0(C(F_1), \ldots, C(F_{2d})) Q_1(C(\partial A)) \prod_{i=1}^{2d} dC(F_i),$$

where the integral on the right-hand side is standard and f is given as in (61). We have thus added considerable "insight" to (62), knowledge of the Haar measure on $U(1)$ and the explicit form of the semigroup Q_t gives us, in principle, full control of the expectation $E_{L(\mu_{\Gamma_\delta})}$.

As a further illustration let A be the union in Γ_1 of three unit cells A^1, A^2, A^3 such that A^1 has a common face with both A^2 and A^3, but A^2 and A^3 have no common face. Let f be a function depending only on faces $F \in \partial A$, i.e.,

$$f(C) = {}^*f_0(\langle C(F) \rangle_{F \in \partial A});$$

notice that this is in complete analogy with (61). If we remark that for cells A^1 and A^2 which are adjacent,

$$Q_1(C(\partial A^1)) \cdot Q_1(C(\partial A^2)) = Q_2(C(\partial(A^1 \cup A^2))),$$

using the semigroup property of Q_t and understanding the product in sense of convolution, we see that

$$(64) \qquad E_{\mu_{\Gamma_\delta}}(f) \approx \int f_0(\langle C(F) \rangle_{F \in \partial A}) Q_{|A|}(C(\partial A)) \prod_{F \in \partial A} C(F),$$

where $|A|$ denotes the Cartesian volume of A, i.e., in this case $|A| = 3$.

The reasoning behind (64) can be extended, even to the hyperfinite case. Let A in Γ_δ be a bounded hyperfinite union of cells; i.e., we assume that the set of standard points of A is bounded. In this case we may think of ∂A as a hyperfinite approximation to a "nice" closed curve in $\mathbb{R}^d$ with an interior of finite volume; see Fig. 7.3. For technical reasons we have added two "curves" S_1 and S_2 so that the resulting partition of Γ_δ, A, B_1, B_2 consists of three connected and simply connected pieces.

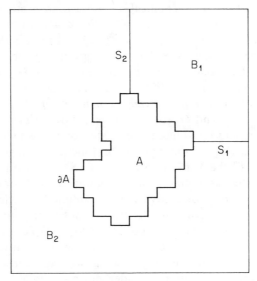

Figure 7.3

Let $K = \{\partial A, S_1, S_2\}$ and $D_K = \{A, B_1, B_2\}$. K is an example of what is called a *regular saturated complex* and D_K is the associated partition of the underlying space. Generalizing (64), we can introduce a probability measure $\nu_{K,\delta}$ associated with K and D_K by

$$(65) \qquad d\nu_{K,\delta}(C) = \prod_{B \in D_K} Q_{|B|}\left(\prod_{F \in \partial B} C(F)\right) \prod_{F \in \partial B} C(F),$$

where Q_t in our examples is given by (63) but could, in principle, be any Markovian semigroup on G satisfying $Q_t(gh) = Q_t(hg)$. We further assume that $Q_{|B|} \approx 1$ if the volume $|B|$ is hyperfinite but not finite. Then, generalizing (64), we may write

$$(66) \qquad\qquad E_{\mu_{\Gamma_\delta}}(f) \approx E_{\nu_{K,\delta}}(f),$$

assuming that f as a function depends only on the "curve" ∂A, i.e.,

$$f(C) = {}^*f_0(\langle C(F)\rangle_{F\in\partial A}).$$

The approximation in (66) will be exact if we add the Loeb construction. Thus we have our final interpretation of (60) for δ infinitesimal in terms of the semigroup Q_t, the volume measure $|B|$, and the Haar measure on G. Notice that in the standard approach the measure $L(\nu_{K,\delta})$ would have to be obtained by a projective limit construction.

The Loeb measure $L(\mu_{\Gamma_\delta})$ can be taken to be the underlying probability measure of a field $C(S)$ of canonical random variables attached to smooth $(d-1)$-dimensional hypersurfaces on $\mathbb{R}^d$. For $d = 2$ and $S = \partial\hat{A}$, this random field coincides essentially, by Remark 2 and (63), with the flow $\int_S A$ of the gauge field A across S. The gauge group was chosen here to be $G = U(1)$, but the same is true in other examples, e.g., $G = SU(2)$ or $G = \mathbb{Z}^2$ [see Albeverio *et al.* (1984b)]. Thus in these situations *the random field $C(S)$ is a realization of the "continuum limit" of "lattice gauge fields."*

For general $d > 2$ and G Abelian one can show (Albeverio *et al.*, 1986) that the random field $C(S)$ has a global Markov property and satisfies the d-dimensional analog of the axioms for Euclidean gauge fields (Seiler, 1982). For $G = \mathbf{T}^n$ the Markov cosurfaces $C(S)$ can also be looked upon as solutions of stochastic differential equations $DA = \xi$ for covector fields A, with D the covariant derivative and ξ an infinitely divisible generalized random field over $\mathbb{R}^d$. This introduces new and interesting mathematical structures, but it would take us too far afield to explore these here. We hope to have convinced the reader that the Loeb construction has once more been put to good use to give a "geometric" construction of the "continuum" case by exploiting the underlying hyperfinite "discrete" model.

7.5. FIELDS AND POLYMERS

In this final section we shall study the close connections which exist between quantum fields and polymer measures. Since their discovery by Symanzik (1969) in the mid-1960s, these relations have played an important part in quantum field theory and helped strengthen the already strong bonds between probability and mathematical physics. As we shall soon see, the basic ingredient in the theory is the representation of the square of the free field Φ as a Poisson random field of local times of Brownian loops. If we add an interaction to Φ, this is reflected in the probabilistic representation as a perturbation involving intersections and self-intersections of the Brownian paths, and we are thus led directly to the problems we studied in Section 6.4. Toward the end of this section, we shall use this relationship to shed new light on Φ_4^4 and (especially) $\Phi_1^2\Phi_2^2$ fields with infinitesimal coupling constants.

A. Poisson Fields of Brownian Bridges

Throughout this section Poisson fields of Brownian loops and bridges will appear in various guises, and it is convenient to begin by studying them in an abstract setting.

Given an internal, $*$-bounded subset Λ of $*\mathbb{R}^d$ and a positive $\delta \in *\mathbb{R}$, let Λ_δ be the lattice

$$\Lambda_\delta = \{n\delta \in \Lambda \mid n \in *\mathbb{Z}^d\}.$$

For each $i \in \Lambda_\delta$, we shall let

$$N_i = \{j \in \Lambda_\delta \mid |i - j| = \delta\}$$

denote the set of next neighbors of i in Λ_δ, and write $|N_i|$ for the number of elements in N_i—obviously, $|N_i| = 2d$ unless i is on the boundary of Λ_δ.

The discrete Laplacian in Λ_δ with Neumann boundary conditions is given by

(1)
$$\Delta_\delta f(i) = \delta^{-2}\left(\sum_{j \in N_i} f(j) - |N_i| f(i) \right).$$

We shall be interested in Markov processes on Λ_δ with infinitesimal generator

$$-\tfrac{1}{2}\Delta_\delta + m^2,$$

where the "mass" m is a positive real number. These processes will have a hyperdiscrete time line $T = \{k\,\Delta t \mid k \in *\mathbb{N}\}$ for some positive infinitesimal Δt that we keep fixed all through the section (in contrast to δ, which we shall allow to be standard on some occasions and infinitesimal on others). We shall have to add a "trap" or "cemetery state" $\square$ to Δ_δ; hence our state space will be

$$\bar\Lambda_\delta = \Lambda_\delta \cup \{\square\}$$

(the trap $\square$ plays exactly the same role here as did s_0 in Chapter 5).

To describe the Markov process more precisely, assume that it is in a state $i \in \Lambda_\delta$ at time t. At time $t + \Delta t$, it will then be in the trap $\square$ with probability $m^2\,\Delta t$; it will be in each one of the neighboring states $j \in N_i$

with probability $\Delta t/2\delta^2$; and it will remain in i with probability $(1 - ((|N_i|/2\delta^2) + m^2)\,\Delta t)$. For this to make sense $((|N_i|/2\delta^2) + m^2)\,\Delta t \leq 1$, i.e.,

$$\delta \geq \sqrt{\frac{|N_i|\,\Delta t}{2(1 - m^2\,\Delta t)}},$$

and we shall always assume that this is the case. If the process is in the trap $\square$ at time t, it will remain in $\square$ at all later times.

Let $X : \Omega \times T \to \bar{\Lambda}_\delta$ be a hyperfinite Markov process which fits the description we have just given. Computing its infinitesimal generator, we get

$$Af(i) = \frac{f(i) - \sum_{j \in N_i} f(j)(\Delta t/2\delta^2) - f(i)(1 - ((|N_i|/2\delta^2) + m^2)\,\Delta t))}{\Delta t}$$

$$= -\frac{1}{2}\delta^{-2}\left[\sum_{j \in N_i} f(j) - |N_i|f(i)\right] + m^2 f(i).$$

Thus $A = -\frac{1}{2}\Delta_\delta + m^2$, exactly as we wanted.

Note that when δ is infinitesimal, then in the interior of Λ_δ, the standard part of X behaves like a Brownian motion which is killed at rate m^2. Observe also that if X is uniformly distributed at time zero, say

$$P\{X(0) = i\} = \delta^d,$$

then X is uniformly distributed at all later times and

$$P\{X(t) = i\} = \delta^d(1 - m^2\,\Delta t)^{t/\Delta t}.$$

It was to achieve this effect that we chose to work with Neumann boundary conditions.

If $i \in \Lambda_\delta$, we let as usual P_i be the probability measure of the process started at i. Thus if P is the probability measure of the process started with uniform distribution $P\{X(0) = j\} = \delta^d$, we have

$$P_i(A) = \delta^{-d}P\{\omega \in A \mid X(\omega, 0) = i\}.$$

If we only count those paths which are not yet trapped at time t, we are led to the probability measures

$$P_i^t(A) = \frac{P_i\{\omega \in A \mid X(\omega, t) \neq \square\}}{(1 - m^2\,\Delta t)^{t/\Delta t}} = \frac{P\{\omega \in A \mid X(\omega, 0) = i \wedge X(\omega, t) \neq \square\}}{\delta^d(1 - m^2\,\Delta t)^{t/\Delta t}}.$$

Given two elements $i, j \in \Lambda_\delta$ which are connected by at least one path of length t, we shall use the measure

$$P_{i,j}^t(A) = \frac{P_i^t\{\omega \in A \mid X(\omega, t) = j\}}{P_i^t\{\omega \mid X(\omega, t) = j\}}$$

to discuss the set of all such Brownian bridges. If there is no such connection between i and j, we let $P_{i,j}^t$ be the point measure on the constant path $X(\omega, t) = \square$. To denote expectations with respect to P_i, P_i^t, and $P_{i,j}^t$, we shall use E_i, E_i^t, and $E_{i,j}^t$, respectively.

So far we have only discussed the second half of our announced topic, "Poisson fields of Brownian bridges." Turning to the Poisson fields, we let α be an internal measure on the space $\Lambda_\delta^2 \times T$. If Σ is the set of all internal maps

$$\sigma : \Lambda_\delta^2 \times T \to \{0, 1\},$$

α induces a measure Q on Σ as follows: let $q_{i,j,t}(\sigma) = \alpha(i, j, t)$ if $\sigma(i, j, t) = 1$ and $q_{i,j,t}(\sigma) = 1 - \alpha(i, j, t)$ otherwise, and put

$$Q(\sigma) = \prod q_{i,j,t}(\sigma),$$

where the product is over all $(i, j, t) \in \Lambda_\delta^2 \times T$.

It is easy to check that (σ, Q) is a Poisson random field with parameter α in the following sense.

7.5.1. LEMMA. Assume that α is nonatomic, i.e., $\alpha\{x\} \approx 0$ for all $x \in \Lambda_\delta^2 \times T$. If $C \subset \Lambda_\delta^2 \times T$ is Loeb measurable with $L(\alpha)(C) < \infty$, then for all $m \in \mathbb{N}$

$$L(Q)\left\{ \sum_{x \in C} \sigma(x) = m \right\} = \frac{L(\alpha)(C)^m}{m!} e^{-L(\alpha)(C)}.$$

PROOF. It is clearly sufficient enough to show that if C is internal and has finite but noninfinitesimal measure, then

$$Q\left\{ \sum_{x \in C} \sigma(x) = m \right\} \approx \frac{\alpha(C)^m}{m!} e^{-\alpha(C)}.$$

Pick a subset $C_0 = \{x_1, x_2, \ldots, x_m\}$ of C of cardinality m. The probability that C_0 is the subset of C where σ equals one is

$$\alpha(x_1)\alpha(x_2) \cdots \alpha(x_m) \cdot \prod_{x \in C - C_0} (1 - \alpha(x)).$$

Since α is nonatomic

$$\prod_{x \in C - C_0} (1 - \alpha(x)) = \exp\left[\sum_{x \in C - C_0} \ln(1 - \alpha(x)) \right] \approx \exp\left[- \sum_{x \in C - C_0} \alpha(x) \right]$$

$$\approx \exp\left[- \sum_{x \in C} \alpha(x) \right] = \exp[-\alpha(C)],$$

where we have used the Taylor expansion of $\ln(1 - \alpha(x))$. The probability that $\sum_{x \in C} \sigma(x) = m$ is thus infinitely close to

$$e^{-\alpha(C)} \sum \alpha(x_1) \alpha(x_2) \cdots \alpha(x_m),$$

where the sum is over all subsets $C_0 = \{x_1, \ldots, x_m\}$ of C of cardinality m. Since each such subset can be ordered in $m!$ different ways, we get

$$Q\left\{ \sum_{x \in C} \sigma(x) = m \right\} \approx \frac{e^{-\alpha(C)}}{m!} \sum \alpha(x_1) \cdots \alpha(x_m)$$

if we sum over all m-tuples $(x_1, \ldots, x_m)$ with distinct coordinates instead. However, since α is nonatomic and C has finite measure, the result of allowing repeated coordinates is just to change the sum by an infinitesimal amount, and thus

$$Q\left\{ \sum_{x \in C} \sigma(x) = m \right\} \approx \frac{e^{-\alpha(C)}}{m!} \sum_{(x_1, \ldots, x_m) \in C^m} \alpha(x_1) \cdots \alpha(x_m)$$

$$= \frac{e^{-\alpha(C)}}{m!} \alpha(C)^m,$$

which proves the lemma.

We shall now combine the Poisson field and the Brownian bridges in one structure. Let

$$\Theta = \Sigma \times \Omega^{\Lambda_\delta^2 \times T},$$

where as usual $\Omega^{\Lambda_\delta^2 \times T}$ is the set of all internal maps from $\Lambda_\delta^2 \times T$ to Ω. Let Z be the internal product measure on Θ obtained by using the measure Q on Σ and the measure $P_{i,j}^t$ on the (i, j, t)-th component of the product $\Omega^{\Lambda_\delta^2 \times T}$.

If $g : \Lambda_\delta \to {}^*\mathbb{R}_+$ is an internal function, define a random variable $T(g) : \Theta \to {}^*\mathbb{R}_+$ by

$$(2) \qquad T(g)(\sigma, \omega) = \sum_{(i,j,t) \in \Lambda_\delta^2 \times T} \sigma(i, j, t) \sum_{s=0}^{t} g(X(s, \omega_{i,j,t})) \Delta t,$$

where $\omega_{i,j,t}$ is the (i, j, t)-th component of ω. We use the convention that $g(\square) = 0$. We shall refer to T as *the Poisson field of Brownian bridges induced by α*.

It is easy to paraphrase the way in which T operates. Each time we choose an element $\xi = (\sigma, \omega)$ in Θ, the first component σ picks out a set of points $(i_1, j_1, t_1), \ldots, (i_m, j_m, t_m)$ in $\Lambda_\delta^2 \times T$, and the second component ω determines a set of Brownian bridges connecting the i_k's to the j_k's in time t_k. The value of T_g is just the sum of the integrals of g over these paths.

The fact that σ is Poisson distributed makes it easy to compute the Laplace transform of $T(g)$:

7.5.2. LEMMA. Let g be a non-negative, internal function such that

(3)
$$\sum_{\Lambda_\delta^2 \times T} (1 - \exp(-t\|g\|_\infty))^2 \alpha(i, j, t)^2 \approx 0.$$

Then

$$E \exp(-T(g)) \approx \exp\left\{\int_{\Lambda_\delta^2 \times T} E_{i,j}^t \left\{\exp\left[-\int_0^t g(X(s))\, ds\right] - 1\right\} d\alpha(i, j, t)\right\}$$

PROOF. For each $(i, j, t) \in \Lambda_\delta^2 \times T$ and each $\omega \in \Omega^{\Lambda_\delta^2 \times T}$, let

$$G_{i,j,t}(\omega) = \int_0^t g(X(\omega_{i,j,t}, s))\, ds,$$

and write **P** for the product measure $\Pi\, P_{i,j,t}$ on $\Omega^{\Lambda_\delta^2 \times T}$. Then

$$E(e^{-T(g)}) = \sum_\omega \sum_{C_0 \subset \Lambda_\delta^2 \times T} \prod_{(i,j,t) \in C_0} \exp(-G_{i,j,t}(\omega)) \alpha(i, j, t)$$

$$\times \prod_{(i,j,t) \notin C_0} (1 - \alpha(i, j, t)) \mathbf{P}\{\omega\}$$

$$= \sum_\omega \prod_{(i,j,t) \in \Lambda_\delta^2 \times T} [(1 - \alpha(i, j, t)) + \exp(-G_{i,j,t}(\omega)) \alpha(i, j, t)] \mathbf{P}\{\omega\},$$

which is easily seen by multiplying out the binomials on the right-hand side. Continuing, we get

$$E \exp(-T(g)) = \sum_\omega \prod_{(i,j,t)} [1 + \alpha(i, j, t)(\exp(-G_{i,j,t}(\omega)) - 1)] \mathbf{P}\{\omega\}$$

$$= \prod_{(i,j,t)} [1 + \alpha(i, j, t) E_{i,j}^t (\exp(-G_{i,j,t}) - 1)]$$

$$= \exp\left\{\sum_{(i,j,t)} \ln[1 + \alpha(i, j, t) E_{i,j}^t (\exp(-G_{i,j,t}) - 1)]\right\}.$$

Using Taylor's formula, we turn this into

$$E \exp(-T(g)) = \exp\left\{\int_{\Lambda_\delta^2 \times T} E_{i,j}^t (\exp(-G_{i,j,t}) - 1)\, d\alpha + R\right\},$$

where the remainder term is negative, and

$$|R| \leq \frac{1}{2} \sum_{(i,j,t)} E_{i,j}^t (\exp(-G_{i,j,t}) - 1)^2 \alpha(i, j, t)^2.$$

But $G_{i,j,t} \leq \|g\|_\infty t$, and thus the right-hand side of this expression is less than

$$\frac{1}{2} \sum_{(i,j,t)} (1 - \exp(-t\|g\|_\infty))^2 \alpha(i,j,t)^2,$$

which is infinitesimal by assumption. This proves the lemma.

We shall use this result to show that with the appropriate choice of the measure α, the Poisson field T has the same Laplace transform as the square of the free field on Λ_δ, and that these two quantities can thus be identified. If the reader wonders why we have used the somewhat curious condition (3) in Lemma 7.5.2 and not, for example, the slightly stronger but more natural condition

$$\sum \alpha(i,j,t)^2 \approx 0,$$

it is because the representation theorem we just mentioned needs the stronger condition.

REMARK. The idea behind the hyperfinite approach to Poisson random fields is by no means new; in fact, the very first application of Loeb measures in probability theory was a similar construction of Poisson processes in Loeb's (1976) original paper [see also Stroyan and Bayod (1985)].

B. The Square of the Free Field as a Local Time Functional

As we now turn to lattice fields, our first task is to compute the Laplace transform of the square of the free field in order to establish the identification referred to above. But before we can begin our calculations, we need to fix the terminology. Let D_{Λ_δ} be the kernel or matrix of $-\frac{1}{2}\Delta_\delta + m^2$ as an operator from $l^2(\Lambda_\delta, \delta^d)$ to $l^2(\Lambda_\delta, \delta^d)$, i.e.,

$$\left(-\frac{1}{2}\Delta_\delta + m^2\right)f(i) = \sum_{j \in \Lambda_\delta} D_{\Lambda_\delta}(i,j)f(j)\,\delta^{2d}.$$

Denote the inverse matrix by C_{Λ_δ}, and observe that $\delta^{-2d}C_{\Lambda_\delta}$ is the kernel of $(-\frac{1}{2}\Delta_\delta + m^2)^{-1}$. Let $\{\tilde{\Phi}_\delta^N(i)\}_{i \in \Lambda_\delta}$ be a Gaussian random vector with mean zero and covariance matrix C_{Λ_δ}, and let $\Phi_\delta^N = \delta^{-d}\tilde{\Phi}_\delta^N$ be its density with respect to the measure δ^d on Λ_δ. Recall from the last section that Φ_δ^N is the free lattice field of mass m on Λ_δ with Neumann boundary conditions. We introduce the square of the free field $\Psi_\delta^N(i) = \Phi_\delta^N(i)^2$, and write

$$\Phi_\delta^N(g) = \sum_{i \in \Lambda_\delta} g(i)\tilde{\Phi}_\delta^N(i) = \sum_{i \in \Lambda_\delta} g(i)\Phi_\delta^N(i)\,\delta^d$$

$$\Psi_\delta^N(g) = \sum_{i \in \Lambda_\delta} g(i)\Psi_\delta^N(i)\,\delta^d$$

for all internal functions g. It will be useful to keep in mind that

$$E(\Phi_\delta^N(f)\Phi_\delta^N(g)) = \sum f(i)g(j)C_{\Lambda_\delta}(i,j) = \langle(-\Delta_\delta + m^2)^{-1}f, g\rangle,$$

where the inner product is in $l^2(\Lambda_\delta, \delta^d)$.

We are now ready to compute $E \exp(-\frac{1}{2}\psi_\delta^N(f))$ (the factor $\frac{1}{2}$ is included for technical convenience) for all positive, internal functions f. By definition of $\{\Phi_\delta^N\}$

$$E \exp(-\tfrac{1}{2}\psi_\delta^N(f)) = (2\pi)^{-|\Lambda_\delta|/2}(\operatorname{Det} C_{\Lambda_\delta})^{-1/2}$$

$$\times \int_{\mathbb{R}^{|\Lambda_\delta|}} \exp\left[-\frac{1}{2}\sum_{i,j} C_{\Lambda_\delta}^{-1}(i,j)q_iq_j\right]$$

$$\times \exp\left[-\frac{\delta^{-d}}{2}\sum_i f(i)q_i^2\right] dq_1\, dq_2 \cdots dq_{|\Lambda_\delta|}.$$

Using the fact that $D_{\Lambda_\delta} = C_{\Lambda_\delta}^{-1}$ and that

$$(4) \qquad \int_{\mathbb{R}^n} \exp(-\tfrac{1}{2}\langle Aq, q\rangle)\, dq = (2\pi)^{n/2} \operatorname{Det} A^{-1/2}$$

for all positive, symmetric matrices A, we get

$$(5) \qquad E \exp(-\tfrac{1}{2}\Psi_\delta^N(f)) = \operatorname{Det} D_{\Lambda_\delta}^{1/2} \operatorname{Det}(D_{\Lambda_\delta} + \delta^{-d}f)^{-1/2}.$$

To compute these determinants, we first translate the problem into the language of operators on $l^2(\Lambda_\delta, \delta^d)$. If we write H_δ for $-\frac{1}{2}\Delta_\delta + m^2$, we have

$$\operatorname{Det}(D_{\Lambda_\delta}) = \delta^{-d|\Lambda_\delta|} \exp[\operatorname{tr} \log H_\delta]$$

and

$$\operatorname{Det}(D_{\Lambda_\delta} + \delta^{-d}f) = \delta^{-d|\Lambda_\delta|} \exp[\operatorname{tr} \log(H_\delta + f)],$$

where tr denotes the trace of an operator. Thus

$$(6) \qquad E \exp(-\tfrac{1}{2}\Psi_\delta^N(f)) = \exp\{-\tfrac{1}{2}\operatorname{tr}[\log(H_\delta + f) - \log(H_\delta)]\}.$$

Our next task is to get rid of the logarithms. Applying the well-known series expansion

$$\log(x) = -\sum_{k=1}^{\infty} \frac{1}{k}(1 - x)^k$$

to a positive operator $A\,\Delta t$, we get $\log(A) + \log(\Delta t) = \log(A\,\Delta t) = -\sum_{k=1}^{\infty}(1/k)(I - A\,\Delta t)^k$. Hence

$$(7) \qquad E \exp(-\tfrac{1}{2}\Psi_\delta^N(f))$$

$$= \exp\left\{\frac{1}{2}\operatorname{tr}\sum_{k=1}^{\infty}\frac{1}{k}[(I - (H_\delta + f)\,\Delta t)^k - (I - H_\delta\,\Delta t)^k]\right\}.$$

To compute the trace, we shall use the Feynman–Kac formula and the orthonormal basis $\{e_i\}_{i \in \Lambda_\delta}$, where

$$e_i(j) = \begin{cases} \delta^{-d/2} & \text{if } i = j, \\ 0 & \text{otherwise.} \end{cases}$$

In order to have the conditions of the hyperfinite Feynman–Kac formula satisfied, it will be convenient to assume that $\|f\|_\infty \le (\ln(1/\Delta t))^{1/2}$. By Corollary 5.3.12, we get

$$\left\langle \sum_{k=1}^\infty \frac{1}{k}[I - (H_\delta + f)\,\Delta t]^k e_i, e_i \right\rangle$$

$$\approx \sum_{k=1}^\infty \frac{1}{k} E\left\{ e_i(X(k\,\Delta t))e_i(X(0)) \exp\left(-\sum_0^k f(X(j\,\Delta t))\,\Delta t\right) \right\},$$

where X is the Markov process generated by H_δ. In terms of the conditional measures P_i^t, $P_{i,j}^t$, this can be rewritten as

$$\left\langle \sum_{k=1}^\infty \frac{1}{k}[I - (H_\delta + f)\,\Delta t]^k e_i, e_i \right\rangle$$

$$\approx \sum_{t=\Delta t}^\infty \frac{1}{t}(1 - m^2\,\Delta t)^{t/\Delta t} P_i^t\{X(t) = i\} E_{i,i}^t \exp\left(-\sum_0^t f(X(s))\,\Delta t\right) \Delta t.$$

When t is noninfinitesimal, $P_i^t\{X(t) = j\}$ is of order of magnitude δ^d, and it is convenient to rescale P_i^t by introducing

$$p(i, j, t) = \delta^{-d} P_i^t\{X(t) = j\}.$$

Note that $p(i, j, t)$ is nothing but the "kernel" of the discrete heat equation in Λ_δ with Neumann boundary conditions. We now get

$$(8) \quad \left\langle \sum_{k=1}^\infty \frac{1}{k}[I - (H_\delta + f)\,\Delta t]^k e_i, e_i \right\rangle$$

$$\approx \sum_{t=\Delta t}^\infty \frac{1}{t}(1 - m^2\,\Delta t)^{t/\Delta t} p(i, i, t) E_{i,i}^t \exp\left(-\sum_0^t f(X(s))\,\Delta t\right) \delta^d\,\Delta t.$$

The trace in our expression (7) for $E \exp(-\frac{1}{2}\Psi_\delta^N(f))$ is the sum of the left-hand side of (8) over all $i \in \Lambda_\delta$. Thus if Λ_δ is finite, we have

$$(9) \quad \operatorname{tr} \sum_{k=1}^\infty \frac{1}{k}[(I - (H_\delta + f)\,\Delta t)^k - (I - H_\delta\,\Delta t)^k]$$

$$\approx \sum_{i \in \Lambda_\delta} \sum_{t=\Delta t}^\infty \frac{1}{t}(1 - m^2\,\Delta t)^{t/\Delta t} p(i, i, t)$$

$$\times E_{i,i}^t \left\{ \exp\left(-\sum_0^t f(X(s))\,\Delta t\right) - 1 \right\} \delta^d\,\Delta t.$$

But if this formula holds for all finite Λ_δ, it must also hold for all hyperfinite Λ_δ with internal cardinality less than some hyperinteger $N_0 \in {}^*\mathbb{N} - \mathbb{N}$. For such Λ_δ we can combine (7) and (9) to get

$$(10) \qquad E \exp\left(-\frac{1}{2}\Psi_\delta^N(f)\right)$$

$$\approx \exp\left\{\frac{1}{2}\sum_{\Lambda_\delta \times T}\frac{1}{t}(1 - m^2\,\Delta t)^{t/\Delta t}p(i, j, t)\right.$$

$$\left. \times E_{i,i}^t\left(\exp\left(-\int_0^t f(X(s))\,ds\right) - 1\right)\delta^d\,\Delta t\right\}$$

(here, and in the remainder of this section, we use the convention that $1/0 = 0$). Comparing (10) and Lemma 7.5.2, we see that if we choose the measure α in 7.5.2 as

$$(11) \qquad \alpha(i, j, t) = (\delta_{ij}/2t)(1 - m^2\,\Delta t)^{t/\Delta t}p(i, j, t)\,\delta^d\,\Delta t,$$

where δ_{ij} is the Kronecker symbol, then $\frac{1}{2}\Psi_\delta^N$ and T have the same distribution (provided, of course, that the condition in 7.5.2 is satisfied). Since α is concentrated on the diagonal $\{(i, i, t) \,|\, i \in \Lambda_\delta, t \in T\}$, the Brownian bridges from which T is constructed will be Brownian loops starting and ending at the same point.

We summarize our findings in the following theorem.

7.5.3. THEOREM. Let Ψ_δ^N be the square of the free lattice field on Λ_δ with Neumann boundary conditions, and let T be the Poisson random field of Brownian loops induced by the measure α in (11). There is a hyperfinite integer $N \in {}^*\mathbb{N} - \mathbb{N}$ such that whenever $|\Lambda_\delta| \leq N$, then

$$E \exp(-\tfrac{1}{2}\Psi(f)) \approx E \exp(-T(f))$$

for all non-negative, internal functions $f: \Lambda_\delta \to {}^*\mathbb{R}$ with $\|f\|_\infty \leq (\log(1/\Delta t))^{1/2}$.

PROOF. We have already carried out all the necessary calculations, and what remains is only a certain amount of bookkeeping; i.e., keeping track of the various conditions we have explicitly or implicitly assumed. As observed above, the condition

$$\|f\|_\infty \leq \left(\log\frac{1}{\Delta t}\right)^{1/2}$$

justifies our use of the Feynman-Kac formula, and if we also choose N no larger than the constant N_0 appearing in the calculations, formula (10) clearly holds.

The rest of the argument is just an appeal to Lemma 7.5.2, and only requires that the condition

$$\sum (1 - \exp(-t\|f\|_\infty))^2 \alpha(i, j, t)^2 \approx 0$$

is satisfied. Using the fact that

$$1 - \exp(-t\|f\|_\infty) \le t\|f\|_\infty,$$

and that the sum

$$S = \frac{1}{4} \sum_{t \in T} (1 - m^2 \Delta t)^{2t/\Delta t} \Delta t = \frac{1}{8m^2 - 4m^4 \Delta t}$$

is finite, we see that if we choose $N \le 1/\sqrt{\Delta t}$, then

$$\sum (1 - \exp(-t\|f\|_\infty))^2 \alpha(i, j, t)^2 \le \sum_{i \in \Lambda_\delta} \sum_{t \in T} t^2 \|f\|_\infty^2 \frac{1}{4t^2} (1 - m^2 \Delta t)^{2t/\Delta t} \Delta t^2$$

$$= S \sum_{i \in \Lambda_\delta} \|f\|_\infty^2 \Delta t \le S \frac{1}{\sqrt{\Delta t}} \log\left(\frac{1}{\Delta t}\right) \Delta t \approx 0,$$

and the proof is complete.

Picking an element $k \in \Lambda_\delta$ and applying T to the function $f_k(i) = \delta^{-d} \delta_{ik}$, we have

$$(12) \qquad T(k)(\sigma, \omega) = T(f_k)(\sigma, \omega)$$

$$= \sum \sigma(i, j, t) \delta^{-d} |\{s < t \,|\, X(s, \omega_{i,j,t}) = k\}| \Delta t.$$

Since $|\{s < t \,|\, X(s, \omega_{i,j,t}) = k\}|$ measures the time X spends at k, Theorem 7.5.3 gives a representation of Φ^2 as a Poisson random field of Brownian local times.

C. Local Time Representations for Interactions Which Are Functions of Φ^2

The next step in our program is to use the representation we just found to study interacting scalar fields. An internal function $u: {}^*\mathbb{R} \to {}^*\mathbb{R}$ defines an interaction U by

$$(13) \qquad U = \lambda \sum_{i \in \Lambda_\delta} u(\Phi_\delta^N(i)) \delta^d,$$

where $\lambda \in {}^*\mathbb{R}$ is a coupling constant. If μ_0 is the free measure (i.e., the probability measure we have been using all along and which makes Φ_δ^N a Gaussian random vector with mean zero and covariance matrix $\delta^{-2d} C_{\Lambda_\delta}$), the interaction measure μ_U is given by

$$(14) \qquad d\mu_U = \frac{\exp(-U) \, d\mu_0}{\int \exp(-U) \, d\mu_0}.$$

Recall from the last section that the interacting field is just Φ_δ^N considered as a random variable with respect to μ_U.

To find the Laplace transform of the interacting field, we must compute

$$(15) \qquad E \exp(-\Phi_\delta^N(g)) \exp(-U),$$

where the expectation is with respect to μ_0. We shall not be able to do this for general interactions, but by using Theorem 7.5.3 we shall obtain the expression

$$(16) \qquad E \exp(-\Phi_\delta^N(g)) F(\tfrac{1}{2}\Psi_\delta^N) \approx E(F(T + T_g)) \exp(\tfrac{1}{2}\langle H_\delta^{-1}g, g\rangle),$$

where $F: {}^*\mathbb{R}^{\Lambda_\delta} \to {}^*\mathbb{R}$ is an internal functional, and T and T_g are independent Poisson fields of Brownian bridges (the measure α inducing T_g depends on the function g). If the interaction is a function of Ψ_δ^N, i.e.,

$$U = \lambda \sum_{i \in \Lambda_\delta} v(\Psi_\delta^N(i)) \, \delta^d,$$

we can then apply (16) to $F(\Psi_\delta^N) = e^{-U}$ to obtain information about (15).

In order to prove (16), we shall first consider the special case where $F(\Psi_\delta^N) = \exp(-\tfrac{1}{2}\Psi_\delta^N(f))$ for some function f, and show that

$$(17) \qquad E(\exp(-\Phi_\delta^N(g)) \exp(-\tfrac{1}{2}\Psi_\delta^N(f)))$$

$$\approx E \exp[-T(f) - T_g(f)] \exp(\tfrac{1}{2}\langle g, H_\delta^{-1}g\rangle).$$

Once this has been established, an easy Stone–Weierstrass-type argument will take care of general F's.

The main tool in computing (17) is simply the formula

$$(18) \qquad (2\pi)^{-n/2} \operatorname{Det} A^{1/2} \int_{\mathbb{R}^n} \exp(-\tfrac{1}{2}(Ax, x)) \exp(-(y, x)) \, dx$$

$$= \exp(\tfrac{1}{2}(A^{-1}y, y))$$

for the Laplace transform of a Gaussian measure. Recalling that Φ_δ^N is Gaussian with covariance matrix $\delta^{-2d}C_{\Lambda_\delta}$ and that $D_{\Lambda_\delta} = C_{\Lambda_\delta}^{-1}$, we get

$$(19) \quad E[\exp(-\Phi_\delta^N(g)) \exp(-\tfrac{1}{2}\Psi_\delta^N(f))]$$

$$= (2\pi)^{-|\Lambda_\delta|/2} \operatorname{Det} D_{\Lambda_\delta}^{1/2} \int_{\mathbb{R}^{|\Lambda_\delta|}} \exp[-\tfrac{1}{2}((D_{\Lambda_\delta} + \delta^{-d}fI)x, x)]$$

$$\times \exp(-(g, x)) \, dx$$

$$= \operatorname{Det} D_{\Lambda_\delta}^{1/2} \operatorname{Det}(D_{\Lambda_\delta} + \delta^{-d}fI)^{-1/2} \exp[\tfrac{1}{2}((D_{\Lambda_\delta} + \delta^{-d}fI)^{-1}g, g)],$$

where all inner products are in $\mathbb{R}^{|\Lambda_\delta|}$, i.e., $(x, y) = \sum_{i \in \Lambda_\delta} x_i y_i$. Since we are elsewhere using the inner product in $l^2(\Lambda_\delta, \delta^d)$, i.e., $\langle f, g \rangle = \sum_{i \in \Lambda_\delta} f(i)g(i)\delta^d$, it is convenient to rewrite (19) as

$$(20) \quad E[\exp(-\Phi_\delta^N(g)) \exp(-\tfrac{1}{2}\Psi_\delta^N(f))]$$
$$= \text{Det } D_{\Lambda_\delta}^{1/2} \text{Det}(D_{\Lambda_\delta} + \delta^{-d}fI)^{-1/2} \exp(\tfrac{1}{2}\langle g, (H_\delta + f)^{-1}g \rangle),$$

where the inner product is in $l^2(\Lambda_\delta, \delta^d)$. Recall that $H_\delta = -\tfrac{1}{2}\Delta_\delta + m^2$.

We have already evaluated the product of the two determinants in (20):

$$\text{Det } D_{\Lambda_\delta}^{1/2} \text{Det}(D_{\Lambda_\delta} + \delta^{-d}fI)^{-1/2} \approx E \exp(-T(f)),$$

where T is the Poisson field of Brownian loops induced by the measure α in (11).

To compute the exponential factor in (20), note that

$$(21) \qquad (H_\delta + f)^{-1} = \sum_{k=0}^{\infty} (I - (H_\delta + f)\,\Delta t)^k\,\Delta t$$

since $H_\delta + f$ is strictly positive. By the Feynman–Kac formula

$$(22) \quad \sum_{k=0}^{\infty} \langle (I - (H_\delta + f)\,\Delta t)^k g, g \rangle$$
$$\approx \sum_{t=0}^{\infty} E\left(g(X(t))g(X(0)) \exp\left(-\int_0^t f(X(s))\, ds\right) \right) \Delta t,$$

where X is the Markov process generated by H_δ. By introducing the conditional measures $p_{i,j}^t$ and their expection $E_{i,j}^t$, the last formula can be written as

$$(23) \quad \sum_{k=0}^{\infty} \langle (I - (H_\delta + f)\,\Delta t)^k g, g \rangle$$
$$\approx \sum_{(i,j,t) \in \Lambda_\delta^2 \times T} g(i)g(j)(1 - m^2\,\Delta t)^{t/\Delta t} p(i, j, t)$$
$$\times E_{i,j}^t\left[\exp\left(-\int_0^t f(X(s))\, ds\right) \delta^{2d}\,\Delta t \right].$$

If we define an internal measure α_g on $\Lambda_\delta^2 \times T$ by

$$(24) \qquad \alpha_g(i, j, t) = \tfrac{1}{2}g(i)g(j)(1 - m^2\,\Delta t)^{t/\Delta t} p(i, j, t)\,\delta^{2d}\,\Delta t,$$

we can combine (21)–(24) to get

$$(25) \quad \exp\left(\frac{1}{2}\langle (H_\delta + f)^{-1}g, g \rangle\right)$$
$$\approx \exp\left\{ \int E_{i,j}^t\left[\exp\left(-\int_0^t f(X(s))\, ds\right) \right] d\alpha_g(i, j, t) \right\}.$$

Note that if $f \equiv 0$, this formula reads

(26)
$$\exp\!\left(\frac{1}{2}\langle H_\delta^{-1} g, g\rangle\right) \approx \exp\!\left\{\int E_{i,j}^t(1)\, d\alpha_g(i, j, t)\right\}$$

(in fact, we have exact equality in this case), and thus

(27)
$$\exp\!\left(\frac{1}{2}\langle (H_\delta + f)^{-1} g, g\rangle\right)$$
$$\approx \exp\!\left\{\int E_{i,j}^t\!\left[\exp\!\left(-\int_0^t f(X(s))\, ds\right) - 1\right] d\alpha_g\right\}$$
$$\times \exp\!\left(\frac{1}{2}\langle H_\delta^{-1} g, g\rangle\right).$$

The reason for this last, rather mysterious maneuver is simply that if T_g is the Poisson random field of Brownian bridges induced by α_g, then

(28)
$$E\,\exp(-T_g(f)) \approx \exp\!\left\{\int E_{i,j}^t\!\left[\exp\!\left(-\int_0^t f(X(s))\, ds\right) - 1\right] d\alpha_g\right\}$$

by Lemma 7.5.2, and thus (27) tells us that

(29) $\quad \exp(\frac{1}{2}\langle (H_\delta + f)^{-1} g, g\rangle) \approx E[\exp(-T_g(f))] \exp(\frac{1}{2}\langle H_\delta^{-1} g, g\rangle).$

Combining (20) and (29) we get

(30)
$$E[\exp(-\Phi_\delta^N(g))\exp(-\tfrac{1}{2}\Psi_\delta^N(f))]$$
$$\approx E\,\exp(-T(f))E\,\exp(-T_g(f))\exp(\tfrac{1}{2}\langle H_\delta^{-1} g, g\rangle),$$

or—provided the random fields $T(f)$ and $T_g(f)$ are constructed independently of each other—

(31)
$$E[\exp(-\Phi_\delta^N(g))\exp(-\tfrac{1}{2}\Psi_\delta^N(f))]$$
$$\approx E\,\exp(-(T + T_g)(f))\exp(\tfrac{1}{2}\langle H_\delta^{-1} g, g\rangle).$$

If $F_f : {}^*\mathbb{R}^{\Lambda_\delta} \to {}^*\mathbb{R}$ is the internal functional
$$F_f(g) = \exp(-\langle g, f\rangle),$$

we can rewrite (30) as

(32) $\quad E[\exp(-\Phi_\delta^N(g))F_f(\tfrac{1}{2}\Psi_\delta^N)] \approx E(F_f(T + T_g))\exp(\tfrac{1}{2}\langle H_\delta^{-1} g, g\rangle),$

where we consider $T + T_g$ as a function from Λ_δ to ${}^*\mathbb{R}$ given by

(33)
$$(T + T_g)(i) = (T + T_g)(f_i),$$

f_i denoting the function $f_i(j) = \delta^{-d}\delta_{ij}$.

Before we extend (32) to more general functionals, let us formulate our results so far as a lemma.

7.5.4. LEMMA. There is an infinite integer $N_1 \in {}^*\mathbb{N}$ such that if $|\Lambda_\delta| \le N_1$, then
$$E[\exp(-\Phi_\delta^N(g))F_f(\tfrac{1}{2}\Psi_\delta^N)] \approx E[F_f(T + T_g)]\exp(\tfrac{1}{2}\langle H_\delta^{-1} g, g\rangle)$$

for all non-negative, internal functions f and g such that $\|f\|_\infty \leq (\log(1/\Delta t))^{1/2}$ and $\|g\|_2/\sqrt{\Delta t} \approx 0$.

PROOF. As in the proof of 7.5.3, we only have to check that the conditions are sufficient for the calculations we have already carried out. Clearly they are strong enough for the use we have made of Theorem 7.5.3, and it is easy to check that since $\|f\|_\infty \leq (\log(1/\Delta t))^{1/2}$, our application of the Feynman–Kac formula in (22) is valid. Hence (27) holds.

To justify the appeal to Lemma 7.5.2, it suffices to check that $\sum \alpha_g(i, j, t)^2 \approx 0$. But

$$\sum \alpha_g(i, j, t)^2 \leq \sum_{(i,j,t)} \frac{1}{4} g(i)^2 g(j)^2 (1 - m^2 \Delta t)^{2t/\Delta t} \delta^{2d} \Delta t^2$$

$$\leq \frac{1}{8m^2 - 4m^4 \Delta t} \left(\sum_i \frac{1}{2} g(i)^2 \delta^d \right)^2 \Delta t \approx 0,$$

and the lemma is proved.

To generalize Lemma 7.5.4, it is necessary to show first that the special functionals F_f span the set of all reasonable functionals in an appropriate sense. We shall use the following terminology. For each $H \in {}^*\mathbb{N}$, an H functional is an internal function

$$F : l^2(\Lambda_\delta, \Lambda^d) \to {}^*\mathbb{R}$$

such that

$$(34) \qquad\qquad |F(f)| \leq H \exp(-\|f\|^2/H)$$

and

$$(35) \qquad\qquad |F(f) - F(g)| \leq H\|f - g\|$$

for all $f, g \in l^2(\Lambda_\delta, \delta^d)$. A hyperfinite sequence (α_n, f_n) consisting of numbers $\alpha_n \in {}^*\mathbb{R}$ and non-negative, internal functions $f_n : \Lambda_\delta \to {}^*\mathbb{R}$ is called a $(K, \Delta t)$ sequence if $\sum |\alpha_n| \leq K$ and $\|f_n\|_\infty \leq (\log(1/\Delta t))^{1/2}$ for all n.

7.5.5. LEMMA. For each $K \in {}^*\mathbb{N} - \mathbb{N}$, there is an $H \in {}^*\mathbb{N} - \mathbb{N}$ with the property that if $|\Lambda_\delta| \leq H$, then for each H functional $F : l^2(\Lambda_\delta, \delta^d) \to {}^*\mathbb{R}$, there is a $(K, \Delta t)$ sequence (α_n, f_n) such that

$$\left| F(g) - \sum \alpha_n F_{f_n}(g) \right| \leq \frac{1}{H}$$

for all non-negative, internal $g : \Lambda_\delta \to {}^*\mathbb{R}$.

PROOF. Assume first that H is finite, and let $\overline{\mathbb{R}_+^{\Lambda_\delta}} = \mathbb{R}_+^{\Lambda_\delta} \cup \{\infty\}$ be the one-point compactification of $\mathbb{R}^{\Lambda_\delta}$. Since Λ_δ is finite, $\mathbb{R}_+^{\Lambda_\delta}$ is a subset of

$l^2(\Lambda_\delta, \delta^d)$, and we can turn F into a standard function $\tilde{F}: \overline{\mathbb{R}_+^{\Lambda_\delta}} \to \mathbb{R}$ by letting

$$\tilde{F}(g) = {}^\circ F(g)$$

for all $g \in \mathbb{R}_+^{\Lambda_\delta}$, and setting

$$\tilde{F}(\infty) = 0.$$

Note that the condition $|F(g)| \le He^{-\|g\|^2/H}$ implies that $\tilde{F}$ is continuous at infinity.

Each strictly positive function $f: \Lambda_\delta \to \mathbb{R}$ defines a continuous function $\tilde{F}_f$ on $\overline{\mathbb{R}_+^{\Lambda_\delta}}$ by

$$\tilde{F}_f(g) = \exp(-\langle g, f \rangle)$$

for $g \in \mathbb{R}_+^{\Lambda_\delta}$, and $\tilde{F}_f(\infty) = 0$; the *strict* positivity is needed to get continuity at infinity. The algebra $\mathscr{A}$ consisting of all finite, linear combinations

$$\beta_0 + \sum_{i=1}^m \beta_i \tilde{F}_{f_i}$$

separates points and contains the constant functions, and since $\overline{\mathbb{R}_+^{\Lambda_\delta}}$ is compact, the Stone-Weierstrass theorem tells us that we can find finite sequences $\{\beta_i\}$, $\{f_i\}$ such that

$$(36) \qquad |\tilde{F}(g) - (\beta_0 + \sum \beta_i \tilde{F}_{f_i}(g))| \le \frac{1}{2H}$$

for all $g \in \overline{\mathbb{R}_+^{\Lambda_\delta}}$.

Considering the f_i's as elements of $l^2(\Lambda_\delta, \delta^d)$, define functionals $F_{f_i}: l^2(\Lambda_\delta, \delta^d) \to {}^*\mathbb{R}$ by

$$F_{f_i}(g) = \exp(-\langle g, f_i \rangle).$$

Using the decay and continuity conditions (34), (35) on F, and the facts that β_i, f_i are standard and f_i is strictly positive, we get from (36) that

$$|F(g) - (\beta_0 + \sum \beta_i F_{f_i}(g))| \le \frac{1}{H}$$

for all non-negative, internal $g: \Lambda_\delta \to {}^*\mathbb{R}$. As (β_i, f_i) is obviously a $(K, \Delta t)$ sequence, this proves the lemma for all finite H, and hence for all sufficiently small, infinite H.

REMARK. The condition $|F(f)| \le H \exp(-\|f\|^2/H)$, which we have used to govern F's behavior at infinity, is not in any sense canonical. In fact, we can replace it by

$$(37) \qquad |F(f)| \le {}^*G(H, \|f\|)$$

for any standard function $G: \mathbb{N} \times \mathbb{R}_+ \to \mathbb{R}_+$ such that $\lim_{x \to \infty} G(n, x) = 0$ for all $n \in \mathbb{N}$.

Combining 7.5.4 and 7.5.5, we get the following theorem.

7.5.6. THEOREM. There is an $H \in {}^*\mathbb{N} - \mathbb{N}$ such that if $|\Lambda_\delta| \leq H$, then for all
H functionals $F: l^2(\Lambda_\delta, \delta^d) \to \mathbb{R}$ and all $g \in l^2(\Lambda_\delta, \delta^d)$ with ${}^\circ\|g\| < \infty$, $g > 0$,

$$(38) \quad E[\exp(-\Phi_\delta^N(g))F(\tfrac{1}{2}\Psi_\delta^N)] \approx E[F(T+T_g)]\exp(\tfrac{1}{2}\langle H_\delta^{-1}g, g\rangle)$$

and, provided ${}^\circ F$ is not identically zero,

$$(39) \quad \frac{E[\exp(-\Phi_\delta^N(g))F(\tfrac{1}{2}\Psi_\delta^N)]}{E[F(\tfrac{1}{2}\Psi_\delta^N)]} \approx \frac{E[F(T+T_g)]\exp(\tfrac{1}{2}\langle H_\delta^{-1}g, g\rangle)}{E[F(T)]}.$$

PROOF. If $|\Lambda_\delta|$ is less than the constant $N_1 \in {}^*\mathbb{N} - \mathbb{N}$ occurring in Lemma
7.5.4, there must be a $K \in {}^*\mathbb{N} - \mathbb{N}$ such that

$$\left| E\left[\exp(-\Phi_\delta^N(g)) \sum \alpha_n F_{f_n}\left(\frac{1}{2}\Psi_\delta^N\right) \right] \right.$$
$$\left. - E[\sum \alpha_n F_{f_n}(T+T_g)] \exp\left(\frac{1}{2}\langle H_\delta^{-1}g, g\rangle\right) \right| \leq \frac{1}{K}$$

for all $(K, \Delta t)$ sequences (α_n, f_n) and all $g \in l^2(\Lambda_\delta, \delta^d)$ with finite norm. If
$|\Lambda_\delta|$ is also less than the constant H in Lemma 7.5.5, we can find a $(K, \Delta t)$
sequence (α_n, f_n) such that the difference $F - \sum \alpha_n F_{f_n}$ is less than H^{-1} in
supremum norm. Consequently,

$$\left| E\left[\exp(-\Phi_\delta^N(g))\left(F\left(\frac{1}{2}\Psi_\delta^N\right) - \sum \alpha_n F_{f_n}\left(\frac{1}{2}\Psi_\delta^n\right) \right) \right] \right|$$
$$\leq \frac{1}{H} E \exp(-\Phi_\delta^N(g)) = \frac{1}{H} \exp\left(\frac{1}{2}\langle H_\delta^{-1}g, g\rangle\right)$$

and

$$\left| E[F(T+T_g) - \sum \alpha_n F_{f_n}(T+T_g)] \exp\left(\frac{1}{2}\langle H_\delta^{-1}g, g\rangle\right) \right|$$
$$\leq \frac{1}{H} \exp\left(\frac{1}{2}\langle H_\delta^{-1}g, g\rangle\right).$$

Since the operator H_δ is bounded from below by m^2, we have

$$\langle H_\delta^{-1}g, g\rangle \leq \|g\|^2 / m^2,$$

and thus

$$\left| E\left[\exp(-\Phi_\delta^N(g))F\left(\frac{1}{2}\Psi_\delta^N\right) \right] - E[F(T+T_g)]\exp\left(\frac{1}{2}\langle H_\delta^{-1}g, g\rangle\right) \right|$$
$$\leq \frac{1}{K} + \frac{2}{H} \exp\left(\frac{\|g\|^2}{2m^2}\right),$$

which proves (38).

Turning to (39), note that if H is finite, $E(F(\tfrac{1}{2}\Psi_\delta^N))$ and $E(F(T))$ are
infinitely close and noninfinitesimal. Hence (39) holds for all finite H, and
the extension to sufficiently small, infinite H is straightforward.

REMARK. The theorem above is a hyperfinite version of Dynkin's representation formula [Dynkin (1984a,b); see also Brydges *et al.* (1982)]. Note that the class of internal functionals F that we allow is sufficiently rich to represent all standard functionals from $l^2(\Lambda, dx)$ to $\mathbb{R}$.

D. Φ^4 and Polymer Measures

Recall from Section 7.4 that Φ^4 fields are given by interactions of the form

(40)
$$U(\Phi_\delta^N) = \frac{\lambda}{4} \sum_{i \in \Lambda_\delta} \Phi_\delta^N(i)^4 \, \delta^d + \frac{a}{2} \sum_{i \in \Lambda_\delta} \Phi_\delta^N(i)^2 \, \delta^d$$

$$= \frac{\lambda}{4} \sum_{i \in \Lambda_\delta} \Psi_\delta^N(i)^2 \, \delta^d + \frac{a}{2} \sum_{i \in \Lambda_\delta} \Psi_\delta^N(i) \, \delta^d,$$

where λ and a are constants. To see how this field can be represented in terms of Brownian local times, note that its generating functional can be written as

(41)
$$E \, \exp(-\Phi_\delta^N(g)) \exp(-U)$$

$$= E \left[\exp(-\Phi_\delta^N(g)) \exp\left(-\frac{\lambda}{4} \sum_{i \in \Lambda_\delta} \Psi_\delta^N(i)^2 \, \delta^d \right. \right.$$

$$\left. \left. -\frac{a}{2} \sum_{i \in \Lambda_\delta} \Psi_\delta^N(i) \, \delta^d \right) \right]$$

$$= E \, \exp(-\Phi_\delta^N(g)) F\left(\frac{1}{2} \Psi_\delta^N \right)$$

for

$$F(f) = \exp\left(-\lambda \sum_{\Lambda_\delta} f^2 \, \delta^d - a \sum_{\Lambda_\delta} f \delta^d \right).$$

When the conditions of Theorem 7.5.6 are satisfied, this means that

(42) $E \, \exp(-\Phi_\delta^N(g)) \exp(-U) \approx E(F(T + T_g)) \exp(\frac{1}{2}\langle H_\delta^{-1} g, g \rangle),$

or, if we spell it out,

(43)
$$E \, \exp(-\Phi_\delta^N(g)) \exp(-U)$$

$$\approx E \, \exp\left\{ -\lambda \sum_{\Lambda_\delta} [T^2(i) + 2T(i) T_g(i) + T_g^2(i)] \delta^d \right.$$

$$\left. - a \sum_{\Lambda_\delta} [T(i) + T_g(i)] \, \delta^d \right\}$$

$$\times \exp\left(\frac{1}{2} \langle H_\delta^{-1} g, g \rangle \right).$$

Before we discuss the terms of the last exponent, we must make our notation a little more explicit. Recall from (12) that T is defined by

$$(44) \qquad T(k)(\sigma, \omega) = \sum_{(i,j,t)} \sigma(i, j, t) \sum_{s=0}^{t} \delta^{-d} 1_{\{k\}}(X(\omega_{i,j,t}, s)) \, \Delta t,$$

where σ is the hyperfinite Poisson field induced by the measure α in (11), and X is the Markov process generated by H_δ. Similarly,

$$(45) \qquad T_g(k)(\sigma_g, \omega_g) = \sum_{(i,j,t)} \sigma_g(i, j, t) \sum_{s=0}^{t} \delta^{-d} 1_{\{k\}}(X_g(\omega_{g,i,j,t}, s)) \, \Delta t,$$

where σ_g is the hyperfinite Poisson field induced by the α_g in (24), and X_g is another copy of the Markov process generated by H_δ. Since T_g is assumed to be independent of T, we let σ, X and σ_g, X_g be chosen independent.

Quadratic terms in T and T_g can now be computed as follows:

$$(46) \qquad \lambda \sum_{k \in \Lambda_\delta} T(k) T_g(k) \, \delta^d$$

$$= \lambda \sum_{k \in \Lambda_\delta} \sum_{(i,j,t)} \sum_{(\tilde{i}, \tilde{j}, \tilde{t})} \sigma(i, j, t) \sigma_g(i, j, t)$$

$$\times \sum_{s=0}^{t} \sum_{\tilde{s}=0}^{t} \delta^{-d} 1_{\{k\}}(X(s)) 1_{\{k\}}(X_g(\tilde{s})) \, \Delta t^2$$

$$= \lambda \sum_{(i,j,t)} \sum_{(\tilde{i}, \tilde{j}, \tilde{t})} \sigma(i, j, t) \sigma_g(\tilde{i}, \tilde{j}, \tilde{t})$$

$$\times \int_0^t \int_0^{\tilde{t}} \tilde{\delta}(X(s) - X_g(\tilde{s})) \, d\tilde{s} \, ds,$$

where $\tilde{\delta}$ is the hyperfinite version of the d-dimensional delta function; $\tilde{\delta}(0) = \delta^{-d}$ and $\tilde{\delta}(i) = 0$ for $i \neq 0$. Similarly,

$$(47) \qquad \lambda \sum_{k \in \Lambda_\delta} T(k)^2 \, \delta^d = \lambda \sum_{(i,j,t)} \sum_{(\tilde{i}, \tilde{j}, \tilde{t})} \sigma(i, j, t) \sigma(\tilde{i}, \tilde{j}, \tilde{t})$$

$$\times \int_0^t \int_0^{\tilde{t}} \tilde{\delta}(X(\omega_{i,j,t}, s) - X(\omega_{\tilde{i},\tilde{j},\tilde{t}}, \tilde{s})) \, d\tilde{s} \, ds$$

and

$$(48) \qquad \lambda \sum_{k \in \Lambda_\delta} T_g(k)^2 \, \delta^d = \lambda \sum_{(i,j,t)} \sum_{(\tilde{i}, \tilde{j}, \tilde{t})} \sigma_g(i, j, t) \sigma_g(\tilde{i}, \tilde{j}, \tilde{t})$$

$$\times \int_0^t \int_0^{\tilde{t}} \tilde{\delta}(X_g(\omega_{g,i,j,t}, s) - X_g(\omega_{g,\tilde{i},\tilde{j},\tilde{t}}, \tilde{s})) \, d\tilde{s} \, ds,$$

where we have displayed the ω's on the right-hand sides to emphasize that unless $(i, j, t) = (\tilde{i}, \tilde{j}, \tilde{t})$, the processes $X(\omega_{i,j,t}, s)$ and $X(\omega_{\tilde{i},\tilde{j},\tilde{t}}, \tilde{s})$ are independent [and so are $X_g(\omega_{g,i,j,t}, s)$ and $X_g(\omega_{g,\tilde{i},\tilde{j},\tilde{t}}, \tilde{s})$].

Linear terms in T and T_g can also be computed:

$$(49) \quad a \sum_{k \in \Lambda_\delta} T(k) \, \delta^d = a \sum_{k \in \Lambda_\delta} \sum_{(i,j,t)} \sigma(i,j,t) \sum_{s=0}^{t} \delta^{-d} 1_{\{k\}}(X(s)) \, \Delta t \, \delta^d$$

$$= a \sum_{(i,j,t)} \sigma(i,j,t) \int_0^1 1_{\Lambda_\delta}(X(s)) \, ds,$$

and analogously

$$(50) \quad a \sum_{k \in \Lambda_\delta} T_g(k) \, \delta^d = a \sum_{(i,j,t)} \sigma_g(i,j,t) \int_0^1 1_{\Lambda_\delta}(X_g(s)) \, ds.$$

Substituting the right-hand sides of (46)–(50) in (43), we arrive at the expression

$$(51) \quad E \exp(-\Phi_\delta^N(g)) \exp(-U)$$

$$\approx E \exp\left\{ -\lambda \sum_{(i,j,t)} \sum_{(\tilde{i},\tilde{j},\tilde{t})} \sigma(i,j,t) \sigma(\tilde{i},\tilde{j},\tilde{t}) \right.$$

$$\times \int_0^t \int_0^{\tilde{t}} \tilde{\delta}(X(\omega_{i,j,t},s) - X(\omega_{\tilde{i},\tilde{j},\tilde{t}},\tilde{s})) \, d\tilde{s} \, ds$$

$$- 2\lambda \sum_{(i,j,t)} \sum_{(\tilde{i},\tilde{j},\tilde{t})} \sigma(i,j,t) \sigma_g(\tilde{i},\tilde{j},\tilde{t})$$

$$\times \int_0^t \int_0^{\tilde{t}} \tilde{\delta}(X(\tilde{s}) - X_g(\tilde{s})) \, d\tilde{s} \, ds$$

$$- \lambda \sum_{(i,j,t)} \sum_{(\tilde{i},\tilde{j},\tilde{t})} \sigma_g(i,j,t) \sigma_g(\tilde{i},\tilde{j},\tilde{t})$$

$$\times \int_0^t \int_0^{\tilde{t}} \tilde{\delta}(X_g(\omega_{g,i,j,t},s) - X_g(\omega_{g,\tilde{i},\tilde{j},\tilde{t}},\tilde{s})) \, d\tilde{s} \, ds$$

$$- a \sum_{(i,j,t)} \sigma(i,j,t) \int_0^t 1_{\Lambda_\delta}(X(s)) \, ds$$

$$\left. - a \sum_{(i,j,t)} \sigma_g(i,j,t) \int_0^t 1_{\Lambda_\delta}(X_g(s)) \, ds \right\}$$

$$\times \exp\left(\frac{1}{2} \langle H_\delta^{-1} g, g \rangle \right).$$

In this formula the last two terms [coming from (49) and (50)] are easy to handle, while the first three are similar to the ones we studied in Section 6.4.C. As long as $(i, j, t) \neq (\tilde{i}, \tilde{j}, \tilde{t})$, the processes occurring inside the same delta functions are independent, and we are in the situation discussed in formulas (6.4.35)–(6.4.37). When $(i, j, t) = (\tilde{i}, \tilde{j}, \tilde{t})$—and note that this event has noninfinitesimal probability—we are dealing with the more complicated polymer measures described in (6.4.39) and (6.4.40). Thus the triviality or nontriviality of a Φ^4 scalar field is intimately tied up with the behavior of polymer measures and perturbations of the Laplacian along Brownian paths.

It should be pointed out at this stage that it is not only our inability to control four-dimensional polymer measures which prevents us from saying anything definite about Φ_4^4. Recall from Section 6.4 that we could only make standard sense of the expressions

$$\exp\left(-\int_0^t \int_0^{\tilde{t}} \lambda \, \delta(b_1(s) - b_2(\tilde{s}))\right) d\tilde{s} \, ds$$

for certain negative, infinitesimal λ. Now it turns out that the result on which we have based our discussion, Theorem 7.5.6, does not hold for negative λ's! In fact, both the numerator and the denominator on the left-hand side of (39) diverge when λ is negative. However, since T and T_g are basically *-finite quantities, the right-hand side does make sense, and it is relatively easy to reinterpret the left-hand side by means of a truncation argument in such a way that the result extends to negative λ's. A much more serious problem is caused by the fact that the infinitesimal λ's we found in Section 6.4 were not constants, but functions of ω and t. It might, of course, be that one can choose them to be constant without violating the argument in Section 6.4, but we have not been able to prove this. On the other hand, there is no obvious way of making sense of the right-hand side of (51) as a quantum field when λ is not constant.

To sum up, let us emphasize that although the computations above and the results of Section 6.4 may seem to suggest that Φ_4^4 could be nontrivial for certain negative and infinitesimal choices of the coupling constant λ, there is a long way to go before such a claim can be either proved or disproved. Work is already in progress; at the time of writing, Andreas Stoll is developing a hyperfinite approach to polymer measures based on (self-repellent) random walks [see Stoll (1985)].

For other discussions of infinitesimal coupling constants and related topics—such as triviality results for repulsive Φ^4 models—see Aizenman (1981), Albeverio *et al.* (1982, 1984a,c), Fröhlich (1982), Gallavotti and Nicolò (1985), Sokal (1982), and Westwater (1980, 1985).

E. $\Phi_1^2\Phi_2^2$ Fields and Local Time Perturbations

As we have just seen, the two major problems we encounter in trying to control Φ^4 through its local time representation are the occurrence of polymer measures on the right-hand side of (51) and the need to choose λ

constant. We shall take a brief look at another kind of interacting quantum fields where the first problem is avoided, but where, unfortunately, the second still is present. What we have in mind are the so-called $(\Phi_1^2\Phi_2^2)_d$ models describing two d-dimensional quantum fields Φ_1 and Φ_2 interacting through a term

$$(52) \qquad \frac{\lambda}{4} \sum_{i \in \Lambda_\delta} \Phi_1^2(i)\Phi_2^2(i) \, \delta^d,$$

where λ is a coupling constant.

To describe the situation mathematically, let m_1 and m_2 be two positive, real numbers representing the masses of the two fields. Let D_k, $k = 1$ or 2, be the matrix of $H_k = -\frac{1}{2}\Delta_d + m_k^2$ as an operator from $l^2(\Lambda_\delta, \delta^d)$ to $l^2(\Lambda_\delta, \delta^d)$, and let C_k be the inverse matrix. Fix a probability measure μ_0 and two independent random vectors $\{\tilde{\Phi}_1(i)\}_{i \in \Lambda_\delta}$, $\{\tilde{\Phi}_2(i)\}_{i \in \Lambda_\delta}$ with covariance matrices C_1 and C_2, respectively. As in Section 7.5.B above, the quantum fields Φ_1 and Φ_2 are simply defined by $\Phi_k = \delta^{-d}\tilde{\Phi}_k$ for $k = 1, 2$. With respect to μ_0, Φ_1 and Φ_2 are clearly independent, free scalar fields.

The interaction measure μ is defined by

$$(53) \qquad d\mu = Z^{-1}\exp\left(-\frac{\lambda}{4}\sum_{i \in \Lambda_\delta}\Phi_1^2(i)\Phi_2^2(i)\,\delta^d\right)d\mu_0,$$

where

$$Z = \int \exp\left(-\frac{\lambda}{4}\sum_{i \in \Lambda_\delta}\Phi_1^2(i)\Phi_2^2(i)\,\delta^d\right)d\mu_0$$

is just the normalizing constant. Assume, for the time being, that λ is non-negative. As usual we want to compute the Laplace transform of μ:

$$(54) \qquad F(f, g) = \int \exp(-\Phi_1(f) - \Phi_2(g))\,d\mu$$

$$= Z^{-1}\int \exp\left(-\sum_i\left\{\frac{\lambda}{4}\Phi_1^2(i)\Phi_2^2(i) + \Phi_1(i)f(i)\right.\right.$$

$$\left.\left. + \Phi_2(i)g(i)\right\}\delta^d\right)d\mu_0.$$

Dealing with two interacting quantum fields, this problem does not fit into the framework discussed in Section 7.5.C, and we shall have to start our calculations from scratch. It turns out, however, that along the way we will be able to make use of some of the formulas we have already derived, and that this will simplify our task considerably. In particular, we shall use the formula

(55) $(2\pi)^{-n/2} \operatorname{Det} A^{1/2} \int_{\mathbb{R}^n} \exp\left(-\frac{1}{2}(Ax, x)\right) \exp(-(y, x)) \, dx$

$$= \exp\left(\frac{1}{2}(A^{-1}y, y)\right)$$

for the Laplace transform of a Gaussian measure; the expression

(56) $\operatorname{Det} D_{\Lambda_\delta}^{1/2} \operatorname{Det}(D_{\Lambda_\delta} + \delta^{-d}f)^{-1/2}$

$$\approx \exp\left\{\int_{\Lambda_\delta^2 \times T} E_{i,j}^t \left(\exp\left(-\int_0^t f(X(s)) \, ds\right) - 1\right) d\alpha\right\}$$

$$= \exp(-\alpha(\Delta_\delta^2 \times T)) \exp\left\{\int_{\Lambda_\delta^2 \times T} E_{i,j}^t \exp\left(-\int_0^t f(X(s)) \, ds\right) d\alpha\right\}$$

obtained by combining (5), (10), and (11), and formula (25), i.e.,

(57) $\exp\left(\frac{1}{2}\langle(H_\delta + f)^{-1}g, g\rangle\right)$

$$\approx \exp\left\{\int_{\Lambda_\delta^2 \times T} E_{i,j}^t \exp\left(-\int_0^t f(X(s)) \, ds\right) d\alpha_g\right\}.$$

We begin the computations by observing that since $\tilde{\Phi}_1$ and $\tilde{\Phi}_2$ are independent and Gaussian with covariance matrices C_1 and C_2

(58) $\tilde{F}(f, g) = ZF(f, g) = (2\pi)^{-|\Lambda_\delta|} \operatorname{Det} C_1^{-1/2} \operatorname{Det} C_2^{-1/2}$

$$\times \int_{\mathbb{R}^{2|\Lambda_\delta|}} \exp\left[-\frac{1}{2} \sum_{i,j \in \Lambda_\delta} C_1^{-1}(i, j)p_i p_j\right]$$

$$\times \exp\left[-\frac{1}{2} \sum_{i,j \in \Lambda_\delta} C_2^{-1}(i, j)q_i q_j\right]$$

$$\times \exp\left[-\sum_{i \in \Lambda_\delta} \left\{\frac{\lambda}{4} p_i^2 q_i^2 \, \delta^{-3d} + p_i f(i) + q_i g(i)\right\}\right]$$

$$\times \prod_{i \in \Lambda_\delta} dp_i \prod_{i \in \Lambda_\delta} dq_i.$$

Let us carry out the integration with respect to the p_i variables. Recalling that $D_k = C_k^{-1}$, $k = 1$ or 2, we get from (55):

(59) $\tilde{F}(f, g) = (2\pi)^{-|\Lambda_\delta|/2} \operatorname{Det} D_2^{1/2}$

$$\times \int_{\mathbb{R}^{|\Lambda_\delta|}} \operatorname{Det} D_1^{1/2} \cdot \operatorname{Det}\left(D_1 + \frac{\lambda}{2} q^2 \, \delta^{-3d}\right)^{-1/2}$$

$$\times \exp\left(\frac{1}{2}\left(\left(D_1 + \frac{\lambda}{2} q^2 \, \delta^{-3d}\right)^{-1} f, f\right)\right)$$

$$\times \exp\left(-\frac{1}{2}\sum D_2(i,j)q_iq_j\right) \cdot \exp\left(-\sum q_i g(i)\right) \prod_{i\in\Lambda_\delta} dq_i.$$

By (56)

(60) $\quad \text{Det } D_1^{1/2} \text{ Det}\left(D_1 + \frac{\lambda}{2}q^2\,\delta^{-3d}\right)^{-1/2}$

$$\approx \exp(-\alpha_1(\Lambda_\delta^2 \times T))$$

$$\times \exp\left\{\int_{\Lambda_\delta^2 \times T} E_{i,j}^t\left(\exp\left(-\int_0^t \frac{\lambda}{2}\delta^{-2d}q(X_1(s))^2\,ds\right)\right)d\alpha_1\right\},$$

where X_1 is the Markov process generated by $H_1 = -\frac{1}{2}\Delta_d + m_1^2$, and α_1 is the measure obtained from (11) by substituting m_1 for m. Similarly, by (57)

(61) $\quad \exp\left(\frac{1}{2}\left(\left(D_1 + \frac{\lambda}{2}q^2\,\delta^{-3d}\right)^{-1}f,f\right)\right)$

$$= \exp\left(\frac{1}{2}\left\langle\left(H_1 + \frac{\lambda}{2}q^2\,\delta^{-2d}\right)^{-1}f,f\right\rangle\right)$$

$$\approx \exp\left\{\int_{\Lambda_\delta^2 \times T} E_{i,j}^t\left(\exp\left(-\int_0^t \frac{\lambda}{2}\delta^{-2d}q(X_1(s))^2\,ds\right)\right)d\alpha_f\right\},$$

where α_f is the measure in (24) with g and m replaced by f and m_1.

Substituting (60) and (61) in (59) and pulling everything which does not depend on q outside the integral, we get

(62) $\quad \tilde{F}(f,g) \approx (2\pi)^{-|\Lambda_\delta|/2} \text{ Det } D_2^{-1/2}$

$$\times \exp(-\alpha_1(\Lambda_\delta^2 \times T)) \int_{\mathbb{R}^{\Lambda_\delta}} \exp\left\{\int_{\Lambda_\delta^2 \times T} E_{i,j}^t\right.$$

$$\times \left(\exp\left(-\int_0^t \frac{\lambda}{2}\delta^{-2d}q(X_1(s))^2\,ds\right)\right)d(\alpha_1 + \alpha_f)\right\}$$

$$\times \exp\left(-\frac{1}{2}\sum D_2(i,j)q_iq_j\right)\exp(-\sum q_i g(i)) \prod_{i\in\Lambda_\delta} dq_i.$$

Strictly speaking, we are cheating here; there are conditions on (56) and (57) which we ought to check, and we cannot just substitute (60) and (61) into (59) and be certain that the result is valid. However, it is not hard to prove that (62) holds when $|\Lambda_\delta|$ and δ^{-1} are finite, and hence when they are infinite and sufficiently small. Since we are trying to communicate an idea rather than prove a theorem, we shall leave all bookkeeping of this sort to the reader from now on.

To compute the q_i-integrals in (62), we first expand the exponential term in a power series

(63) $\quad \exp\left\{ \int_{\Lambda_\delta^2 \times T} E_{i,j}^t \left(\exp\left(-\int_0^t \frac{\lambda}{2} \delta^{-2d} q(X_1(s))^2 \, ds \right) \right) d(\alpha_1 + \alpha_f) \right\}$

$\qquad = \sum_{n=0}^\infty \frac{1}{n!} \left(E_{i,j}^t \left(\exp\left(-\int_0^t \frac{\lambda}{2} \delta^{-2d} q(X_1(s))^2 \, ds \right) \right) d(\alpha_1 + \alpha_f) \right)^n.$

Let β_f^n be the measure on $\Lambda_\delta^{2n} \times T$ obtained by multiplying $\alpha_1 + \alpha_f$ by itself n times. Assume that $X_1^1(\omega_1, \cdot), \dots, X_1^n(\omega_n, \cdot)$ are independent copies of X_1, and let

$$E_{i_1,\dots,i_n,j_1,\dots,j_n}^{t_1,\dots,t_n}$$

denote expectation with respect to the product measure $\Pi_{k=1}^n P_{i_k j_k}^{t_k}(\omega_k)$. We can rewrite (63) as

(64) $\quad \exp\left\{ \int_{\Lambda_\delta^2 \times T} E_{i,j}^t \left(\exp\left(-\int_0^t \frac{\lambda}{2} \delta^{-2d} q(X_1(s))^2 \, ds \right) \right) d(\alpha_1 + \alpha_f) \right\}$

$\qquad = \sum_{n=0}^\infty \frac{1}{n!} \int_{\Lambda_\delta^{2n} \times T^n} E_{i_1,\dots,i_n,j_1,\dots,j_n}^{t_1,\dots,t_n}$

$\qquad\qquad \times \exp\left(-\frac{1}{2} \lambda \delta^{-2d} \sum_{k=1}^n \int_0^{t_k} q(X_1^k(s))^2 \, ds \right) d\beta_f^n.$

Putting this into (62) and interchanging the order of integration, we see that

(65) $\quad \tilde{F}(f, g) \approx (2\pi)^{-|\Lambda_\delta|/2} \operatorname{Det} D_2^{1/2} \exp(-\alpha_1(\Lambda_\delta^2 \times T))$

$\qquad \times \sum_{n=0}^\infty \frac{1}{n!} \int_{\Lambda_\delta^{2n} \times T^n} E_{i_1,\dots,i_n,j_1,\dots,j_n}^{t_1,\dots,t_n}$

$\qquad \times \left(\int_{\mathbb{R}^{|\Lambda_\delta|}} \exp\left(-\frac{1}{2} \sum D_2(i,j) q_i q_j - \sum q_i g(i) \right) \right.$

$\qquad \left. \times \exp\left(-\frac{1}{2} \lambda \delta^{-2d} \sum_{k=1}^n \int_0^{t_k} q(X_1^k(s))^2 \, ds \right) \prod_{i \in \Lambda_\delta} dq_i \right) d\beta_f^n.$

The idea is that we can now compute the innermost integral by using the formula for the Laplace transform of Gaussian measures. To see this, let L_n be the "local time"

(66) $\qquad\qquad L_n(i) = \delta^{-d} \sum_{k=1}^n |\{s \le t_k \,|\, X_1^k(s) = i\}| \, \Delta t,$

and observe that

$$\exp\left(-\frac{1}{2}\lambda\,\delta^{-2d}\sum_{k=1}^{n}\int_{0}^{t_k}q(X_1^k(s))^2\,ds\right)=\exp\left(-\frac{1}{2}\lambda\,\delta^{-d}\sum_{i\in\Lambda_\delta}L_n(i)q_i^2\right).$$

By (55)

$$(2\pi)^{-|\Lambda_\delta|/2}\int_{\mathbb{R}^{|\Lambda_\delta|}}\exp\left(-\frac{1}{2}((D_2+\lambda\delta^{-d}L_n)q,q)\right)\exp(-(g,q))\,dq$$

$$=\operatorname{Det}(D_2+\lambda\,\delta^{-d}L_n)^{-1/2}\exp\left(\frac{1}{2}((D_2+\lambda\,\delta^{-d}L_n)^{-1}g,g)\right),$$

and thus (65) becomes

(67) $$\tilde{F}(f,g)\approx\exp(-\alpha_1(\Lambda_\delta^2\times T))\operatorname{Det}D_2^{1/2}$$

$$\times\sum_{n=0}^{\infty}\frac{1}{n!}\int_{\Lambda_\delta^{2n}\times T^n}E_{i_1,\dots,i_n,j_1,\dots,j_n}^{t_1,\dots,t_n}$$

$$\times\left(\operatorname{Det}(D_2+\lambda\,\delta^{-d}L_n)^{-1/2}\right.$$

$$\left.\times\exp\left(\frac{1}{2}\langle(H_2+\lambda L_n)^{-1}g,g\rangle\right)\right)d\beta_f^n.$$

If X_2 is a copy of the Markov process generated by $H_2=-\frac{1}{2}\Delta_d+m_2^2$, which is independent of all the X_1^k's, we get from (56) that

(68) $$\operatorname{Det}D_2^{1/2}\operatorname{Det}(D_2+\lambda\,\delta^{-d}L_n)^{-1/2}$$

$$\approx\exp(-\alpha_2(\Lambda_\delta^2\times T))$$

$$\times\exp\left\{\int_{\Lambda_\delta^2\times T}\tilde{E}_{i,j}^t\left(\exp\left(-\lambda\int_0^t L_n(X_2(s))\,ds\right)\right)d\alpha_2\right\},$$

where $\tilde{E}$ is expectation with respect to the measure governing X_2, and α_2 is the measure (11) with m replaced by m_2. Similarly, by (57)

(69) $$\exp\left(\frac{1}{2}\langle(H_2+\lambda L_n)^{-1}g,g\rangle\right)$$

$$\approx\exp\left\{\int_{\Lambda_\delta^2\times T}\tilde{E}_{i,j}^t\left(\exp\left(-\lambda\int_0^t L_n(X_2(s))\,ds\right)\right)d\alpha_g\right\},$$

where α_g is the measure in (24) with m replaced by m_2.

Substituting (68) and (69) into (67)—and leaving the bookkeeping to the reader—we get

(70) $\tilde{F}(f, g) \approx \exp(-(\alpha_1 + \alpha_2)(\Lambda_\delta^2 \times T))$

$$\times \sum_{n=0}^{\infty} \frac{1}{n!} \int_{\Lambda_\delta^{2n} \times T^n} E_{i_1,\ldots,i_n,j_1,\ldots,j_n}^{t_1,\ldots,t_n}$$

$$\times \left\{ \exp \int_{\Lambda_\delta^2 \times T} \tilde{E}_{i,j}^t \left(\exp\left(-\lambda \int_0^t L_n(X_2(s))\, ds \right) \right) \right.$$

$$\left. \times d(\alpha_2 + \alpha_g) \right\} d\beta_f^n.$$

Recalling the definition of L_n, we may rewrite this as

(71) $\tilde{F}(f, g) \approx \exp(-(\alpha_1 + \alpha_2)(\Lambda_\delta^2 \times T))$

$$\times \sum_{n=0}^{\infty} \frac{1}{n!} \int_{\Lambda_\delta^{2n} \times T^n} E_{i_1,\ldots,i_n,j_1,\ldots,j_n}^{t_1,\ldots,t_n}$$

$$\times \left\{ \exp \int_{\Lambda_\delta^2 \times T} \tilde{E}_{i,j}^t \left(\exp\left(-\lambda \sum_{k=1}^n \int_0^t \int_0^{t_k} \right. \right. \right.$$

$$\left. \left. \left. \times \tilde{\delta}(X_2(s) - X_1^k(s_k))\, ds_k\, ds \right) \right) d(\alpha_2 + \alpha_g) \right\} d\beta_f^n,$$

where as before $\tilde{\delta}$ is the hyperfinite version of the delta function given by $\tilde{\delta}(0) = \delta^{-d}$ and $\delta(i) = 0$ for $i \neq 0$.

Formula (71) may seem curiously unsymmetric in Φ_1 and Φ_2, but if we expand the exponential term in a power series, we see that

(72) $\tilde{F}(f, g) \approx \exp(-(\alpha_1 + \alpha_2)(\Lambda_\delta^2 \times T))$

$$\times \sum_{n=0}^{\infty} \sum_{m=0}^{\infty} \frac{1}{n!} \frac{1}{m!} \int_{\Lambda_\delta^{2n} \times T^n} \int_{\Lambda_\delta^{2m} \times T^m}$$

$$\times E_{i_1,\ldots,i_n,j_1,\ldots,j_n}^{t_1,\ldots,t_n} \tilde{E}_{i_1,\ldots,i_m,j_1,\ldots,j_m}^{\tilde{t}_1,\ldots,\tilde{t}_m}$$

$$\times \exp\left(-\lambda \sum_{k=1}^n \sum_{l=1}^m \int_0^{t_k} \int_0^{\tilde{t}_l} \right.$$

$$\left. \times \tilde{\delta}(X_2^l(\tilde{s}_l) - X_1^k(s_k))\, d\tilde{s}_l\, ds_k \right) d\tilde{\beta}_g^m\, d\beta_f^n,$$

where X_2^1, X_2^2, ... are independent copies of X_2, and where $\tilde{E}_{i_1,...,i_m,j_1,...,j_m}^{\tilde{i}_1,...,\tilde{i}_m}$ and $\tilde{\beta}_g^m$ are constructed from the X_2^l's and $\alpha_2 + \alpha_g$ in the same way as $E_{i_1,...,i_n,j_1,...,j_n}^{t_1,...,t_n}$ and β_f^n were constructed from the X_1^k's and $\alpha_1 + \alpha_f$.

Since the normalization constant Z in (58) equals $\tilde{F}(0, 0)$, we finally arrive at the expression

$$(73) \quad F(f, g) \approx \left(\sum_{n=0}^{\infty} \sum_{m=0}^{\infty} \frac{1}{n!} \frac{1}{m!} \int_{\Lambda_\delta^{2n} \times T^n} \int_{\Lambda_\delta^{2m} \times T^m} E_{i_1,...,i_n,j_1,...,j_n}^{t_1,...,t_n} \tilde{E}_{i_1,...,i_m,j_1,...,j_m}^{\tilde{i}_1,...,\tilde{i}_m} \right.$$

$$\times \exp\left(-\lambda \sum_{k=1}^{n} \sum_{l=1}^{m} \int_0^{t_k} \int_0^{\tilde{t}_l} \tilde{\delta}(X_2^l(\tilde{s}_l) - X_1^k(s_k)) \, d\tilde{s}_l \, ds_k \right)$$

$$\times \left. d\tilde{\beta}_g^m \, d\beta_f^n \right) \Bigg/ \left(\sum_{n=0}^{\infty} \sum_{m=0}^{\infty} \frac{1}{n!} \frac{1}{m!} \int_{\Lambda_\delta^{2n} \times T^n} \int_{\Lambda_\delta^{2m} \times T^m} \right.$$

$$\times E_{i_1,...,i_n,j_1,...,j_n}^{t_1,...,t_n} \tilde{E}_{i_1,...,i_m,j_1,...,j_m}^{\tilde{i}_1,...,\tilde{i}_m}$$

$$\times \exp\left(-\lambda \sum_{k=1}^{n} \sum_{l=1}^{m} \int_0^{t_k} \int_0^{t_l} \right.$$

$$\left. \times \tilde{\delta}(X_2^l(\tilde{s}_l) - X_1^k(s_k)) \, d\tilde{s}_l \, ds_k \right) d\alpha_2^m \, d\alpha_1^n \bigg),$$

provided $|\Lambda_\delta|$ and δ^{-1} are less than some infinite bound, and f and g are "reasonable" functions.

This, then, is our representation of $\Phi_1^2 \Phi_2^2$ in terms of local time functionals. Note that since X_2^l and X_1^k are independent, the exponential terms are always of the kind studied in Section 6.4, and no polymer measures enter into the discussion. But we are still faced with two problems. The first is that while in 6.4 we were forced to choose λ negative and infinitesimal in order to obtain something nontrivial, the calculations above apply only to non-negative λ's. In fact, the interaction term $e^{-(\lambda/4)\Phi_1^2 \Phi_2^2}$ is not integrable with respect to the free measure μ_0 when $\lambda < 0$, and our calculations thus make no sense for such choices of λ. But since the right-hand side of (73) always exists, it is not difficult to make sense of the left-hand side by a suitable truncation argument in such a way that (73) holds (or, more opportunistically, we can simply define the right-hand side to be the interpretation of the left-hand side when $\lambda < 0$).

The second problem is more difficult—and of much greater importance: Is there a negative, infinitesimal λ which makes (73) different from the Laplace transform of the free field; i.e., is $\Phi_1^2 \Phi_2^2$ nontrivial for $d = 4$ or $d = 5$? The results of Section 6.4 indicate that the answer would be yes if λ were allowed to vary with ω and t, but we simply do not have sufficient control to answer the question for constant λ's. All we know from Chapter 6 is that the permissible choices of λ depend on the local behavior of the

Brownian path, and that we thus are dealing with a question concerning self-similarity of Brownian motions. We shall leave the problem open-ended, only emphasizing once again what this section is meant to illustrate—the ease and flexibility with which hyperfinite methods can handle some of the intricate conceptual and computational questions of quantum field theory.

REMARK. As a further illustration let us remark that the exponential model discussed in Section 7.4.C recently has attracted much attention due to the connection of its zero mass version with *the theory of relativistic strings*. The massless exponential interaction model is called the *Liouville model*, since it was studied originally by Liouville as a classical field theory. The connection between the quantum version of this field theory and relativistic strings was observed by Polyakov in 1981. In fact, it turns out that the theory of relativistic strings in space–time dimension $D \le 13$ (corresponding to $\alpha < \sqrt{4\pi}$, see 7.4.C for the choice of α) can be expressed in terms of expectations with respect to the pathspace measure of the Liouville model, see Albeverio *et al.* (1986a). The reduction of the string model to the (hyperfinite) exponential Euclidean model also reduces the "singularities" of the string approach to a discussion of singularities of the Euclidean model, and can thus be discussed by the hyperfinite methods developed in this chapter.

REFERENCES

M. Aizenman (1981). Proof of the triviality of Φ_d^4 field theory and some mean field features of using models for $d > 4$. *Phys. Rev. Lett.* **47**.

S. Albeverio and R. Høegh-Krohn (1974). The Wightman axioms and the mass gap for strong interactions of exponential type in two-dimensional space-time. *J. Funct. Anal.* **16**.

S. Albeverio and R. Høegh-Krohn (1979). Uniqueness and the global Markov property for Euclidean fields. The case of trigonometric interaction. *Comm. Math. Phys.* **68**.

S. Albeverio and R. Høegh-Krohn (1980). Martingale convergence and the exponential interaction in $\mathbb{R}^n$. *In* (L. Streit, ed.), *Quantum Fields—Algebras and Processes*. Springer-Verlag, Berlin and New York.

S. Albeverio and R. Høegh-Krohn (1984a). Local and global Markov fields. *Rep. Math. Phys.* **19**.

S. Albeverio and R. Høegh-Krohn (1984b). Diffusion fields, quantum fields and fields with values in Lie groups. *In* (M. Pinsky, ed.), *Stochastic Analysis and Applications*. Dekker, New York.

S. Albeverio, G. Gallavotti, and R. Høegh-Krohn (1979). Some results for the exponential interaction in two or more dimensions. *Comm. Math. Phys.* **70**.

S. Albeverio, R. Høegh-Krohn, and G. Olsen (1981). The global Markov property for lattice systems. *J. Multivariate Anal.* **11**.

S. Albeverio, Ph. Blanchard, and R. Høegh-Krohn (1982). Some applications of functional integration. *In* (R. Schrader, R. Seiler, and D. A. Uhlenbrock, eds.), *Mathematical Problems in Theoretical Physics*. Lect. Notes Phys. 153, Springer-Verlag, Berlin and New York.

S. Albeverio, J. E. Fenstad, R. Høegh-Krohn, W. Karwowski, and T. Lindstrøm (1984a). Perturbations of the Laplacian supported by null sets, with applications to polymer measures and quantum fields. *Phys. Lett.* **104.**

S. Albeverio, R. Høegh-Krohn, and H. Holden (1984b). Markov cosurfaces and gauge fields. *Acta Phys. Austriaca, Suppl.* XXVI.

S. Albeverio, Ph. Blanchard, and R. Høegh-Krohn (1984c). Newtonian diffusions and planets, with a remark on nonstandard Dirichlet forms and polymers. *In* (A. Truman and D. Williams, eds.), *Proceedings of the LMS-Symp. on Stochastic Analysis and Applications (Swansea, 1983).* Springer-Verlag, Berlin, and New York.

S. Albeverio, R. Høegh-Krohn, and H. Holden (1985). Markov processes in infinite dimensional spaces, Markov fields, and Markov cosurfaces. *In* (L. Arnold and P. Kotelenz, eds.) *Stochastic Space–Time Models, Limit Theorems.* Reidel, Dordrecht.

S. Albeverio, R. Høegh-Krohn, and H. Holden (1986). Markov cosurfaces and quantum fields (in preparation).

S. Albeverio, R. Høegh-Krohn, S. Paycha, and S. Scarlatti (1986a). Pathspace measure for the Liouville quantum field theory and the construction of relativistic strings. *Phys. Lett.* (to appear)

J. Bellissard and R. Høegh-Krohn (1982). Compactness and the maximal Gibbs state for random Gibbs fields on a lattice. *Comm. Math. Phys.* **84.**

J. Bellissard and P. Picco (1979). Lattice quantum fields: uniqueness and Markov property. (preprint) Marseille.

P. Blanchard and J. Tarski (1978). Renormalizable interactions in two dimensions and sharp-time fields. *Acta Phys. Austriaca* **19.**

D. Brydges, J. Fröhlich, and T. Spencer (1982). The random walk representation of classical spin systems and their correlation inequalities. *Comm. Math. Phys.* **83.**

D. C. Brydges, J. Fröhlich, and A. D. Sokal (1983). A new proof of the existence and nontriviality of the continuum φ_2^4 and φ_3^4 quantum field theories. *Comm. Math. Phys.* **91.**

R. L. Dobrushin (1968a). Gibbsian random fields for lattice systems with pairwise interactions. *Functional. Anal. Appl.* **2.**

R. L. Dobrushin (1968b). Description of a random field by means of conditional probabilities and the conditions governing its regularity. *Theory Probab. Its Appl.* **13.**

E. B. Dynkin (1984a). Gaussian and nongaussian random fields associated with Markov processes. *J. Funct. Anal.* **55.**

E. B. Dynkin (1984b). Polynomials of the occupation field and related random fields. *J. Funct. Anal.* **58.**

W. G. Faris (1979). The stochastic Heisenberg model. *J. Funct. Anal.* **32.**

R. Fittler (1984). Some nonstandard quantum electrodynamics. *Helv. Phys. Acta* **57.**

H. Föllmer (1975). Phase transition and Martin boundary. *Sem. Probab., Strasbourg* IX. *Lecture Notes in Math.* **465,** Springer-Verlag, Berlin and New York.

H. Föllmer (1980). On the global Markov property. *In* L. Streit (ed.), *Quantum Fields—Algebras, Processes.* Springer-Verlag, Berlin and New York.

J. Fröhlich (1980). Some results and comments on quantized gauge fields. In *Recent Developments in Gauge Theory, Cargese Summer Inst., 1979.* Plenum Press, New York.

J. Fröhlich (1982). On the triviality of $\lambda \Phi_d^4$-theories and the approach to the critical point in $d \geq 4$ dimensions. *Nucl. Phys. B* **200.**

G. Gallavotti and F. Nicolò (1985). Renormalization theory in four dimensional scalar fields II. *Comm. Math. Phys.* **101.**

R. Gielerak (1983). Verification of the global Markov property in some classes of strongly coupled exponential interactions. *J. Math. Phys.* **24.**

R. J. Glauber (1963). Time dependent statistics of the Ising model. *J. Math. Phys.* **4.**

J. Glimm and A. Jaffe (1981). *Quantum Physics, a Functional Integral Point of View*. Springer-Verlag, Berlin and New York.

S. Goldstein (1980). Remarks on the global Markov property. *Comm. Math. Phys.* **74**.

L. Gross (1979). Decay of correlation in classical lattice models at high temperature. *Comm. Math. Phys.* **68**.

L. Gross (1982). Thermodynamics, statistical mechanics and random fields. In *Ecole d'Eté de Probabilités de Saint-Flour. X-1980. Lecture Notes Math.* **929**, Springer-Verlag, Berlin and New York.

L. L. Helms and P. A. Loeb (1979). Applications of nonstandard analysis to spin models. *J. Math. Anal. Appl.* **69**.

L. L. Helms and P. A. Loeb (1982). Bounds on the oscillation of spin systems. *J. Math. Anal. Appl.* **86**.

Y. Higuchi (1984). A remark on the global Markov property for the d-dimensional Ising model. *Proc. Jpn. Acad. Ser. A* **60**.

R. Holley (1970). A class of interactions in an infinite spin system. *Adv. Math.* **5**.

R. Holley (1972). Markovian interaction processes with finite range interactions. *Ann. Math. Statist.* **43**.

R. Holley and D. W. Stroock (1976). L_2 theory for the stochastic Ising model. *Z. Wahrsch. Verw. Gebiete* **35**.

R. Holley and D. W. Stroock (1977). In one and two dimensions every stationary measure for a stochastic Ising model is a Gibbs state. *Comm. Math. Phys.* **55**.

A. E. Hurd (1981). Nonstandard analysis and lattice statistical mechanics: a variational principle. *Trans. Amer. Math. Soc.* **263**.

R. B. Israel (1979). *Convexity in the Theory of Lattice Gases*. Princeton Univ. Press, Princeton, New Jersey.

P. J. Kelemen and A. Robinson (1972). The nonstandard $\lambda : \varphi_2^4(x)$: model. *J. Math. Phys.* **13**.

C. Kessler (1984). Nonstandard methods in random fields. Thesis, Bochum.

C. Kessler (1985). Examples of extremal lattice fields without the global Markov property. Publ. RIMS 21.

R. Kindermann and J. L. Snell (1980). *Markov Random Fields and their Applications*. Amer. Math. Soc. Providence, Rhode Island.

D. Laugwitz (1978). *Infinitesimalkalkül*. Bibl. Inst., Mannheim.

J. L. Lebowitz and A. Martin-Löf (1972). On the uniqueness of the equilibrium state for Ising spin systems. *Comm. Math. Phys.* **25**.

Li Bang-He (1978). Nonstandard analysis and multiplication of distributions. *Sci. Sinica* **21**.

T. M. Liggett (1985). *Interacting Particle Systems*. Springer-Verlag, Berlin and New York.

P. A. Loeb (1976). Conversion from nonstandard to standard measure spaces and applications in probability theory. *Israel J. Math.* **25**.

P. A. Martin (1977). On the stochastic dynamics of Ising models. *J. Statist. Phys.* **16**.

J. Mikusinski and R. Sikorski (1973). *Theory of Distributions. The Sequential Approach*. Elsevier, Amsterdam.

S. Nagamachi and T. Nishimura (1984). Linear canonical transformations on Fermion Fock space with indefinite metric (preprint). Tokushima.

E. Nelson (1973). Construction of quantum fields from Markov fields. *J. Funct. Anal.* **12**.

A. Ostebee, P. Gambardella, and M. Dresden (1976). A "nonstandard" approach to the thermodynamic limit. II. Weakly tempered potentials and neutral Coulomb systems. *J. Math. Phys.* **17**.

C. Preston (1976). *Random fields. Lecture Notes in Math.* **534**, Springer-Verlag, Berlin and New York.

M. M. Richter (1982). *Ideale Punkte, Monaden und Nichtstandard-Methoden*. Vieweg, Wiesbaden.

A. Robinson (1966). *Nonstandard Analysis.* North-Holland Publ., Amsterdam.

M. Röckner (1985). A Dirichlet problem for distributions and specifications for random fields. *Mem. Amer. Math. Soc.* **54.**

D. Ruelle (1983). *Statistical Mechanics: Rigorous Results.* Benjamin, New York.

E. Seiler (1982). *Gauge theories as a problem of constructive quantum field theory and statistical mechanics. Lecture Notes in Phys.* **159,** Springer-Verlag, Berlin and New York.

B. Simon (1974). *The $P(\varphi)_2$ Euclidean (Quantum) Field Theory.* Princeton Univ. Press, Princeton, New Jersey.

Y. Sinai (1982). *Theory of Phase Transitions: Rigorous Results.* Pergamon, Oxford.

A. Sokal (1982). An alternate constructive approach to the Φ_3^4 quantum field theory, and a possible destructive approach to Φ_4^4. *Ann. Inst. H. Poincaré Sect. A* **37.**

F. Spitzer (1970). Interaction of Markov processes. *Adv. Math.* **5.**

A. Stoll (1985). *Selfrepellent random walks and polymer measures in two dimensions.* Doctoral dissertation. Bochum.

D. W. Stroock (1978). Lectures on infinite interacting systems. *Lectures in Math.* **11.**

K. Stroyan and J. Bayod (1985). *Foundations of Infinitesimal Stochastic Analysis.* North-Holland Publ., Amsterdam (to appear).

W. G. Sullivan (1975). Markov processes for random fields. *Comm. Dublin Inst. Adv. Stud. Ser. A* **23.**

K. Symanzik (1969). Euclidean quantum field theory. *In* (R. Jost, ed.), *Local Quantum Theory.* Academic Press, New York and London.

H. von Weizsäcker (1980). A simple example concerning the global Markov property of lattice random fields. *Proc. Winter School Abstr. Anal., 8th, Praha.*

J. Westwater (1980). On Edwards' model for long polymer chains. *Comm. Math. Phys.* **72.**

J. Westwater (1985). On Edwards' model for long polymer chains. *In* (S. Albeverio and Ph. Blanchard, eds.) *Trends and Developments in the Eighties.* Proc. Bielefield Enc. Math. Phys. IV. World Scientific, Singapore.

B. Zegarliński (1985). Uniqueness and the global Markov property for Euclidean fields: the case of general exponential interaction. *Comm. Math. Phys.*

B. Zegarliński (1984). Extremality and the global Markov property, II: global Markov property for non-FKG maximal Gibbs measures. BiBoS (preprint). Bielefeld.

INDEX

509